environmental science

understanding our changing earth

environmental science

understanding our changing earth

AMERICAN GEOLOGICAL INSTITUTE

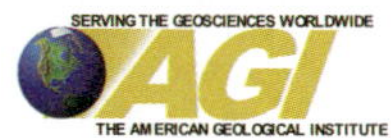

CENGAGE Learning™

Australia • Brazil • Japan • Korea • Mexico • Singapore • Spain • United Kingdom • United States

Environmental Science: Understanding Our Changing Earth
American Geological Institute

High School Environmental Science was developed by the American Geological Institute, in collaboration with the Biological Sciences Curriculum Study, with funding from the National Science Foundation (ESI- 0242636).

Executive Director: P. Patrick Leahy, American Geological Institute

Principal Investigator: Ann E. Benbow, American Geological Institute

Co-Principal Investigator: Christopher M. Keane, American Geological Institute

Writers:
Margaret E. Coleman
Lisa M. Damian-Marvin
John Field
Patti Geis
David Hanych
Matthew L. Hoover

Project Evaluators:
Robert A. Bernoff
Do Yong Park

National Advisory Board:
P. Geoffrey Feiss, The College of William and Mary, Chair
Rodger Bybee, Biological Sciences Curriculum Study
Frank Darytichen, Science Supervisor; Woodbridge Township Schools
Carl F. Katsu, Fairfield Area School District
Bruce F. Molnia, U.S. Geological Survey
Barbara Moss, National Speleological Society
Stephen Prensky, Society of Petrophysicists and Well Log Analysts
Glenn A. Richard, Mineral Physics Institute
Robert Ridky, U.S. Geological Survey
Parvinder S. Sethi , Radford University
J. Thomas Sims, Soil Science Society of America
Lynn Sironen, North Kingstown High School
Myra Thayer, Alan Leis Instructional Center
William Tucci, Mathematics and Science North Carolina Department of Public Instruction
Pam van Scotter, Biological Sciences Curriculum Study
Richard S. Williams, Jr., PhD, Woods Hole Science Center; U.S. Geological Survey

Content Reviewers:
Emmett Wright, GLOBE
BSCS folks Steve Getty, Dave Hanych, Anne Westbrook, Pamela Van Scotter

Delmar content reviewers:
Jessica L. Rettagliata, Livingston High School, NJ
Audrey Carmosino, Medford High School, MA
Erik Quissell, Boise High School, ID
Brian Roodbeen, Richmond High School, MI
Carie L. Callan, Edgewater High School, FL
Michael Lopatka, Edgewater High School, FL
Rick Beal, SUNY ESF
Ted Endreny, SUNY ESF

Photographic Research:
Jessica Ball, American Geological Institute
Jason Betzner, American Geological Institute
Laura Middaugh Rios, American Geological Institute

Teacher Coordinators:
Amy Spaziani, American Geological Institute
Monica Umstead, American Geological Institute

Pilot and Field Test Teachers:
Malissa Attebery
Katie Barlas
Amit Basu
Erin Binns
Deborah Buffington
Philip Carmichael
Patsy Cicala
Rosemary Davidson
David Gale
Sharon Gallant
Christine Gregory
Eric Groshoff
Joseph Halloran
Robert Henry
Nina Hike-Teague
Joseph Jakupcak
Susan Kelly
Steven Lacombe
Linda Leyva
Yolanda Luick
Cheryl Manning
Laura Petit
Glenn Pickett
Mark Powers
Stephanie Ramon
Jonathan Rice
Roy Ryman
Mary Ann Schnieders
Caroline Singler
Jo Ann J. Staiti
Adrien Tanquay
Lesley Urasky
Robert Weinfurtner
Tracie Wood

Library of Congress Control Number: 2009935914

ISBN-13: 978-1-4283-1170-1
ISBN-10: 1-4283-1170-X

Delmar
5 Maxwell Drive
Clifton Park, NY 12065-2919
USA

Cengage Learning products are represented in Canada by Nelson Education, Ltd.

For your lifelong learning solutions, visit **delmar.cengage.com**

Visit our corporate website at **cengage.com.**

Printed in the United States of America
1 2 3 4 5 6 7 13 12 11 10 09

contents

unit one earth and its environments

unit two living at earth system interfaces

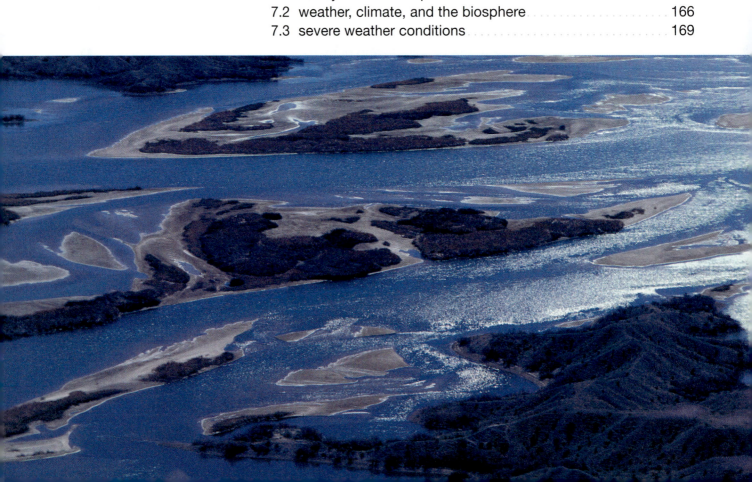

unit three living with a dynamic earth

unit four depending on the earth

extensive teaching/learning materials

A complete supplemental package is provided together with this textbook. It is intended to assist teachers as they plan their teaching strategies:

- **Instructor's Guide** (ISBN: 1-4283-1172-6) featuring chapter objectives, key term list, answer key for chapter review questions, and lesson plans.
- **Classmaster CD-ROM** (ISBN: 1-4283-1171-8) contains a PDF file of the Instructor's Guide, an instructor slide presentation in PowerPoint®, an Image Library made up of the informative illustrations and diagrams within the textbook, and an ExamView test bank.
- **Laboratory Manual** (ISBN: 1-4283-1173-4) and **Lab Manual CD-ROM** (ISBN: 1-4283-1175-0) is made up of over 100 laboratory experiments and activities that correspond with the material presented in each chapter.
- **Lab Manual Instructor's Guide** provides answers to lab manual exercises and additional guidance for the instructor on how to guide students through the activity.
- **Classroom Interactivity CD-ROM** (ISBN: 1-4283-1176-9) allows instructors to create a dynamic learning environment while engaging their students in active participation. This CD-ROM contains four different "game-show-themed" applications to be run by the instructor. All questions are taken directly from the text and serve as a great tool in reinforcing the main concepts of each unit.

Also available

VisionsofEarth™ DVD Series (ISBN 1-4283-7663-1)

Developed by the American Geological Institute, this four disk series is aligned to the National Science Education Standards and is specifically designed to fit current and future Earth Science curricula in schools around the world. Drawing upon the American Geological Institute's huge geoscientific archive of HD television footage, this series provides captivating high definition footage and Hollywood-style animations to show the Earth's processes in action. Each disk is broken up into separate units that discuss specific processes. A correlation guide is available for use with the *Environmental Science* text to take you to the specific spot within the series.

unit one

earth and its environments

chapter 1

earth systems

© Laura Petit

Figure 1-1 Archbishop Shaw High School students.

On Saturday August 27, 2005, it became clear that Hurricane Katrina was headed toward New Orleans. Students at Archbishop Shaw High School (Figure 1-1) knew the drill; pack up your most precious possessions, put the TV and other items up on shelves in case of flooding, gather up pets and family members, and leave. Go north, east, or west but get out of the path of the hurricane. It had happened before and it was happening again. But this time would be the worst.

The students and their families left New Orleans on Saturday and Sunday along with hundreds of thousands of other people. The roads were clogged—Interstate 10 became a long parking lot and trips that normally took 2 hours would take 10 or more as the evacuation got underway. The students left just in time.

Hurricane Katrina came ashore on Monday, August 29, and New Orleans was devastated. Winds reached 225.3 kph (140 mph) in southeast Louisiana and 161 kph (100 mph) in New Orleans. The wind blew the ocean ashore in a giant storm surge that filled up rivers, canals, and lakes. Rain fell at rates of over 2.5 cm (1 inch) an hour and the land was drenched. Then the levees—the earthen banks constructed to protect New Orleans from rising flood waters—broke. Over 80 percent of New Orleans became flooded (Figure 1-2). People who could not evacuate were trapped. Many lost everything: homes, pets, and, in some cases, family members. It would be weeks or even months before electricity, water, and cable were restored. Some parts of New Orleans have still not recovered (Figure 1-3).

The people of New Orleans were caught in the dynamic, churning, and violent interface between land, air, and water. The wind and flood waters combined to smash buildings, farms, forests, bridges, and lives. All living things in New Orleans were threatened by the violent interaction of land, air, and water. Severe storms like Katrina show how amazingly dynamic the Earth can be—and how it can change people's lives overnight.

The Archbishop Shaw High School students did not see Katrina come ashore. They moved about for weeks and months waiting for New Orleans to become habitable again. They went to new schools,

Figure 1-2 New Orleans before and after Hurricane Katrina struck.

Figure 1-3 Part of New Orleans that still has not recovered from Katrina damage.

moved from place to place, and lived in crowded quarters with extended family members. Throughout this time, they did not know what the storm had done to their homes or friends. When they returned to New Orleans, many felt lucky. Most of them lived on higher ground west of downtown New Orleans. Their homes were not extensively flooded or destroyed. However, the city was devastated and they will never forget the impact of Hurricane Katrina on their lives and the lives of others living in New Orleans.

Earth is a special place (Figure 1-4). It is a distinctive planet with abundant water, and there is life almost everywhere near its surface. But the Earth is constantly changing, and many of these changes affect people. Some changes, as the effects of Hurricane Katrina show, can happen within hours. People can best understand how and why Earth changes by investigating its systems. A **system** *is a group of things that regularly interact or are interdependent and that together form a unified whole.*

There are four major Earth systems:

- *The* **geosphere** *is the solid Earth and all its related parts.*
- *The* **atmosphere** *is the envelope of gases that surround Earth.*
- *The* **hydrosphere** *is the water in the oceans, lakes, rivers, underground, and in permanent ice accumulations.*
- *The* **biosphere** *is the life that inhabits Earth and evidence of past life.*

This book is about the Earth, how it changes, and how people live with the Earth. People do things and make choices every day that affect Earth systems, the environments they define, and the many other organisms that call Earth home. Understanding how and why the Earth changes helps people live more safely and more harmoniously with Earth.

Figure 1-4 Satellite view of Earth—the classic "blue marble" photo. *(NASA Goddard Space Flight Center Image by Reto Stöckli (land surface, shallow water, clouds). Enhancements by Robert Simmon (ocean color, compositing, 3D globes, animation). Data and technical support: MODIS Land Group; MODIS Science Data Support Team; MODIS Atmosphere Group; MODIS Ocean Group; Additional data: USGS EROS Data Center (topography); USGS Terrestrial Remote Sensing Flagstaff Field Center (Antarctica); Defense Meteorological Satellite Program (city lights))*

AFTER COMPLETING THIS CHAPTER, YOU WILL UNDERSTAND:
The principles that define how systems work, including matter and energy transfers, reservoirs and fluxes, residence times, and system types.
What Earth systems are, including the geosphere, hydrosphere, atmosphere, and biosphere.
The carbon and water cycles as examples of the interactions of Earth systems.
Why Earth systems and their interactions are important to people.

 1.1 energy, matter, and systems

There are some aspects of **systems** that will help you understand how they work. One key aspect is how energy and matter are transferred between them. How much energy and matter are stored, moved around, and transferred in and out of systems is also important. The transfers of energy and movement of matter in systems commonly occur in ways that make systems appear to be unchanging—to be in a **steady state**. In nature this is mostly an appearance because natural systems are commonly changing.

Energy and Systems

Energy enables systems to change and interact. The main sources of energy that drive Earth's systems are the sun's radiation, chemical reactions, and Earth's internal heat. The force of **gravity** also plays a major role in

Earth's systems. Some examples of the role of energy in Earth's systems include:

- Radiation from the sun heats the atmosphere and ocean water.
- Radioactive elements in minerals decay and heat Earth's interior.
- Chemical reactions convert radiation from the sun into stored chemical energy in green plants, called photosynthesis.

You will learn more about energy in Chapter 15. From a systems perspective, energy is important because it drives changes and movement of matter and because it can be transferred between systems.

Matter and Systems

Matter is all physical substances. Like energy, matter is stored in systems, moved around in them, and transferred among them. Take the element carbon, for example.

Carbon is stored in systems. It is stored in the body as a component of the molecules (such as proteins) that make life possible. But carbon is also stored in rocks of the geosphere, in gases of the atmosphere (especially carbon dioxide), and in water as dissolved carbonate ions.

Carbon is moved around in systems. The carbon in food is used by the body to help make new tissue. Wind and currents move carbon around in the atmosphere and in water bodies. Rocks once deeply buried in the geosphere, such as those containing diamonds (a form of pure carbon), are uplifted and exposed on Earth's surface.

Carbon is transferred among systems in the **carbon cycle** (Figure 1-5). Plants transfer carbon from the atmosphere to the biosphere by absorbing carbon dioxide. The geosphere releases carbon dioxide to the atmosphere through volcanoes and, currently, the combustion of fossil fuels. Organisms in the biosphere make shells and other structures from carbonate ions in seawater. In addition, carbon-bearing materials accumulate on the seafloor and become rocks in the geosphere.

Carbon is an element with many essential roles in Earth systems. The transfer of carbon in the carbon cycle (see Figure 1-5) takes place at local to global scales. Many other elements, nitrogen and sulfur for example, are involved in similar cycles. These cycles are examples of how matter is involved in the interactions of Earth systems.

Reservoirs, Flux, and Residence Time

Places where energy or matter is stored in systems are called **reservoirs**. Rocks are Earth's largest carbon reservoir. A **tonne** is a metric ton, equal to 1,000 kilograms, or 2204.6 pounds. All the carbon in land plants totals 560 billion tonnes and all the carbon in the atmosphere totals about 720 billion tonnes. These are large reservoirs, but sediments and sedimentary rocks contain 150,000 to 200,000 times more carbon! Need some more examples of reservoirs? Earth's oceans are a water reservoir, Earth's atmosphere is an oxygen reservoir, and Earth's interior is a heat reservoir.

You now know that matter and energy can be transferred from one reservoir to another. **Flux** is the rate at which these transfers take place. Land plants absorb carbon from the atmosphere—the flux of carbon from the atmosphere to plants—at a rate of 120 billion tonnes per year. However, the land plant reservoir contains 560 billion tonnes of carbon. Why do you think land plants are not a much larger carbon reservoir if they receive 120 billion tonnes per year from the atmosphere?

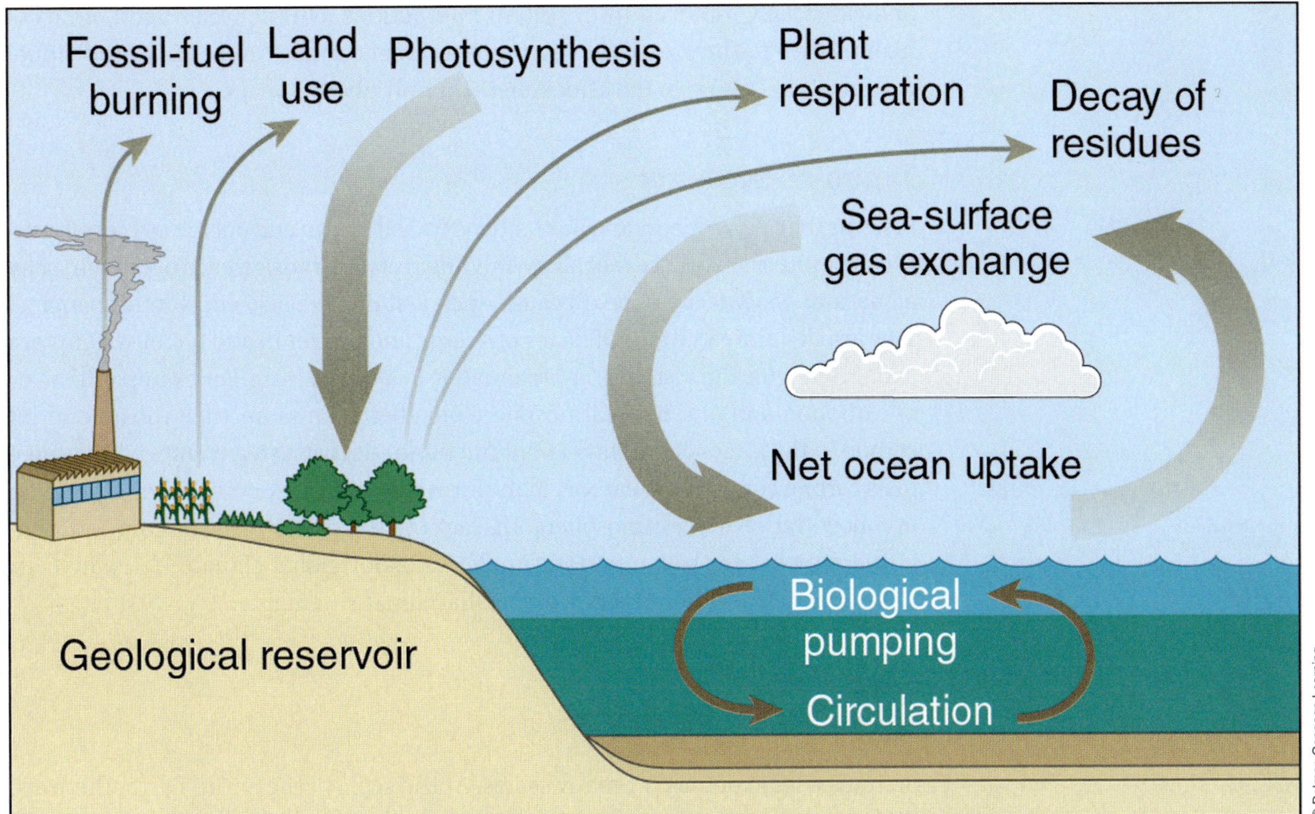

Figure 1-5 The carbon cycle.

The answer is that carbon is lost from land plants each year too. For example, plants die, decompose, and transfer about 60 billion tonnes of carbon to soils each year. In the process of converting food to energy (respiration [see Chapter 2]), living plants release carbon (in the form of carbon dioxide) back to the atmosphere. This carbon flux is about 60 billion tonnes per year. Overall, land plants receive about as much carbon as they give up each year.

An important concept that helps us understand the C cycle is the average amount of time it is contained in a specific reservoir. This is called its **residence time**. Residence time is readily calculated for steady-state systems in which fluxes in (inflows) and out (outflows) of a reservoir are about the same. In these cases, residence time can be calculated using the following equation:

$$\text{Residence Time} = \frac{\text{Reservoir Size}}{\text{Flux In}} = \frac{\text{Reservoir Size}}{\text{Flux Out}}$$

Using the values presented in the previous paragraph, the residence time of carbon in plants can be calculated:

$$\text{Residence Time} = \frac{\text{560 billion tonnes}}{\text{120 billion tonnes per year}} = \text{4.7 years}$$

Where flux is low relative to the size of the reservoir, residence times can be very long: Carbon can reside in rocks for millions or even billions of years. It essentially becomes trapped in rocks compared to other reservoirs. Large reservoirs with long residence times are called **sinks**. Sinks are places where matter gets trapped and essentially removed from system interactions. On the other hand, if flux is high relative to the size of the reservoir, residence times can be short. Individual carbon atoms reside in the atmosphere for only about 3 years.

Types of Systems

Systems that allow energy and matter to be transferred in and out are **open systems**. **Closed systems** are those that allow only energy to be transferred in or out. Virtually all natural systems are open. Because open systems are also constantly changing, they are **dynamic systems**. If fluxes of energy and matter in and out of systems are about the same, the system is approximately in a steady state. For example, if fluxes of carbon in and out of the atmosphere are about the same, then the amount of carbon in the atmosphere stays about the same—it is at a steady state with respect to its carbon content. But carbon, in the form of carbon dioxide, has been increasing in concentration in the atmosphere. The lack of a steady-state carbon concentration in the atmosphere has important implications for global climate. You will learn more about atmospheric carbon and global climate in Chapter 4.

1.2 earth systems

Systems can be defined at many scales. At the scale of the entire Earth, there are four major systems: the geosphere, hydrosphere, atmosphere, and biosphere. The character of each of these major systems is described in this section. Although several examples of how these systems interact are presented here, the interactions of the major Earth systems is a main focus of other chapters in this book.

The Geosphere

The geosphere is the solid Earth and all its related parts. In detail, the geosphere is fairly complicated and even includes liquid components deep within the Earth. You may think that the geosphere is "rock solid," but it is a very dynamic and open system. Active processes, evidenced by volcanoes and earthquakes, for example, change the geosphere every day. The core, mantle, and crust are the three basic parts of the geosphere.

The Core

The innermost part of the geosphere is the core (Figure 1-6). The **inner core** is very hot, but it is solid because the pressures there are tremendous. It is mostly made up of metallic iron and nickel. The **outer core** is also composed of iron and nickel, but because the pressures are lower it is liquid. Earth's magnetism originates in its core.

The Mantle

The **mantle** makes up most of the geosphere (see Figure 1-6). Overall it has a moderate density because it contains iron and magnesium along with some lighter elements, especially silicon and oxygen. Most of the mantle is solid but there is a part in the upper mantle, called the asthenosphere (see Figure 1-6), that can slowly flow, like very thick syrup. Temperature variations make the asthenosphere

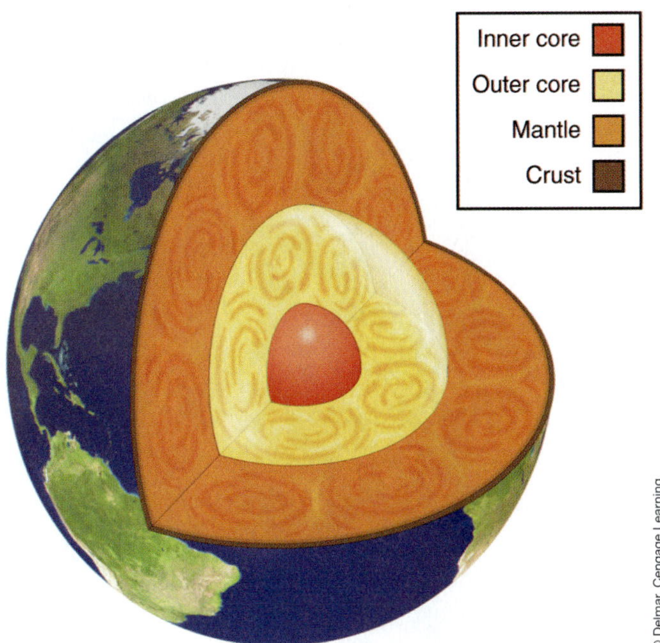

Figure 1-6 Earth structure: core, mantle, asthenosphere, and crust.

unstable—material slowly sinks, rises, and moves from place to place in ways that are still being investigated by scientists. These movements have very important influences on other parts of the geosphere and the other Earth systems.

The Crust

The **crust** is the thinnest layer in Earth's internal structure (see Figure 1-6). There are two types of crust: that which developed under oceans (oceanic crust) and that which developed on continents (continental crust). Oceanic crust is only 5 to 8 km (a few miles) thick and mostly composed of iron, magnesium, calcium, silicon, and oxygen. Continental crust on average is about 50 km (30 miles) thick and contains a wide variety of light elements; silicon and aluminum are very abundant in continental crust. The differences between oceanic crust and continental crust reflect their different origins (see Chapter 4).

The Lithosphere

Oceanic crust and continental crust are rigid. However, rigid rocks are also present in the uppermost mantle below the base of the crust. Scientists recognize the rigid rocks of the crust and uppermost mantle as the lithosphere (Figure 1-7). Because of its rigidity, the lithosphere is the part of the geosphere that can—and does—break. As you will see, how the lithosphere breaks is a major control on where volcanoes and earthquakes occur.

The Rock Cycle

Rocks are the solid part of the geosphere that people can climb on or pick up and throw across a pond. Technically, **rocks** are aggregates of one or more minerals. A **mineral** is a naturally occurring inorganic solid made up of an element or a combination of elements that has an ordered arrangement of atoms and a characteristic chemical composition. In some cases rocks do not contain minerals; they are masses of noncrystalline material such as coal or volcanic glass. Most rocks are hard, solid mineral aggregates. Regardless of how hard

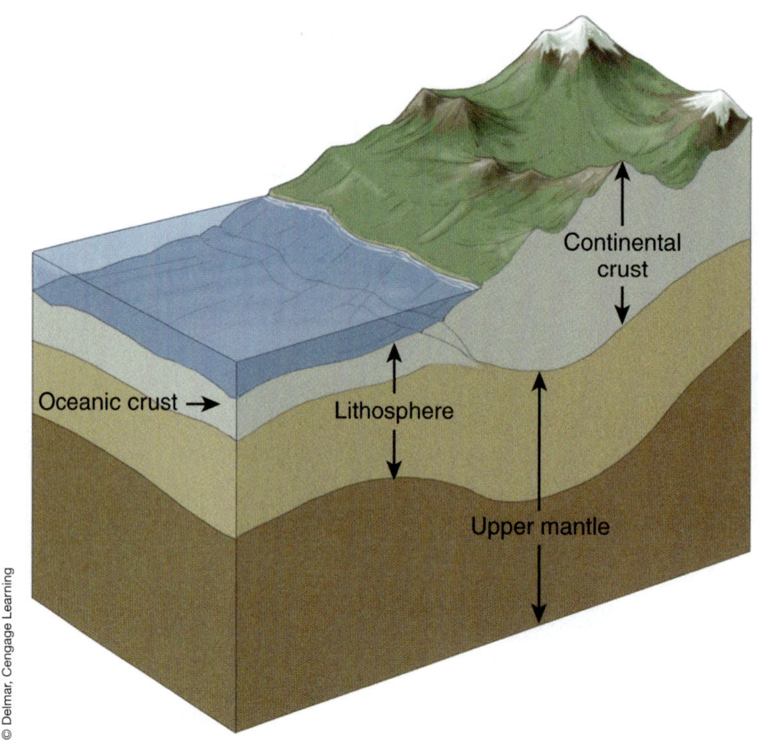

Figure 1-7 The lithosphere and its relation to the mantle and both oceanic and continental crust.

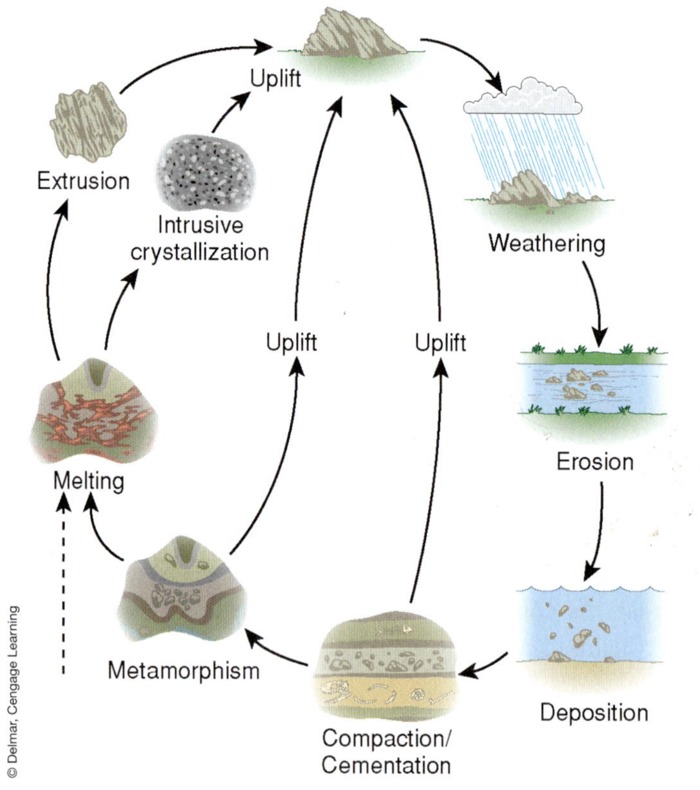

Figure 1-8 The rock cycle.

and solid rocks are though, they are all constantly changing. How rocks change is best illustrated by the **rock cycle** (Figure 1-8).

There are three major kinds of rocks involved in the rock cycle: igneous, sedimentary, and metamorphic (Figure 1-9). **Igneous** rocks are those that form from molten material. Molten material is called **magma** if it is inside Earth and

lava if it is erupted onto Earth's surface. Solidifying magma forms bodies of crystalline rock that vary in size from small injected sheets only centimeters (inches) wide to very large irregular masses tens of kilometers (miles) long and at least a few kilometers (miles) thick. When magma solidifies slowly, as when it is emplaced in the hot, deep parts of the crust, the minerals that form have time to grow to larger sizes. The large crystals are visible to the eye and impart a coarse texture to the new rock. Granite is a common example (see Figure 1-9).

Just the opposite happens when lava is erupted onto the Earth's surface at volcanoes. In volcanic settings, molten material cools and solidifies very quickly. As a result, the minerals that crystallize are very small and volcanic rocks characteristically have a fine texture. In some cases, solidification is so rapid that minerals do not have time to form at all and the solidified lava becomes a glass.

When Earth's initial crust formed, all the new rocks on Earth's surface were igneous. Today igneous rocks continue to form. They form in the crust and on Earth's surface at volcanoes. All rocks at Earth's surface are there only temporarily because they are constantly being changed by weathering and erosion processes.

Weathering processes are the physical, chemical, and biological processes that naturally break rocks into smaller and smaller pieces, change their minerals, and eventually create soils (see Chapter 14). Once rocks are broken into pieces and changed on Earth's surface, gravity starts them in motion and erosion begins.

Erosion moves broken rock from higher to lower elevations. At first, rocks literally fall down steep mountain slopes, but they soon become carried along by the running water of streams and rivers. Rock debris—either rock fragments or individual mineral grains—that is moved by flowing water or wind is called **sediment**. The sand and gravel along the shores of rivers, lakes, and oceans is sediment. Muddy rivers such as the Mississippi carry tremendous amounts (loads) of sediment to the oceans.

(a)

(b)

(c)

Figure 1-9 (a) Typical igneous (granite); (b) sedimentary (sandstone); and (c) metamorphic (gneiss) rocks. *(Courtesy the Earth Science World Image Bank)*

Sediment is in motion until it arrives at a place where it can accumulate, or be deposited, rather than continue being moved by running water or wind. The places that accumulate sediments are called **basins**. When organisms living in water die, their remains sink to the bottom and become incorporated in other sediment. Minerals precipitated from seawater also become sediment.

Sediment accumulates in basins in layers. Over long periods, commonly many millions of years, thick sequences of sediment layers form. Sedimentary sequences can be several kilometers (miles) thick in places. Changes take place in the deeper parts of thick sedimentary sections. The originally loose sediment changes into solid **sedimentary rock**, sand becomes sandstone (see Figure 1-9), clay-rich sediment becomes shale, and calcium carbonate sediment becomes limestone.

Sedimentary basins commonly do not last forever. At very deep levels, all rocks experience such high pressures and temperatures that they change into new mineral aggregates; they are metamorphosed. **Metamorphic rocks** are the

(a)

(b)

(c)

Figure 1-10 The dynamic geosphere.
((a) Courtesy the ESW Image Bank Photo by USGS Cascade Volcano Observatory
(b) Courtesy the ESW Image Bank Photo © Oklahoma University
(c) Courtesy the ESW Image Bank Photo © Bruce Molnia, Terra Photographics)

result of these changes (see Figure 1-9). They may have originally been any type of rock—igneous, sedimentary, or even previously metamorphosed rock—but at deep crustal pressures and temperatures they all change into various types of metamorphic rocks. If the heating of these rocks continues, they will eventually become hot enough to melt. The new melt is magma that can rise and solidify to form new igneous rocks. The generation of new magma and igneous rocks starts the rock cycle over again.

(a)

(b)

(c)

Figure 1-11 The geosphere provides people with resources.
((a) Courtesy the ESW Image Bank Photo Bruce Molnia, Terra Photographics
(b) Courtesy the ESW Image Bank Photo Courtesy United States Geological Survey
(c) Courtesy the ESW Image Bank Photo © Marli Miller, University of Oregon)

The Geosphere and People

The geosphere is in the news every day because of its interactions with people. Volcanoes erupt, earthquakes devastate cities, or landslides wipe out communities; in other words, the geosphere's effects on people make it news (Figure 1-10). You will learn about volcanoes in Chapter 10, earthquakes and tsunamis in Chapter 11, and unstable land in Chapter 12. These are natural hazards produced by a dynamic and changing geosphere. Although these geosphere processes cannot be controlled by people, they can be understood. Understanding can help people live more safely with them. For example, understanding helps people make better choices about where they live, how they construct their buildings, and how they will respond in a natural hazard emergency. The geosphere is not just a source of natural hazards, it is also the source of natural resources that people need.

The geosphere provides people with soil, energy, and mineral resources that are the foundation of modern society (Figure 1-11). You will learn about soils in Chapter 14, energy in Chapter 15, and minerals in Chapter 16. Soils are produced from surface weathering of the geosphere. Energy is provided by fuels such as oil, coal, and the radioactive element uranium—all formed in and recovered from the geosphere. In addition, the geosphere provides metals such as iron, copper, and lead as well as a host of other materials people use in everything from bridges to iPods. These resources are essential to modern society, and Americans use more of them per person than any other society on Earth.

The geosphere affects people but people affect the geosphere, too. As they recover resources, the geosphere is changed through developments such as mines (see Chapter 16). People change land as they farm (see Chapter 10) or develop cities (see Chapter 8). In addition, people create wastes (garbage) that may be disposed of in ways that pollute the geosphere (see Chapter 8). Everyone, including you, interacts with the geosphere every day.

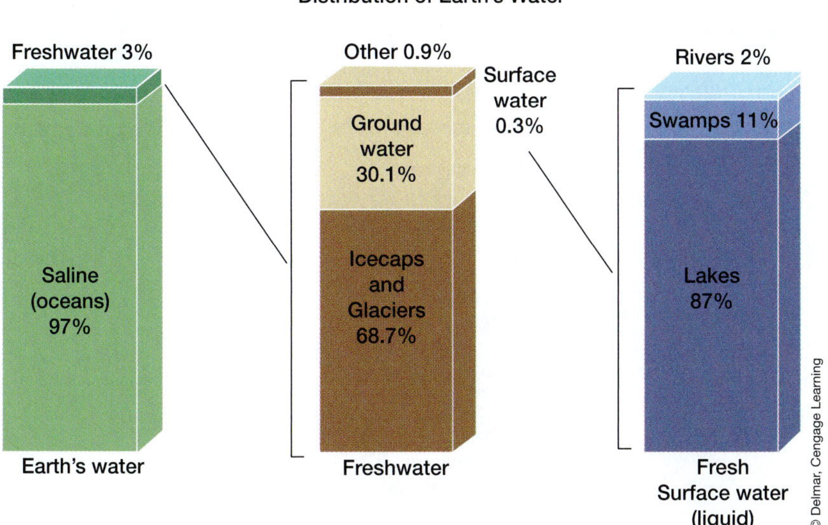

Figure 1-12 Earth's major water reservoirs and their sizes in the hydrosphere.

The Hydrosphere

The hydrosphere is all the water on and near the Earth's surface. It includes water frozen in glaciers and the polar ice caps, water in streams and lakes, and water in the oceans (Figure 1-12). It even includes the water in the geosphere near Earth's surface—ground water.

Earth is a special place when it comes to water. Earth is the only planet in the solar system where water can exist in all three common states of matter (solid, liquid, or gas) at or near its surface. Earth would be just another rocky planet if it were smaller and closer to the sun. The Earth's orbit keeps it at just the right distance from the sun's heat source. It is not too hot (water would evaporate), not too cold (water would be frozen), but just right. Earth is just the right size, too. Earth's gravity keeps water in the atmosphere from escaping into space. This is why Earth is called the water planet—the only planet that can support life as we know it.

The hydrosphere has many roles in Earth systems. It is essential for life (see Chapter 2), it effectively stores heat from the sun and transfers it around the world (see Chapter 7), and it helps wear down the geosphere's mountains. The water cycle illustrates many of these Earth system interactions.

The Water Cycle

The hydrosphere is a very dynamic and open system. Water is constantly moving among its major reservoirs, including the oceans, the atmosphere, surface water (streams and lakes), and ground water. How water moves among these reservoirs defines Earth's **water cycle** (Figure 1-13).

The largest reservoir in the hydrosphere is the world ocean. Oceans cover 71 percent of Earth's surface and contain 97.3 percent of all the water in the hydrosphere; that is, 1.23 billion cubic km (295 million cubic miles). Oceans average 3.7 km (2.3 miles) in depth and reach maximum depths of 11 km (6.9 miles) in places called **trenches**. This depth is greater than the height of Mt. Everest, the tallest mountain above sea level. However you look at oceans, they are huge.

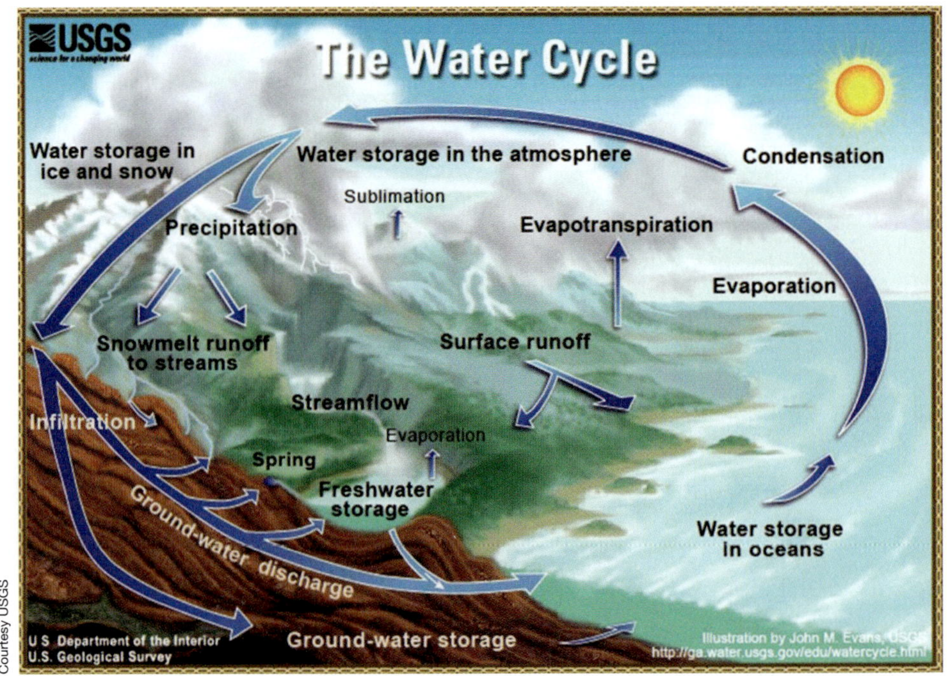

Courtesy USGS

Figure 1-13 The water cycle.

Water that evaporates from the oceans is the principal source of water in the atmosphere. About 86 percent of the atmospheric water is evaporated from the oceans; the remainder is evaporated from the land and **transpired** by plants (water gathered in plant roots is released to the atmosphere through their leaves). For all its importance in the water cycle, the atmosphere is a very small water reservoir. It contains only about 0.001 percent of the hydrosphere (12,700 cubic km or 3,047 cubic miles).

Water in the atmosphere is water vapor, which is the gaseous form of water. However, air can hold only so much water vapor before it condenses into tiny liquid droplets or, if cold enough, into snowflakes or ice particles. Condensed water forms clouds and, when droplets become large enough, precipitates as rain, snow, or ice (hail). The precipitation of water from the atmosphere is a critical part of the water cycle. It transfers water to the land that is needed by people and other living things.

When water precipitates from the atmosphere on land it becomes available to support life, percolate through soils, run off the surface into streams and lakes, or be added to glaciers and ice caps. Glaciers and permanent ice caps are the largest reservoir of nonsalty (fresh) water on Earth. They contain 26,500,000 cubic km (6.36 million cubic miles) or 2.1 percent of all the water in the hydrosphere. If all this ice were to melt, the water added to the world ocean would cause sea level to rise over 70 m (230 feet)!

The water that fills the cracks and voids in the near-surface part of the geosphere, the water that wells tap into for people's use, is ground water. Some of the precipitation on land infiltrates the geosphere and becomes part of the ground water reservoir. This reservoir contains 7,600,000 cubic km or 0.6 percent of all the water in the hydrosphere. The rest of the precipitation on land runs off on the surface and moves to the oceans. By returning water to the oceans, surface runoff completes the water cycle.

Although the paths of water movement are complicated in detail, there is a general cycling of water through the hydrosphere—from the oceans to the atmosphere to land then back to the oceans. In every part of this cycle, water movements transfer energy and interact with all other Earth systems, but the atmosphere is the key link that makes the water cycle possible.

The Hydrosphere and People

Can you imagine living without water? People die in about 7 days if they do not have water to drink. Of course you knew that. It is locked in our collective memory by the many movie scenes of a lost traveler, crawling across a desert, dying as his search for water fails. However, most Americans do not worry much about getting the water they need. They just turn on the faucet and get a drink whenever they want. For all its importance to life, it is remarkable how we take the availability of this valuable resource for granted.

Consider the amount of Earth's water that is even potentially good for people to drink (see Figure 1-12). Only 3 percent is fresh water and most of this is trapped in glaciers and the polar ice caps. This leaves less than 1 percent for people to use. You will learn how people use, and misuse, water in Chapter 13. For now just remember that:

- The hydrosphere is essential for life—people and all other living things.
- Fresh water is a limited resource. It should not be taken for granted.
- People affect water resources. They have been depleted and polluted by misuse. Sound water management is an ongoing challenge in many places in the world.

(a)

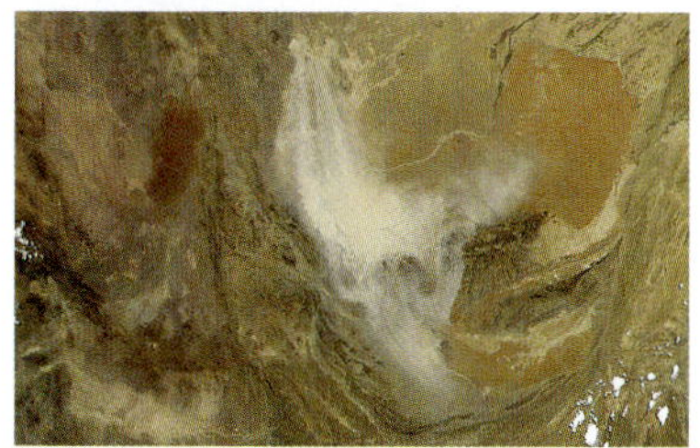

(b)

(c)

Figure 1-14 The dynamic atmosphere. *((a) Courtesy ESW Image Bank Photo by National Oceanic and Atmospheric Administration (b) Courtesy ESW Image Bank © NASA SeaWiFS Project (c) Courtesy ESW Image Bank Photo by National Oceanic and Atmospheric Administration)*

The Atmosphere

The atmosphere is the envelope of gases (air) surrounding Earth. The atmosphere is not very dense and people commonly cannot see or smell it, but they sure can feel it. Moving air (wind) can be so strong that it blows people and buildings to the ground. Birds and airplanes fly through the atmosphere. Clouds float about in the atmosphere and falling meteoroids (shooting stars) burn up in the atmosphere. It extends from within soils and near-surface rocks upward for many hundreds of kilometers (hundreds of miles). Its upper limit, though, is a moving target because it does not have a distinct top; it just becomes less dense and gradually merges into space. The atmosphere is Earth's most dynamic system (Figure 1-14). Each day's changing weather is the most striking example of the atmosphere's very dynamic nature.

Earth's atmosphere is mostly a mixture of nitrogen and oxygen (Table 1-1). Other components, although present in small amounts, can be very important. One example is carbon dioxide. It is only about 0.035 percent of the atmosphere, but even this small amount is an important influence on global climate (see Chapter 4). Gases like carbon dioxide are called greenhouse gases because they trap heat energy and keep Earth's surface from being entirely frozen. They make Earth habitable for people.

Structure of the Atmosphere

Anyone who has hiked around in high mountains or just walked up some stairs too fast in Denver knows that the atmosphere is different at higher elevations. The huffing and puffing that people do to catch their breath shows that there just is not as much oxygen up there as there is at lower elevations. Next time you

Table 1-1 *Composition of the atmosphere, including minor and variable components.*

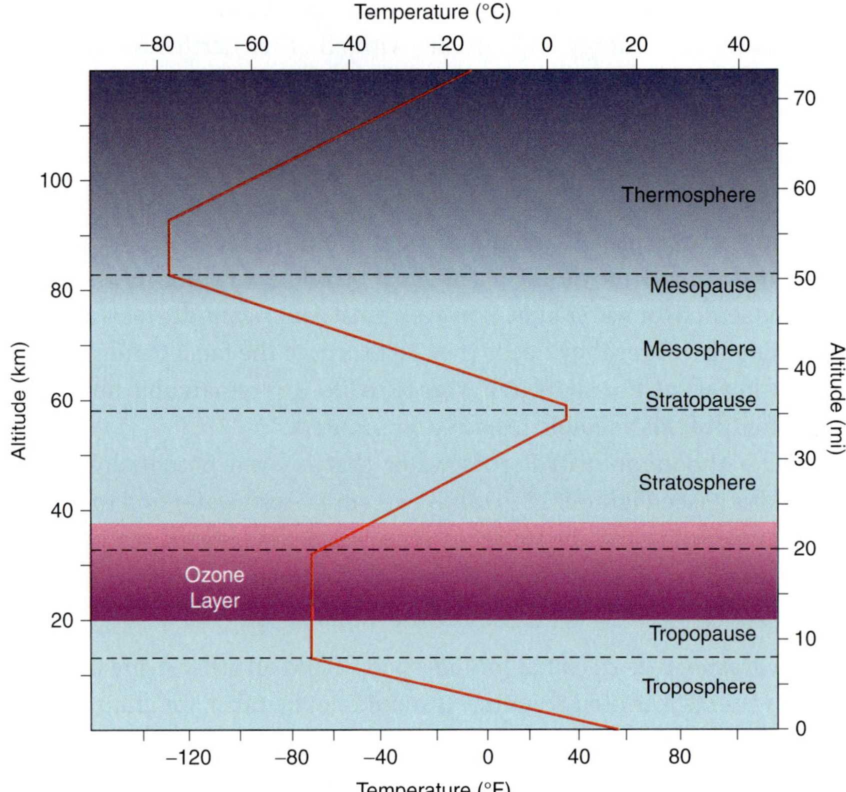

watch a sports team play in Denver, perhaps against the Broncos, Nuggets, or Avalanche, think about this. Do the visiting players get tired first?

Lots of visitors to Denver are just not adjusted to the atmosphere's lower oxygen content at a mile-high elevation. The atmosphere gets less dense (has lesser amounts of gases) the farther from Earth's surface you go. The temperature of the atmosphere changes upward from Earth's surface in some predictable ways too. Temperature changes are what define the atmosphere's structure. The atmosphere contains four principal layers. From the surface upward these are the troposphere, stratosphere, mesosphere, and thermosphere (Figure 1-15).

The **troposphere**, the lowest layer of the atmosphere, is where most of the action is in the atmosphere. It is where life exists, where weather happens (see Chapter 7), and where most of the clouds are. About half the mass of the entire atmosphere is in the lower 5 km (3 miles) of the troposphere. Temperatures gradually decrease upward through the troposphere but at its top, the **tropopause**, they reverse this trend and start to increase (see Figure 1-15).

The layer above the troposphere is the **stratosphere**. The stratosphere is much less dense than the troposphere and, because it contains little water vapor, clouds are not typically present within it. The stratosphere has a stratified character caused by the successively warmer air temperatures at higher levels (see Figure 1-15). As a result, the stratosphere is a clear part of the atmosphere where winds are generally parallel to Earth's surface. Pilots love this and the lower stratosphere is where airliners commonly fly. If you get a chance to fly across the country, look out the window. You will be in the stratosphere and the clouds far below will be in the troposphere. At the top of the stratosphere, temperatures start to decrease at the **stratopause** (see Figure 1-15).

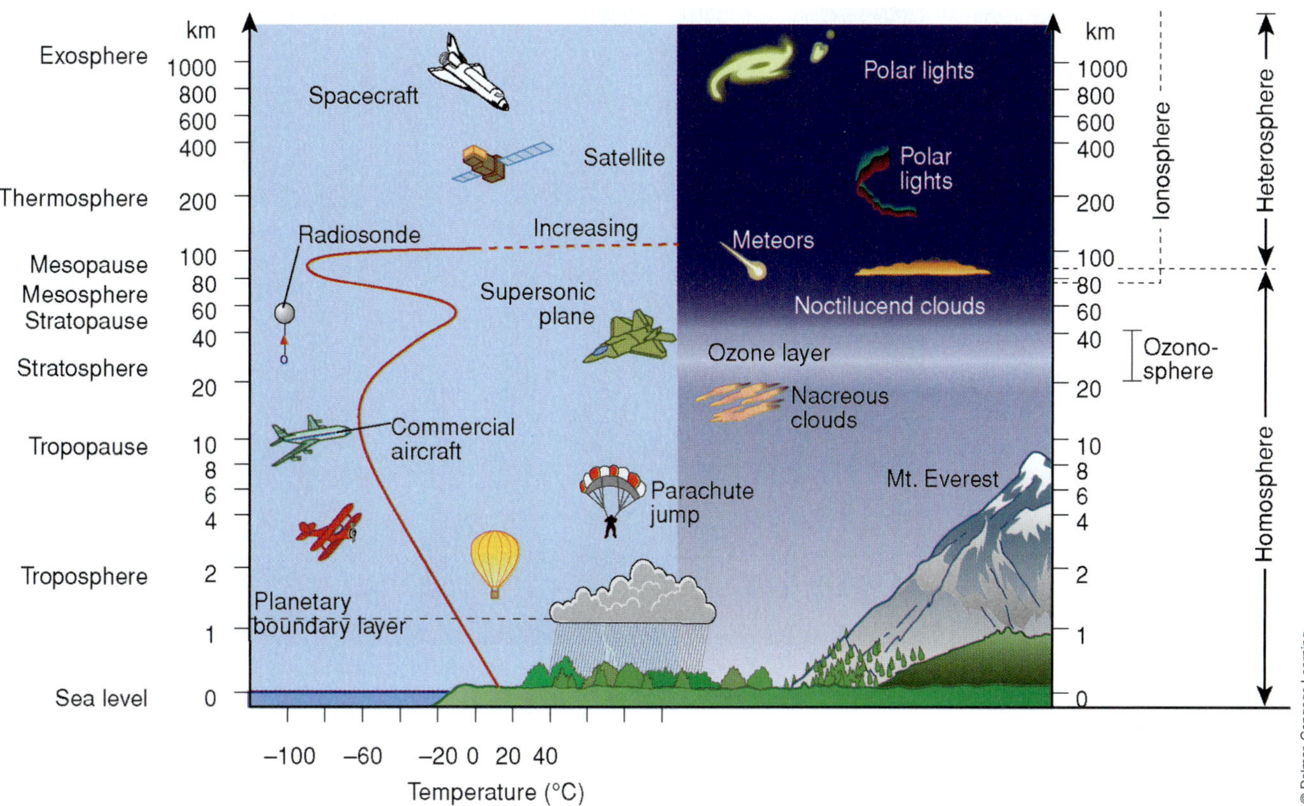

Figure 1-15 The structure of the atmosphere.

The stratosphere has a special part, a sublayer, called the **ozonosphere** or just **ozone layer** (see Figure 1-15). This is where naturally occurring ozone is concentrated in the atmosphere. Ozone is a type of oxygen molecule that contains three atoms (O_3) instead of the regular two (O_2). It is formed when ultraviolet radiation in sunlight interacts with O_2. The ozone layer decreases the amount of ultraviolet radiation that reaches Earth's surface. This shielding is a good thing because ultraviolet radiation is dangerous. It can cause skin cancer among other things. You will learn more about the ozone layer in Chapter 4.

The layer above the stratosphere is the **mesosphere** (see Figure 1-15). Sunlight interacts with sparse gas molecules and changes them into electrically charged particles, called **ions**, in parts of the mesosphere. These particles reflect radio waves sent from Earth. There is enough air in the mesosphere to frictionally heat falling meteoroids and create the bright, fast-moving shooting stars (meteors) visible on clear night skies (Figure 1-16). The mesosphere is also where the space shuttle starts flying upon reentry to the atmosphere. Temperatures gradually decrease upward to the top of the mesosphere at the **mesopause** (see Figure 1-15).

The **thermosphere** is the atmospheric layer above the mesosphere. Gas molecules become fewer and fewer upward through the thermosphere as it gradually merges into space. The molecules floating around in the thermosphere commonly interact with sunlight to form electrically charged particles (ions). These ions make the thermosphere home to the wonderful northern (aurora borealis) and southern (aurora australis) lights—the moving sheets and wisps of colored lights visible at high latitudes on clear nights (Figure 1-17).

Figure 1-16 A meteorite shower in the mesosphere.

Figure 1-17 Northern lights.
(Courtesy ESW Image Bank Photo by Dr. Yohsuke Kamide in 1977. Courtesy National Oceanic and Atmospheric Administration)

The Atmosphere and People

The atmosphere is easy for people to relate to. Perhaps it is because it is so dynamic, we see it moving and changing every day. Maybe it is because every breath emphasizes how much people depend on it. The atmosphere:

- Provides resources essential to life—oxygen for living things and the carbon dioxide that plants use in photosynthesis.
- Protects organisms from harmful radiation from the sun.
- Controls Earth's temperature and global climate.
- Transfers energy and matter between Earth systems as in the water and carbon cycles.

The Biosphere

The biosphere is all life on Earth (Figure 1-18). It is present underground, in soils and rocks, scattered through all parts of the hydrosphere, and floats or flies around in the atmosphere. For the obvious organisms, from ants to zebras, people know life when they see it. Life reproduces itself and controls aspects of itself (self-regulates) that help it get along in its environment. For more obscure life forms that seem to be everywhere, including in people, making sure that they reproduce and self-regulate is important to defining life. The biosphere is an important part of what makes Earth a unique planet in the solar system.

Life's Organization

After hundreds of years of study, much focused on discovering and describing the character of organisms, scientists have come to understand how life is organized. New organisms, species, continue to be discovered each year.

Figure 1-18 Diversity of life. *((a) Photo by Sharon Franklin. Colorization by Stephen Ausmus. Image courtesy of ARS Photo Library USDA (b) Courtesy National Oceanic and Atmospheric Administration (c) Photo by Scott Bauer. Image courtesy of ARS Photo Library, USDA (d) Photo by Scott Bauer. Image courtesy of ARS Photo Library, USDA (e) Courtesy ESW Image Bank © Marli Miller, University of Oregon)*

Species are groups of interbreeding populations that cannot make fertile offspring with other similar groups. For example, a turtle in a freshwater pond that cannot breed with other turtles such as giant sea turtles is a member of one species. People around the world, regardless of color, shape, or size, belong to one species. Many scientists classify all the known species into five categories, or kingdoms, of life (Figure 1-19).

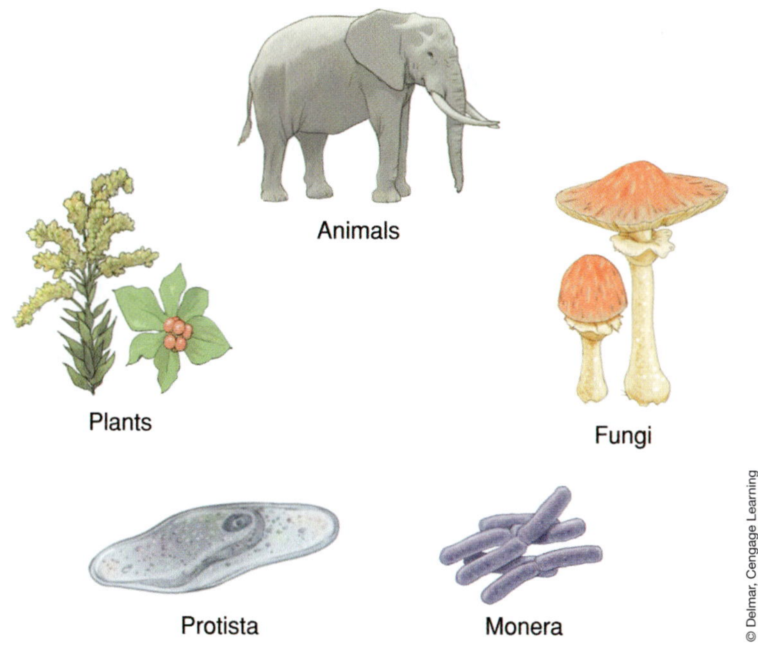

Figure 1-19 Fivefold organization (kingdoms) of life.

- **Monera**: These include the Archaebacteria and Eubacteria. Archaebacteria are one-celled microscopic organisms, otherwise known as the primitive "old" bacteria. The earliest life forms on Earth were Archaebacteria. They have some distinctive chemical and genetic characteristics, but are best known for being able to live in a variety of harsh conditions. The Eubacteria are common bacteria. They include a tremendous variety of one-celled species, from those that use photosynthesis (cyanobactera) to those that cause disease. Note that some scientists consider the Eubacteria and Archaebacteria to be their own separate kingdoms.
- **Protista**: These are mostly one-celled organisms, but they have internal cell structures such as a nucleus that separate them from bacteria. There are untold numbers of protista in the world. They include algae and tiny organisms (protozoans) that can move around on their own. They eat each other a lot. Some have hard parts, skeletons, and these commonly become fossils.
- **Fungi**: Fungi cannot move around, so they grow embedded in a food-bearing medium like soil. They secrete enzymes that help decompose food and make nutrients available to be absorbed into their bodies. Mushrooms, molds, and yeast are fungi.
- **Plantae**: These are plants, like oak trees or poison ivy, that make their food by photosynthesis.
- **Animalia**: These are multicelled organisms with cell membranes instead of cell walls. They have a nervous system and include fish, frogs, and you.

The genetic ties between the kingdoms of life (and their many subdivisions) back to a common ancestor can be illustrated in a diagram, sometimes called the "tree of life" (Figure 1-20). This diagram shows that animals, an entire kingdom in the grouping of life, are a small part of the biosphere's diversity.

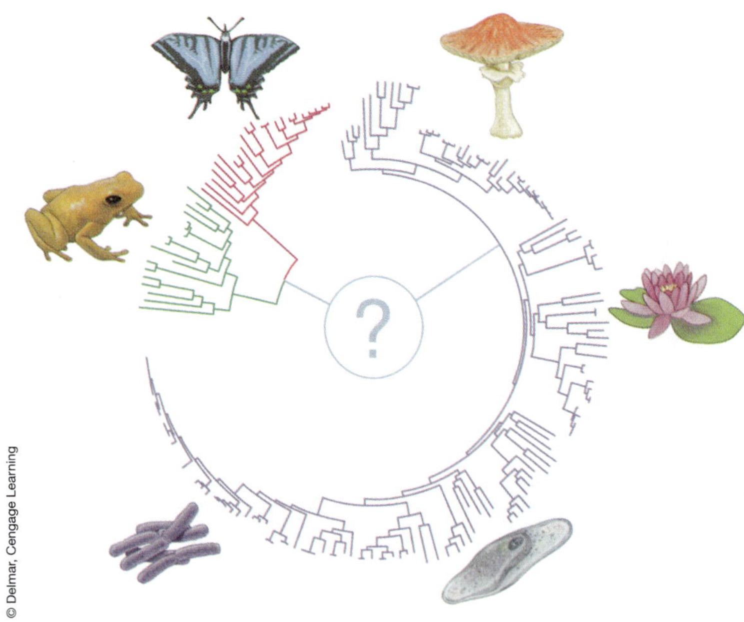

Figure 1-20 The "tree-of-life."

The Biosphere and People

Can you count the ways you use or depend on the biosphere? There are some obvious ways and some less obvious ways. Perhaps the most obvious is that people eat other parts of the biosphere to survive. People eat just about everything you can imagine that has nutritional value. The biosphere also returns oxygen to the atmosphere through photosynthesis (see Chapter 2), supplies wood and fiber for buildings and clothes, and provides chemicals for drugs that fight serious diseases. The biosphere is a natural resource that people need. A way to measure the health of the biosphere is by its biodiversity.

Biodiversity is a general term for the variability of life. It includes the number of species, the variety of species types and interactions, and the genetic diversity within species. Greater biodiversity is thought to make ecosystems more resilient to stress; there are more alternative ways to capture energy, produce food, and recycle nutrients (see Chapter 2). Biodiversity is also thought to be a resource that people can use. For example, ecosystems with genetic and species diversity could be a source of new or better medicines and food crops in the future. Conserving, rather than decreasing, biodiversity appears to be in people's best interest.

Biodiversity is commonly measured by the number of species in an ecosystem. It is estimated that there are more than 1.5 million species that have been described in Earth's biosphere. New species are constantly being discovered and the actual number of species is probably much higher than has been described. The number of species in the biosphere has fluctuated widely. In fact, millions and millions of species—many more than are alive today—have died out (become extinct) in Earth's history (see Chapter 4).

Species are always becoming extinct. Some die out gradually and some evolve into different species. Sometimes large numbers of species from all around the world die out in just a few millions of years. These are very notable times in the history of the biosphere because life was drastically changed at these times. They are called mass extinctions (see Chapter 4).

Drastic, global-scale changes in Earth's physical environment has driven past mass extinctions although the exact causes are not always well understood. Scientists believe that several species are becoming extinct every day, but that the long-term average rate of natural extinction is only about one species per day. It appears that another mass extinction is now underway.

Today's extinctions seem to be caused by people's effects on the biosphere. Other chapters in this book clarify these effects more completely, but some typical things that people do that cause extinctions are destruction of habitat (such as deforestation) and overpredation (people have actually killed off entire species [see Chapter 2]). Preserving species and biodiversity is important to people. It would not be so good if the only species able to live with people in the long term turned out to be bacteria and cockroaches!

 1.3 how science studies earth systems

Science sometimes seems to have little relevance to you or your daily activities. This is far from the case when it comes to understanding Earth systems and how they affect people, or how people affect Earth systems. The scientific method is an approach to asking and answering questions that scientists use to explore and explain the way Earth systems interact and change. It is also a way of solving problems that everyone can use in their daily lives. This process includes five steps:

Questioning and defining problems: The first step for a scientist is to ask questions about the Earth and the problem that is being encountered. It is very important to accurately define the problem. What causes a volcano to erupt? How do we determine where earthquakes will occur? Why do floods recur in some places? What happens if people dispose of their waste in rivers?

Developing hypotheses—answering the questions: In the second step, scientists develop a tentative explanation or answer to their questions, called a hypothesis. A hypothesis is a possible solution to the problem. It is a testable explanation consistent with all we know about the situation. In fact, Earth scientists have found it very helpful to develop more than one hypothesis if possible. They call these "multiple working hypotheses" and use them to help understand the range of answers that need to be considered. It helps them avoid bias or an attachment to one particular solution.

Making predictions: The hypothesis is used to predict outcomes or relationships about the situation. The outcomes or relationships implied by the hypothesis can then be investigated to determine if they are as predicted.

Acquiring and using new data and information: Next, data and information are gathered to test and evaluate the hypothesis and its predictions. A scientist may gather data in a laboratory such as measuring the type and amount of a pollutant produced by a certain chemical reaction. Sometimes the gathering of data may come from making observations about the problem such as monitoring the locations and magnitudes of earthquakes in a region over a period.

Evaluating results: In the final step, the original hypothesis is reevaluated. If it is not well supported by new data and information, then a new hypothesis is developed and the process is repeated. On the other hand, if the original hypothesis is supported, it becomes more firmly established and is used to address the problem.

This method of investigating scientific questions is a repeated process. Even if the initial tests are positive, a hypothesis can continue to be tested in more detail and further refined. The process enables people to develop better and better answers to questions about the Earth.

Applying this method to understand Earth systems is a continuous process. The ongoing observation of Earth gives us the opportunity to continually refine hypotheses. In some cases, many extensively tested hypotheses become integrated in a well-accepted statement of relationships called a theory.

1.4 conclusion

Earth systems and people are closely connected. They provide resources that people need to fuel, build, or grow virtually everything they use each day. In turn, almost everything people do affects Earth systems in some way, and Earth systems directly affect people. Where people choose to live can be dangerous. Floods (see Chapter 6), storms (see Chapter 7), earthquakes (see Chapter 11), and volcanic eruptions (see Chapter 10)—all the result of Earth system processes—can have dire consequences for people. It pays to understand Earth systems.

IN THIS CHAPTER YOU HAVE LEARNED:

- How systems work. Energy and matter are stored in system reservoirs and transferred between them at certain rates, their flux. Systems in which energy and matter are at approximately constant levels are in a steady state. Residence time is the average length of time that energy or matter stays in a reservoir under steady state conditions.

- Both energy and matter are transferred in and out of open systems. Natural systems are almost all open and changing (dynamic).

- The carbon cycle illustrates how matter is stored in Earth systems, moved about within them, and transferred among them.

- Earth's major systems are the geosphere, hydrosphere, atmosphere, and biosphere. All Earth systems interact with one another. These interactions cause Earth to be dynamic, or constantly changing.

- The geosphere is the solid Earth and all its related parts, including a dense and partly molten core; a thick mantle; and a relatively thin, light crust. The upper mantle (asthenosphere) has parts that slowly rise, sink, and flow. There are two types of crust: thin oceanic crust and thicker (and lighter) continental crust. The crust and the uppermost part of the mantle are rigid and together are called the lithosphere. The rigid lithosphere can be broken.

- The hydrosphere is all the water on and near the Earth's surface. It includes water frozen in glaciers and the polar ice caps, water in the streams and lakes on land, and water in the oceans. It even includes the water in the geosphere near Earth's surface, known as ground water. Oceans cover 71 percent of Earth's surface and contain 97.5 percent of all the water in the hydrosphere. The amount of fresh water that people can use is only about 1 percent of the hydrosphere.

- Earth's water is in motion from the oceans to land and back again in the water cycle. In this cycle, water evaporates from the oceans and other places where water exists, moves through the atmosphere, and then precipitates as rain and snow on land. Surface runoff from rain and melting snow forms the many creeks, lakes, and rivers through which surface water continually flows back to the oceans. Significant amounts of rain and melting snow soak into the ground, fill up open spaces in soil, and percolate deeper to become part of a vast, interconnected ground water system.

- The atmosphere is the envelope of gases (mostly nitrogen and oxygen) surrounding Earth. It contains four principal layers defined by temperature changes through them. The atmosphere is Earth's most dynamic system as evidenced by the changing weather of each day. Although it is constantly changing, the atmosphere is a major control on climate and it provides essential support and protection of the biosphere.

- The biosphere is all life on Earth. It is present underground, in soil, scattered through all parts of the hydrosphere, and floats or flies around in the atmosphere. Life reproduces itself and controls aspects of its self (self-regulates) that help it get along in its environment. The biodiversity of the biosphere is valuable to people.

- People affect Earth systems and Earth systems affect people.

KEY TERMS

Animalia
Archaebacteria
Atmosphere
Basins
Biodiversity
Biosphere
Carbon cycle
Closed systems
Crust
Dynamic systems
Energy
Erosion
Eubacteria
Flux
Fungi
Geosphere
Gravity
Hydrosphere
Igneous
Inner core
Ions
Lava
Magma
Mantle
Matter
Mesopause
Mesosphere
Metamorphic rocks
Mineral
Open systems
Outer core
Ozone layer
Ozonosphere
Plantae
Protista
Reservoir
Residence time
Rock cycle
Rocks
Sediment
Sedimentary rock
Sinks
Species
Steady state
Stratopause
Stratosphere
System
Thermosphere
Tonne
Transpired
Trenches
Tropopause
Troposphere
Water cycle
Weathering processes

REVIEW QUESTIONS

1. What is a system? Give an example of a system and explain how it is a system.
2. Define the Earth's systems.
3. What is one example of a closed system on Earth? An open system?
4. How are igneous, sedimentary, and metamorphic rocks formed?

5. Explain the roles of weathering and erosion in the rock cycle.

6. Explain the role of reservoirs in the water cycle.

7. How is the biosphere organized?

8. In what ways does air temperature change as you move from sea level up through the layers of the atmosphere? Why does this happen?

9. In what ways is a tsunami an example of interactions between Earth's systems?

10. Give two examples of how humans are affected by Earth's systems.

chapter 2

ecology

As evening falls in a suburb of Boston, Massachusetts, a group of students from Revere High School walk quietly and listen. Beep…beep…beep…their receiver locates the direction of the signal. It is coming from a clearing between some houses where a dog-like animal is running. The animal has thick, brown fur; a long and slender snout; pointed ears; and a tail with a black tip. It is Maple, a wild coyote wearing a radio transmitter collar (Figure 2-1). The students are conducting a field study to investigate the interactions between urban coyotes and their environment. By tracking Maple's location and observing her activities, the Revere High students are able to determine where Maple lives (her range), what she eats, and the general patterns in her daily life. Are you surprised that coyotes live in Boston?

Figure 2-1 Maple, a wild coyote wearing a radio transmitter collar.

*Coyotes now live just about everywhere in North America from Alaska to Mexico and east to the Atlantic Ocean. These collie-sized relatives of domestic dogs may be the most adaptable animals (other than humans) in the world. They adapt by changing how they interact with each other, with other organisms (especially what they eat!), and with their physical environment. To understand coyotes requires understanding their interactions with the physical (abiotic) and living (***biotic***) world around them.* ***Ecology*** *is the scientific study of the interactions between organisms, like the coyote, and their environment.*

Ecology helps people understand Earth's biosphere, how it changes, and how it interacts with other Earth systems. Ecology also helps people better understand their roles in the biosphere.

AFTER COMPLETING THIS CHAPTER, YOU WILL UNDERSTAND:
How the biosphere is organized.
How ecology helps us understand the biosphere.
How organisms interact with other organisms and their physical environment.
How the biosphere is connected to, and interacts with, other Earth systems.
The place of humans in the biosphere.
How humans affect the rest of the biosphere.

 understanding the biosphere

Understanding the biosphere starts with understanding how it is organized. Ecologists have organized the biosphere to help them identify and understand its interactions and changes within it. But understanding these interactions and changes also requires knowledge of the physical environment in which organisms live, their habitat, and the role of the organisms in their habitats—their niche.

The Biosphere's Organization

The biosphere, all organisms and where they live on Earth, is organized at many different levels. Identifying these levels is not something ecologists do to have lots of things to talk about. They need to understand the biosphere's organization to investigate how organisms interact and change. Ecologists study the biosphere at six levels: individuals, populations, communities, ecosystems, biomes, and the entire biosphere itself (Figure 2-2).

- An **individual** is one organism—a turtle in a pond or a person on a street.
- A **population** is a group of organisms of the same species that live in a specific area—all the turtles in Walden Pond.
- A **community** is made up of all the populations of various species that live in an area—all the turtles, fish, insects, snails, plants, and so forth that live in Walden Pond.
- A **biome** is a community that covers a large geographical area—the temperate forests of the northwestern United States are an example.
- An **ecosystem** is a community and its abiotic—all the organisms, water, bottom sediments, and shoreline of Walden Pond.
- The **biosphere** is all of Earth's ecosystems.

Habitat—Where Organisms Live

Ecologists call the environmental surroundings of an organism its **habitat**. These surroundings are an organism's physical "address." Habitat provides an organism with the environmental conditions and **resources** such as food, water, shelter, and space it needs to survive and reproduce. There are two basic kinds of habitat. A terrestrial habitat is on land and aquatic habitat is on or in water (Figure 2-3). Terrestrial and aquatic habitats can be further divided; examples are coastal beaches, streams, wetlands, meadows, forests, and so forth.

Ecologists also divide habitats into **microhabitats**. Caves are a terrestrial habitat (Figure 2-4) but within a cave may be many microhabitats including the cave floor, walls, ceiling, underneath rubble, or perhaps along the roof where roots penetrate down from the ground above. Ecologists find it essential to identify an organism's habitat if they are to understand its abiotic and biotic interactions within an ecosystem.

Niche—What Organisms Do

Just as you can think of an organism's habitat as its address, you can think of an organism's **niche** as its "profession" or role in an ecosystem. A simple way to define an animal's niche is to describe what the animal eats and what eats the animal. It is common, though, that ecologists need to consider a more complete

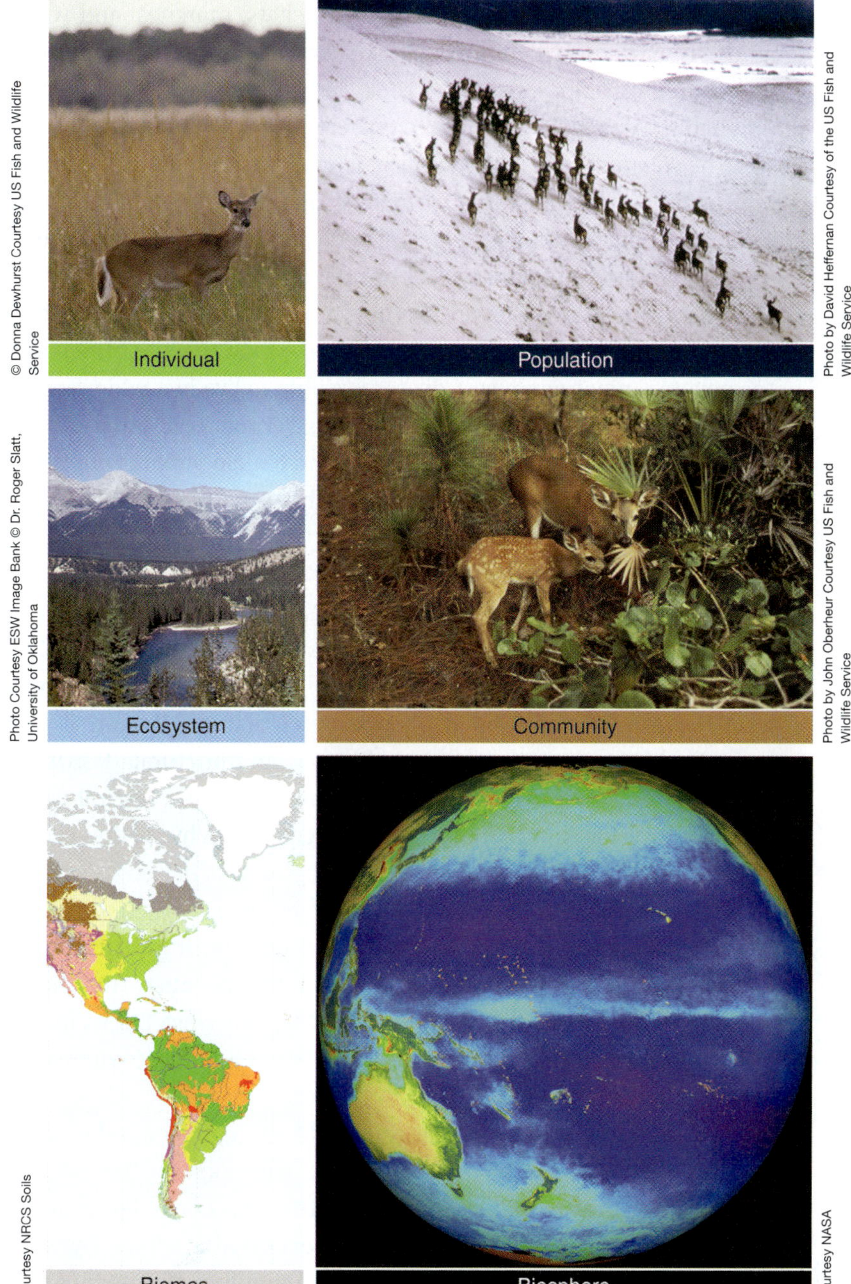

Figure 2-2 Six levels of the biosphere: individuals, populations, communities, ecosystems, biomes, and the biosphere as a whole.

description of an organism's relationships to the physical environment and to other organisms in defining its niche. These relationships are sometimes called **niche dimensions**. Every organism has a niche with many dimensions. These dimensions identify the physical conditions an organism can tolerate, the resources it uses, and other aspects of its lifestyle. For example, rainbow trout occupy the colder, clean, and non-silty parts of a stream where they feast on insects, invertebrates, and smaller fish. Other fish occupy warmer or dirtier parts of the stream.

If the Revere High School students were to define the habitat and niche of the coyotes they are studying, what would they say? Because coyotes live on land, they would define the habitat as terrestrial. The students have probably observed that their coyotes live in a suburban habit,

Figure 2-3 Terrestrial and aquatic habitats. *(Photo by Hans Stuart, Courtesy of the US Fish and Wildlife Service)*

Figure 2-4 Caves are a terrestrial habitat.

Figure 2-5 Wildfires can be devastating to organisms and their habitats. (*Image courtesy ESW Image Bank. Photographer: John McColgan Courtesy Alaska Forest Service*)

Figure 2-6 Suppressing fire in jack pine forests almost led to the extinction of the Kirtland's warbler. (*Photo by Lou George, Courtesy of the US Fish and Wildlife Service*)

spending much of their time in local thickets, marshes, and woodlands. Their dens, which are occupied for the purpose of giving birth, are located on slopes, banks, or rocky ledges and are often hidden by downed trees or stumps, or are in culverts.

The coyote's niche in this suburban habitat is a predatory nocturnal mammal, eating a wide variety of foods including meat, garbage, insects, rodents, rabbits, birds, deer, carrion, and even berries and fruits. Bears, mountain lions, and wolves used to prey on coyotes, but due to their declining populations, they are no longer much of a threat to the coyotes. Because coyotes are considered a nuisance by many people, humans are the coyote's major enemies. Coyotes are able to live in a wide range of climates and conditions. They breed during February or March and give birth in April or May, and both the male and female remain with the litter until the offspring are able to hunt for themselves.

Fire—Friend or Foe?

Wildfires can be devastating to organisms and their habitats (Figure 2-5). Should we prevent them from happening? Why do you have this opinion? Does any scientific evidence support your opinion? If so, what is it?

Ecologists once thought that fire was an unnatural, destructive force in ecosystems. Today, scientists think that it is a natural part of some ecosystems and a useful management tool. Why did they change their minds? They changed their minds because of new evidence.

Scientific thinking is based on evidence. Scientists collect evidence by making observations and conducting experiments. Scientists changed their minds about fire because they began gathering evidence that supported a different view. Let us look at some of the evidence that made scientists change their view of fire and people's fire management policies.

One of the first things that ecologists do to understand wildfires is determine the fire history of an area. They look for the evidence of previous fires in plants, soil, and historical records. For example, ecologists can determine how often fire occurred in a forest by taking core samples out of trees. By counting the number of tree rings from the bark to the center, they can determine the age of the tree. By looking to see which rings have charcoal present, they can tell when fires burned the tree. What evidence of fire would you see in the soil?

Ecologists also study the interactions of animals and plants in burned and unburned areas. One of these studies showed that suppression of fire in jack pine forests in Michigan almost resulted in the extinction of the Kirtland's warbler (Figure 2-6). The Kirtland warbler is an endangered bird that lives in a small part of northern Michigan. It has very specific nesting requirements. It builds its nests only in jack pines that have dead branches at the ground level. The only jack pine trees that have these branches are 6 to 21 years old. The bird will not nest in younger or older trees.

In 1961, ecologists found about 1,000 Kirtland's warblers in their study area. Ten years later, there were fewer than 500. A survey revealed that there were not enough suitable jack pines for nesting. Most of the trees in the forest were too old because fire suppression by people had reduced jack pine reproduction.

To save the bird from extinction, ecologists began burning small sections of the jack pine forest to create new habitat for the warbler. Unfortunately, the number of birds remained below 500 until 1991. It then increased to 800 in 1992. Why did it increase? Evidence suggests that the increase resulted from a planned burn that went out of control in 1980. Instead of burning 200 acres as planned, the fire burned 25,000 acres and damaged 44 houses. Ten years later, the jack pine trees that started growing after the massive fire were suitable for nesting.

Now that you have looked at some evidence, is fire a friend or a foe? What do you mean when you say that fire is a friend or a foe? What human values are you using to make such judgments? Answering these questions is difficult without a complete understanding of the interactions of organisms and their physical environment before and after wildfires.

2.2 understanding interactions in the biosphere

Ecologists study three types of interactions that occur within ecosystems. They study the effects that the physical environment has on organisms, the effects that organisms have on the physical environment, and the effects that organisms have on each other.

Earth Systems' Influences on Organisms

The physical environment is a strong influence on the distribution and abundance of organisms. **Distribution** refers to where organisms are found; **abundance** refers to the number of organisms present. For example, rooted aquatic plants grow only in shallow areas of lakes, sharks only live in saltwater, snakes cannot live in the Arctic, and polar bears cannot live in Florida (well, maybe they can if they live in a zoo that provides them appropriate habitat!). Almost everyone knows that physical environmental factors limit the distribution and abundance of organisms. Gardeners, for example, know that some crops do not grow well if the soil is too dry, so they water their plants. If the soil is low in nitrogen, they add fertilizer. Here are a few of the ways that the hydrosphere, atmosphere, and geosphere influence organisms.

The hydrosphere not only provides water for cellular processes, it also has important physical and chemical characteristics, especially its quality and temperature, that influence organisms. Water quality depends on the amount and nature of dissolved constituents. These create conditions that organisms must adapt to. Water's saltiness (salinity) is an example. Another is the presence of pollutants such as small amounts of dissolved metal that can be tolerated by some organisms but not others. In addition water temperature is a very important physical characteristic of aquatic environments. Water cannot be too hot or too cold for many organisms to be comfortable and survive (Figure 2-7).

Courtesy USFWS

Figure 2-7 Water temperature is important for organisms' survival.

Some important characteristics of the atmosphere that influence organisms include temperature, moisture content, and the presence of pollutants. The temperature of the atmosphere adjacent to Earth's

Courtesy USGS

Figure 2-8 Acid rain can harm plants over large areas.

surface can make life bearable or unbearable. Almost everyone has experienced cold north winds in the winter—time to head south! But what if you are an organism that cannot head south? You will have to be able to adapt or you will not make it through the winter. If the air is too dry, then precipitation decreases, droughts occur, and living things begin to struggle. An example of an atmospheric pollutant that is hard on the biosphere is sulfur dioxide. This gas is emitted by the burning of coal (see Chapter 15) and other industrial activities, but it is also released in large amounts during some volcanic eruptions (see Chapter 10). The emitted sulfur dioxide reacts with water vapor in the atmosphere to form tiny droplets of sulfuric acid. This causes acidic rain that can harm plants over large areas (Figure 2-8).

The principal way that the geosphere influences organisms is through the physical properties of soils; for example, its mineral and moisture content. You will learn a lot about soils and the biosphere in Chapter 14. Other physical characteristics of the geosphere that are important are its elevation (or depth in lakes and oceans) and relief (topography). Have you ever driven or hiked up a mountain to treeline and beyond? Some plants and animals cannot live at high elevations where temperatures are colder and snow may accumulate during the winter. Does this environment sound like the result of Earth system interactions? It sure is. Hydrosphere (snow), atmosphere (temperature), and geosphere (elevation) interact to create different habitats at different elevations in mountainous terrain.

Tolerance

Ecologists need to know the requirements of species to explain some of the patterns in nature. Abiotic factors influence the distribution and abundance of species because every species is adapted to live within a certain range of physical environmental conditions. This is called a species' **range of tolerance**. Outside of this range, the species cannot survive.

Every species has a tolerance range for many abiotic factors. For example, creek chubs (a fish) can tolerate water temperatures between 4.5 and 30.3°C (40.1 and 86.5°F). Brook trout can tolerate temperatures between 0.5 and 25.3°C (32.9 and 77.5°F). If you had a pond on your property that reached water temperatures of 26°C (78.8°F) during the summer, what would happen if you tried to create a fishing pond by stocking it with brook trout?

Within a species' range of tolerance for a factor, there is a smaller range called the **optimal range**, within which the condition of the factor is ideally suited for that organism. Above and below this optimal range, the species does not do as well. For example, brook trout may be able to survive in water temperatures outside its optimum, but it may not reproduce at these temperatures.

Traits

To survive and thrive in their habitats, species need to be suited to the environmental conditions that exist around them. Species have **traits** (distinguishing features or characteristics) that enhance their survival and reproductive success in a particular environment. These traits can be physical (such as an animal having warm fur to survive cold temperatures), behavioral (such as animals burrowing underground to avoid extremely high temperatures), or physiological (such as chemical processes that allow certain trees to photosynthesize at low temperatures).

Biomes

Figure 2-9 Terrestrial biomes of the world.
((a) Courtesy ESW Image Bank © Bruce Molnia, Terra Photographics
(b) Courtesy ESW Image Bank © Marli Miller, University of Oregon
(c) Courtesy ESW Image Bank © Marli Miller, University of Oregon
(d) Courtesy ESW Image Bank © Bruce Molnia, Terra Photographics
(e) Courtesy of Jason Betzner. (RAINFOREST)
(f) Courtesy ESW Image Bank © Michael Collier
(g) Courtesy ESW Image Bank © Bruce Molnia, Terra Photographics
(h) Courtesy NRCS Soils)

The fact that different species have different tolerance ranges and different adaptations is one reason why various habitats support diverse communities of organisms. It is also a reason why communities vary geographically. In fact, different regions of the world have different biomes because **abiotic** factors such as temperature, rainfall, and soil composition vary geographically.

Biomes

In the early 1930s, Frederic Clements and Victor Shelford developed a system for classifying the world's biotic communities into biomes (Figure 2-9). A biome is a community that covers a large geographic area. It is distinguished by its dominant vegetation and climate. Ecologists recognize from 6 to 14 or more biomes, depending on which classification system they use. Let us take a close look at several of these biomes and how the organisms that live in each have adapted to the conditions that are present.

Organism Influences on the Physical Environment

As you have just seen, the physical environment affects organisms. The opposite is also true. Organisms affect the physical environment. For example, earthworms loosen the soil when they crawl through it. Coyotes dig holes in the ground and use them for shelter. In the deep parts of some lakes, microbes living in the sediment use up the oxygen in bottom water, making it unfit for fish to survive.

Some remarkable effects of organisms on the physical environment come from their role in developing the geosphere's soils. You will learn more about

Figure 2-10 Algal "blooms." (*Image from ESW Image Bank. Courtesy MODIS NASA Moderate Resolution Imaging Spectroradiometer, NASA*)

Photo by: Susanne Miller Courtesy of the USFWS

Figure 2-11 An example of predation.

Courtesy USFWS

Figure 2-12 Bluebird houses avoided by starlings and house sparrows.

organisms and soil in Chapter 14. Soils are full of life and it is not just plants. Millions of species live in soil: bacteria, protozoa, fungi, worms of all kinds, and insects. They help decompose organic matter by converting it back to its molecular and elemental building blocks. Some capture nitrogen from the atmosphere and convert it to forms that plants can use as nutrients. In addition the physical breakdown of rocks and minerals in soil is enhanced by the organisms living in it. As you will see in Chapter 14, soil is essential for life and the role of organisms in creating and changing soil is very important.

Another example of organisms affecting the physical environment occurs in the hydrosphere. Where nutrient levels are high in aquatic habitats, algae (microscopic plants) can drastically increase in abundance and form algal "blooms." Under dark and cloudy conditions algae use up much dissolved oxygen in the water. Algae can lower the dissolved oxygen so much that fish find themselves trying to survive outside their range of tolerance but they cannot. Algal blooms are notorious for causing fish kills (Figure 2-10).

Organisms' Influences on Other Organisms

It is important to remember that living things are part of the environment just like nonliving things. Furthermore, living things can influence the distribution and number of other organisms just like nonliving things. To understand how, let us examine some of the interactions that take place between organisms.

Scientists have observed that every organism interacts with other organisms in various ways. For example, when an organism of one species eats an organism of another species, the interaction is called **predation** (Figure 2-11). Predation can affect the distribution and abundance of **prey** organisms in an area. Predation also results in the transfer of energy from one organism to another. At what organizational level of the biosphere do you think predatory interactions occur?

Can you remember a time when you were on a sports team competing with others to get control of a loose ball? Such **competition** occurs in nature, too, when species use resources that are in short supply. A resource is something that a species needs to survive and reproduce. Competition is another interaction that can affect a species' distribution or abundance. When competition occurs, there are several possible outcomes.

One possible outcome of competition is that one of the competing species will find alternative resources that they can use. For example, North American bluebirds nest in holes in trees. When starlings and house sparrows were introduced from Europe, they began competing with bluebirds for the same tree holes. Because the number of suitable holes is limited, the number of bluebirds declined drastically. To save the bluebirds, conservationists began building artificial bluebird houses that starling and house sparrows do not like to use. In this case, one of the species (the bluebirds) found an alternative resource to use—one that the conservationists provided (Figure 2-12)!

Another possible outcome of competition is **resource partitioning**, where two species competing for a single resource adapt so that parts of the same resource can be used by both species. For example, if a habitat contained bats and frogs, both of which eat insects, they could share the insects through resource partitioning: Because bats can fly, they would eat the flying insects, and the frogs, which are terrestrial, would eat the crawling insects. Resource partitioning can be thought of as "splitting the niche." If neither of these two possibilities occurs, one of the competing species will emigrate to a new habitat to find the resources it needs, or one or both of the species will suffer a population decline, possibly leading to one of the species becoming locally extinct.

Courtesy Peter Kolbl

Figure 2-13 An example of mutualism.

Another type of interaction involves two organisms from different species living in direct contact with one another. This type of interaction is called **symbiotic**. When a symbiotic interaction between two organisms benefits both organisms, it is called **mutualism** (Figure 2-13). When a symbiotic interaction benefits one organism but hurts the other, it is called **parasitism**. When a symbiotic interaction benefits one organism but does not help or harm the other, it is called **commensalism**.

Making Sense of Interactions in the Biosphere

In the mid-19th century, Charles Darwin (Figure 2-14) revolutionized people's understanding of the living world. In his 1859 book, *On the Origin of Species*, he presented evidence that species change across time (that is, evolve). He also said that natural selection was the most important mechanism of evolutionary change. His evidence and reasoning were so persuasive that most biologists accepted his ideas by the end of the century. Before 1859, most people did not think that species changed.

Evolution refers to change in the genetic makeup and traits of a population, not individuals. As a population evolves, it can change so much across time that it becomes a new species. It is a new species when individuals that make up the population no longer mate with individuals from other populations.

Armed with these ideas, Darwin was able to explain how species form, why they have traits in common, and how they become suited to their environment. Today, the concept of biological evolution is one of the foundations of biology. In fact, the famous biologist Theodosius Dobzhansky once wrote that "nothing in biology makes sense except in the light of evolution." That includes our understanding of ecology.

Scientists now look at everything they learn about the interrelationships between organisms and their environment in the light of evolution. This is because organisms evolve in environmental settings. Both living and nonliving environmental factors influence how populations evolve.

For example, natural selection in prey species favors the evolution of faster running, better camouflage, and other antipredator traits. In predators, such as coyotes, it favors the evolution of more effective hunting behaviors or keener senses. It also accounts for the origin of the mutualistic interactions and evolutionary changes that

Figure 2-14 Charles Darwin.

occur because of competition. Finally, evolution by natural selection explains how organisms have become suited to live in the many different physical environments in Earth's systems.

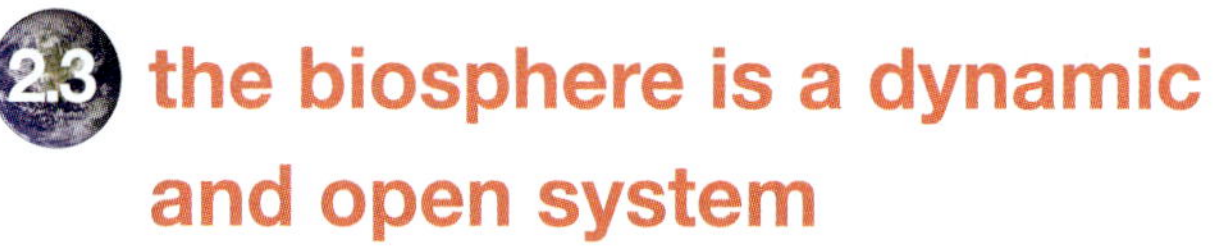

2.3 the biosphere is a dynamic and open system

The biosphere is constantly changing and evolving through the interactions of systems. Energy and matter are transferred within the biosphere and between the biosphere and other Earth systems. Here are some examples.

Energy Transfer in Ecosystems

Sunlight is the source of energy in almost all ecosystems. It is converted to energy (food) that organisms can use by **photosynthesis**. Through photosynthesis, plants, algae, and some bacteria use part of the sunlight they receive to change carbon dioxide and water into carbohydrate (sugar) molecules ($C_6H_{12}O_6$) that they use for food. Oxygen is released through photosynthesis (Figure 2-15).

The equation in Figure 2-15 shows that a plant that begins with six molecules of carbon dioxide, six molecules of water, and radiant energy from the sun can rearrange those atoms to produce one molecule of sugar and six molecules of oxygen. Some of the energy that is absorbed from the sun is stored in the chemical bonds of the sugar molecule and can be released and used either by the plant itself or by a consumer that eats the plant through the process of cellular respiration.

Cellular respiration breaks down the sugar molecule in a reaction that is the reverse of the photosynthesis reaction:

$$C_6H_{12}O_6 + 6\ O_2 \rightarrow 6\ CO_2 + 6\ H_2O + \text{energy}$$

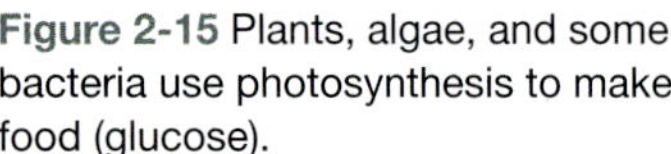

Figure 2-15 Plants, algae, and some bacteria use photosynthesis to make food (glucose).

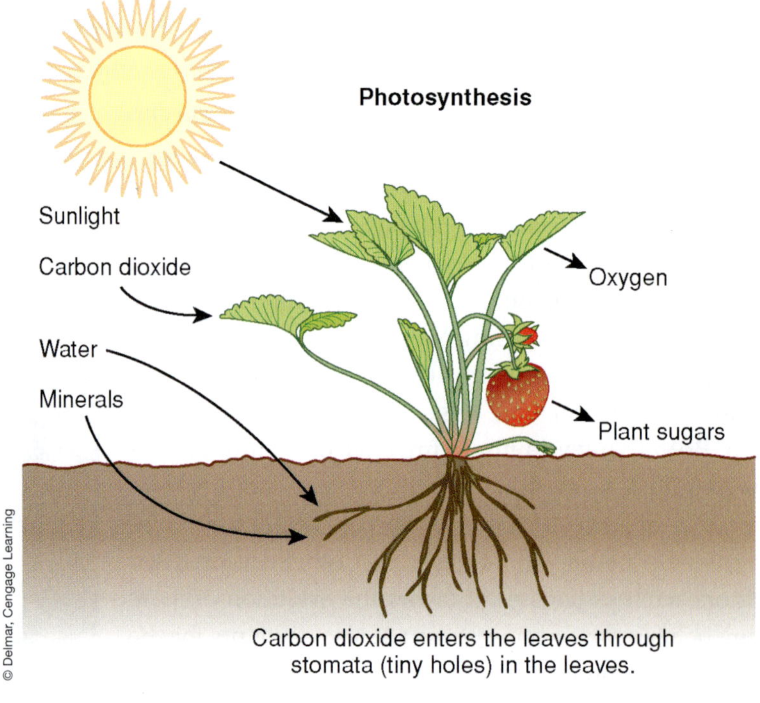

Cellular respiration breaks the chemical bonds within the sugar molecule, releasing the molecule's stored energy so that it can be used by the organism. When plants carry out cellular respiration, they use most of the energy to stay alive; this energy is eventually discharged into the environment as heat. The small amount of leftover energy is used to make new plant tissue—tissue that is made up of many kinds of energy-rich molecules that are available for herbivores to eat.

(a)

(b)

Figure 2-16 Herbivores gain energy from plants, whereas carnivores rely on meat for their energy. ((a) *Photo by Jesse Achtenberg, Courtesy U.S. Fish and Wildlife Service* (b) *Photo by Gary M. Stolz, Courtesy U.S. Fish and Wildlife Service*)

Organisms that create their own food through photosynthesis are called **producers**. They capture energy, store it, and set the stage for its transfer among other organisms in an ecosystem. They are the foundations of food webs. The organisms that cannot make their own food and therefore survive by eating other organisms are called **consumers**. There are three different types of consumers: herbivores, carnivores, and omnivores.

Herbivores, varying from insects to elephants, eat plants. Herbivores use the process of cellular respiration to break down the energy-rich molecules that form the plant's digestible tissues. Herbivores use most of the energy released through cellular respiration to stay alive; this energy is eventually discharged into the environment as heat. Like plants, they use the small amount of energy that is left over to make new body tissue—tissue that is made up of many kinds of energy-rich molecules and that is available for **carnivores** (meat eaters) to consume (Figure 2-16).

Carnivores also use cellular respiration to break down the energy-rich molecules that form the herbivore's digestible tissues. Carnivores use most of the energy released through cellular respiration to stay alive; this energy is eventually discharged into the environment as heat. They use the small amount of energy that is left over to make new body tissue—tissue that is made up of many kinds of energy-rich molecules and that is available for other carnivores (tertiary consumers) to eat.

Omnivores are animals that eat both plants and animals. They satisfy their energy needs by using cellular respiration in the same way as herbivores and carnivores (Figure 2-17).

Decomposers are the last step of an ecosystem's path of energy transfers. They participate in several ways:

- Any plants or plant parts left uneaten eventually die. The dead plant tissue becomes food for decomposers.
- Any part of an animal that remains uneaten by carnivores becomes food for the decomposers.
- Any tissue—plant or animal—that is not digested by consumers is released as feces. Feces become food for decomposers.

Figure 2-17 Omnivores gain energy from consuming both plants and meat. (*Photo by Chris Sevheen Courtesy U.S. Fish and Wildlife Service*)

By breaking down the organic material in wastes and dead organisms, decomposers obtain the energy they need to live. Decomposers are the last link of the ecosystem energy flow. All of the energy provided to them through feeding on waste and dead organisms is eventually discharged into the environment as heat.

Energy moves the same way in all ecosystems even though different species make up the food web in different ecosystems

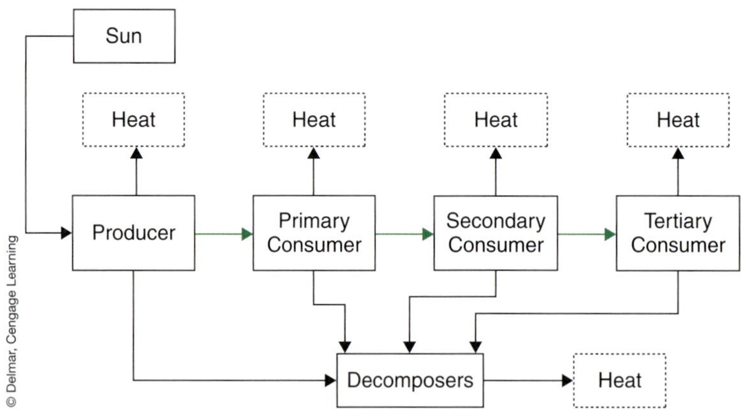

Figure 2-18 Energy moves the same way in all ecosystems.

Figure 2-19 The source of energy for organisms that live near hot, deep-sea vents is hydrogen sulfide. *(Courtesy NOAA Photo by C. Van Dover OAR/National Undersea Research Program (NURP); College of William & Mary)*

(Figure 2-18). Notice that all energy is provided to the community through the producers, which, in most cases, is from the sun. At each link of the web, energy is lost to the environment as heat. The fate of all the energy provided to the food web is to be returned to the environment as heat. Energy flows in one direction through all ecosystems and is not recycled.

Ultimately, the sun provides the energy for life in most, but not all, ecosystems. For example, the source of energy for organisms that live near hot, deep-sea vents is hydrogen sulfide (Figure 2-19). Bacteria extract energy from hydrogen sulfide molecules present in vent water and use it to make sugar molecules. Other species in the community live off the bacteria. Giant clams and crabs eat them. Tube worms live in a mutualistic relationship with them.

The Biomass Pyramid

Now that you know how energy moves through an ecosystem let us examine how many organisms, how much biomass, and how much energy is present at each level of a food web.

If you travel to Africa to see the plants and animals that live on the African plains, you will notice a pattern that ecologists see in almost all natural communities. The community contains lots of plants (producers), far fewer herbivores (primary consumers), even fewer carnivores (secondary consumers), and very few top carnivores (tertiary consumers). Ecologists see the same pattern if they measure the amount of biomass of organisms at each feeding level. This pattern is represented by a biomass pyramid (Figure 2-20).

So why is there less biomass at higher **trophic levels**? The answer lies in the transfer of energy that you just learned about. All organisms depend on the energy stored by producers: The primary consumers get their energy from the producers, the secondary consumers get their energy from the primary consumers, and so on.

A basic law of physics states that energy can never be created or destroyed but it can be converted from one form to another. You might think then that all of the energy absorbed by the producers eventually makes its way to the animals at higher trophic levels. This is not true for several reasons. First, not everything in the lower levels gets eaten by higher trophic organisms. An herbivore might eat the tender leaves of a tree but not the bark. A carnivore might eat another

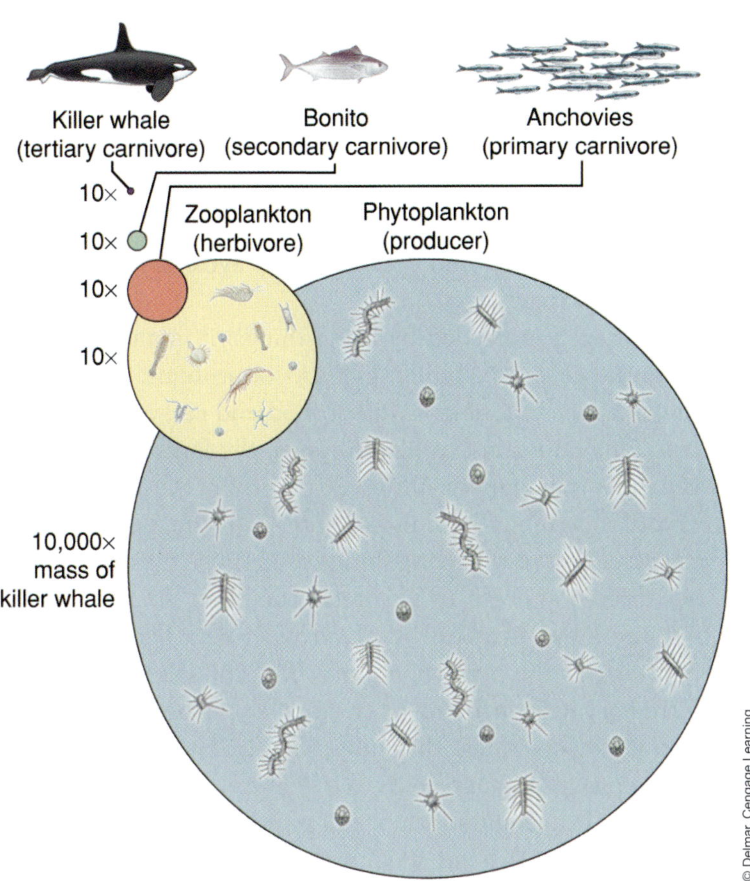

Figure 2-20 Biomass pyramid.

animal's meat but not its bones. Some organisms die and decompose without being eaten by organisms at higher trophic levels. Because all parts of all organism store some of the energy of that organism, any uneaten part is "lost energy" to higher trophic levels. In addition, not everything that an organism eats is digested; any undigested food part is also "lost energy."

Finally, consider the food that *is* eaten and digested. Remember that organisms at each trophic level use about 90 percent of the food energy they obtain just to stay alive. This energy is eventually released as heat into the atmosphere. Only 10 percent of the energy that an organism obtains from its food is converted into new body tissue that organisms at the next feeding level can eat. This means that each trophic level contains only one-tenth of the energy of the level below it. The fewer organisms and less biomass at each level reflect the smaller amount of energy that is present. Most ecosystems have fewer than five trophic levels because there is not enough energy in the form of edible tissue at the top to support another feeding level.

Why are **predators** (Figure 2-21) less abundant than producers? Predators are less abundant because there is not much energy available at the top of a food web to support lots of them. Most of the energy in an ecosystem is used up as it moves up the food web from one trophic level to the next.

Now that you know how energy moves through ecosystems, let us examine how certain life-sustaining chemical elements cycle between the living and nonliving parts of an ecosystem.

Figure 2-21 Predators.
(Courtesy NASA, The Met Office, UK/SAFARI 2000/ ORNL)

Figure 2-22 Specialized bacteria are responsible for nitrogen fixation in some plants.

Material Transfer in Ecosystems

Energy moves through ecosystems in one direction—from producers to herbivores to carnivores. Life-sustaining chemical elements and molecules, however, cycle through ecosystems. They are continually circulated through the atmosphere, hydrosphere, geosphere, and biosphere. You learned about a very important example in Chapter 1—the carbon cycle. Another cycle that is essential to life involves nitrogen.

Nitrogen is needed by living things. It is an essential element in amino acids, the building blocks of proteins. In addition it is in the enzymes that help us digest food and carry out other chemical reactions important to life. However, the largest reservoir of nitrogen is the atmosphere. Almost 80 percent of the atmosphere is made up of the gaseous form of nitrogen—N_2—the form that is not very reactive. As a result, living things cannot use the atmospheric nitrogen directly. It needs to be changed or "fixed" first.

There are two ways that nitrogen gets fixed in nature. First, lightning in the atmosphere provides enough energy to break the nitrogen-nitrogen bond in N_2 and let nitrogen atoms react with oxygen to form nitrogen oxides. This conversion makes the nitrogen available for reactions that produce amino acids (see Chapter 1). However, nitrogen fixation by lightning only accounts for a small part of the total nitrogen fixed in nature. The really hard work when it comes to nitrogen fixation is done by specialized bacteria (Figure 2-22).

Many different bacteria fix nitrogen but most live in soil. Some are free living, but some live symbiotically with the roots of certain plants: for example, soybeans and alfalfa. Photosynthetic cyanobacteria (blue green algae) that live in aquatic environments also fix nitrogen. Therefore, the microscopic world of bacteria is very helpful because they fix nitrogen from the atmosphere and make it available, especially to plants.

Nitrogen in soil is an essential and limiting nutrient for plants. If plants use it up faster than it gets fixed in soil they will stop growing; in other words, they will die if they do not get fixed nitrogen. Animals get their nitrogen from plants or from eating each other. The path that nitrogen takes from the atmosphere to the biosphere is, therefore, largely through bacteria in soil.

Guess where else the soil gets fixed nitrogen—from people. People make nitrogen-based fertilizer by reacting atmospheric nitrogen with hydrogen. The amount of nitrogen fixed in this way is about the same as the amount fixed by lightning and bacteria. Therefore, people play a significant part in the nitrogen cycle. Without this source of fertilizer, people could not grow the food and fiber they need.

Many nitrogen compounds are readily dissolved in water and they get flushed from fields and other soils into rivers, lakes, and the oceans. The oceans are a large nitrogen reservoir even though they contain 95 percent less nitrogen than the atmosphere. Even so, the oceans contain 100,000 times the amount of nitrogen that is in all of Earth's soils and plant life. The biggest nitrogen reservoir on Earth is the geosphere. Nitrogen accumulates in the sediments of the oceans and becomes incorporated in rocks through the rock cycle. Rocks of the geosphere contain about 100 times the amount of nitrogen in the atmosphere.

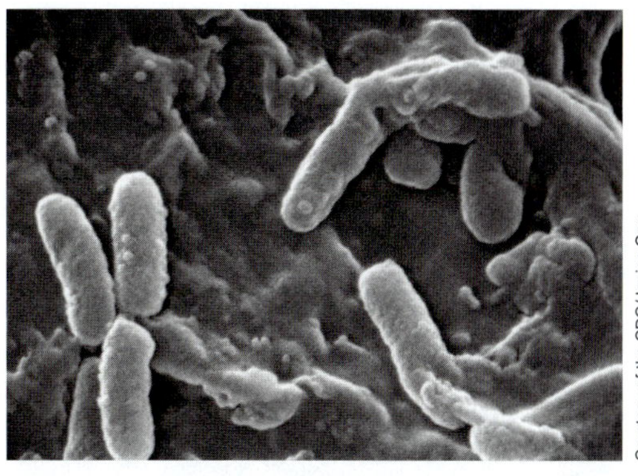

Figure 2-23 Some types of anaerobic bacteria are able to fix nitrogen.

This nitrogen is essentially trapped in the geosphere—only a small amount gets released to other Earth systems from the geosphere.

What happens to the nitrogen in the biosphere? When living things die, decay processes—most involving bacteria again—break down nitrogen by changing organic molecules into inorganic ones like ammonia (NH_3). This nitrogen can then be taken up by plants again or, you guessed it, be "denitrified." Denitrification converts nitrogen compounds back to the nonreactive gaseous form, N_2. Bacteria living in places where oxygen is not readily available—anaerobic environments that develop in certain sediments and deep soils, for example—do the job (Figure 2-23). They use nitrogen compounds as their energy source (food) and produce N_2 in the process. This nonreactive nitrogen can then pass back to the atmosphere from soils or the oceans. Therefore, nitrogen is an important example of a material cycle involving all Earth's systems. But it is a cycle in which bacteria in the biosphere play very important roles (Figure 2-24).

 the biosphere and humans

What do you think are some of the more important roles for humans in the biosphere? Humans, the species known as *Homo sapiens*, are just one of 1.6 million animal species in the biosphere. Today there are over 6 billion *Homo sapiens*, 50 years ago there were only half as many. The population of humans

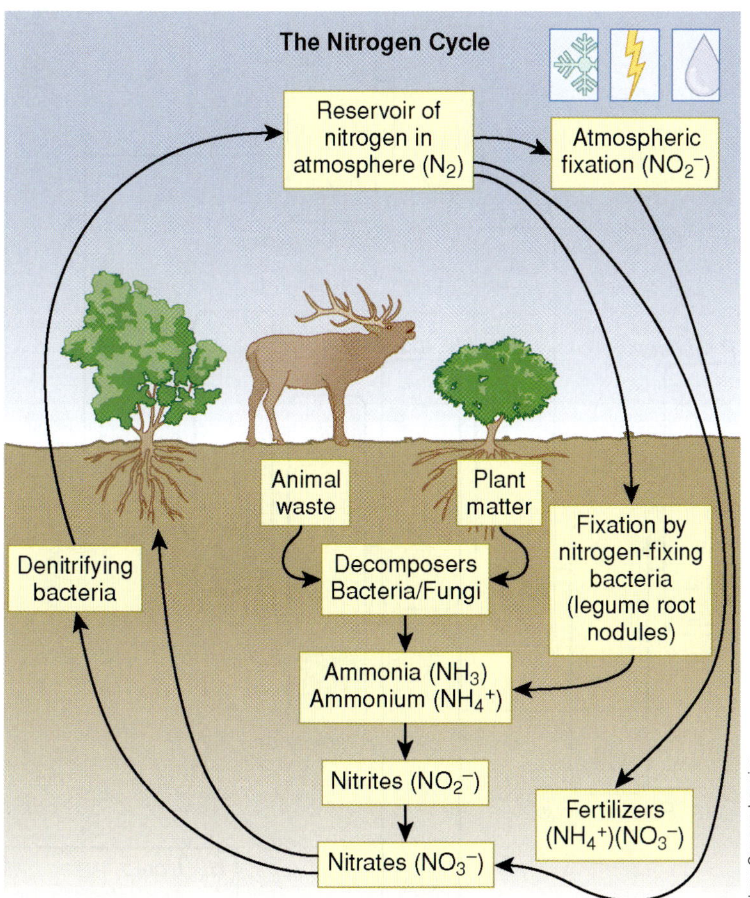

Figure 2-24 The nitrogen cycle.

Figure 2-25 *Australopithecus afarensis.*

is growing at the rate of over 100 million each year. You will learn more about populations and how they change in Chapter 3. For now, the important point is that there are a lot of people in the world. Where did all these people come from?

Human Origins

In the animal kingdom, *Homo sapiens* have a special characteristic in that their brains are the largest for their size. This makes humans the most intelligent animal species and literally capable of changing the world. This very important characteristic has taken several million years to develop.

People's genetic roots are in a family of species called the *Hominidae. Hominidae* are part of a larger group—a superfamily—that includes apes. Fossils show that some ancestral apes lived as long as 15 to 20 million years ago. The species *Australopithecus afarensis* (Figure 2-25), the oldest known *Hominidae,* evolved about 4 million years ago. This species could walk upright, but it had legs and arms adapted to living in trees. It appears to have been a creature intermediate between apes and humans. Since 4 million years ago several species have evolved in the *Hominidae* family (Figure 2-26).

Figure 2-26 The Hominidae family.

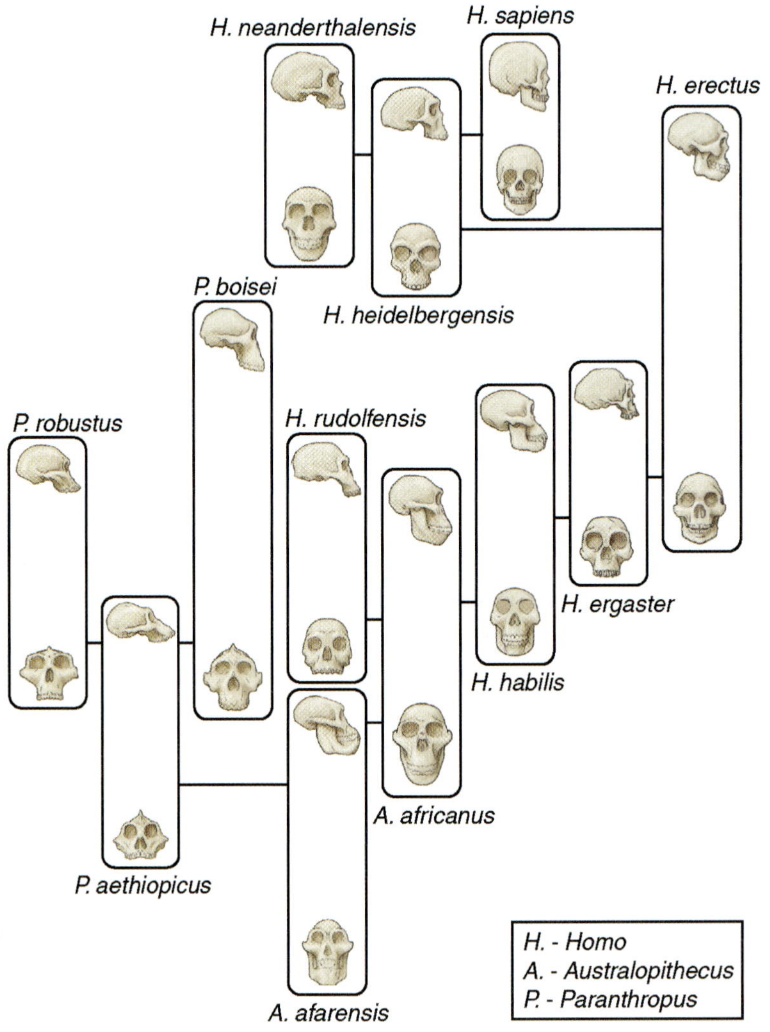

About 2.3 million years ago the ancestors of modern humans evolved from *Australopithicus.* The first of these, called early Homo in Figure 2-26, had a very human characteristic—a large brain. Early Homo's brain was almost twice as large as that of its *Australopithicus* ancestors. *Homo erectus,* who lived from about 1.6 million to 300,000 years ago, made excellent stone tools (Figure 2-27) and was the first human ancestor to migrate widely outside Africa. It continued the increasing brain size trend but its brain was still about 30 percent smaller than that of *Homo sapiens.* Neanderthals were a separate species (*Homo neanderthalensis*) that overlapped in time with *Homo sapiens.* Neanderthals had large brains but they also had large bulky bodies. They developed a distinctive culture but died out soon after *Homo sapiens* migrated from Africa into their European habitat about 35,000 years ago. *Homo sapiens,* or modern humans, evolved in Africa about 150,000 years ago and are the only surviving species in the *Hominidae* family. *Homo sapiens* have survived and increased their population to over 6 billion because their intelligence enables them to be supremely adaptive: They have literally come to inherit the Earth.

Figure 2-27 *Homo erectus.*

Humans and Ecosystems

Humans are the most adaptive animals on Earth; they beat the coyote hands down. They have learned how to survive in parched deserts, dense rainforests, and the frozen Arctic. They have done more than survive. People have developed rich cultures, graced their lives with the arts, and used science and technology to understand the world around them. There seems no limit to what *Homo sapiens* can do, and therein lies the rub. People are capable of drastically changing the biosphere through their interactions in ecosystems.

People significantly affect all of Earth's systems. They pollute the atmosphere, dam rivers, convert forests to farmland, and mine minerals from the Earth. The many ways that people interact with and affect the atmosphere, hydrosphere, and geosphere are discussed more completely in the other chapters of this book. All of these impacts affect ecosystems. After all, ecosystems are the combination of the living and nonliving parts of the environment. Let us examine how people have specifically affected the biosphere.

People have displaced or destroyed ecosystems. Perhaps the most widespread example is the conversion of forests, grasslands, and even desert to farmland. This is the principal way that people have provided the food that sustains their large and growing population. In the process, ecosystems are changed, displaced, or even destroyed. Examples of people destroying ecosystems are the cutting of virgin forests and the overgrazing of grasslands by flocks of domesticated sheep, goats, or other herbivores. Overgrazing is a major factor in the desertification of large regions.

People also destroy ecosystems to build houses, roads, buildings, recreation areas, and other developments. They build dams on rivers to meet household water needs, irrigate fields, control flooding, and generate power. They drain wetlands to create additional land for human use. Fragmentation occurs when ecosystems are broken apart by human development into smaller and smaller areas of undisturbed land, pushing the native plant and animal species into correspondingly smaller and smaller habitats.

Figure 2-28 The Carolina parakeet.

These activities clearly have a positive effect on human quality of life, but how are they affecting ecosystems? As the area of natural habitats decreases, the number of organisms living in those habitats must decrease, too. This results in a decrease in the populations of the native species, including the possibility of some species becoming locally extinct. If the entire habitat of an endangered species is destroyed, that species may become globally extinct. As you have learned, a change such as this within a community affects the entire system, possibly even the natural processes that people depend on.

People also affect ecosystems by introducing competing species. Species within a community develop their niche based on the biotic and abiotic conditions that surround them. All of the species of a community compete or cooperate as they share the available resources. Occasionally, either accidentally or on purpose, humans introduce a new species into these complex communities. For example, humans import farm animals, bring predators into an area to control native pests, and allow some species to accidentally stow away on ships and other forms of transportation. Some of these species' introductions do not seriously disrupt the local ecosystem, but they occasionally upset the local ecological equilibrium and cause harm to native flora and fauna. In some cases, the introduced species overpower the native species, causing local extinctions.

Actions such as these have made Hawaii one of the extinction capitals of the world. Twenty-four birds and dozens of snails, insects, and plants have become extinct in Hawaii since the first European contact in 1778. However, extinction can happen anywhere. Did you know that a wild parakeet used to live in the southeastern United States? The Carolina parakeet (Figure 2-28) has been extinct since about 1920.

People have hunted species for food or other purposes to the point that they have become extinct. These species were overharvested—they were depleted at a rate faster than nature could replace them. Entire ecosystems can be permanently altered if overharvesting occurs. Once again, short-term gains can cause long-term negative consequences. You will learn more about one example, the passenger pigeon, in Chapter 4.

People have also polluted ecosystems. The release of chemicals to the atmosphere, hydrosphere, and geosphere—as when fertilizers are used in agriculture, industrial facilities emit gases, accidental spills occur, or wastes are improperly discarded—finds many ways to affect the biosphere. Entire rivers, lakes, and estuaries have become poisoned in these ways. Fish have become contaminated and unfit to eat. Certain species, for example, carnivorous birds such as falcons, have been on the verge of extinction due to environmental pollution.

Many of humans' impacts on ecosystems have the general effect of decreasing biodiversity. You learned in Chapter 1 that biodiversity is a natural resource that people can benefit from. In general, being considerate members of the biosphere is in people's best interest, but accomplishing this is a tremendous challenge, especially as the population of people increases. Human population growth is such an important factor in people's impacts on the environment that it is the focus of Chapter 3.

2.5 conclusion

This chapter has focused on interactions within the biosphere and between the biosphere and other Earth systems. Understanding biosphere interactions is the realm of ecology, the study of ecosystems.

IN THIS CHAPTER YOU HAVE LEARNED:

- Life is organized into five kingdoms and three larger groups called domains. The animal kingdom includes people.
- The biosphere, all organisms and where they live on Earth, is organized at six levels: individuals, populations, communities, biomes, ecosystems, and the entire biosphere itself.
- Ecology is the study of the interactions of organisms and their physical environment. It is the science that recognizes and studies the interactions of the biosphere with the atmosphere, hydrosphere, and geosphere.
- Organisms influence the physical environment. They can change the character of the atmosphere, hydrosphere, and geosphere.
- The physical environment influences where organisms can live and how well they can live.
- Organisms influence other organisms. Producers convert the energy of sunlight to the stored energy of food and are the foundation of food webs that transfer energy through ecosystems. Organisms can compete, be predator or prey, or live in symbiotic relationships.
- The biosphere cycles materials through the other Earth systems. Materials essential for life, such as carbon, are stored in all of Earth's systems. Fluxes of these materials (nutrients) to and from these reservoirs and the biosphere are essential to support life.
- Humans are a late-developing part of the animal kingdom. They have evolved from their genetic ancestors over the last 4 million years. Humans are an especially important part of the biosphere because of their characteristic high intelligence (large brain for their size).
- People have an impact on ecosystems in almost all their activities. They have caused species to become extinct, displaced or destroyed ecosystems, and polluted ecosystems.
- Most of people's impacts on the biosphere cause a decrease in biodiversity.

Although people's impacts can be thought of as negative in many cases, the good news is that people can live in harmony with the world around them. The key need is to understand the environmental consequences of one's activities. As you study the following chapters, you will learn that people have choices about how they affect the environment. Earth's systems are resilient and people are supremely adaptive. If people recognize their impacts and accept their responsibilities, they can create, sustain, and protect a healthy biosphere.

KEY TERMS

Abiotic
Abundance
Biome
Biosphere
Biotic
Carnivores
Cellular respiration
Commensalism
Community
Competition
Consumers
Distribution
Ecology
Ecosystem
Habitat
Herbivores
Individual
Microhabitats
Mutualism
Niche
Niche dimensions
Omnivores
Optimal range
Parasitism
Photosynthesis
Population
Predation
Predator
Prey
Producers
Range of tolerance
Resource partitioning
Resources
Symbiotic
Traits
Trophic level

REVIEW QUESTIONS

1. At what levels is the biosphere organized?
2. What does an ecologist study about the Earth?
3. What is the difference between biotic and abiotic factors in an ecosystem?
4. What is the respective role of producers and consumers in the biosphere?
5. List and explain two examples of symbiotic types of relationships.
6. Select one organism with which you are familiar and explain how it is suited to its niche.
7. What can happen in a food web if a predator becomes extinct? Explain your reasoning.
8. Why is carbon such an important element in the biosphere?
9. Explain two ways in which nitrogen moves through the biosphere.
10. Describe two effects that humans have had on a particular ecosystem.

chapter 3

human population

Alaska. Do you think of Alaska as a distant, cold, and perhaps dark place where few people live? If so, you are partly right. It is the farthest northwest part of North America, it does get cold and dark in the winter, and, yes, compared to most states few people live in Alaska. Alaska's population is only about 600,000 people or only about one person for every square mile of surface area. With so much land and so few people, how can population be much of a concern? Well, like other parts of America, there are places where people prefer to live and even Alaska has urban areas that are rapidly growing.

Students at Palmer High School in Palmer, Alaska, have experienced residential growth in their community and are concerned about it. Palmer is in the Matanuska ("Mat") Valley. This valley is about 20 miles north of Anchorage, Alaska's largest

Figure 3-1 The Matanuska Valley, Alaska. (*by James Stewart Creative Commons License Source*)

city (Figure 3-1). Palmer and other nearby communities are included in the Matanuska Susitna Borough; a borough is like a county "outside" (Alaska-speak for the rest of America). The Matanuska Susitna Borough is about the size of West Virginia. Dust and silt blown into the Mat Valley from the flood plains of nearby glacial rivers have helped create excellent soils. The soils are so good for farming that the U.S. government helped move entire families here from drought-stricken Midwest farms during the Great Depression of the 1930s. More recently, the Mat Valley has become a suburban mecca for people from Anchorage. It is the fastest growing area in Alaska. Its population has increased almost 50 percent in the last 10 years (Figure 3-2).

Suburban growth in the Mat Valley means more housing subdivisions, fewer farmlands, and increasing encroachment on sensitive wetlands and

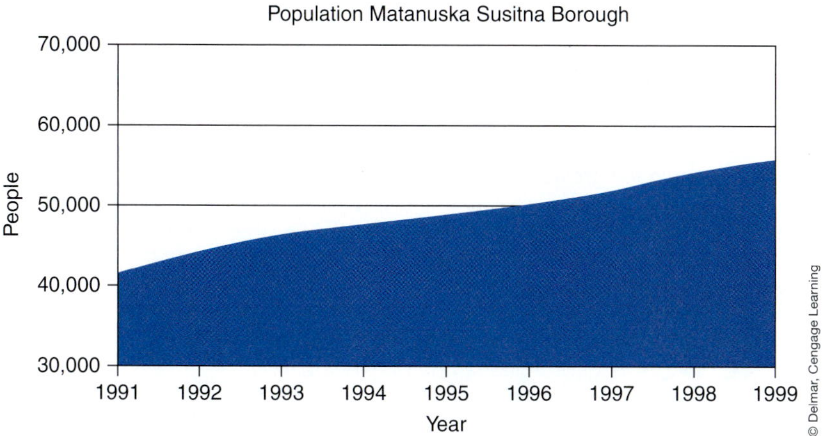

Figure 3-2 Population growth in the Matanuska Susitna Borough, Alaska.

coastal ecosystems. Students at Palmer High investigated the impacts of increasing residential development in their area. They concluded that wetlands and coastal ecosystems were especially vulnerable to increasing residential growth. The students developed a plan to protect these lands. Their plan would include expanding wildlife refuges along natural boundaries such as streams rather than arbitrary property lines, creating corridors for wildlife between refuges and other sensitive areas, and delineating buffer zones around the most important ecosystems where development is restricted. So even in remote, sparsely populated Alaska, increasing numbers of people can lead to concerns about changes to habitat and displacement of wildlife. People have an impact on the environment wherever they live, but special concerns develop where population growth happens, whether it be a specific community or the world as a whole.

A population can be thought of as a dynamic system. It can change in size and composition (characteristics such as people's ages and the proportion of males to females) depending on a wide range of influences within the population as well as the biotic and abiotic conditions around them. If a population is stable in its size and composition, then it can be considered as being in a steady state. But many natural populations change in size and composition over time. As you learned in Chapter 2, coyote populations in America have increased over the last 100 years as this species adapted to living around and close to people. Over a period of perhaps several thousand years or more, the population of every species of dinosaur decreased to zero, becoming extinct 65 million years ago. People are a very important nonstable population. The number of people on Earth is now doubling about every 50 years! In addition people are a special case when trying to understand the biosphere and its interactions with other Earth systems.

People are supremely adaptive. They eat just about anything. Well, maybe you do not eat just about anything but people around the world have come to eat all sorts of things. People also have conquered diseases and on average are living longer and longer. They change the world around them to their benefit whenever and wherever they can. As a result, people have become the dominating influence on the environmental health of Earth. The more people there are, the more influence on Earth's environment they have. This is why human population is the focus of this chapter.

Every chapter in this book will help you better understand how people interact with Earth's systems. This chapter focuses on how the number of

AFTER COMPLETING THIS CHAPTER, YOU WILL UNDERSTAND:
The basics of population dynamics, including how birth and death rates regulate population size, the role of migration, and how populations develop different age characteristics.
How human population growth has varied over time, factors influencing the dramatic increase in human population, and what it will likely be in 2050.
Where the human population is distributed, how this distribution is changing, and where most people are likely to live in 2050.
The relation of human population to the general environmental health of the atmosphere, biosphere, hydrosphere, and geosphere.
What the concept of sustainability is and why the number of people is important to achieving it.
What carrying capacity is and how people change Earth's carrying capacity for themselves and other living things.
Why the study of human population dynamics is important to you and your future prosperity.

people and their needs are especially important influences on the environment. People as individuals and as societies make choices every day that affect the environment. One of these choices is how many children a family will have—a key factor in determining how large the human population will become.

 population basics

What makes populations change in size and composition? There are three basic controls: birth rate, death rate, and migration. These controls combine to define a population's **growth rate**. Growth rate is the key to understanding how populations change.

Growth Rate

The **birth rate** is defined as the number of live births per 1000 people per year. It is calculated using the total number of live births in a year and the mid-year number of people in a population. The number of births per person is multiplied by 1000 to determine the birth rate. The **death rate** is defined as the number of deaths per 1000 people per year. By using the total number of deaths in a year and the mid-year number of people in the population, the death rate is calculated the same way as the birth rate. Note that these birth and death rates are for the population as a whole. They are not specific to a particular age group (such as ages 15 to 49) so they are called "**crude rates**."

Migration takes into account the number of people who move from one area to another (*immigration* is migration into and *emigration* is migration out of an area). Migration helps understand population increases or decreases in specific areas. For example, the recent increase in the population of the Mat Valley in Alaska is due largely to immigration, not a big increase in the number of births in the local hospital! Since about 1990, 30 percent or more of the United States population growth has been from migration of people into the country. Migration can make a big difference in the population of a specific area. But migration means that for every place that gains population, another place loses population. For the Earth as a whole, the difference between birth rate and death rate is the human population's "**natural growth rate**."

Growth rate is commonly expressed as the percentage population change. If the birth rate in a population is 30 (per 1000 people per year) and the death rate is 20 (per 1000 people per year), then the natural growth rate is 10 (per 1000 people per year). This example population is growing at the rate of 1 percent per year (10/1000 = 0.01 or 1 percent). If migration is taken into account, the average annual growth rate for a given population can be determined (average annual growth rate = birth rate − death rate + immigration − emigration).

Natural growth rates vary between regions of the world. Developed countries tend to have low to negative natural growth rates. Germany, for example, has a growth rate of −0.2 percent. Less developed countries tend to have higher growth rates. The average natural growth rate for less developed countries, excluding China, is 1.8 percent; Niger in Africa has one of the highest at 3.5 percent. If the growth rate is zero (birth rate is equal to death rate) and migration is not a factor, there is said to be **zero population growth**, or **ZPG**. ZPG means the population is stabilized at a specific size and composition. ZPG is considered a worthy long-term goal for Earth's human population by many who are concerned about people's impacts on the environment.

Growth rates are a key to predicting future populations. Population growth has another aspect that helps us understand why the number of people can increase rapidly. If a population has a positive growth rate for many years, it will experience what is called **exponential growth**. Even if the growth rate stays the same, the number of people added to the population will increase exponentially each year. This is because the base or beginning population each year is greater than it was the previous year. If last year a population had 1,000,000 people and it grew at a rate of 1 percent, then the starting population for the coming year is 1,000,000 plus 10,000 (1 percent), or 1,010,000. So, in the second year the growth rate of 1 percent will add 10,100 people, not 10,000 like it did the year before (Figure 3-3).

Such small additional increases do not seem like much from year to year but over time they really add up. Growth rates are often used to calculate **doubling time**—the amount of time it will take for a given population to

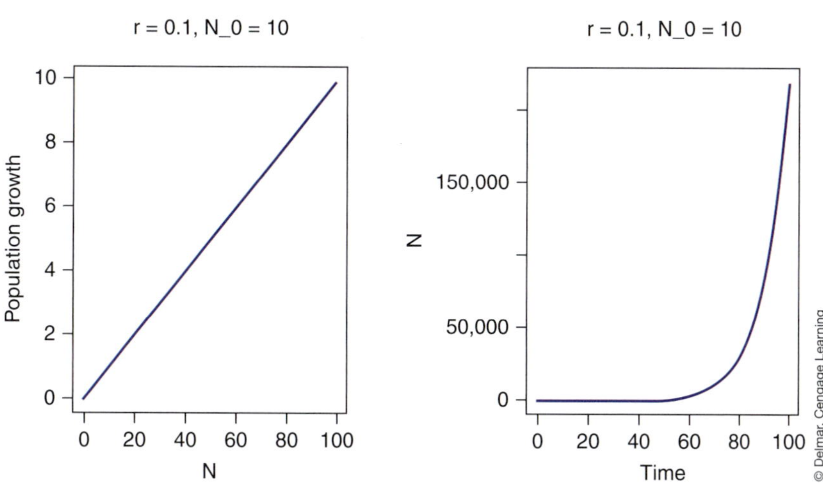

Figure 3-3 Exponential population growth.

double. At a growth rate of 1 percent, the doubling time is only 70 years. Over tens to hundreds of years, just a few human generations, exponential growth can lead to tremendous population increases. It is the same type of growth that your savings account can have as long as you leave all your money (starting population) plus the interest it earns each year (yearly population increase) in the bank. It is because of exponential growth that the size of the human population can now double in about 50 years, reflecting an average global natural growth rate of about 1.3 percent.

Differences between birth and death rates tell more about populations than just their growth rates. Birth and death rates that are sustained for several generations also control the age composition of a population. The age structure of populations and their relation to birth and death rates is illustrated in Figure 3-4. About equal numbers of younger and older people are present in a stable ZPG population. Belgium, for example, has a growth rate of 0.1 percent; 17 percent of its 10.4 million people are less than 15 years old and 17 percent are over 65. Populations where birth rates are much higher than death rates become dominated by younger people. Let us compare two countries with about the same population, Nigeria and Japan.

Comparing Nigeria and Japan

Nigeria and Japan have about the same number of people; Nigeria has 137.4 million and Japan has 127.6 million. However, in Nigeria, where the 2004 birth rate was 42 and the death rate was 13, 44 percent of the 137.4 million people are less than 15 years old (Figure 3-5). Japan, on the other hand, had a 2004 birth rate of 9 and a death rate of 9. Only 14 percent of Japan's 127.6 million people are less than 15 years old, whereas 19 percent are over the age of 65 (Figure 3-6).

Nigeria faces tremendous challenges in educating and employing its large number of young people. Japan needs to provide for an aging population with many retired people and is increasingly concerned about maintaining an adequate workforce. European countries with aging populations have addressed this problem by augmenting their workforce through immigration. Many

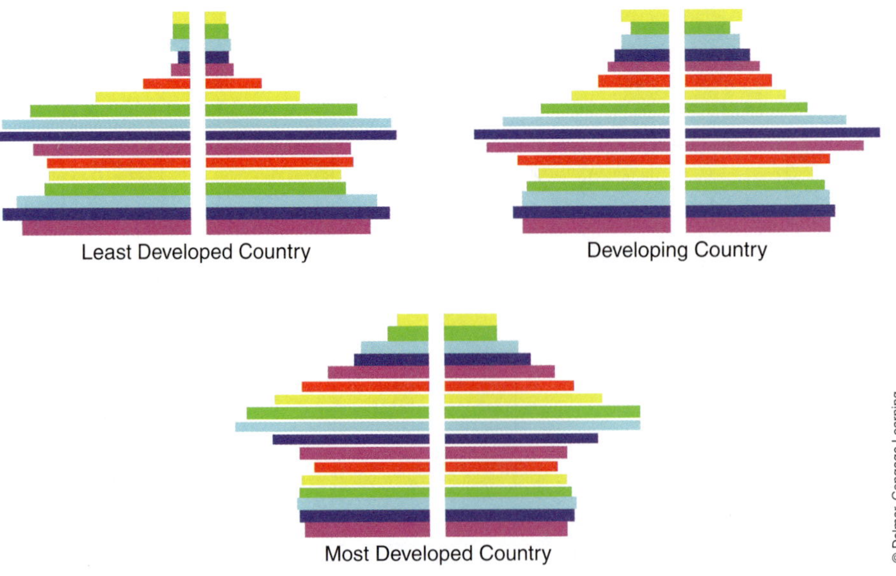

Figure 3-4 Population age structures for countries with different birth and death rates.

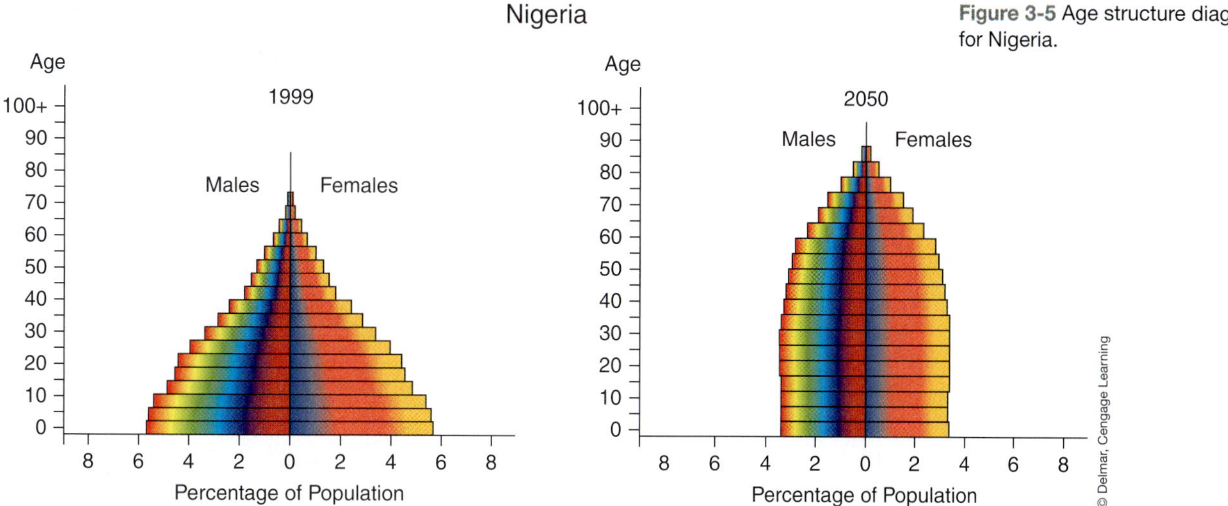

Figure 3-5 Age structure diagram for Nigeria.

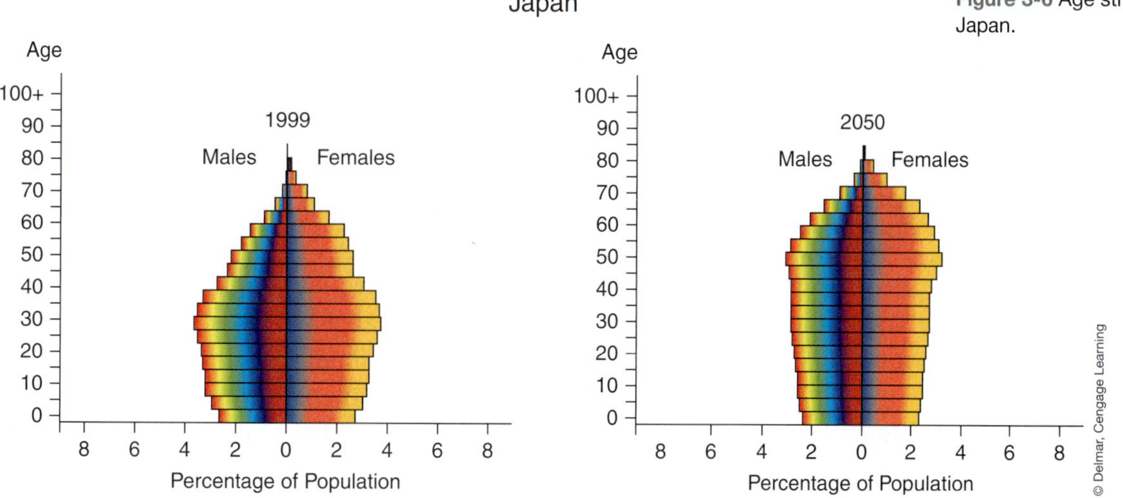

Figure 3-6 Age structure diagram for Japan.

people from less developed countries immigrate to Europe to find jobs. The age structure differences between Japan and Nigeria emphasize how sustained differences in birth and death rates lead to very different population compositions and very different economic and social challenges.

Some people who study populations theorize that populations evolve in four stages (Figure 3-7). The theory of this evolving character is called **demographic transition**. (Demography is the science dealing with the distribution and vital statistics of populations.) Examples of demographic transition have been identified by studying industrialized countries like those in Europe. In these countries, both birth and death rates were initially high, but as industrialization progressed, the death rate fell faster than the birth rate. Eventually both birth and death rates reached stable lower levels and the number of people stopped rapidly increasing as the population approached ZPG. If this theory applies to the world as a whole, then sometime in the future—after the world's countries have socially and economically progressed to some degree—the world's population is predicted to stabilize. How people get from here to there is uncertain and possibly unlikely if conscious efforts to control population are not maintained or expanded.

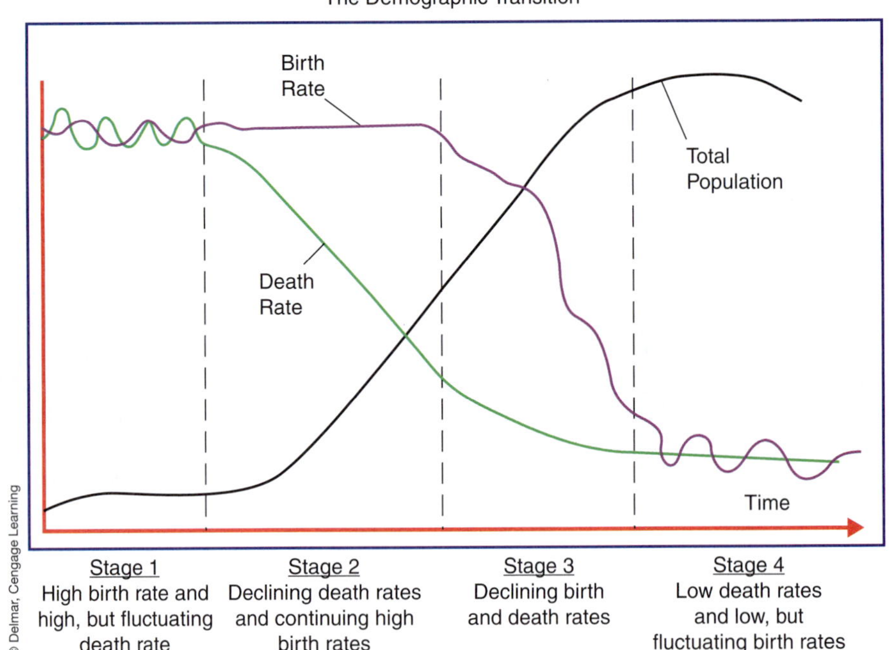

Figure 3-7 Theory of demographic transition.

3.2 population history and trends

Predicting the future of human population requires understanding its past and its present. The history of population changes helps explain how people got to this point and identifies the future-defining trends that are in place. History shows that human population has increased exponentially. Even though disease, famine, war, and natural disaster have taken tremendous tolls on human population through history, the relentless increase in the number of people on Earth has continued. We know that in the near future, that is, within your lifetime, there will be a more densely populated world with the most growth occurring in less developed countries and an increasing concentration of people living in cities—the megacities of the future.

Human Population Growth

Understanding the history of human population growth gets tougher the farther back in time one investigates. Even contemporary census figures are estimates. However, the general picture is clear: Human population has changed in three general stages (Figure 3-8).

Prior to 10,000 B.C., people were hunter-gatherers. They fed themselves with the animals they could catch and the fruits and plants they could collect. The food resources of the land were a limit on how many people could be fed. This limit, a short life expectancy of perhaps 20 years, and low birth rates kept the population from rapidly increasing. At this time it is estimated that the total number of people on Earth was about 6 million.

It was about 10,000 B.C. when people first domesticated plants and animals. This was the beginning of agriculture and the agrarian stage of human population development. Agriculture allowed people to produce more food and

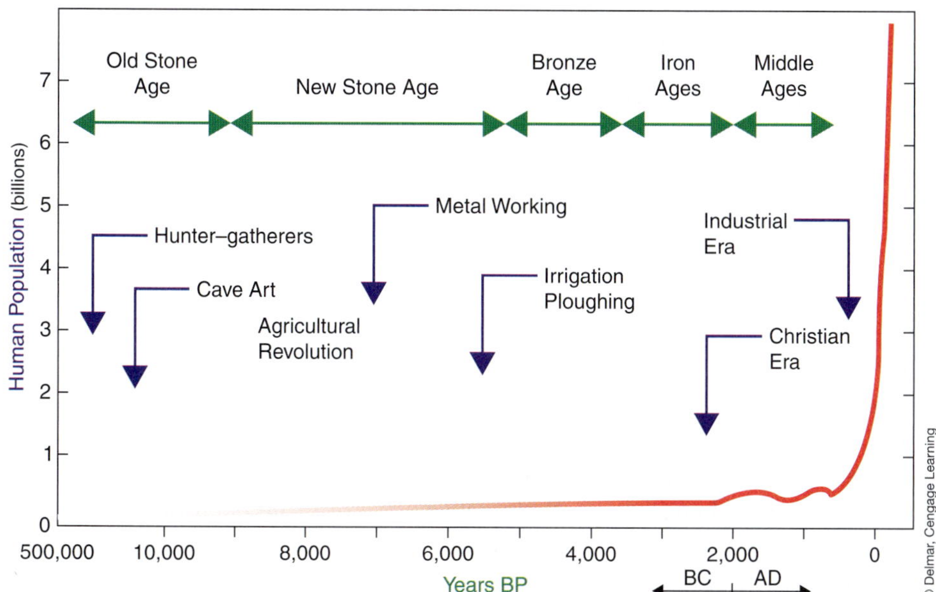

Figure 3-8 World's population growth through history.

to reside far longer in specific places. Life expectancy and birth rates increased. Growth rates were low though, perhaps 0.1 percent, and it took 10,000 years—to 0 A.D.—for worldwide population to reach 250,000 million. From 0 A.D. to the end of the agrarian stage in 1750 A.D., however, worldwide population increased to 750 to 800 million people. So when America's original 13 colonies were going full speed, just 255 years ago, world population was still less than 1 billion people.

Things changed dramatically with the onset of the industrial revolution. As people learned to use new forms of energy—especially coal and petroleum (see Chapter 15)—they became better food producers, started controlling diseases, and lived more safely. The world population hit a landmark 1 billion people in the year 1804, doubled to 2 billion by 1927, doubled again to 4 billion by the early 1970s, and reached 5 billion people by 1987. Most remarkably, the world went from 5 billion to 6 billion people in just 12 years from 1987 to 1999!

Life expectancies increased to about 50 years in 1850 and to 60 or more today. If you are a male in the United States, your life expectancy is now 75 years; for females, it is now 80 years. The better nutrition and health care that have come with economic development have added years to people's lives.

The world's population has had some setbacks from time to time as disease, famine, war, and natural disaster increased death rates. For example:

- Natural disasters such as the Indian Ocean tsunami of December 26, 2004, have killed hundreds of thousands of people in just hours.
- Famines such as the Irish Potato Famine of 1845 to 1849 have killed hundreds of thousands to a few million people in a few years.
- Wars have led to the deaths of millions; over 50 million people died between 1939 and 1945 in World War II.
- Diseases, such as the bubonic (black) plague, which killed about 25 million people or one-third the population of Europe in 5 years (1347–1352), have been devastating to particular areas or societies.

In reviewing the historical impacts of disease, disaster, war, and famine on population, countries and regions have been very severely affected, but these factors have not significantly changed the continuing growth of global population. The large number of people and their wide distribution around the world has kept the negative impacts of these factors from significantly slowing global population growth.

Disease

Even though human population growth has only temporarily been set back by disease, it has been absolutely devastating at times. Did you know that it was not too long ago that people did not understand the connection between bacteria and disease? Today most people understand that unclean water can be bad for their health. This is especially true for water that is contaminated with germs people pass on in their sewage. Waterborne bacteria have caused epidemic after epidemic of diseases such as cholera and typhoid fever. However, people did not get the connection between bacteria-bearing water and disease until about 150 years ago. Before then the world's devastating disease outbreaks such as the plague that devastated medieval Europe were thought to be caused by bad air or just "God's will." Disease seemed to come out of nowhere and kill previously healthy people in a day or two.

Understanding that unclean water was bad for human health started with the preserving work of John Snow, a doctor in London, England, during the mid-nineteenth century. Dr. Snow and a few other doctors and scientists of the time suspected that diseases were caused and spread by something in the water. During a cholera outbreak in 1854 in London, Dr. Snow rigorously tested his hypothesis that this disease was being spread through unclean water. He interviewed people and recorded where the sick lived and where they got their water. He found that almost all the people who died lived near a community well on Broad Street. Two people who died but lived outside the Broad Street area were found to have bottled and used water from the Broad Street well. Furthermore, people who lived near the well but got their water from other sources—such as workers at a local brewery who received free beer and did not use the well—did not contract the disease. Altogether, Dr. Snow showed that, in a 10-day span, over 500 people who lived within 230 meters (250 yards) of the Broad Street well contracted cholera.

Image taken by Hannah Mably, 2008

Figure 3-9 Broad Street pump replica in London.

What made the Broad Street well the likely source of cholera? Dr. Snow found that a house at 40 Broad Street, where four people died of cholera, had a cesspool that overflowed into a drain that passed near the well. The well water was being contaminated with human waste. Because the actual cholera-causing bacteria in the water were not identified at this time, it was an uphill battle for Dr. Snow to convince the city to disable the Broad Street well (Figure 3-9). However, his rigorous investigations and diligent evaluation of alternative causes left little doubt that the Broad Street well was the source of cholera. The city finally became convinced and the cholera epidemic died out soon after the well was disabled. Dr. Snow's studies, marked by thorough data gathering and the testing of multiple hypotheses, have gone down in history as a landmark scientific study of disease. He is known as the "Father of Epidemiology," the study of diseases in a population.

Disease Today

Today a disease that is having a significant impact on some populations is HIV/AIDS. Over 20 million people have already died of this disease; about 1.9 million died in 2003. Southern Africa was the hardest hit. Here the percentage of the population with HIV/AIDS is commonly greater than 20 percent. Almost 39 percent of the people in Swaziland have HIV/AIDS. There are not very many old people in Swaziland; 43 percent of the population is less than 15 years old and only 3 percent are 65 or older.

HIV/AIDS may be the most deadly disease now attacking large segments of the human population. Other diseases have been identified that could also be devastating if they were not recognized and adequately addressed. Severe acute respiratory syndrome (SARS) is a recent example of limiting the effects of a deadly disease. SARS is a virus that attacks the human respiratory system. About 10 percent of the people who contract SARS die of its pneumonia-like effects; up to 50 percent of older people who get SARS may die. Within a few months of its first reported occurrence in China (February 2003), SARS had spread to more than 24 countries in North and South America, Europe, and Asia. However, national and world health organizations quickly responded and integrated their efforts to identify and isolate the disease and its carriers. The spread of this contagious disease was halted and the total deaths from the outbreak are thought to have been in the several hundreds. The international effort to contain the 2003 SARS outbreak is a success story for global disease control. Currently the world is concerned about the potential spread of avian (bird) influenza (flu) through large parts of the human population. A few people in local areas have already contracted this disease and birds are spreading it among themselves around the world. It looks like bird flu could be the next big international challenge to controlling disease.

Trends

Changes in human population over the past 50 years define several trends that help to predict how this population will change in the future. Organizations such as the United Nations and the Population Reference Bureau in Washington D.C. make special efforts to understand ongoing changes in human population. Their studies help show what the size, distribution, and age of the human population is likely to be in 2050—about when your grandchildren could be taking this class. Although predictions of future population require several assumptions, especially about future birth and death rates, the established trends provide some assurances that the future is generally predictable.

The fastest growing period of human population occurred in the last half of the twentieth century. The highest average growth rate, 2.1 percent, was in the 1960s, and the highest number of people added in a year, 86 million, was in the 1980s. At the end of 2004, the world population totaled 6.4 billion, and it was increasing by about 80 million people per year. The number of people in 2050 will be about 9.3 billion or about 50 percent more than today.

Although there will continue to be tremendous disparities in economic opportunity and access to education and health care between the less developed and developed parts of the world, family planning is having an impact: Over 60 percent of the married women of childbearing age (15–49) in Latin America

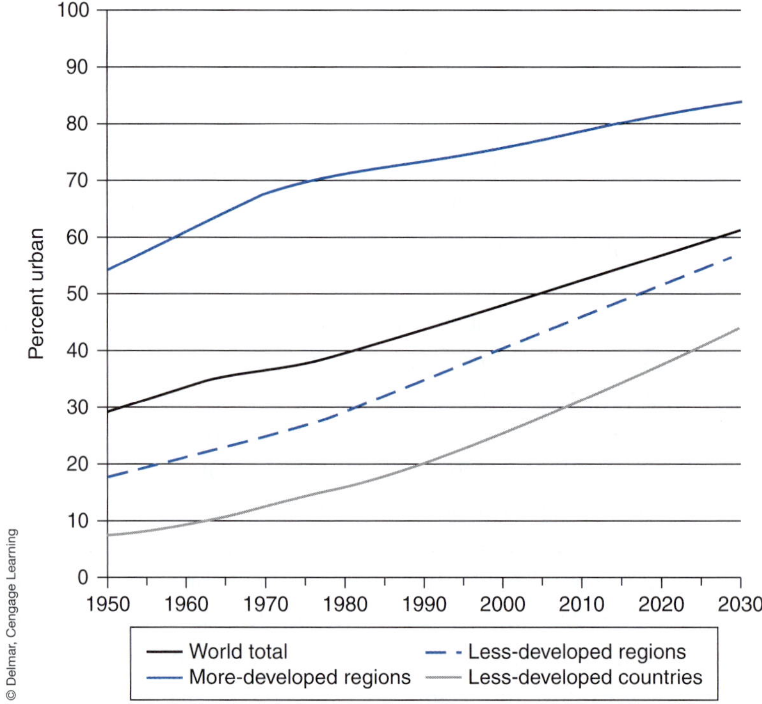

Figure 3-10 Expected urbanization growth.

use modern methods of contraception. In China, national policies, some pretty heavy-handed such as mandatory sterilization if a couple had more than one child, has led to a 2004 growth rate of only 0.6 percent. Birth rates have generally decreased during the last 40 years, and some studies conclude that the world's population will stabilize at about 10 billion people by the end of this century.

Where will all these new people live? Between now and 2050, over 99 percent of the world's 3 plus billion new people will live in less developed countries. The United States, with a growth rate of 0.6 percent and higher levels of immigration, will be the only developed country with significant population growth between now and 2050. By 2050, 90 percent of the world's people will be living in less developed countries.

In addition more and more people will be living in cities. By 2030, as much as 60 percent of the people will live in cities (Figure 3-10). In 1960, there were only two cities in the world with more than 10 million people: New York and Tokyo. By 2015 there are expected to be 26 of these megacities, most of them being in less developed countries. Although you may be thinking of cities as dirty and crowded, as some parts can be, people living in urban areas generally have better access to education and health care than their rural counterparts. Historically, urban people have come to have longer life expectancies than rural people. As you will learn in Chapter 8, well-managed cities can be part of the solution when trying to balance population needs with environmental protection.

Another change underway is the aging of the human population. In 2000, the median age of the world's population was 26.1 years, but by 2050 it is expected to be about 38 years. Aging of the population is most pronounced in developed countries but it is underway in less developed countries too. By 2050, over 30 percent of the population in developed countries will be 65 or older. In less developed countries, which now only have about 5 percent of their population above the age of 65, the proportion of people 65 or older will be as much as 20 percent or more. These are significant changes with a number of related economic and social impacts within particular societies like those encountered by Nigeria and Japan. From the viewpoint of global population dynamics, one of the important implications of the world's aging population is the increased migration that may occur. Immigration could help replenish workforces in older societies. The movement of these people could help further global economic and cultural integration.

Meet the Baby Boomers

You may not think that aging is much of an issue yet, but let us look at the example of the baby boomers in the United States. World War II was a time of tremendous economic and social stress for Americans. Women staffed the factories to build ships, tanks, and airplanes. Rationing of food and fuel was necessary. Millions of men and women went to war overseas; about 500,000 Americans died in this conflict. As you can imagine, many people put normal activities such as starting and raising a family on hold during the war. But when the war ended, things got back to normal fast. Men came home, women stopped working in the factories, and babies started to be born—lots of babies (Figure 3-11). From 1946 to 1964, the baby boom generation, or just "boomers," was added to the population. The number of births went from less than 3,000,000 per year during the war to over 4,000,000 per year in the late 1950s and early 1960s. It seems that as soon as normal activities of family life were possible, Americans really got with the program.

The boomers, some 76,000,000 in total, have had an impact on all aspects of American life. They stressed hospital facilities when they were babies and educational resources when they were children. They have dominated the workforce for most of their lives. Some 85 percent of boomers hold jobs—mothers and fathers both. Dual-income families became the norm with boomers, and now they are starting to retire.

Boomers comprise 28 percent of the U.S. population and some 4 million of them turned 50 in 2005. As they plan for retirement, they expect the federal Social Security program to help them. Taxes paid by future workers—you—will provide more and more of the Social Security dollars used by boomers. So an aging population is important to you. The future care and feeding of boomers is an example of what all aging societies will be facing as their populations get older and older.

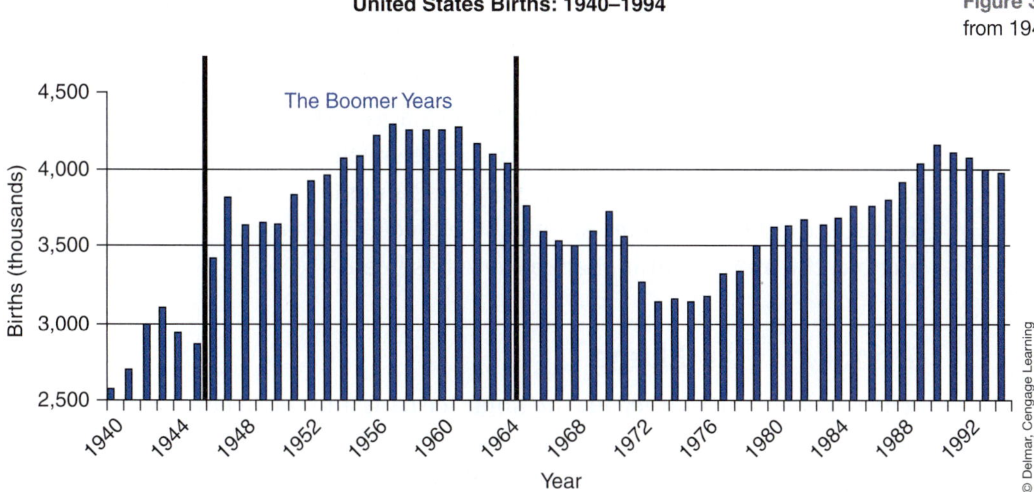

Figure 3-11 Number of births per year from 1940 to 1994.

 population and the environment

People have an impact on the environment, and the more people there are, the greater their impact. With a 50 percent increase in population by 2050, or 3 billion more people, it is important to understand what their environmental effects are and how the negative ones can be mitigated or prevented. Other chapters in this book focus more specifically and more completely on many of people's impacts. In this chapter it is the tie between large numbers of people and some global environmental consequences for the biosphere, atmosphere, hydrosphere, and geosphere that is important.

Population and Biodiversity

In Chapter 1, you learned that biodiversity, which refers to the entire variability of life, including species, ecosystems, and genetic diversity, is a good thing. There are 1.4 million described or known species, but estimates of the total number of species range from 10 million to as many as 30 million. Species naturally become extinct. Most species that have evolved on Earth are now extinct, having lived on average for 2 to 10 million years. Some scientists think that the rate of extinction is now high, perhaps as high as during some of the great mass extinctions of the distant past (see Chapter 1). Estimating contemporary extinction rates is a hard job. It requires many assumptions about the actual number of species and the rates of habitat loss. But some are convinced that extinction rates rise as human population increases. Here is a key reason why.

Another Island Ecology Example

As you learned in Chapter 1, islands are a great place to study populations and their evolution. Darwin's studies about species made the Galápagos Islands famous (see Chapter 2). Other ecology studies on islands have shown a key relation between habitat area and the number of species; in other words, the more the habitat area, the greater the number of species (Figure 3-12). Figure 3-12 shows the number of reptile and amphibian species on the Caribbean Islands. If an island's area is 90 percent smaller than another one, it will have half as many species. This is an important relationship because species diversity is proportional to habitat area. This relationship is the foundation of the conclusion that biodiversity is seriously jeopardized by human developments. The more people change the land and displace or destroy habitat, the fewer species there will be. Of course, people can decrease biodiversity in other ways such as by hunting (see Chapter 4) or by introducing nonindigenous species that out-compete native species (see Chapter 5). The relationship shown in Figure 3-12 is very fundamental: the more habitat people change, the fewer species should be expected to survive. So far, the more people, the more habitat change there has been.

Population and Land

The Earth is pretty large, but the polar ice caps and the oceans are not places where humans can live. This leaves about 25 percent of Earth's surface—49 million square miles—to even potentially be habitable, including some

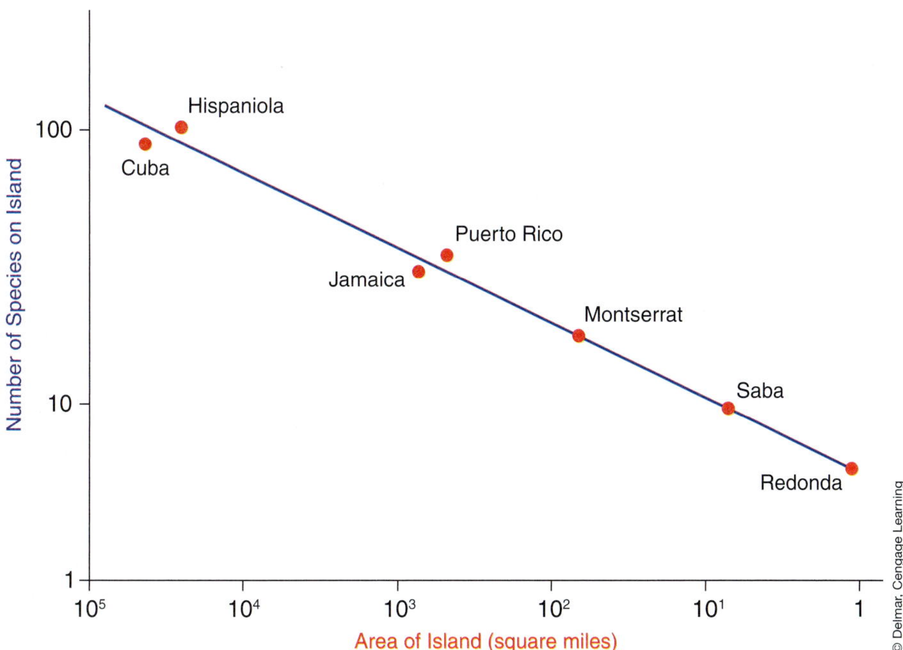

Figure 3-12 Diagram showing the relation of island size to species number on Caribbean Islands.

pretty marginal areas like the large deserts. With 6.4 billion people, the current population density is 50 people per square kilometer (130 people per square mile) of potentially habitable land. When the number of people reaches 9.3 billion in 2050, the population density will be 73 people per square kilometer (190 people per square mile). This is about 3.4 acres per person.

But you know that people are not spread out equally around the world. Population densities vary tremendously. Some islands are very densely populated. Malta, an island in the Mediterranean Sea, has 3,229 people per square mile. Cities have the highest population densities. Monaco (a city that is also a country) has 44,000 people per square mile, and Macao, China, has 58, 100 people per square mile—that is packed! The big question though is how much land do people actually use or, in some cases, use up.

People use land for crops, pasture, forests, and other resource developments and they have an impact on land when it becomes polluted or degraded, especially by erosion (see Chapter 14). For example, it is estimated that 75 billion tons of soil are lost globally each year through the effects of land degradation and erosion (Figure 3-13). The value of this land resource has been placed at $400 billion per year, or over $60 per year for every person on Earth! However, increasing numbers of people do not need to lead to increasing land degradation and soil loss. It depends on what the people choose to do to the land. If land use choices continue for the next 50 years as they have in the past, the negative impacts will increase as population increases. In some cases, people actually affect land and soil to the point that the land's productivity as farmland is irreparably harmed. Soil, commonly thought of as a renewable resource that can be used over and over, can actually become a nonrenewable resource due to people's actions. People's impacts on the hydrosphere, especially fresh water resources, are another example.

Figure 3-13 Soil erosion. (*Courtesy ESW Image Bank © Bruce Molnia, Terra Photographics*)

Population and Fresh Water

Water is one of Earth's resources that is essential to humans. It makes up 60 percent of the weight of people's bodies and they will die in about a week without it. The human body needs about eight glasses of water (about ½ gallon) a day to be healthy. Of course people use water in many other ways, too, but drinking clean water every day is essential to sustaining a healthy life. Therefore, more population growth means more need for clean, fresh water. Today people need over 3 billion gallons a day of Earth's fresh water to meet their drinking needs. If the population increases to 9.3 billion people in 2050, then an additional 1.5 to 2 billion gallons per day will be needed—50 percent more than is being used for drinking today. Where will this water come from?

Water resources are already being stretched around the world. Even in the United States water resources are being depleted, some in ways that cannot be reversed. People pollute water resources, too. For an Earth resource as essential as water, it is surprising how much people take it for granted. As the world's population grows and water becomes more scarce and valuable, perhaps people will be more concerned about how it is used.

Population and Air Quality

If you live in a big city or have visited one, you have probably experienced air pollution. The exhaust emissions from cars and other vehicles are a major cause, although industrial sources of pollutants are also contributors. Emissions that degrade air quality include several gases, but carbon monoxide (CO) and nitrogen dioxide (NO_2) are two primary examples. NO_2 is involved in reactions that need energy from sunlight (photochemical reactions) that create low-level ozone and ultimately help make smog. Smog can cause respiratory problems for people. In some cities smog alerts warn people to stay inside during times of particularly bad air quality. Smog can get so thick that it is difficult to see through it.

Smog and polluted air are not just a city problem. CO and NO_2 form whenever hydrocarbon fuels, grasslands, or forests are burned. Lightning in the atmosphere also creates NO_2 and some is generated by bacteria in soils. Thanks to new space technology, we now know where the high concentrations of NO_2 are around the world and where they originated. People's emissions of NO_2 have caused air pollution around the world (Figure 3-14).

New instruments onboard Earth-monitoring satellites can measure the concentration of air pollutants in the atmosphere. They can survey the entire world in 6 days. By averaging out measurements over periods up to 18 months, they are able to separate natural variations and define the concentrations of pollutants generated by people. The resulting global maps (Figure 3-15) show where anomalous concentrations of pollutants are. Using NO_2, for example (see Figure 3-15), the global satellite monitoring surveys show that cities, industrialized areas, coal-fired power plants, and even shipping lanes are sources of higher NO_2 concentrations in the atmosphere. They also show that air pollutants from these sources are dispersed in gigantic plumes around the world. For example, CO plumes from the burning of grass and forest land in South America make their way across the South Atlantic and Indian Ocean all the way to Australia (see Figure 3-15).

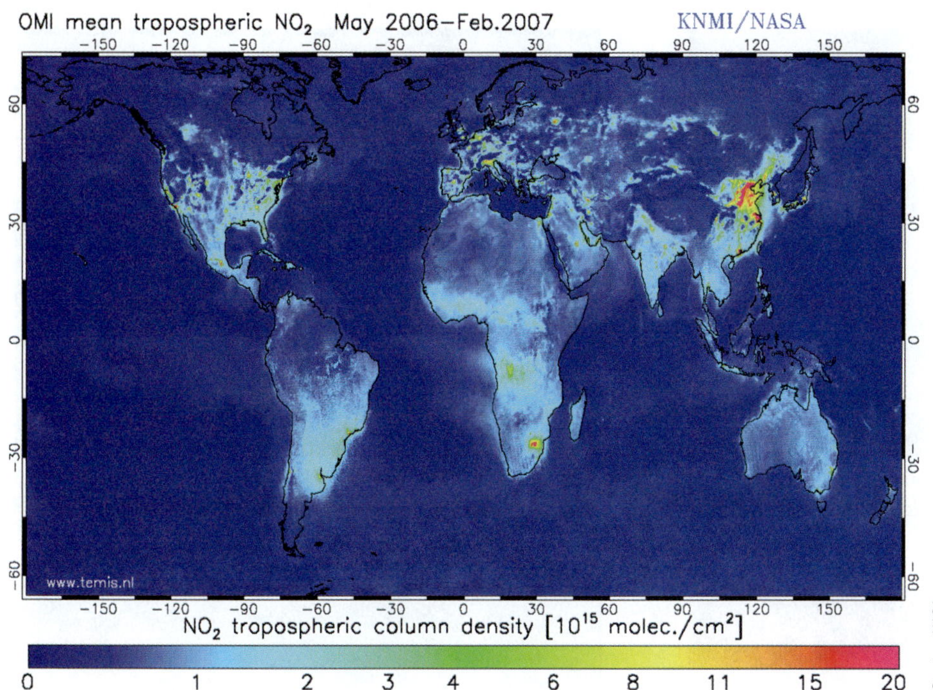

Figure 3-14 NO_2 distribution in the atmosphere.

For the first time people have tools to help them measure and map their impact on air quality everywhere on Earth. As we expect people's needs for energy to continue and increase as the world's population increases, the continued global degradation of air quality is likely without efforts to control potentially harmful emissions. Fortunately, technologies exist that can help do this, and stringent air quality standards in many U.S. cities have shown how these technologies can reverse pollution trends and improve air quality. Being able to do this and actually doing it—especially in a globally meaningful way—are two different things however. As satellites are showing, controlling air pollution has become a global problem that will require global-scale efforts to reverse.

Sustainability

Understanding how people interact with Earth is a major theme of this book. From the preceding examples, it is clear that people's efforts to meet their needs have come to affect all of Earth's systems, not only in general but in specific ways as well. The enormous and growing human population is using land, decreasing natural habitat and biodiversity, depleting water resources, and lowering air quality around the world. These changes in Earth systems follow directly from people's efforts to provide themselves and their families with shelter, food, water, clothing, and fuel. Some people, like those in the United States, contribute to these changes by their efforts to provide a very high standard of living that includes everything from iPods to SUVs. In Chapters 13, 14, 15, and 16 you will learn specifically how people have come to need and rely on Earth's resources and especially how people's use of Earth's resources are having an impact on the environment. Some of these resources are not renewable; in other words, they are used up when they are consumed. Coal, oil, and natural gas energy resources are the key examples, but people have also used other resources, such

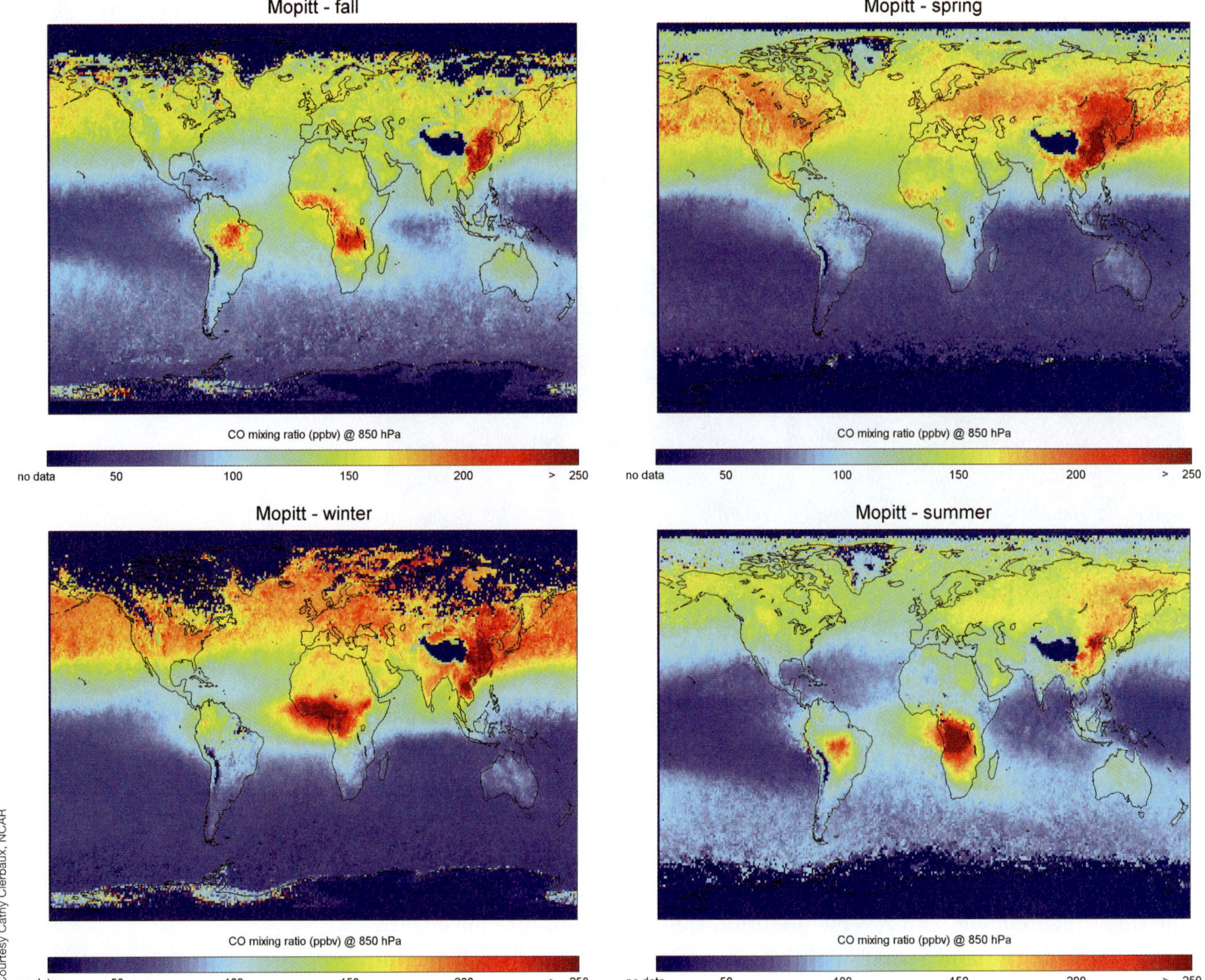

Figure 3-15 Global map showing the distribution of CO in the atmosphere.

as soil and water, in ways that have essentially made them nonrenewable, too. For example, it takes hundreds to thousands of years for soil to redevelop if it has been severely degraded. Some depleted ground water resources may not be recharged for hundreds of years as well.

The concept of **sustainability** has been developed to help people better understand how their actions are important to both theirs and Earth's future condition. Sustainability is most commonly defined as providing for the present needs of people in ways that do not jeopardize the needs of people in the future. This is easily said but not so easily converted into specifics. What are the specific needs to be met in the future? At what level? Well, for sure the basics need to be provided: things like shelter, food, and clothing.

How about ensuring adequate biodiversity for the future? Is this a need to be anticipated and provided for future generations? To live sustainably does include the wide spectrum of needs from clothing to protected habitat and biodiversity. As a result, achieving worldwide sustainability is a very challenging goal. It implies that all people should have the opportunity for a healthy and safe life in balance with the natural world. This is not the case today and providing it for 9 or more billion people in the future is unlikely without some serious resolve to make changes.

One of the factors related to understanding sustainability is Earth's "**carrying capacity**." Carrying capacity is the number of people Earth can sustainably support at some reasonable level of economic and social well-being; that is, standard of living. What is reasonable? Again it is a tough call and depends tremendously on a wide range of cultural and social perspectives. Carrying capacity could be looked upon as just a physical problem, such as how much food and other resources can Earth supply at sustainable levels and how many people this supply will support. Many efforts to estimate Earth's sustainable carrying capacity with respect to people have been made. As you might expect, there is a very wide range of estimates. After all, they depend on numerous assumptions about how much people need, what technology they will develop, and many social and economic factors. Some estimates conclude that the human population has already exceeded Earth's carrying capacity. The estimates are as low as a few billion and as high as 100 billion people. With the estimates so wide ranging, you might be wondering how useful they are in the first place, but it is only prudent to be tackling these tough questions about sustainability and Earth's carrying capacity. Consider the example of Easter Island that follows.

Easter Island

Easter Island is a 64-square mile pile of volcanic rocks about as far from other places in the world as you can get. It is in the South Pacific Ocean 2,300 miles west of South America. Easter Island is famous for its large stone statues, called *moai*, that are as much as 21.3 meters (70 feet) tall and weigh as much as 63.5 metric tons (70 tons) (Figure 3-16).

Moai were carved in a quarry on one of the island's volcanoes, moved several miles to coastal settings, and erected on large stone platforms. There were 393 moai moved from the quarry to platforms, 97 left lying along transport roads, and 397 under construction in the quarry. The Easter Island people were really into making moai. They were used to honor ancestors and helped confirm relations between gods and the island people's chiefs.

The Easter Island people were Polynesians who came from other islands far to the west; the closest are the Pitcairn Islands 2,092 km (1,300 miles) away. The exact date of their arrival on Easter Island is uncertain, but archeological studies indicate that they probably arrived by 900 A.D. The early inhabitants found a forest-covered island.

Large palms and as many as 21 other tree species, now extinct, welcomed the new inhabitants. These trees were the key resources that the people needed to make sea-going canoes, roofs for their homes, and ropes (made from fibrous tree bark). The canoes were needed for fishing and the ropes and logs were needed to transport and erect moai. However, when the first European, Dutch explorer Jacob Roggeveen, arrived on April 5, 1722, the forests were gone. Today there are only 48 native plant species on Easter Island, and the tallest reaches the great height of 2.1 meters (7 feet). What happened to the valuable forests?

Courtesy NOAA

Figure 3-16 Large stone statues on Easter Island.

Easter Island's population grew; they became an agrarian society, and the forests were cut down. Estimates of the peak population size on Easter Island vary from 6,000 to 30,000 or from population densities of about 100 to 450 per square mile. Considering the intense agricultural use of the island that developed, estimates toward the high end of this range seem reasonable. A lot of people lived on Easter Island up to about 1400 A.D. These are the people who carved, transported, and erected the many moai. These statues were erected from about 1100 A.D. to 1600 A.D. They became larger and some were capped by cylindrical blocks of red rock (*pukao*) later in this period.

The time of larger moai construction coincided with a couple of other important changes on Easter Island. First, in 1400 A.D. is about when big trees started to become scarce. The forests were pretty much gone by the 1600s. The second change was the expansion of agriculture from the lowlands to just about everywhere possible on the island, starting about 1300 A.D. Agricultural areas can be mapped and studied because the islanders developed gardens and planting areas surrounded by rock walls to protect them from the wind and help maintain soil temperature. Remnants of rock-protected gardens are located throughout the interior of Easter Island. So from about 1400 to 1600 A.D., the forests were disappearing and much of the island's population was either growing food or building moai.

The forests were disappearing because the trees were being used up. They were cut down to make canoes, provide daily fuel, cremate the deceased (Easter Island crematoria contain the remains of thousands of people), provide more land for gardens, and make timbers and ropes needed to transport and erect moai. The forest's large trees, which are a renewable resource if properly managed, instead became extinct. Along with the forests went the stability of Easter Island society.

The loss of the forests led to erosion and soil degradation that lowered agricultural productivity. Without large canoes, the ocean could no longer be a source of offshore seafood such as dolphin and tuna. All the chief's efforts to ensure abundant food through construction of moai did not pay off. It was the forests, not the gods, that needed the Easter Islanders' respect and consideration. The people lost faith. Social turmoil arose—moai were toppled, chiefs were replaced by military leaders, and a civil war erupted about 1680 A.D. Starving islanders eventually resorted to cannibalism. By the 1700s, cracked human bones were common in the middens (garbage dumps).

The surviving people developed new social structures and became an island of chicken farmers to help meet their food needs. But the abundance of resources they initially discovered on and around Easter Island were either gone or no longer accessible to them. When Roggeveen first visited the island, the only boats left were leaky 10-footers that could only carry one or two people. There were no standing moai and the culture that had created them had become extinct along with the forests needed to sustain it.

3.4 conclusion

As you continue with this book, you will encounter many examples of how people are interacting with Earth and its resources in ways that may or may not be helping achieve sustainability—in ways that are stressing Earth's carrying capacity. There are a tremendous number of people on Earth and there are going to be more. The good news appears to be that population growth will not go on indefinitely. However, the very large and growing number of people is a concern if Earth's carrying capacity is to be respected. The world will become a smaller and smaller place as more resources are needed and competed for by increasing numbers of people. Regardless of what Earth's carrying capacity is ultimately determined to be, it is clear that the more people there are, the more challenging it will be to achieve sustainability. It is only prudent for people—including you—to be concerned about the future of human population growth.

IN THIS CHAPTER YOU HAVE LEARNED:

- A population's birth and death rates determine its growth rate and its age structure. Growth rates can be positive or negative. A growth rate of zero means that the population has stabilized at a certain size and age composition—that it has reached zero population growth, or ZPG.
- Human population growth has happened in three stages: the hunter-gatherer, agrarian, and industrial stages. Most of the rapid population growth has happened in the industrial stage since about 1750. Exponential growth explains how populations increase rapidly even with low growth rates.
- There will be about 9.3 billion people in 2050, an increase of 50 percent over the 6.4 billion people at the end of 2004. Most of these new people will live in cities in less developed countries.
- Catastrophes such as disease, famine, war, and natural disasters have devastated populations in certain areas, but they have not significantly slowed human population growth as a whole.
- The size of the human population is a significant factor in determining the environmental health of Earth. People affect the environment and the more people there are, the greater their impact. Examples include loss of biodiversity through habitat disruption, depletion of water resources, degradation of soil, and global air pollution.
- Sustainability provides for people's needs in ways that do not jeopardize the needs of people in the future. Sustainability includes providing basic resources for shelter, food, energy, and clothing as well as general environmental health such as biodiversity.
- Earth's carrying capacity is the number of people Earth can sustainably support at some reasonable standard of living. Estimating Earth's carrying capacity is difficult and depends on a wide range of assumptions about people and their future needs and accomplishments. Identifying and respecting Earth's sustainable carrying capacity is important.
- The more people on Earth, the more challenging it will be to respect Earth's carrying capacity and achieve sustainability. The more people there are, the more competition for Earth's resources there will be. People—including you—can help determine what Earth's future human population will be.

KEY TERM LIST

Birth rate
Carrying capacity
Crude rates
Death rate
Demographic transition
Doubling time
Exponential growth
Growth rate
Migration
Natural growth rate
Sustainability
Zero population growth (ZPG)

REVIEW QUESTIONS

1. Explain how growth rate is calculated.
2. What factors affect population size?
3. What is zero population growth?
4. Explain what carrying capacity is and its effect on population size.
5. What resources do you think will most affect population size in developing countries in the next 20 years? What is your evidence?
6. What factors have increased life expectancies in Western populations? How has an increased life expectancy changed the demographics of Western countries?
7. Describe the factors that had a negative impact on the Easter Island population. How do you think the Easter Island population could have prevented what happened to it?
8. Explain why the population demographics are so different in Japan and Nigeria—two countries with roughly the same size populations.
9. What effect will increasing human world populations have on the atmosphere?
10. What effect will increasing human world populations have on the diversity of the biosphere?

chapter 4

environmental change

Would you like to monitor environmental change in the Bahamas? Seems like a nice place to do a lot of things but what does it mean to monitor the environment? What changes? What is measured? Students from Forest Hill High School in West Palm Beach, Florida, learned the answers to these questions the classic way—they went to the "field" and made their own observations. The students visited reefs around Lee Stocking Island. There they learned about types of corals, the ecology of coral reefs, and population interactions among various reefs. The students also visited the RV Kristina, an environmental monitoring buoy near Lee Stocking Island. The buoy collects data, including ocean temperature, which is sent by satellite to the Atlantic Oceanographic & Meteorological Laboratory.

What did Forest Hill students learn about monitoring the environment? They learned how scientists use the RV Kristina data to determine when ocean conditions can cause coral bleaching. Coral is colorful, but when photosynthetic organisms that live in the coral die or are expelled, the coral turns white. Coral bleaching is caused by several factors, including high ocean temperatures. Monitoring ocean temperatures can guide scientists to areas where coral bleaching is more likely. Understanding where coral bleaching occurs is important because it can be an indicator of longer term temperature increases around the world's oceans.

Monitoring ocean conditions is one way to identify and measure environmental change. In this chapter you will study how the environment changes, the **rates** *and* **scales** *at which these changes occur, and how the consequences of environmental change may not be as local or simple as they first appear.*

AFTER COMPLETING THIS CHAPTER, YOU WILL UNDERSTAND:
How environmental change can be in a steady state, can be shifting back and forth (oscillating), or evolving.
How rates of environmental change vary from those happening in seconds to those taking many millions of years to complete.
How the scale of environmental change varies from local changes in a pond or valley to those that change the entire world.
How environmental change in one Earth system can affect other Earth systems often distant in both space and time.
How science and environmental monitoring help to understand environmental change.
How environmental change can have both positive and negative impacts and be challenging to understand.

 environmental change around you

Environmental change happens all around you. People are very familiar with some environmental changes and not at all familiar with others. For example, people have adapted fairly well to weather changes—we have summer and winter clothes, insulated homes and buildings, and we schedule our vacations around the seasons. Even extreme weather conditions like hurricanes may not surprise us although they can be devastating.

Many environmental changes are not very obvious, or even perceptible to people although over time they can be very significant. Did you know that sea level has been rising for thousands of years and that this is flooding coasts around the world?

The last major **glaciation**, when massive ice sheets covered most of Canada and extended south into the Midwest states, peaked between 35,000 and 15,000 years ago (Figure 4-1). During this time, the water cycle moved tremendous volumes of water from the oceans and stored it on land in extensive glacial ice sheets. As the oceans lost water, sea level became lower and lower. About 18,000 years ago, sea level was 100 meters (330 feet) lower than it is today (Figure 4-2).

When glacial expansion ceased about 18,000 years ago, and the ice began to melt as Earth's climate warmed, more and more water was transferred back

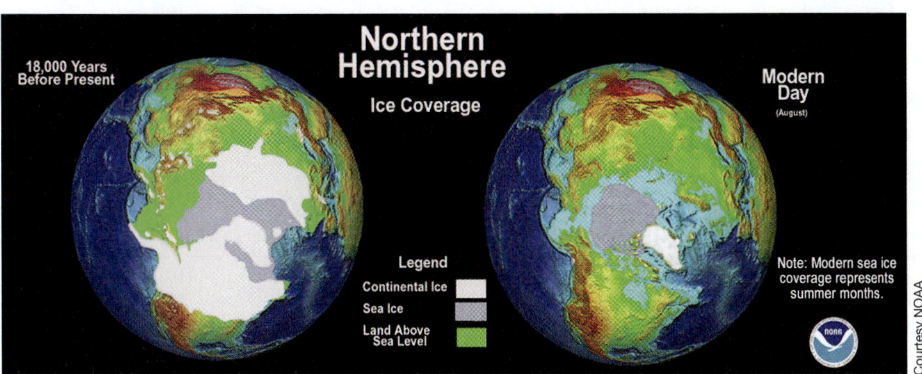

Figure 4-1 The last maximum glacial advance in the upper midwest United States.

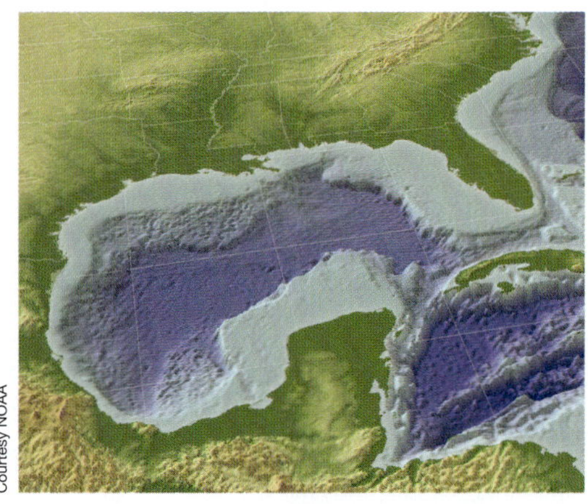

Figure 4-2 Shoreline location now and 18,000 years ago.

to the oceans and sea level began to rise. Rising sea level gradually flooded coastlines until about 6,000 years ago. Since then global sea level has fluctuated up and down over a small range (Figure 4-3). The rates of sea level change are very low and imperceptible to people and even fast changes were only measured in millimeters per year. Over long periods, however, even low rates such as these can create big changes.

Figure 4-3 shows that, in detail, sea level has not changed in a steady way. Instead there were times when sea level actually fell, as between about 9,000 and 8,200 years ago. But from about 10,000 to 6,000 years ago, the average of these small irregular changes indicate a general trend of rising sea level. From about 6,000 years ago to the present, there have been many small sea level rises and falls. The effects of many small irregular changes averaged over longer periods indicate a general trend and suggest that the processes causing the change are more or less continuously happening. There is evidence of an acceleration in the rate of sea level rise in the last 50 years. This is a general characteristic of changes caused by Earth system processes—over short terms they fluctuate but over longer terms they average out to define general trends.

People cause many environmental changes—we call them anthropogenic changes—that we can readily observe. Have you noticed any changing environmental conditions around your school recently? Is more trash appearing on nearby sidewalks? Are more cars showing up in the parking lot? Look at Figure 4-4. How many environmental changes can you identify or infer from this photograph? The photograph suggests that the woodland environment is being encroached on by farmland and urban development. Through history people have converted natural lands to other uses. These land use conversions meet many needs of people but they also change natural habitat, soil character, and the distribution of plants and animals.

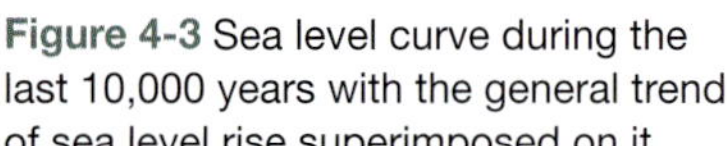

Figure 4-3 Sea level curve during the last 10,000 years with the general trend of sea level rise superimposed on it.

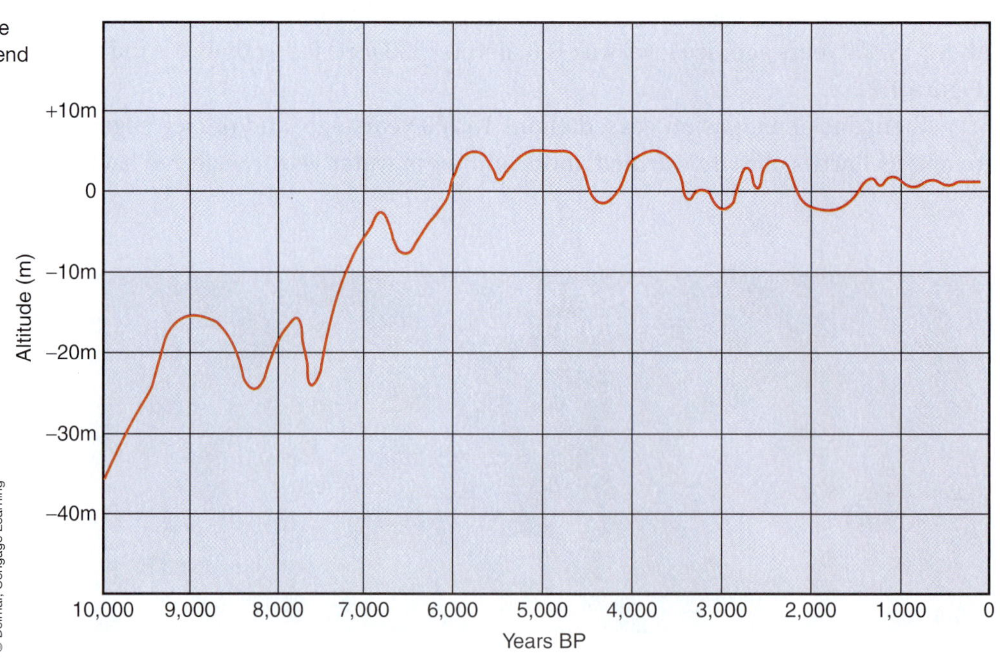

Figure 4-4 Farmland, woodland, and city.

 thinking about environmental change

Earth and its environments commonly leave people with the impression that they are stable and unchanging, but many changes may be occurring simultaneously and we do not perceive the change because opposing natural processes are always in play. For example, plants are growing and dying, soils are gaining and losing components, and water is moving into and out of river systems all at the same time. If the number of plants that grow is about the same as those that die, the components gained by a soil are about the same as those that are lost, and the amount of water entering a river system is about the same as the amount leaving the river system, then the opposing processes are in balance and a steady state is approached.

It is uncommon for opposing natural processes to be in a steady state. In some cases, their interactions lead to shifts or **oscillations** between conditions. The populations of snowshoe hare (prey) and Canada lynx (predator) are an example (Figure 4-5). These populations change dramatically but there is still

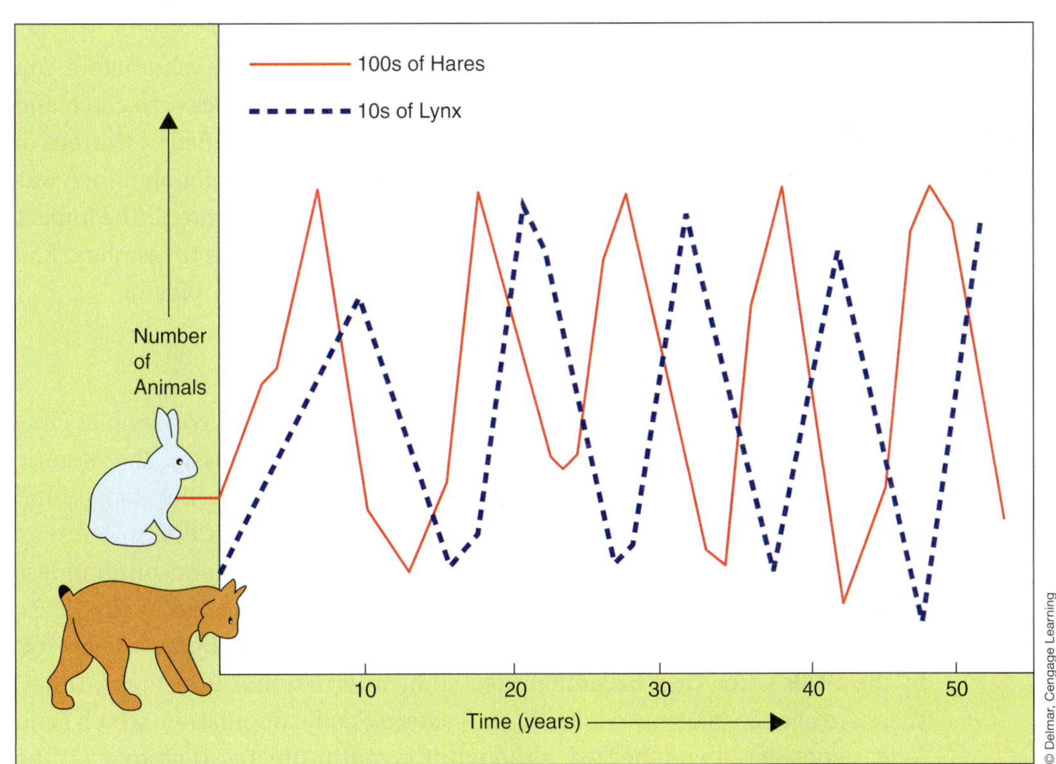

Figure 4-5 Populations of snowshoe hare and Canada lynx for years 1895 to 1924 in a part of Canada.

a consistent pattern between the populations—changes in lynx populations follow similar changes in hare populations in time. Oscillating, rather than stable, populations characterize these species.

It is even more common for environmental change to be evolving rather than be in steady state or oscillating; that is, for the environment to undergo gradual change that is not balanced and is not shifting back and forth. Continents slowly move across Earth's surface, rising sea level floods coastlines even if we do not notice it, and species evolve and the Earth's biosphere changes over time. These are examples of ongoing and evolving environmental changes that will take place no matter what people do. However, people can and do affect many natural processes. A purpose of this book is to help you understand the differences between evolving natural changes and environmental changes caused by people.

Even though many natural environmental changes are inevitable, those caused by people may be controlled or even reversed and prevented. The examples described below, the extinction of the passenger pigeon and the near-extinction of the gray whale, show how environmental changes caused by people can be either permanent or reversed depending on their choices.

Figure 4-6 Martha, the last passenger pigeon.

Demise of the Passenger Pigeon

Several billion passenger pigeons, 15- to 16-inch (38.1–40.6 cm) tall birds resembling mourning doves, lived in the forests of the Midwest when colonists first came to America. Passenger pigeons formed huge flocks that blackened the sky and took days to pass overhead. At one time, they may have been the most numerous birds on Earth. Now they are probably the only species for which we know the exact date of its extinction. The last known passenger pigeon, Martha (Figure 4-6), died at 1:00 P.M. on September 1, 1914, in the Cincinnati Zoological Garden.

The fate of the passenger pigeon was determined years before. You see, passenger pigeons had a problem—they were easy to catch and good to eat. In the 1800s, passenger pigeons were killed by the tens of thousands per day for months at a time. By 1860, their population was in decline and by the 1890s, they had all but disappeared. The impact of hunting and the conversion of their forest habitat to farmland had a permanent or irreversible effect on the passenger pigeon.

Saving Gray Whales

Human effects on natural populations do not need to be irreversible. Gray whales once migrated along both the Atlantic and Pacific coasts. The Atlantic gray whale population was hunted to extinction in the 1700s. The eastern and western Pacific populations were nearly extinct by the 1890s due to hunting. Hunting pressures then declined as whales became sparse and hunting less profitable. The Pacific populations were able to recover some, but by the 1920s, increased hunting resumed. There were only a few hundred gray whales left by the 1940s when they became protected by international agreement. Today, there are over 20,000 gray whales in the eastern Pacific population, which is no longer considered endangered, although it is still protected (Figure 4-7). The

(a)

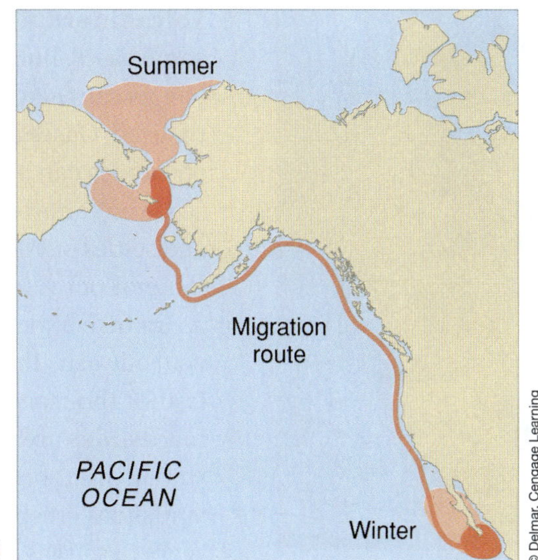

(b)

Figure 4-7 Gray whale and its migration path through the eastern Pacific.

western Pacific gray whale population seems to still be in jeopardy, but recovery of the eastern Pacific population shows that people's negative impacts can be reversed if recognized in time and protection measures are put in place.

4.3 understanding rates and scales of environmental change

One of the biggest challenges that people face in understanding environmental change comes from not being able to relate to the rates and scales at which they occur. Do you find it easy to understand that building mountains takes millions of years? If people cannot directly observe a change happening or record information about the change in a reasonable period (days, weeks, or years), then it is difficult to fully comprehend the change and how important it is. In the following sections, we will examine ways to better understand the rates and scales of environmental change.

Short-Term Environmental Changes

The rates of environmental change easiest to understand are those that people can see. Earthquakes, volcanic eruptions, landslides, oil spills, and floods are examples. They can have major impacts within seconds, hours, or days. But the changes people immediately perceive are commonly a small part of the total change underway. Let us examine some of these short-term examples further.

- **Earthquakes** cause changes in the geosphere (see Chapter 11) within seconds or minutes. Thousands of very minor earthquakes, ones that only sensitive instruments can detect, occur every day. Around the world, people feel about 400 earthquakes each day but only about 100 cause damage in a year. The largest earthquakes, called great earthquakes, only happen on average once a year. Earthquakes are a wonderful example of the active, moving, and changing geosphere, but people can only perceive a small part of the changes they cause.

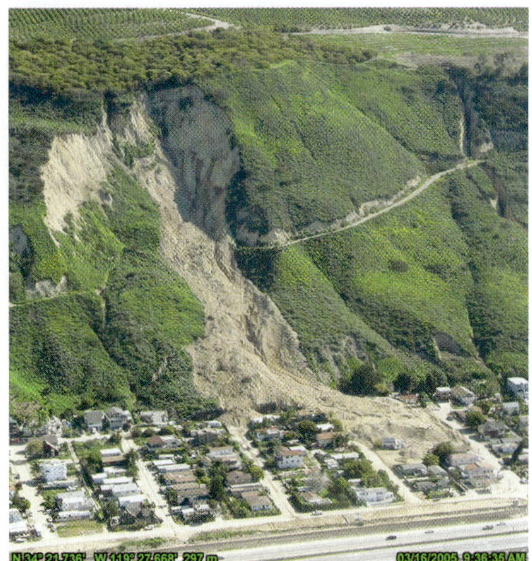

Figure 4-8 The La Conchita landslide in southern California. *(Photographer: Courtesy USGS Photo by Mark Reid and Jonathan Godt, U.S. Geological Survey's Landslide Hazards Program, Pam Irvine, California Geological Survey)*

- **Volcanic eruptions** can last for hours or days and even weeks in some cases. But an interesting thing about volcanoes is that their effects can take from a few years to hundreds of years to be fully realized. Gases and solid particles blasted into the atmosphere at volcanoes can be blown around the world and take a few years to settle out or otherwise be removed. The erupted debris that drifts and settles out downwind of volcanoes (called ash; see Chapter 11) is mineral rich and wonderful beginning material for soils. However, it can take hundreds of years for a fertile soil to develop from volcanic ash. Through soils the effects of volcanic eruptions can last for thousands of years.
- **Landslides** are commonly catastrophic failures of the ground's stability that occur in seconds to minutes (Figure 4-8). Some landslides creep so slowly downhill, called slump, that it takes years to see the results; for example, cracked streets and broken water lines. If a landslide dams a stream and a lake forms, it will take years for new aquatic habitats to fully develop. Some landslides lead to increased erosion and sediment dispersal that also causes habitat changes for many years. In many cases, a specific landslide is just the beginning of environmental changes that take up to years to complete.
- **Oil spills** in the ocean can be complete—all the oil released to the environment—within minutes, hours, or days (Figure 4-9). The released oil eventually, in weeks to years, becomes widely dispersed, evaporates, or accumulates on the sea floor or coast. Animals encounter gooey oil for days or weeks and, if their populations are negatively affected, may take years to recover. The remnants of the oil get covered up in beach gravel, stuck in cracks of coastal rocks, or form tar balls that last tens of years or more. In general, an oil spill is just the beginning of environmental changes that take years to complete. The good news is that nature is typically able to recover from oil spills, although it takes a while.
- **Floods** vary from those occurring within minutes to those that take days to complete to reach maximum high water and recede to normal levels (Figure 4-10). Thunderstorms in the mountains above desert canyons can cause flash floods that descend so quickly that they surprise people, sometimes with deadly consequences. Other floods are ones people can watch rise over days to maximum flow levels. They provide plenty of time to get out of the way. Floods disperse sediment, and this has an effect over a longer time. Flood sediments can become wonderful soils and productive farmland. In places, sediments can be farmed the same year they are deposited but mature, rich soils take hundreds of years to develop and the resulting farmlands can be productive for thousands of years or more.

Figure 4-9 The Exxon Valdez oil spill. *(Photo courtesy of the Exxon Valdez Oil Spill Trustee Council)*

Figure 4-10 Flooding across a large floodplain. *(Courtesy Earth Science World Image Bank. © Michael Collier)*

Therefore, even rapid environmental changes like these examples have consequences that may take many years or longer to complete. What about the changes taking very, very long periods to happen? Mountains can gain elevation at rates of a few to several millimeters per year (Figure 4-11), and sea level changes have varied from millimeters to centimeters per year. These rates seem very slow and insignificant but they are about the same rate that the tire tread on a car wears down (2 mm or 0.08 inch per year) or children grow (0.8–3.9 inches or 2–10 cm per year). Earth system processes operating at these rates develop tremendous environmental changes because they are sustained for tremendous lengths of time—a million, tens of millions, or even hundreds of millions of years. These are the lengths of time needed to understand Earth history and the evolution of Earth's systems.

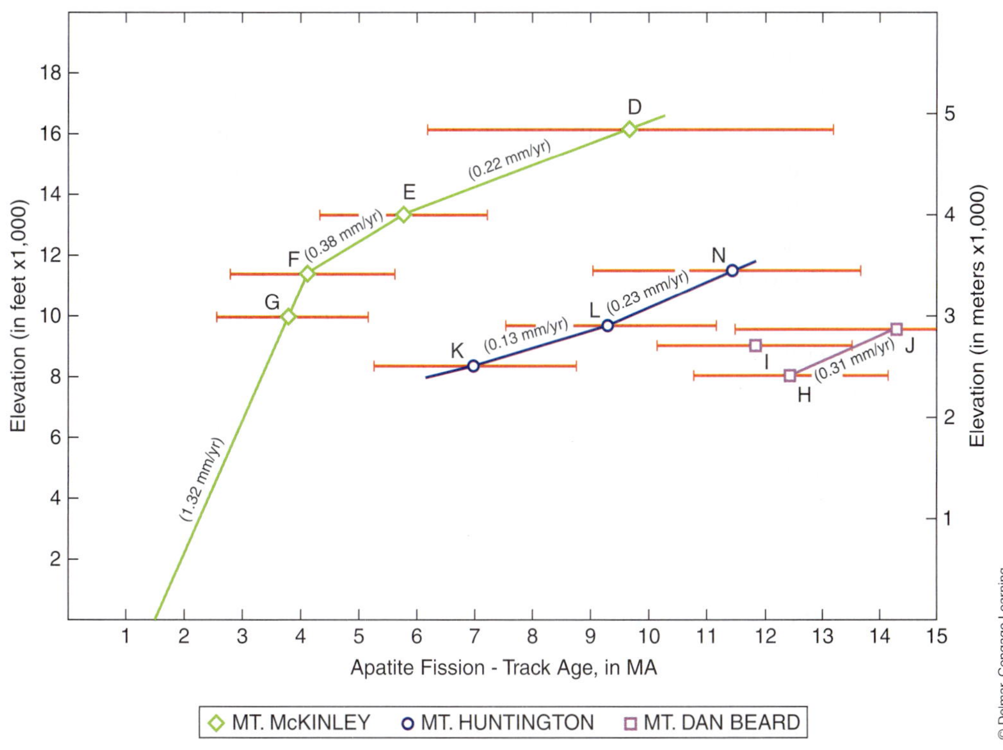

Figure 4-11 Mt. McKinley uplift rate.

Long-Term Environmental Changes

Many processes take place at slow rates and require longer periods to accomplish significant changes. For example, coastal erosion can happen as the result of a sudden storm (at a very fast rate), but it can also happen very slowly, taking tens to hundreds of years to become noticeable. People can come to understand the rates of environmental change like those in gradual coastal erosion because of historical records, even photographs or perhaps stories told by grandparents. But what about processes that need very long periods to accomplish significant changes?

Earth is about 4.5 billion years old and it has taken every one of these years to evolve the world we live in today. Earth scientists have been studying Earth history for about 300 years and still have much to learn about how Earth's systems have evolved. In fact, understanding Earth's dynamic and evolving past is an important way to learn about the processes that are underway and causing environmental changes today. Here are the key long-term changes in Earth systems that have produced our present-day Earth.

- **Evolution of the Atmosphere:** Volcanism, very widespread as the young Earth segregated internally, released gases—especially carbon dioxide and water vapor—that became Earth's early atmosphere. Volcanoes do not emit much oxygen but photosynthetic **cyanobacteria** do. They started releasing oxygen about 3.5 billion years ago (BYA) but it took another billion years before oxygen's concentration in the atmosphere began to increase (and it was extremely toxic to early forms of life). This is because the oxygen produced by cyanobacteria was quickly used up in chemical reactions with other elements, especially abundant and reactive iron and sulfur. Much of the early oxygen that reacted with iron became trapped in iron-rich sediments called banded iron formation (Figure 4-12). Banded iron formation is a chemical sink where a lot of Earth's early-formed oxygen is trapped. This iron was originally suspended in the ocean but precipitated out because of reactions with oxygen. About 2 BYA the readily available chemical sinks for atmospheric oxygen were used up and the oxygen content of the atmosphere began increasing. By about 600 million years ago (MYA), the atmosphere's oxygen levels had increased toward that of today and by 400 MYA the ozone layer appears to have formed. Overall it took about 4 billion years to develop an atmosphere like that of today.
- **Origin of Oceans:** Earth was not covered by oceans like it is today during its earliest stages of development. Water was a component of some minerals in the primordial material that formed the initial Earth. This and other water from ice or gas particles in comets and other interplanetary debris was Earth's original water. Volcanic activity released water from the geosphere into the atmosphere and from there it could be precipitated on Earth's surface. This water formed Earth's first ocean some 3.5 to 3.8 BYA.
- **Evolution of Life:** The biosphere started with very simple single cell organisms. The oldest fossils are of 3.4 billion-year-old bacteria, but evidence of photosynthesis is present in carbonaceous sediments as old as 3.8 billion years. Multicell organisms, including algae, started developing by about 3 BYA. These early components of the biosphere were very important. Their photosynthetic processes produced oxygen

Figure 4-12 Banded iron formation.

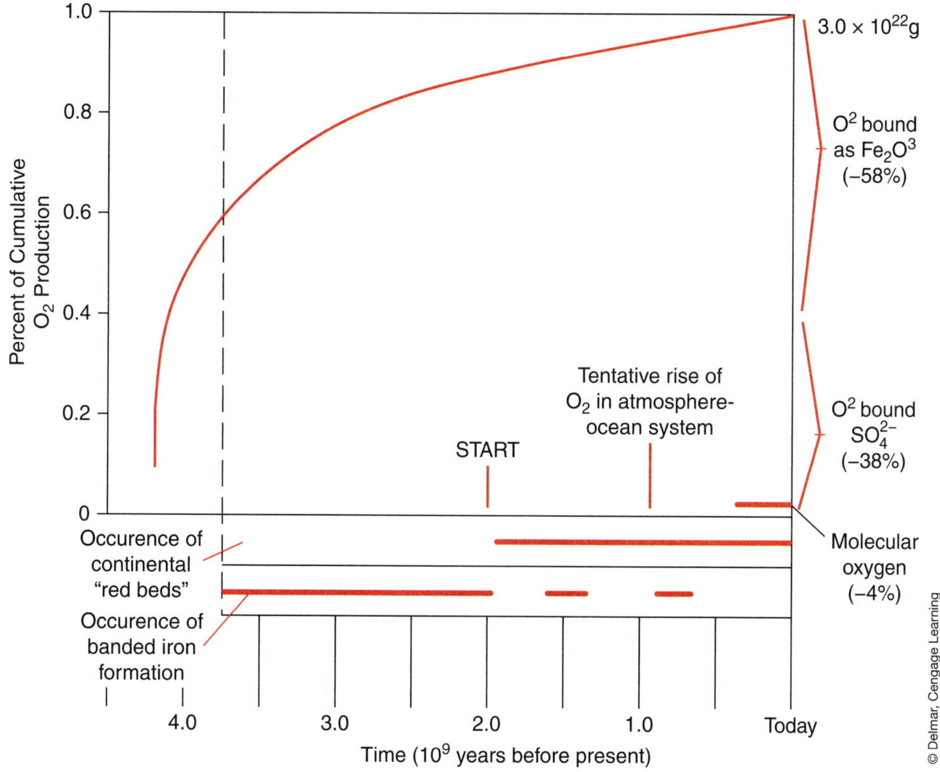

that eventually was able to accumulate in the atmosphere. About 600 MYA life started becoming incredibly diverse. By 550 MYA an explosion of new species, including animals, was underway (Cambrian). This early explosion of diverse life took place in the sea. About 400 MYA, with higher atmospheric oxygen levels and a developed ozone layer in place, plants became the first life on land. By 250 MYA insects and reptiles joined plants in colonizing the terrestrial world.

- **Assembling Continents**: Large continents formed between 2.3 and 3 BYA. Plate tectonic motions (see Chapter 9) have caused continents to collide, split apart, and collide again several times since then. Some collisions aggregated all the continents into supercontinents. Examples of supercontinents are Rodinia, which formed about 1 BYA, and Pangea, which formed 250 MYA (Figure 4-13).

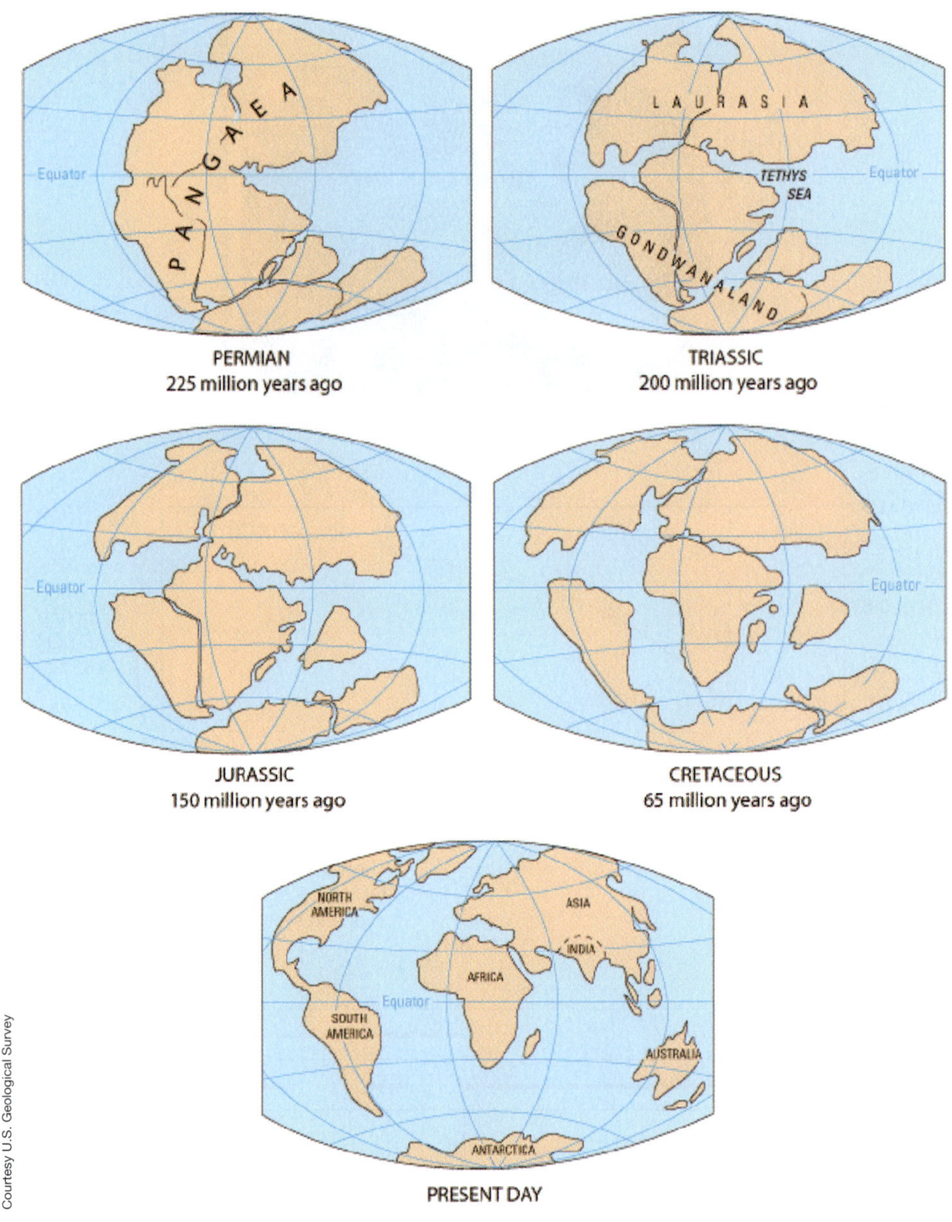

Figure 4-13 Reconstruction of Pangea.

Courtesy U.S. Geological Survey

These examples from Earth's history show that some types of changes can take millions of years and can have global impacts. These can be **episodic** such as the aggregating and breaking apart of continents. They can also be changes that evolve Earth in fundamental ways such as the development of oceans and an oxygen-rich atmosphere. Earth's long history of environmental change can be somewhat sobering for people who think that Earth's environments can be managed and controlled. Earth system processes are ongoing and capable, over time, of significant change regardless of what people do.

Scales of Environmental Change

The scales of environmental changes are just as varied as their rates. Environmental changes can affect a pond, a river, or an ocean and a field, a forest, or a continent. People commonly think that environmental change is local, perhaps in a river, a particular watershed, or an ecosystem. But it is becoming increasingly understood that environmental changes are interrelated and that changes at one scale can have impacts at many others. Here are some examples.

Figure 4-14 Mt. Pinatubo eruption cloud. (*Photo by R.S. Culbreth, U.S. Air Force, June 12, 1991, Courtesy USGS Photo Library*)

Volcanic Eruptions

The virtually immediate effects of violent volcanic eruptions—the masses of lava, hot gases, and rock debris flowing down the sides of volcanoes—are devastating to people and the environment within tens of miles of the volcano. The dark, dense, and huge clouds ejected from volcanoes during violent eruptions can send tremendous amounts of volcanic ash—a mixture of fine rock, mineral, and glass (frozen magma) fragments—to very high altitudes in the atmosphere (Figure 4-14). The ash from the 1980 Mt. St. Helens plume covered an area of about 91,000 square kilometers or 35,135 square miles within 10 hours (Figure 4-15). In very large eruptions, volcanic ash is carried by winds completely around the Earth and can cause cooling in air temperatures.

Mercury in the Environment

Coal from the geosphere contains very small amounts of mercury, an element that the Environmental Protection Agency (EPA) considers potentially very harmful to people. Mercury is a metal, but it is a liquid at room temperature and it easily becomes a gas (volatizes) at higher temperatures (Figure 4-16). This

Figure 4-15 Path of Mt. St. Helens ash plume.

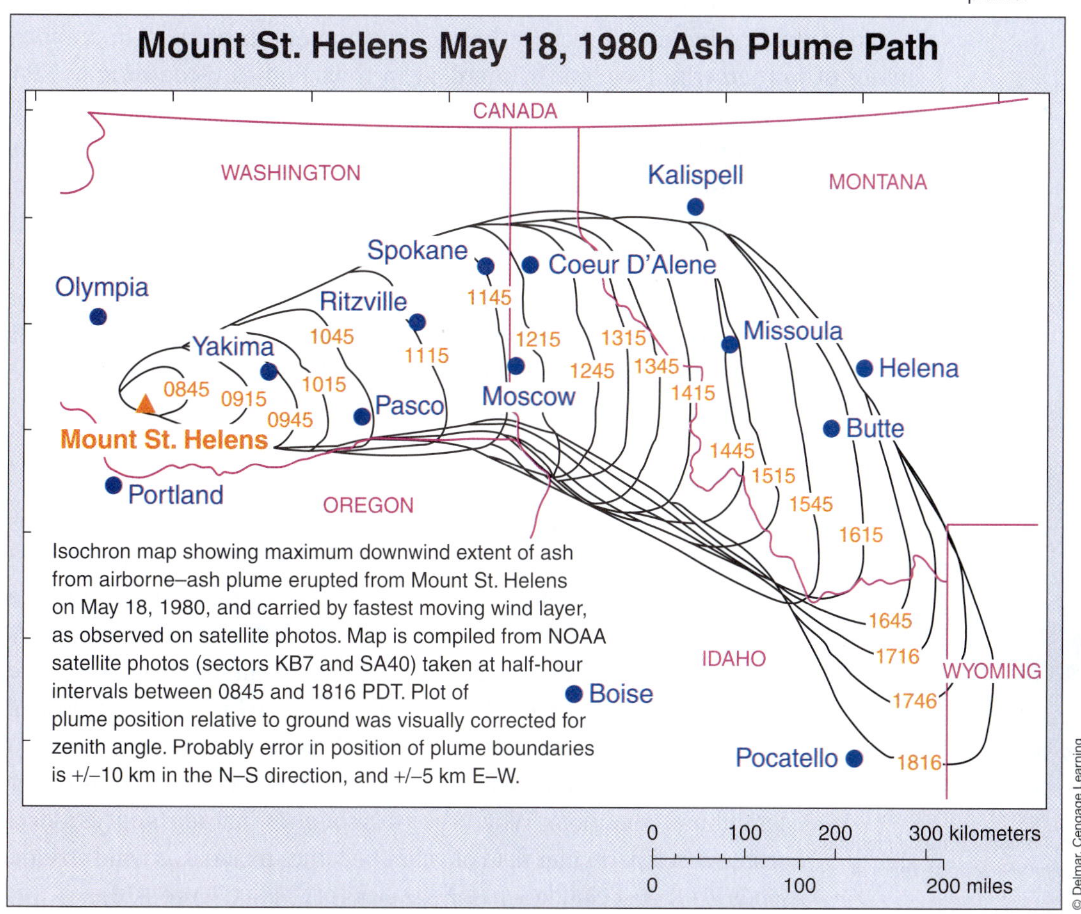

© Delmar, Cengage Learning

Figure 4-16 Mercury at room temperature.

is why coal burning releases mercury to the environment. Mercury volatizes and escapes with other gases up the smoke stacks of coal-burning power plants.

Not a lot of mercury is emitted, but it does not take much to be an environmental concern. Coal-burning power plants are a local, or **point source**, of mercury contamination but the immediate impact is regional in scale. Mercury settles on the landscape over thousands of square kilometers (miles), areas big enough to include parts of several states. Here it undergoes a change and enters the biosphere.

Mercury from coal burning is an inorganic form that is not hazardous, but once it enters aquatic environments, such as wetlands and lakes, microbes change it to another form called methylmercury. Methylmercury is the form that, if ingested in organisms, is not metabolized and accumulates in their bodies. It becomes a long-lasting component of the food chain.

Courtesy NOAA

Figure 4-17 A large tuna.

Fish are a classic example. Little ones that ingest methylmercury get eaten by bigger ones and the mercury is passed on up the food chain. Guess where a lot of the methylmercury is now—in ocean fishes like tuna.

Tuna are a large fish (Figure 4-17) that make the entire Atlantic and Pacific Oceans their home. They migrate great distances and eat tremendous amounts of little fish containing trace amounts of methylmercury. This methylmercury gradually accumulates in their bodies and it is there when tuna are caught and sent to grocery store shelves for people to eat. If it accumulates in people's bodies, it can cause problems, especially brain development problems in unborn babies of women who have methylmercury in their bodies. According to EPA estimates, as many as 1 in 12 women have accumulated enough methylmercury in their bodies for it to be a concern for their babies' health. This is why the EPA recommends a restricted intake of tuna, especially for women having babies.

So what is the scale of the impact of mercury in the environment? It varies from local effects around point sources, regional effects downwind or downstream of point sources, to global effects. The global effects are most apparent after mercury becomes part of the biosphere in aquatic habitats. It becomes passed along food chains until it is present throughout the oceans and on grocery store shelves around the world.

Figure 4-18 Satellite photograph of dust clouds over the northwest Sea of Japan. (*Image courtesy Jacques Descloitres, MODIS Land Rapid Response Team at NASA GSFC*)

Eroding Farmland

Northwest China is an agricultural area that has experienced serious soil erosion. Extensive plowing and grazing, accompanied by dry and hot drought conditions, enables winds to blow away soil from bare fields. It has been estimated that soil is lost from 2,300 square kilometers (900 square miles) each year. Several tons of soil lost from a hectare (0.004 square miles) would only account for 0.04 inch (1 mm) loss of soil. You will not even notice it even though it took many years to form the soil, 2.25 cm (1 inch) takes about 500 years to form under agricultural conditions. Where does this soil go? What starts out as a local problem in a particular field or valley becomes regional as wind erosion sends giant dust clouds across large parts of China (Figure 4-18).

The huge dust clouds obscure the sun and reduce the visibility so much that traffic is slowed and airports close in several of China's cities. However, the dust clouds do not stop there. They continue east and neighboring countries like Japan are affected. Perhaps most surprising though is that dust from China makes its way all the way across the Pacific Ocean to the United States. In early 2001, people in downtown Denver could hardly see the nearby Rocky Mountains because dust from China obscured the view. Dust can also carry invasive pathogens and toxins. From a field in northwest China to downtown Denver—how's that for a range of scale for an environmental impact of people's farming practices?

These examples of rates and scales of environmental change should help you understand a very important aspect of environmental impacts. The immediate observable impact is generally just the beginning of a series of changes that play out over a variety of rates and scales. The impact you directly observe is just part of the overall effect. This fundamental characteristic of environmental change is due to the interrelated character of Earth systems. What affects one may affect them all in some way. For this reason, people have found it helpful to monitor and measure environmental change. This helps them discover environmental changes and relationships that are not that directly observable.

4.4 measuring environmental change

Measuring environmental change is important to understanding how fast, how widespread, and how significant the change is. But what needs to be measured? Environmental change can be represented by factors that reflect the general environmental conditions. Representative factors that can be measured are **indicators** of the general environmental conditions. Measuring and monitoring environmental indicators enable people to track environmental changes over time. They can tell people the status of environmental conditions and whether environmental conditions are getting better or worse.

Environmental indicators vary depending on the environmental conditions that need to be monitored and understood. For example, nitrate levels are a good indicator of water quality, toxic contaminant levels in animals high in food chains can be indicators of environmental pollution, and surface temperature is an indicator of climate change. Here are examples of how environmental indicators are measured and used.

Understanding Coral Reefs

Remember the RV Kristina visited by students from Forest Hill High School? The RV Kristina is one of several environmental monitoring buoys in a network called the Coral Reef Early Warning System (CREWS) that measures ocean conditions that adversely affect coral reef systems. These buoys enable scientists to predict where coral bleaching may occur. Remember coral bleaching? Corals gain their bright colors, and important nutrients, from photosynthetic organisms that live within them (Figure 4-19). If these organisms die, the host corals will turn white or "bleach" and die themselves within a few weeks. The loss of corals and the reefs they build would mean the loss of habitat for hundreds of thousands of species.

Florida Keys National Marine Sanctuary, Courtesy NOAA

Figure 4-19 Brightly colored ocean coral are good photosynthetic organism indicators.

The causes of coral bleaching are not well understood, but the frequency and distribution of bleaching events seem to be increasing. It appears that sea surface temperatures are an important control. If sea surface temperatures of 1°C (33.8°F) warmer than usual are sustained for a month or more, bleaching may happen. Scientists use the RV Kristina and other monitoring buoys to measure sea surface temperature, wind speed, and the intensity of solar radiation. These three environmental indicators are analyzed together to predict where coral bleaching may occur. Knowing where coral bleaching is likely helps scientists prepare for and study these perplexing events.

Osprey and Environmental Pollution

A very different type of environmental indicator can be used to evaluate pollution of the environment. Ospreys, for example, are good indicators of pollution in aquatic ecosystems. Ospreys eat larger fish and are at the top of a fairly complex food chain (Figure 4-20). They also migrate thousands of miles from summer nesting areas in Canada to wintering sites in Mexico and Central America. If **pollutants** are widespread in the environment they are likely to accumulate in the fish that osprey eat and, therefore, in osprey. It also is helpful that ospreys commonly lay an extra egg that does not hatch. This egg can be used as a representative sample of the osprey and analyzed for a number of toxic contaminants such as pesticides and mercury.

Courtesy NPS

Figure 4-20 An osprey catching a fish.

4.5 people and environmental change

People also cause environmental change. This is because they use Earth's resources, they generate wastes, and they use technologies that release materials to the environment. Virtually everything people use to feed, house, and clothe themselves requires Earth's resources. Everyone uses plants, animals, water, soil, rocks, minerals, and energy resources from Earth. People change forests into farmlands, build dams on rivers to produce electricity, and mine materials to build cars and continue expanding our cities.

Specific examples of people's use of Earth's resources already discussed in this chapter include the taking of passenger pigeons and gray whales. While using Earth's resources, people generate things they do not need or cannot use—waste. Waste includes the obvious things like household trash and sewage, but it also includes many less obvious things like the light and heat lost from cities (Figure 4-21).

Other types of waste include things like emissions from vehicles and "used" water containing heat (thermal pollution) or contaminants. People's pollution of the environment pretty much follows from their handling of wastes, but some also comes from using materials such as fertilizers and insecticides.

People's effects on the environment are increasing. The increasing environmental effects reflect a growing population and efforts to advance people's standards of living around the world. These key trends indicate that people will place more and more demands on Earth's resources and increase their potential to create significant environmental change. The biggest challenge will be to fully understand the environmental consequences of people's actions and to make choices that reflect this understanding. Consider people's experience with the pesticide, DDT.

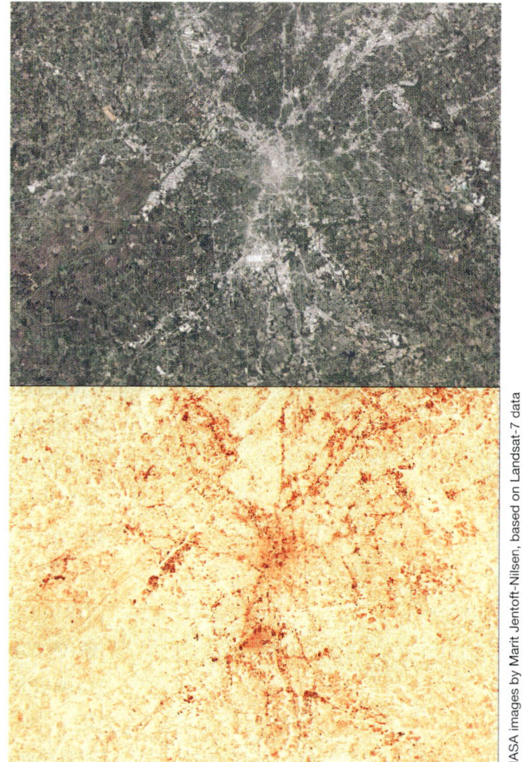

NASA images by Marit Jentoft-Nilsen, based on Landsat-7 data

Figure 4-21 Urban heat island effect.

An Historical Example—DDT

The challenge we face to better understand the environmental consequences of people's actions—and how scientists work to help do this—is well illustrated by an historical example, this being the use of DDT to wipe out malaria in our country. DDT is an organic chemical compound that is an effective insecticide. It was first synthesized in 1874 laboratory experiments by Othmar Zeidler. However, it was not until 1939 that Dr. Paul Muller, working for Geigy Pharmaceutical in Switzerland, discovered a very useful property of DDT—it killed insects, including mosquitoes (Figure 4-22).

In 1943, DDT was used by Allied military forces to combat malaria transmitted by mosquitoes and typhus transmitted by lice. It was very important in WWII, because soldiers found themselves in places around the world where these diseases could be more deadly or debilitating than enemy bullets. DDT's success in fighting disease-spreading insects during WWII quickly led to its general use against malaria and typhus. The benefits of DDT for people were so tremendous that Dr. Muller was awarded the Nobel Prize for its development and application in 1948. In 1955, an international effort to combat malaria based on the use of DDT was initiated, and in a few years malaria was eradicated in developed countries and drastically reduced in several other places. About 25 million people, mostly children, were saved by the use of DDT. It was also extensively used as an agricultural insecticide through the 1950s and 1960s.

As DDT became widely used, its ecological effects began to be recognized. Whereas DDT has a low toxicity in mammals, aquatic species, including many fish, find it very toxic. In addition, fish can accumulate DDT in their bodies and pass it on to animals that eat them like pelicans, eagles, and ospreys. During the 1960s, bird populations such as these became jeopardized because their egg shells were becoming thinner than normal—an effect that many concluded was caused by the DDT that the birds ate in their food. The negative environmental impacts of DDT became widely known to the public, and on the last day of December 1972, the general use of DDT was banned in the United States by the EPA.

Figure 4-22 Insecticide containing DDT.

The use of DDT was already declining by 1972, partly because effective substitutes for its use as an agricultural insecticide were developed. But elsewhere in the world, especially where malaria is a threat to people, DDT is still used and still very helpful to people. However, once DDT is released to the environment it can persist for several years and become widely dispersed in the biosphere; its use in one area can be a concern elsewhere. As a result, discussions about the positive and negative effects of DDT are still occurring. Some argue that misuse of DDT can be avoided while still using it to combat disease, whereas others feel that any release of DDT to the environment should not be allowed. DDT use is a great example of a new scientific discovery helping overcome a dreadful human malady. It is also an example of how science continues to evolve understanding and gives us insight into the complexities surrounding DDT's use and dispersal in the environment. Only 24 years passed between Dr. Muller's Nobel Prize and the banning of DDT's general use in the United States.

The example of DDT illustrates the complexity of many environmental changes. The existence of both positive and negative environmental changes associated with DDT use created a tremendous discussion about how to manage this chemical. It was eventually decided that the negative impacts in the United States were more important than the positive effects, especially after malaria was eradicated. What is positive and what is negative about an environmental change is a moving target in many cases. There are trade-offs associated with many environmental changes and tough decisions need to be made in many cases. Understanding and evaluating the trade-offs is important to making the best decisions. The issues become complex when people's welfare is perceived to be in competition with the environment's welfare. Here are some more examples.

Competing for Land—the Arctic National Wildlife Refuge

America's dependence on fossil fuels requires an ongoing search for new sources of petroleum, and that process affects the ecosystem of the surrounding area. A good example of this process is the Arctic National Wildlife Refuge (ANWR), located in the northeast corner of Alaska along the Arctic Ocean. About 8 percent of the 7.7 million hectares (19 million acres) in ANWR on the coastal plain is prospective for large petroleum accumulations. However, this is also the general area where a large caribou herd has its calves each year (Figure 4-23).

Courtesy USFWS

Figure 4-23 Caribou on the coastal plain of the Arctic National Wildlife Refuge.

Also, ANWR is relatively undeveloped and as close to pristine wilderness as possible in the United States. The oil industry argues that it can explore and develop this area with minimal impact, especially to the caribou. Some local Inuit people and the State of Alaska want oil exploration in hopes of getting needed jobs and income. Another Native American group does not want oil exploration because they fear negative impacts on the caribou they depend on. Environmental groups oppose oil exploration and development because they feel that the value of the wilderness is greater than any oil that might be discovered. These issues have been debated in Congress for many years and they are still unresolved.

Competing for Fish—the North Pacific Steelhead

Steelhead are a large, hard-fighting, beautiful variety of rainbow trout (Figure 4-24). They are born in streams of the Pacific Northwest but they live large parts of their lives in the ocean. They return to the streams to spawn and then go back to the sea. They may make these trips, runs, a few times in their lives. Their time in the ocean, where food is plentiful, is why they can become large—exceptionally large steelhead can weigh 18.14 kilograms (40 pounds).

Steelhead are a prized sport fish. Some people are obsessed with them. These people will venture out in the worst, blustery, cold winter weather in search of steelhead. To support this habit they spend money, lots of money, on things like boats, tackle, fishing guides, food, and lodging. They significantly help the economy of the towns along the steelhead streams. These people also catch fish and over the years they have caught lots of fish. But they are not the only ones catching steelhead. Native American tribes are allowed to net steelhead in the rivers and sell them commercially. Tribes are estimated to catch about half of the annual steelhead harvest in the areas they fish. Fishing and degradation of upstream habitat has led to a serious decline in steelhead populations. These declines have been sufficient to convince government regulators that steelhead populations in some areas need to be listed as threatened under the Endangered Species Act. They are at risk of becoming extinct in these areas.

Steelhead populations have been declining for many years. Hatcheries have attempted to help overcome these declines by releasing thousands of baby steelhead—smolt—into the rivers. Hatcheries, in effect, increase the birth rate of steelhead. Now the steelhead runs are a mix of hatchery and "wild" fish. It is the wild fish that are listed as threatened under the Endangered Species Act. When species are listed as threatened, government regulators take actions to help reverse the population decline. In the case of steelhead, they can protect stream habitat and limit fishing. In some areas, regulators now prohibit the taking—killing—of wild fish by sport fishing. Regulations that limit the use of private land along the streams to protect habitat are also expected.

These new regulations have been met by a very vocal opposition. Businesses are worried that sport fishing will decline, the people sport fishing argue between those who want to keep their fish and those who want to "catch and release," and Native Americans are concerned that resentment of their allotted catch will increase. Concerns are significant enough to have led to a lawsuit to invalidate the threatened status of steelhead. Regardless of the competing arguments about how steelhead populations should be managed, one fact remains: The populations of these beautiful wild fish are much smaller than they once were.

Figure 4-24 A large steelhead. (*Courtesy Pacific Northwest 'National Laboratory, US Department of Energy*)

Figure 4-25 Stillwater mine. (*Courtesy Earth Science World Image Bank. © Marli Miller, University of Oregon*)

Guiding Environmental Impacts of Mining

This example shows how people dealt with the complexities of mine development in Montana, a place where protecting nature and supplying resources and jobs were apparently in conflict. The conflicts arose over expanding the Stillwater platinum and palladium mine (Figure 4-25).

Platinum (Pt) and palladium (Pd), two metals closely associated with each other in nature, have diverse uses, such as in fuel cells, dental restorative materials, and cancer chemotherapy. However, their principal use is to prevent air pollution; they are the key components in the catalytic converters that reduce air pollution from vehicle exhaust emissions. In 1999, almost 50 percent (110,000 kilograms) of the United States' consumption of Pt and Pd was by the automotive industry for catalytic converters. The need for these metals in controlling exhaust emissions is increasing as more cars are produced, as the use of catalytic converters is extended to other types of vehicles (such as SUVs), and as more countries (such as Brazil and China) adopt air pollution regulations like those in the United States.

The Stillwater Mining Company controls a very large Pt/Pd deposit, called the J-M Reef, in Montana. In the 1990s, Stillwater Mining became convinced that they needed to increase their production to help meet the increasing demand for Pt/Pd. To do this, they decided to expand operations at the Stillwater Mine, increase their processing capacity, and start a new mine, the East Boulder Mine, on the western end of the J-M Reef. Montana environmental regulations and permitting requirements guided all these expansion efforts, but many local citizens were concerned that more oversight was needed to ensure that environmental issues were satisfactorily addressed throughout the several decades of expected mine life. Both the Stillwater and East Boulder mines are located near trout streams, high-use recreation areas, and agricultural lands (mention of leaching from mine tailings).

The local citizens groups (Northern Plains Resource Council, Cottonwood Resource Council, and the Stillwater Protective Association) were so concerned that they filed a lawsuit, challenging an environmental impact statement important to Stillwater Mining's expansion plans. However, in this case, both sides wanted to rise above their adversary roles and work together as good neighbors. The result? A legally binding, "Good Neighbor Agreement," was signed by all the parties in early May 2000. Some of the highlights of this agreement are:

- Stillwater Mining will invest in developing and implementing new water treatment and waste reduction technologies to achieve zero discharge of wastewater and eliminate or reduce the need for very large waste dumps.
- Thousands of acres of rangeland and wildlife habitat owned by Stillwater Mining will be dedicated to conservation easements.
- A comprehensive water protection program, including expanded water monitoring, will be put in place to ensure that water quality and fisheries are not degraded.

- Stillwater Mining will limit traffic by busing employees in order to decrease congestion and air pollution.
- The citizens groups are guaranteed access to all environmental compliance and performance information and have the right to inspect mine operations with independent technical and scientific consultants.
- Stillwater Mining will fund periodic, independent environmental performance audits and implement audit recommendations.
- Standing committees, such as the Citizens Oversight Committee and the Responsible Mining and Technology Committee with members from mine management and the citizens groups, will manage all aspects of the agreement's implementation and resolve new issues that arise. Disputes that cannot be resolved by the committees can go to binding arbitration.

It appears that the key to reaching the agreement was the willingness of both sides to meet, be open with each other, and work to define common ground. The negotiation participants included Stillwater Mining senior management and an 11-person team of local ranchers, professionals, retirees, and others representing the citizens groups. The participants can justly claim that they have created a precedent-setting agreement and example for resolving conflicts over the impacts of mining developments.

4.6 conclusion

Environmental change is ongoing everywhere and at all rates and scales. In addition environmental changes are interrelated, involving all Earth systems. Whether natural or caused by humans, environmental change is complex and challenging to understand. People are a cause of environmental change and they face difficult decisions when their wants and needs cause undesirable environmental changes.

IN THIS CHAPTER YOU HAVE LEARNED:

- Environmental change can happen almost instantly or take many millions of years to occur.
- Environmental change can happen at all scales from a local pond to the entire Earth.
- Environmental change can be in a steady state, oscillating, or evolving. Evolving environmental change is common in Earth systems.
- Environmental changes are commonly related to other environmental or Earth system changes.
- People cause environmental change through their use of Earth's resources and their generation of waste.
- Environmental indicators help people identify and monitor environmental changes.
- Science helps people understand and respond to environmental change.

KEY TERMS

Cyanobacteria
Earthquakes
Episodic
Floods
Glaciation
Indicators
Landslides
Oil spills
Oscillation
Point source
Pollutants
Rate
Scale
Volcanic eruptions

REVIEW QUESTIONS

1. The environment can change very rapidly or very slowly. Give an example of what can cause each type of change.
2. Explain how a portion of the environment can be in a steady state.
3. Give an example of a local environmental change and a global change.
4. Describe two ways in which scientists monitor environmental change.
5. What factors can lead to the extinction of a species?
6. What can humans do to prevent the loss of species?
7. In what ways can a volcanic eruption result in short-term environmental change?
8. Describe how methylmercury caused environmental change.
9. What role do coral reefs play in helping scientists to understand environmental change?
10. Discuss what was both good and bad for the biosphere about the use of DDT. Support your answer with evidence from the chapter.

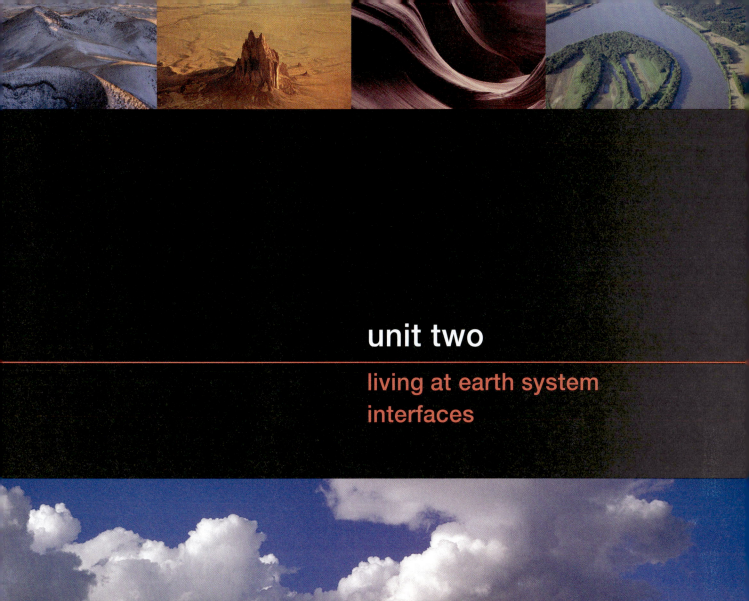

unit two

living at earth system interfaces

chapter 5

coasts

Students at Ball High School in Galveston, Texas, head to the beach during class time. However, rather than sunglasses, towels, and bathing suits, these students take their hip waders, measuring tapes, surveying levels, and sample bags. They are part of the Texas High School Coastal Monitoring Program organized by the Texas Bureau of Economic Geology. Ball High students essentially live on an old beach. The land Galveston is built on was created from beach sand stacked up by coastal waves and currents. These waves and currents continually change the local beach environment, and once in awhile, large storms do, too. In 1900, a hurricane flooded the entire city of Galveston, destroying thousands of homes and killing over 8,000 people.

Observations and measurements obtained by the Ball High School students show how the beach environment is changing. They survey beach profiles (Figure 5-1), map the distribution of vegetation along the shoreline, and collect sediment samples at different times during the school year. They combine these data with observations of wave and weather conditions to explain the changes and compare them to those along other parts of the Texas coast. Several high schools participate in monitoring the coastal changes in Texas. Knowing how the beach environment naturally changes is important information needed to guide sound use of the shoreline not just in Texas, but along any coast.

Lots of people live along coasts, the thin strip of land at the interface between land and ocean. Many more are drawn to coasts for recreation—to

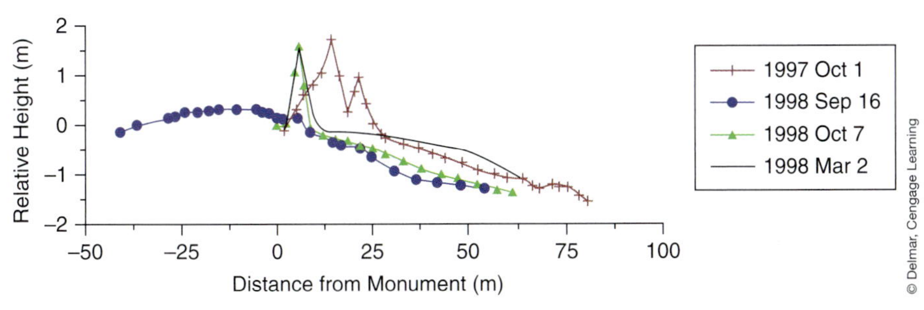

Figure 5-1 Repeat beach profiles from Galveston Island, Texas, from 1997 to 1999.

Figure 5-2 The endangered Florida manatee. (*Photo by Rathburn, Gaylen Courtesy U.S. Fish and Wildlife Service*)

swim in the surf, relax in the sunshine, or picnic in the sand dunes. It is fun going to the seashore.

People also go to the coasts to learn more about them. Like the students at Ball High School, scientists study coasts around the world so that people and coastal ecosystems can be better protected. Coastal habitat is home to many plant and animal species, including some like the Florida manatee (Figure 5-2) that are protected by the U.S. Endangered Species Act. Only through an understanding of coastal processes and coastal environments can people provide the needed protection for themselves and other organisms.

Coasts are also very important economically. People need to exchange goods and services with other nations. Most of this trade occurs on the coast where ocean-going cargo ships come to port and deliver everything from food to cars. Every year over $500 billion worth of goods are imported and $200 billion worth is exported through U.S. ports. This requires a tremendous amount of ship traffic and many workers to load and unload material. It is no wonder then that the biggest cities in the country are port cities—New York, Boston, Seattle, Los Angeles, and New Orleans are examples.

AFTER COMPLETING THIS CHAPTER, YOU WILL UNDERSTAND:
What coasts are.
The Earth system interactions that influence coasts, including waves, tides, sea level changes, geosphere movements, and storm surges.
The nature of the biosphere that inhabits the coast, including on beaches, in estuaries, in salt marshes, and in mangrove swamps.
How people are affected by coastal processes, including flooding, shoreline erosion, and sedimentation.
How people affect the coast through land developments, waste disposal, changes in coastal biota, and attempts to control coastal processes.

5.1 coasts and earth systems

What images come to mind as you think about "heading to the coast"? This image varies, no doubt, depending on your past experiences and where you live (Figure 5-3). Where the land meets the water is perhaps common to all of our thoughts—a boundary between the hydrosphere and the geosphere—a "line" where land and water meet. The terms *shoreline* and *coastline* embody this idea, but the coast is always changing and evolving with each passing storm that reshapes the beach, with each tidal cycle that fills and empties a lagoon, and with each new community of plants and animals that colonize a sand dune. For a hermit crab, it changes with each crashing wave that exposes a shell it can use as a home.

The **coast** is more than a line or boundary, however. It is a transition zone where all of Earth's systems interact. People recognize beaches, tidal flats, estuaries, and other settings as characteristically coastal, not only because of their physical appearance, but also because of the plant and animal species that inhabit them. Where else but a coastal environment can you find species such as chitons, which are specially adapted to withstand severe water loss

Figure 5-3 Various coastal zones around the United States.
((a) Courtesy the Earth Science World Image Bank Copyright Marli Miller University of Oregon
(b) Courtesy USGS
(c) Photo by Mr. William Folsom, NOAA, NMFS Courtesy NOAA)

during periodic exposure at low tide (Figure 5-4)? The variable physical conditions created by the ongoing Earth system interactions along coasts are a major influence on the plants and animals that live there.

The coast is shaped by many Earth system interactions. These interactions produce waves, tides, sea level change, shifting coastlines caused by geosphere (tectonic) movements, and storm surges. Some of these processes act almost continuously on the coast (waves and tides), but others occur only a few times per year (storm surges), very rarely (geosphere movements), or very slowly (sea level change). The reshaping of the coast by frequent processes can mask the impact of less frequent but commonly more powerful and potentially dangerous ones. Earth system interactions combine to create habitat for the biosphere that is unique to coasts. Recognizing the nature, frequency, and magnitude of these processes, their impacts, and when they will occur, is the key to protecting people and ecosystems along the coast.

Figure 5-4 A fuzzy chiton (*Acanthopleura granulate*) from the Cayman Islands.

Figure 5-5 Surfing the "tube" on the North Shore of Oahu, Hawaii. *(Courtesy NOAA, Photo by: Commander John Bortniak, NOAA Corps (ret.))*

Making Waves

Have you ever seen videos or television programs of people surfing huge waves off the coast of Hawaii, California, and other exotic locations around the world (Figure 5-5)? Two questions come to mind when watching surfers. First, how on earth do they stand up on those boards without crashing and killing themselves?

Second, how do waves get so big, curl over on the top, and create a "pipeline" for surfers to ride through? Ocean **waves** are movements of near-surface water caused by wind. Wind transfers energy of movement (kinetic energy; see Chapter 15) to the water as it blows across the water's surface.

Making waves starts by the atmosphere pushing down or lifting up on the water's surface. Downward air movements depress the water surface in places, and upward air movements allow the water surface to rise slightly. Once an uneven water surface is developed, wind pushing against the high spots can transfer lateral movement to the water and start waves. The higher the wind velocity in a given direction, the larger the wave will become. But to create large waves the wind must also blow over a long distance (called fetch). The largest waves occur in oceans because they have the greatest fetch—potentially thousands of kilometers (miles) across which the wind can blow. However, one gust of wind, regardless of the fetch area, will not create a large wave. Long sustained winds, or winds of long duration, are also needed to create large waves. Wind velocity, fetch, and duration, then, are the three primary controls on wave size and the amount of energy contained within them.

Coasts with the largest waves are those most exposed to the open ocean—no islands, inlets, or other barriers to block the path of waves generated by large, powerful, and long-lasting winds at sea. The next time you are standing at the beach on a calm, windless day remember that the waves you see approaching may have originated several days earlier and thousands of kilometers (miles) away. Waves lose little energy and maintain their size and shape as they travel across the open ocean even when the wind that formed them stops.

However, once the wave approaches the coast and moves into shallower water, the wave will change its shape and height as it begins to interact with the seafloor. These changes begin when the depth of water is less than half the distance between successive waves and becomes most pronounced when the water depth is less than this. The interaction of the wave with the ocean bottom at these shallow depths causes the water near the bottom of the wave to move more slowly than the water closer to the surface. Consequently, the top of the wave starts to "outrace" the base. This creates a higher and steeper wave that eventually becomes too steep and unstable to maintain its form (Figure 5-6). The wave eventually breaks and collapses forward. These breaking waves are what surfers ride toward shore.

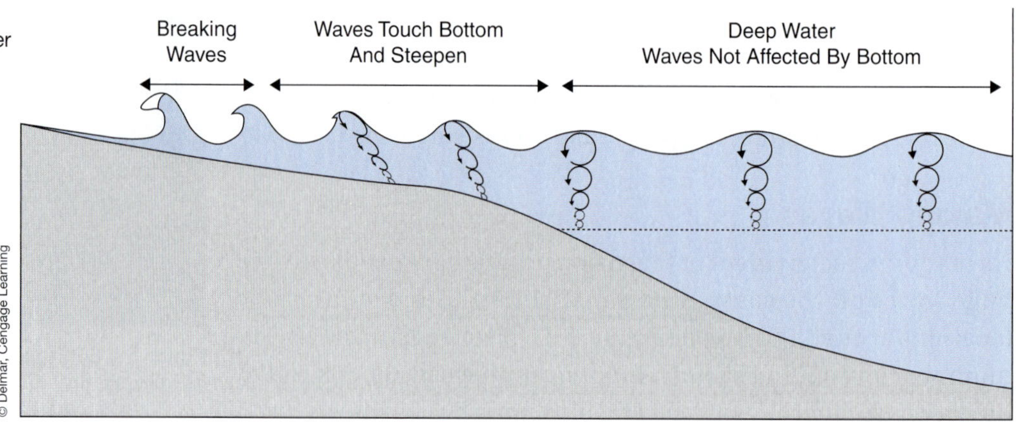

Figure 5-6 Changes in the shape of the wave as it reaches shallower water.

Making Beaches

Surfers and water are not the only things moving toward a sandy beach; sand does too. The water that rushes up the beach from breaking waves (swash) carries sand along with it. This water flows back down the beach as backwash, but due to infiltration backwash contains less water than swash and it commonly flows more slowly too. This allows the sand to accumulate, or be deposited, and form a **beach**.

The geosphere's coastlines are rarely perfectly straight or parallel to approaching waves. This causes waves to break on the coast at an angle, and sand moves up across the beach at an angle with it. Gravity, however, causes the returning water to flow more directly back down the slope of the beach. This happens wave after wave, causing sand to be moved along the length of the beach. The water movements along a beach created by waves breaking at an angle to the coast are called longshore currents (Figure 5-7). Longshore currents are familiar to those who have spent time floating along a beach where the waves approach the shore at an angle. After a while, a swimmer will be moved along parallel to the beach, perhaps quite a distance from where he or she initially entered the water. The long walk back to a beach towel and snacks is the result of longshore currents not only moving sand and water down the beach, but the swimmer as well!

Waves are an essential process in the formation of beaches, but a source of sand or gravel is also needed for beaches to form. Where do this sand and gravel come from that swash moves up the beach? The sand cannot come from far offshore because, remember, waves do not interact with the ocean bottom until the water depth is very shallow near the coast. Sand and gravel on a beach—sediment—must be transported from inland areas, different places along the coast, or the nearby shallow seafloor. High sea cliffs, frequently collapsing under the incessant pounding of the waves, are one source of sediment (Figure 5-8). A very important source of sediment is rivers. Rivers can deliver large amounts of sediment to coasts from mountains hundreds to thousands of kilometers (miles) away. Sediment at the mouths of rivers is initially deposited in deltas (Figure 5-9). The sediment is then carried away from the delta by longshore currents, redistributing it to form beaches. Without sources of sediment, waves, and longshore currents, beaches would not exist.

Making Tides

Tides are the alternating and regular rise and fall in sea level along a coast; there are two rises to a high level (high tide) and two falls to a low level (low tide) about every 24 hours. Whereas waves are formed by a transfer of wind energy from the atmosphere to the hydrosphere, tides are the result of a transfer of gravitational energy to the hydrosphere by solar system interactions. Daily variations in tides are created by the gravitational interaction between the Earth and the moon. Tidal differences that correspond to quarterly phases of the moon during a lunar month (27.3 days) are caused by gravitational interactions between Earth's hydrosphere and the moon and the sun.

High and low tides are produced by a combination of forces acting on the surface of the ocean (Figure 5-10). The first force, the moon's gravity, pulls water in the hydrosphere toward the moon. The second force, which works in opposition to the moon's gravity, is the centrifugal force caused by the Earth and moon revolving around their common center of gravity (a point within the

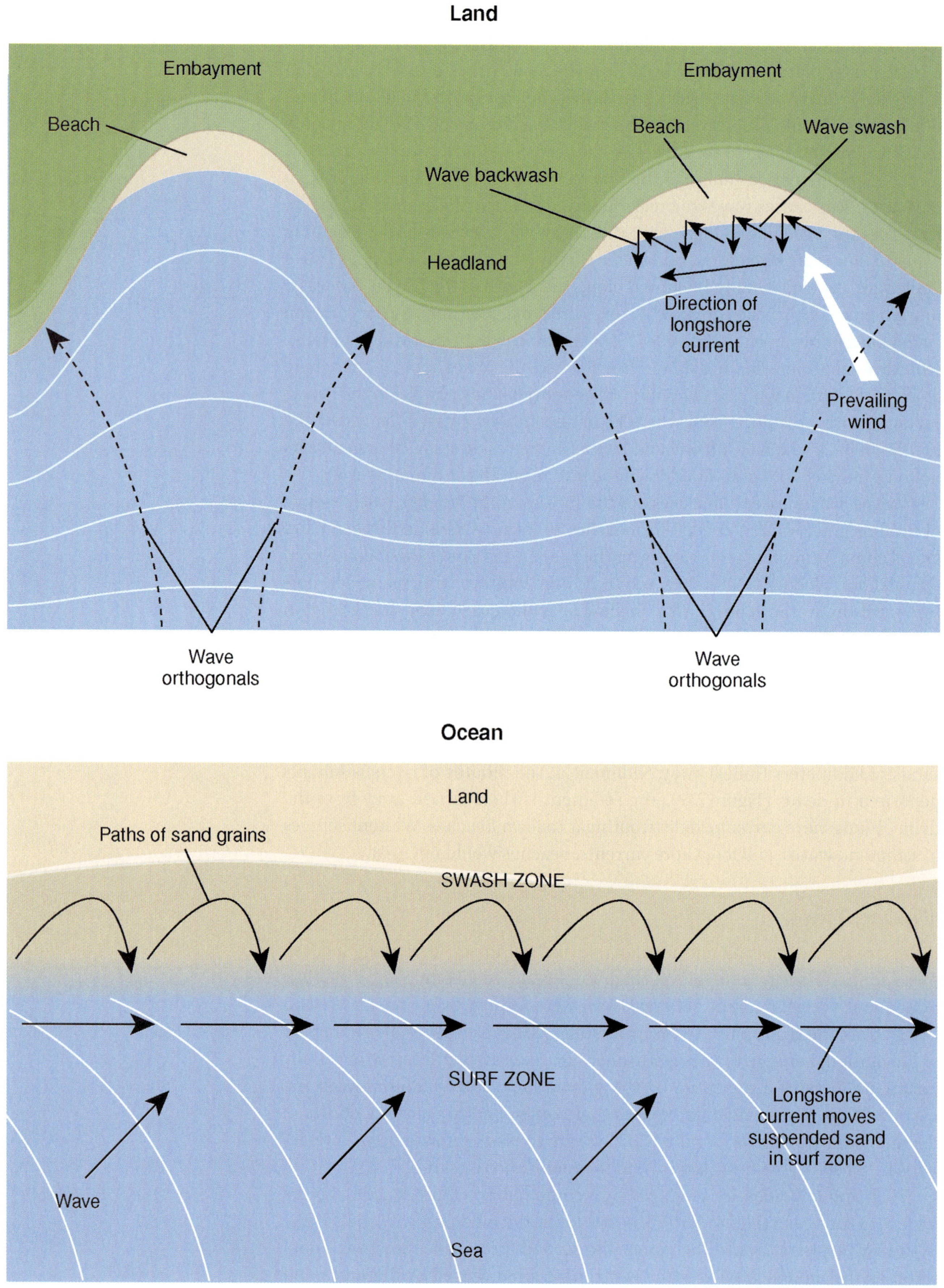

Land
Embayment
Embayment
Beach
Beach
Wave swash
Wave backwash
Headland
Direction of longshore current
Prevailing wind
Wave orthogonals
Wave orthogonals
Ocean
Land
Paths of sand grains
SWASH ZONE
SURF ZONE
Longshore current moves suspended sand in surf zone
Wave
Sea

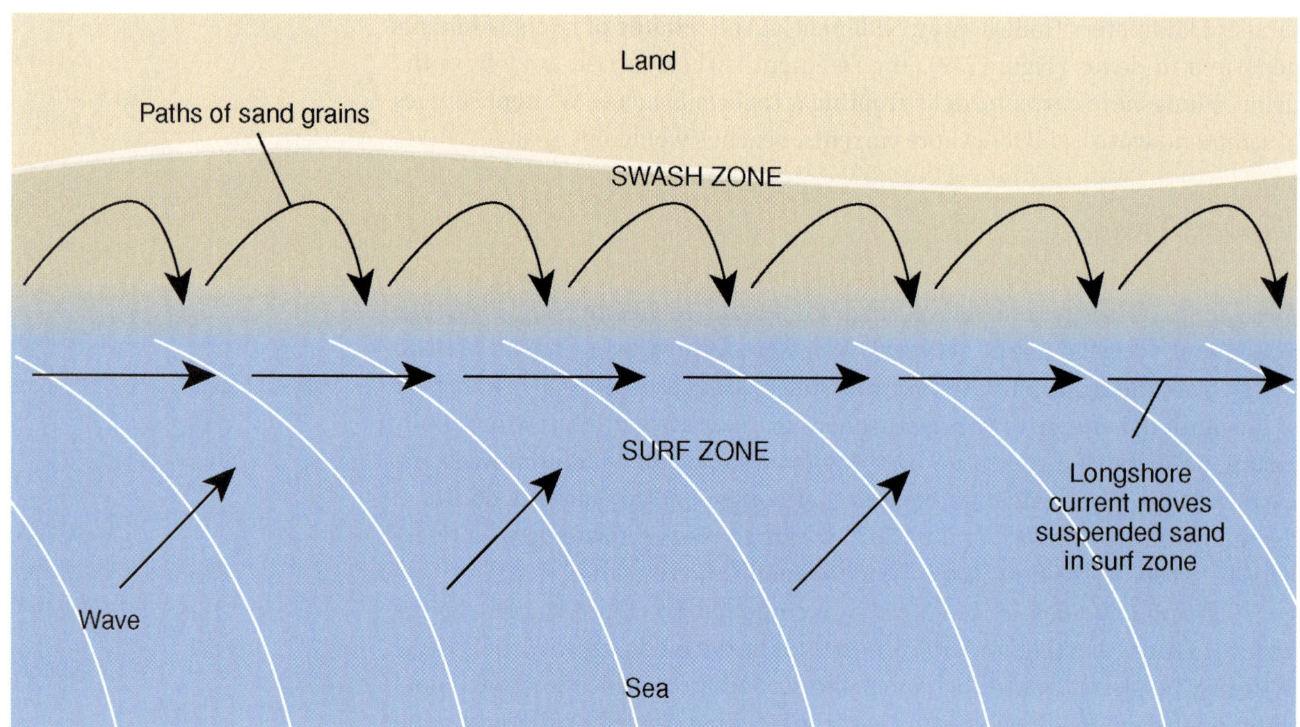

Figure 5-7 Longshore currents.

Earth, about one-quarter the distance from the surface to the center). Centrifugal force is an outward-directed force that causes matter in and on the Earth, including its oceans, to move outward. You feel a similar outward-pulling force when you are spinning on an amusement park ride. The centrifugal force is the same all over the Earth, but the force of the moon's gravity depends on the distance from the moon.

The centrifugal force is the same magnitude in both locations, but the force of the moon's gravity is greater on the side of Earth closest to the moon. Because the moon's gravity is stronger than the centrifugal force at this location, ocean water is pulled to this side of the Earth, forming a tidal bulge (or high tide). On the opposite side of the Earth the centrifugal force is stronger than the moon's gravity, so ocean water is pulled to this side as well, creating another high tide. Therefore, two high tides typically occur each day—one when the moon is directly overhead (closest to a location on Earth's surface) and another when the moon is farthest from a location. Low tides result when the moon is at right angles to a location because the ocean water is being pulled from these areas both toward the moon and toward the location of the most pronounced centrifugal forces opposite the moon.

Gravitational interactions between Earth's hydrosphere and both the moon and the sun combine to create greater elevation differences between high and low tides (tidal range). Although the sun is extremely far from the Earth (150 million km or 93 million miles), it is also extremely massive (333 thousand times the mass of Earth)—massive enough to create a gravitational force that can influence

Figure 5-8 A rocky headland, or seacliff, being eroded by waves. *(Courtesy the Earth Science World Image Bank. © Bruce Molnia, Terra Photographics)*

Figure 5-9 A delta at the mouth of a river. *(Courtesy of Earth Sciences and Image Analysis Laboratory, NASA Johnson Space Center)*

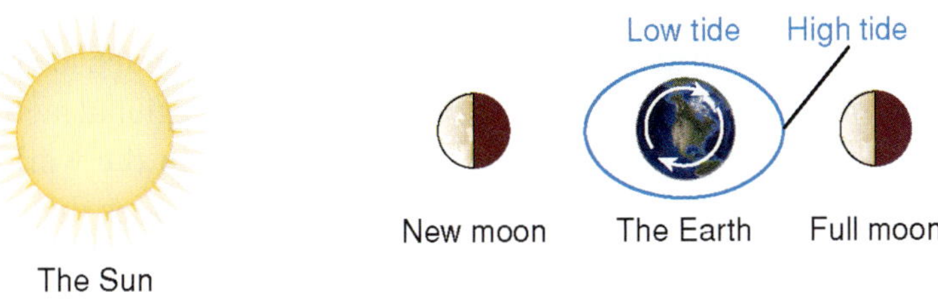

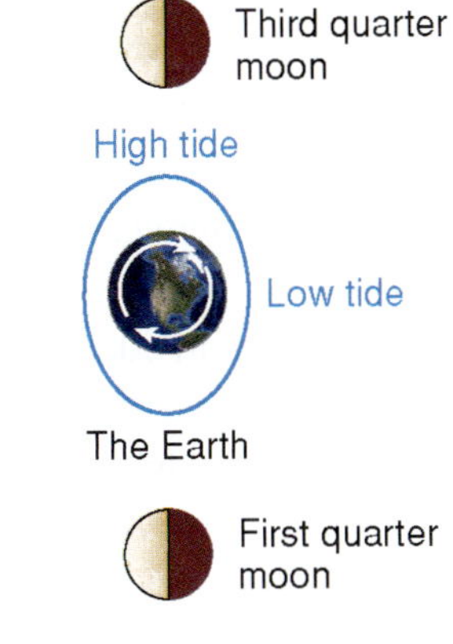

Figure 5-10 Earth-moon-sun interactions responsible for high and low tides.

Figure 5-11 Tidal variations at Canada's Bay of Fundy. (Courtesy NASA/GSFC/METI/ERSDAC/JAROS, and U.S./Japan ASTER Science Team)

monthly variations in the Earth's tidal range. For example, when the Earth, sun, and moon are all in a line, the gravitational pull of the moon and the sun combine to generate a higher-than-normal high tide and a lower-than-normal low tide.

Tides not only vary at a given location through time, but they also vary quite dramatically from place to place, even over short distances. These variations are the result of a multitude of factors that make predicting tidal ranges quite complicated. Perhaps most importantly, the shape of the land and seafloor along the coast affects the tidal range in a specific area. Tidal ranges are most pronounced where the tide enters narrow inlets because the water coming in with the tide flows into a smaller space that forces the water level to rise. This is exactly the setting of the Bay of Fundy in Canada, which experiences the greatest tidal range in the world at 12 meters (40 feet; Figure 5-11). In other words, the sea level rises and falls 12 meters (40 feet) about every 12 hours! The opposite happens at oceanic islands far from continents; they typically experience very small tides of less than 1 meter (3.3 feet).

What effects do tidal changes have on a coast? The main effect is that seawater is moved back and forth across the land each day. As tides rise, it comes ashore and flows up bays and rivers. When the tide falls, muddy or marshy areas called tidal flats become exposed. Exposed tidal flats can be several kilometers (miles) in width. This means that the shoreline can shift laterally several kilometers (miles) in the little more than 6 hours between high and low tides.

Moving Coastlines

Tides shift the shoreline back and forth every day. Other processes can move the location of the coastline for thousands of years or more. Past times of coastlines being higher than today are evidenced by beach deposits, marine fossils, and other coastal features found high above the present shoreline. In other locations, terrestrial features, including human settlements, are found submerged beneath the sea, formed at a time when the coastline was much lower. Changes in coastline location have resulted from both global and regional factors. Global changes are caused by changes in the volume of water in the world's oceans. Regional changes are generally related to geosphere processes that change the elevation of the land along a coast.

You learned about global sea level changes in Chapter 4. They are related to changes in the volume of water stored in the polar ice caps and continental ice sheets that form during times of colder global climate. Water evaporated from the oceans in the water cycle becomes trapped in the ice accumulations and sea level falls. During times of warmer global climates, the ice accumulations melt and send water back to the oceans. Sea level rises at these times. The most recent example was between 18,000 and 6,000 years ago when global sea level rose 120 meters (390 feet; Figure 5-12).

Regional changes in coastline location are generally caused by movements of the geosphere relative to the sea (hydrosphere). In high latitudes, the weight of polar ice can gradually depress Earth's crust as ice builds up during periods of colder global climates. During these times the depressing (subsiding) crust allows the sea to transgress (spread inland) across coastlines to relatively higher

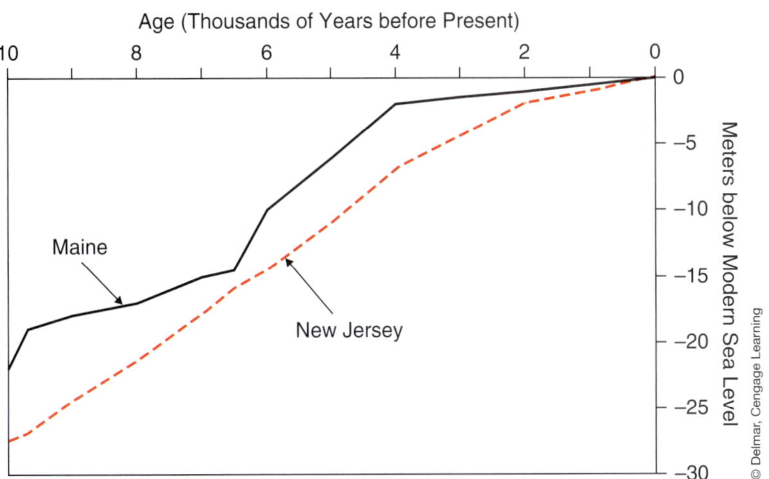

Figure 5-12 Global sea level curve for the last 21,000 years.

levels—the sea essentially floods and submerges the subsiding coastal areas. In these cases, the crust is subsiding (sinking) faster than sea level is falling in response to the cold times.

During periods of warmer global climate, the opposite can happen at high latitudes. Polar ice melts during these times and the decreasing weight of the ice allows the previously depressed crust to rise, or rebound, relative to sea level. In these cases, the shoreline moves back (regresses) toward the oceans as more and more land becomes exposed from uplift. Both of these processes (ice melt and crustal rebound) have caused a gradual shift in some regions' coastlines, like in Glacier Bay, Alaska (Figure 5-13). Rapid movements of the geosphere (tectonic movements) may also change the location of coastlines.

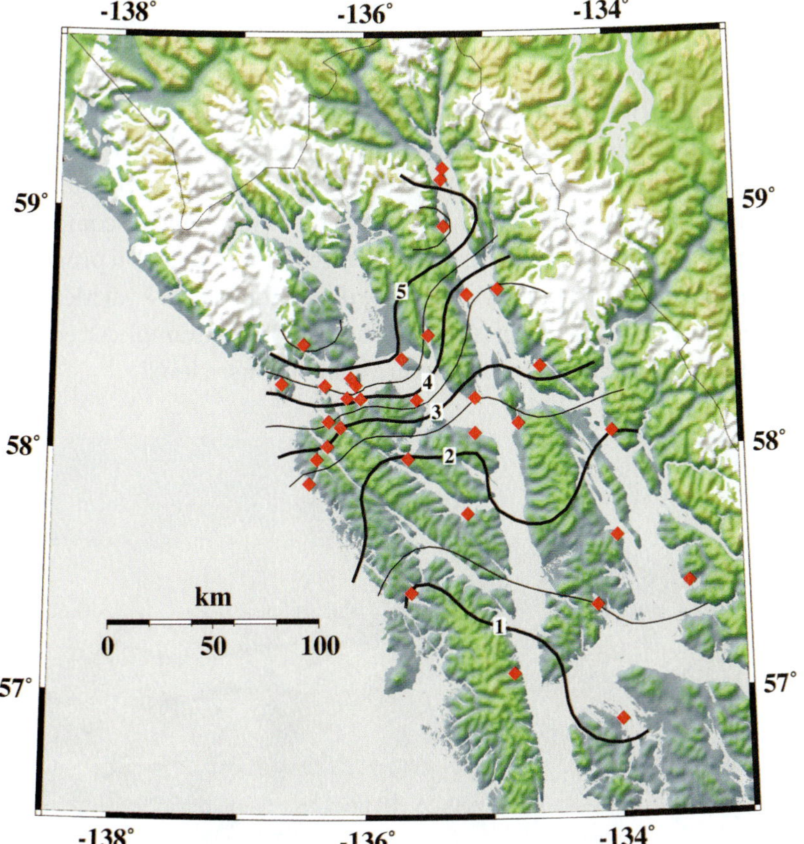

Figure 5-13 Changes in the historical shorelines in Glacier Bay, Alaska. *(Copyright Chris Larsen, Geophysical Institute, University of Alaska, Fairbanks)*

Courtesy USGS

Figure 5-14 Shoreline uplift following the Great Alaska Earthquake, 1964.

Tectonic movements can uplift or submerge coastlines. Geosphere movements during earthquakes commonly include vertical displacement of the Earth's surface. The Great Alaska Earthquake of 1964, lasting just 4 minutes, caused over 11 meters (36 feet) of uplift in some areas with other areas subsiding over 2 meters (6.5 feet) relative to sea level. Subsidence of land near the coast resulted in extensive coastal flooding. Uplift of the sea floor in other areas exposed the ocean floor and created new pieces of coastal land almost 1 km (0.62 miles) wide (Figure 5-14).

Tectonic movements can also gradually shift the geosphere up or down. Gradual uplifts cause the coastline to recede, leaving behind beach deposits (Figure 5-15). Gradual subsidence can lead to the shoreline migrating landward in places eroding into bedrock and forming sea cliffs; in other places rivers and valleys may become extensively submerged (Figure 5-16). Coastlines that move tectonically can have complicated histories as regional factors interact with global sea level changes.

Figure 5-15 Beach ridges on the north coast of Saaremaa, Estonia. *(Photo courtesy Mark A. Wilson, Department of Geology, The College of Wooster)*

Storm Surges—Colliding Earth Systems

Storm surges are a temporary local rise in sea level caused by strong storm winds blowing onshore. They are, in effect, a large dome or bulge of water that the atmosphere blows ashore (Figure 5-17). The height of a storm surge is ultimately determined by interactions of the hydrosphere with the atmosphere and geosphere. The stronger the onshore winds and the shallower the water leading up to the coast, the higher the surge will be. Surge heights are also greater if they occur at high tide or if they flow into bays or inlets.

Storm surges are commonly associated with hurricanes and tropical storms. Such storms annually threaten the Atlantic and Gulf Coasts of the United States. Hurricane Hugo hit the South Carolina coast in October 1989, raising sea level up to 11 meters (36 feet) in places and causing $9 billion dollars in damage. For a house directly on the shore, a rise in sea level of 11 meters (36 feet) would flood the home over the third floor. If it were not that tall, the house could be completely submerged. In many places storm surges are a greater threat to life and property along coasts than the associated storm winds. The devastation that New Orleans experienced from Hurricane Katrina in 2005 (see Chapter 1) is an example. Of course it did not help that large parts of New Orleans are built below sea level!

Figure 5-16 Submerged coastline of Prince William Sound, Alaska.

Courtesy NOAA

People and all other living things along a coast get trapped at the interface between atmosphere, hydrosphere, and geosphere by storm surges. They are capable of dramatic modifications to the shoreline and habitats, including completely destroying sand spits and barrier islands created by longshore currents. The overtopping surges form new tidal inlets and channels where none existed before (Figure 5-18). The slow continuous transport of sand by longshore currents will slowly fill in tidal inlets formed by storm surges if tidal currents are not strong enough to remove sand accumulating in the inlet. Over many years, evidence of the storm surge can be removed by the actions of longshore currents, potentially leaving the landscape with little evidence of previous powerful

Courtesy NOAA

Figure 5-17 A storm surge in action.

July 17, 2001

August 31, 2005

Photo by: Dennis Krohn Courtesy USGS

Figure 5-18 Before and after a storm surge on a beach.

and damaging storms. These changing coastal landscapes pose challenges to organisms adapted to specific habitats and community leaders trying to assess and plan for the full realm of hazards that can occur along a coast.

 coasts and the biosphere

The physical character of the coast—the transition zone where all Earth systems interact—is constantly changing. Waves incessantly pound the shore, tides continually rise and fall, and storm surges periodically alter coastal settings. Once in a while, large parts of a coast can be changed by tectonic displacements. In light of its dynamic nature, it is not surprising that coasts are a habitat for some of the most interesting and versatile organisms in the biosphere—organisms that have adapted to the ever-changing abiotic conditions in places like beaches, estuaries, salt marches, and mangrove swamps.

Beach Habitats

A beach habitat's most unique characteristic may be the constant movement of sand. Sand moves up and down the beach with each passing wave or along the beach with longshore currents. Sand not only moves in small amounts with wave swash but large amounts of sand can be removed to shallow areas offshore by stronger waves during stormier winter periods. Subsequent calmer summer months lead to the replacement of the sand as gentler waves slowly move sand back up the beach. Organisms inhabiting the beach environment must not only adapt to the shifting sands with each wave but also must be able to relocate as a beach's position shifts from season to season or storm to storm.

The force of the water created with the swash and backwash of a wave is the primary abiotic factor influencing the types and distribution of species found on a beach. Distinct habitats are found moving up the beach from the water's edge (Figure 5-19). In the surf zone, moving sand and waves keep rooted plants from taking hold on the surface. Animals that live in the surf zone must either be strong swimmers or able to burrow rapidly into the sand to protect themselves from waves. One such burrowing organism is the sand crab (*Emerita*), which is found on the Atlantic and Pacific coasts of North and South America. Sand crabs burrow into the sand tail first, leaving only part of their head sticking out to grab food floating by (Figure 5-20). Their large antennae trap microscopic food particles as the backwash of a wave rushes past their shallow burrow. Another species adapted to the surf zone is the razor clam. Several species of razor clams exist along both the Atlantic and Pacific coasts: the Pacific razor clam (*Siliqua patula*), Atlantic jackknife clam (*Ensis directus*), and Atlantic razor clam (*Siliqua costata*). The Pacific razor clam can tunnel through a few feet of sand in seconds and large gills allow it to breathe while buried in the moist sand (Figure 5-21). Both the razor clam and sand crab are protected from the surf by a tough external structure—the clam's shell and the crab's exoskeleton.

Another reason for burrowing in the surf zone is to avoid predation. Sand crabs face threats from both the sea and land. When submerged, sand crabs are preyed on by fish such as yellowfin croaker and barred surfperch. This makes sand crabs excellent bait for people fishing at the beach. Sand crabs also

Figure 5-19 Ecological zoning found on a beach face.

become prey for birds, such as gulls, when the waves' backwash can wash away the sand around their burrow and expose them. Sandy beaches do not provide much protective cover for animals that live on them.

Higher on the beach, energy transfer through the ecosystem does not begin with rooted photosynthetic plants like most terrestrial environments; rooted plants cannot take hold where waves even occasionally—as during storms—wash ashore. Nutrients supporting life in the beach habitat must, therefore, come from somewhere else. The major source of nutrients along a beach is the ocean. Nutrients come ashore in three forms: (1) beach "wrack," which consists of large masses of seaweed and algae; (2) sea foam, which consists of microscopic algae; and (3) microscopic plants and animals floating on the ocean's surface. Although the sand on which it rests may appear devoid of life, a complex ecosystem develops within the beach wrack—the base of the food web in this habitat (Figure 5-22). Carnivorous organisms, such as the rove beetle (*Thinopinus*), feast on the millions of sand shrimp, beach hoppers, and

Figure 5-20 A sand crab at home.

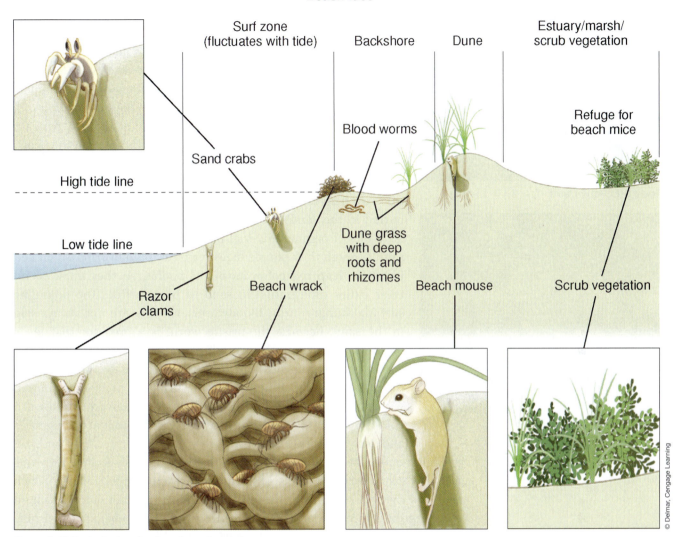

Figure 5-21 Razor clams.

Figure 5-22 A pile of beach wrack. (*Photo by: Captain Albert E. Theberge, NOAA Corps (ret.), Courtesy NOAA*)

other herbivores that both reside in and eat the seaweed. The beach hoppers, sometimes called sand fleas, are actually tiny crustaceans that crawl and hop along the sand and beach wrack. They are well adapted to beach living because they have gills that function almost as lungs. They must occasionally enter the ocean to wet their gills, but they will drown if completely submerged in seawater—an important reason they do not reside permanently in the surf zone.

The beach wrack also supports a mini-ecosystem beneath the sand. As seawater and rainwater percolate through the beach wrack, microscopic organic particles are carried down into the tiny spaces between the sand grains. Wave action, when present, can also help move organic material into the sand. Bloodworms buried in the sand will actually swallow the sand, digesting the little bits of organic material present and passing the remaining sand through their simple digestive tract. Bloodworms, so named for their deep red color, have soft bodies, so they must spend their entire time buried in sand to be safe from the pounding surf and hungry birds. Bloodworms, though, are not always safe in the sand, because the bristle worm is a predator that feeds on them there. The bloodworm's eating habits mix the sand and spread microorganisms that further accelerate decomposition of the organic matter. Whether on the surface of the sand or buried within it, nutrients washing ashore from the ocean are the base of a complex food web consisting of primary and secondary consumers and decomposers (Figure 5-23).

Higher on the beach, **dunes** are found that are originally formed as berms created by strong waves during large storms but are later topped by wind-blown sands (see Figure 5-19). Only the largest storms affect this part of the beach, so years can go by without impacts from waves. However, plants and animals wishing to colonize the dunes must still contend with several abiotic and biotic factors, including surface stability, dryness, and the lack of nutrient-rich soils. Beach grass can take hold on the dune sands but only a tenuous hold because the sand still regularly shifts under the influence of winds blowing off the ocean. The deep roots of the beach grass help anchor it on the unstable dune sands (see Figure 5-19). The deep root systems of the beach grass serve another purpose too. Dune sands dry out very rapidly after a rainstorm because the water quickly percolates downward. Consequently, beach grass must grow deep roots to find adequate freshwater to survive. Beach grass must also retain whatever moisture it finds, resisting the drying effect of the wind, sun, and salt spray typical of the shore. The long, narrow, spiky leaves serve this purpose as do the thick, waxy, or fuzzy leaves characteristic of other plants that grow on beach dunes.

Beach grass is one of the very few plants able to adapt to the harsh conditions of the dunes. Large trees, more typical of a forested hillside, cannot find sufficient nutrients, water, or surface stability to survive on the loose and moving sands of beach dunes.

Some animals are also found on the dunes sometimes in a symbiotic relationship with beach grass. For example, beach mice, the only small mammal living year round on the dunes, dig their burrows near a clump of grass or other vegetation to find protection from predators (Figure 5-24). Eight subspecies of

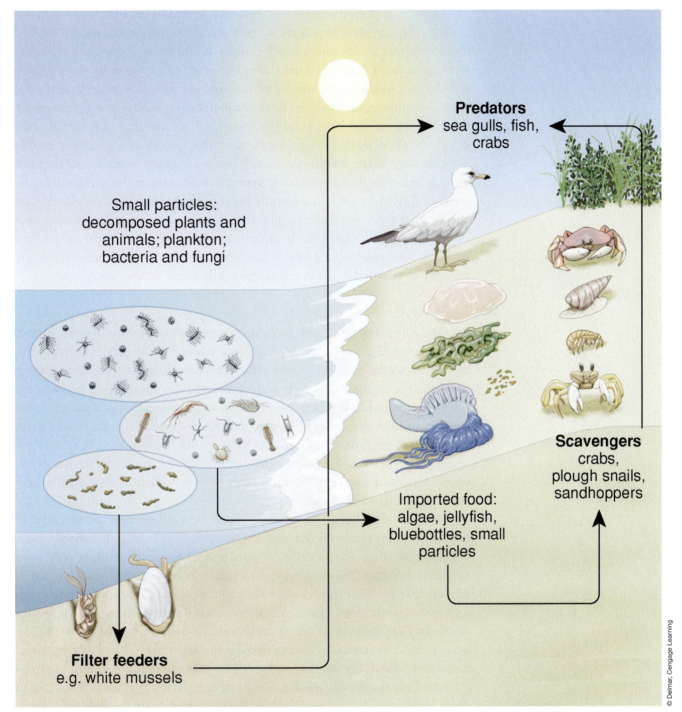

Predators
sea gulls, fish, crabs

Small particles:
decomposed plants and animals; plankton; bacteria and fungi

Imported food:
algae, jellyfish, bluebottles, small particles

Scavengers
crabs, plough snails, sandhoppers

Filter feeders
e.g. white mussels

Figure 5-23 Food web and trophic levels on the beach.

the Oldfield mouse (*Peromyscus polionotus*), collectively known as beach mice, occur on beach dunes on the Gulf Coast from Alabama to Florida and on the Atlantic Coast of Florida. Vegetation in the dune areas occupied by beach mice is primarily sea oats (*Uniola paniculata*) and dune panic grass (*Panicum amarum*), with the seeds of sea oats being an important food source for the mice, as are insects and acorns. The acorns are from oak trees (*Quercus* spp.) growing further inland in the scrub vegetation adjacent to the dunes.

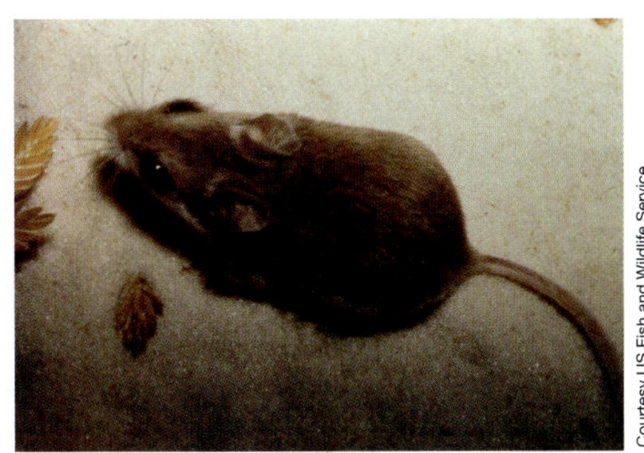

Figure 5-24 A beach mouse.

From an evolution perspective, species adapted to the beach environment must be able to relocate or "pick up and move." Entire beaches can be washed away by storms, and habitats shift locations up and down a beach with seasonal changes in wave strength. Sand crabs, like many beach animals, have pelagic larval stages, meaning the "babies" of the various species float near the surface of the ocean. This ensures that the larvae drift off to new locations with potentially excellent beach habitat. Although individual adults may not survive the destruction of a beach during a strong storm, the species as a whole can survive because some of the floating larvae will land on another beach where they can grow to adulthood and ultimately reproduce. Species without this ability to relocate simply do not survive in the highly dynamic beach environment. The species that do develop the required adaptations are naturally selected over time and survive. Consequently, environments with harsh abiotic factors, like beaches, are characterized by low species diversity. Without competition, the few species adapted to the beach are generally found in great abundance.

Estuary Habitats

Estuaries are partly protected places along coasts where freshwater and saltwater mix (Figure 5-25). The sands moving down a beach by longshore currents can move out into the open ocean when the beach curves into an embayment along the coastline, creating sand spits and a protected quiet water area between the spit and mainland. Whereas the spit produces environments very similar to other beaches, the protected body of water, or bay, formed behind the spit is completely different and supports a wide range of habitats. But the formation of spits is just one way to form bays. Many bays formed when sea level rise during the last 21,000 years submerged valleys in the lower part of river systems. Consequently, bays come in all different sizes and shapes.

Estuaries develop in bays where there is a significant contribution of freshwater from rivers on one end and a significant tidal influence from the ocean on the other. The mix of freshwater and saltwater in an estuary is known as brackish water. Seaward portions of the estuary will be saltier, whereas areas nearer the mouths of rivers will be fresher. Variations in water chemistry characterize estuaries. Organisms able to tolerate both saltwater and freshwater conditions are best suited for estuaries and many have evolved over time to occupy these environments. An example is the Atlantic croaker (*Micropogonias undulatus*), one of the most abundant bottom-dwelling fish found in estuaries from the Chesapeake Bay south to Florida (Figure 5-26). They move in and out of freshwater through the year. In the spring they migrate up rivers to freshwater, in the summer they move back and forth between freshwater and salt water, and in the fall they move out of estuaries to the nearby open ocean to spawn.

Some organisms only live in estuaries for part of their life. Salmon are born in freshwater, their eggs having incubated in gravel at the bottom of rivers, but spend most of their adult life in the salty ocean. Some young salmon depend on estuary environments to ease their transition from freshwater to salt water. Within the estuary, young salmon undergo physiological changes that allow them to survive in the changing water chemistry.

The shores of estuaries and bays are commonly tidal flats, especially where the tidal range is high. Organisms occupying the intertidal zone, the area between the normal high and low tide lines, must contend with a number

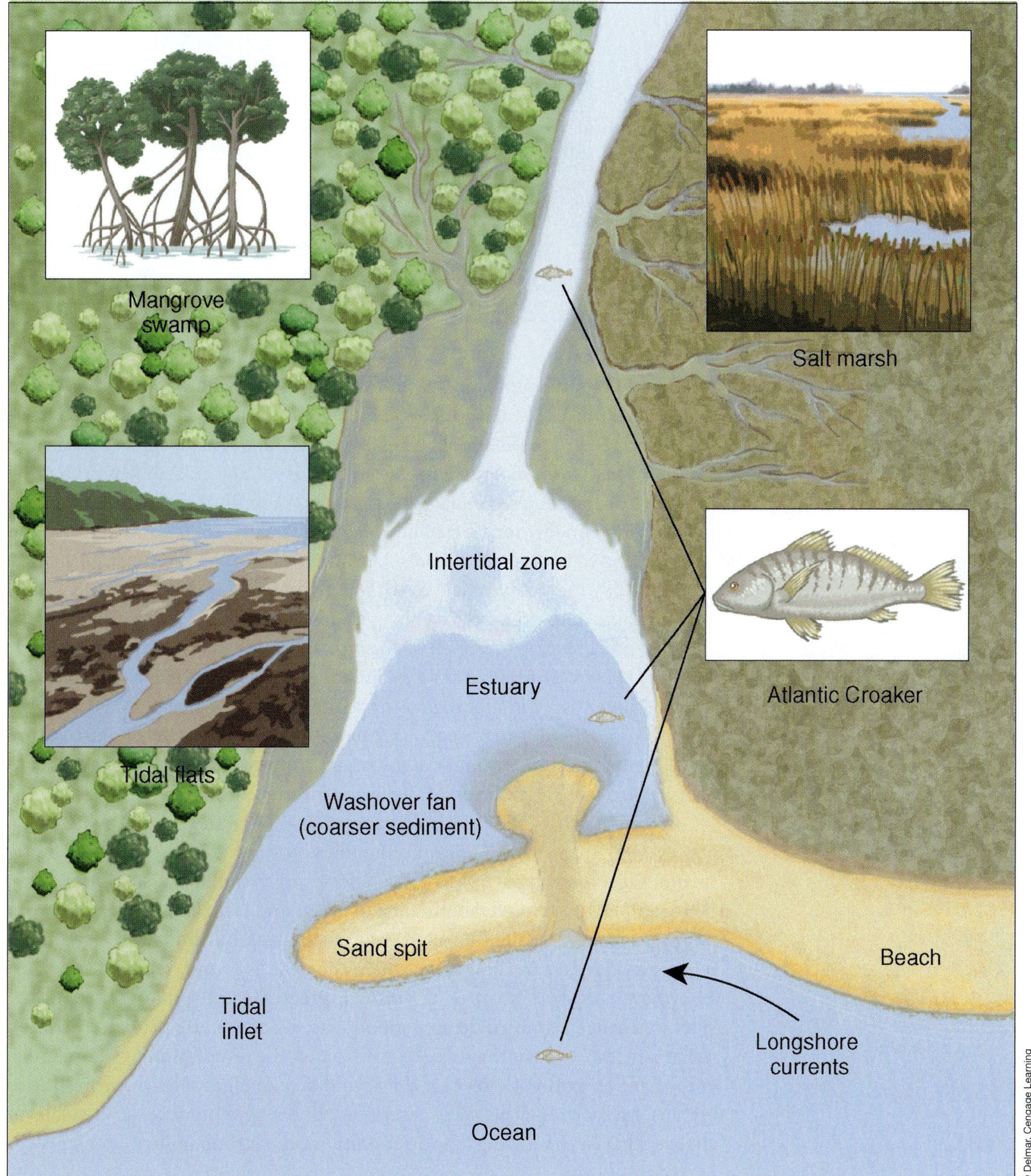

Figure 5-25 Habitats in a typical estuary.

of abiotic factors but the length of time they are exposed to air during low tides is especially important. Intertidal organisms have developed a number of adaptations to tolerate these abiotic conditions. Many species, such as snails, seastars, fish, and chitons, will retreat to tide pools to avoid water loss at low tide, but then move about searching for food while submerged at high tide. Tide pool sculpins, a small fish species, tend to occupy the same

Figure 5-26 The Atlantic croaker is able to live in both freshwater and salt water.

Figure 5-27 Barnacles. (*Courtesy Mark A. Wilson, Department of Geology, The College of Wooster*)

tide pool with each low tide and feed on shrimp and worms. Those animals unable to move about, such as barnacles, tend to concentrate in the tide pools where water is almost always present. Within these tide pools, microscopic organisms are more likely to float by the barnacles (Figure 5-27). Barnacles are filter feeders that capture food on feathery meshlike appendages that sweep through the water. The hard shell of the barnacle attaches strongly to rocky substrates, allowing the barnacle to withstand crashing waves and avoid drying out when exposed for short periods, serving much the same purpose as the hard platy shell of chitons. These shells are also defenses that allow barnacles to survive outside tide pools. Barnacles become less and less abundant higher in the intertidal zone because of the longer time they are exposed to air there. A species' tolerance of temperature fluctuations, drying, water loss, and other abiotic factors largely controls its distribution in the upper intertidal zone.

In lower parts of the intertidal zone, species diversity is higher due to the less extreme exposure conditions. The impacts of predation and other biotic factors exert a stronger influence on species diversity and distribution here. For example, the relative abundance of green and red algae is dependent on the presence of a snail. This snail eats green algae but finds red algae inedible. Consequently, in tidal areas where the snail is found, red algae flourish. Elsewhere red algae must compete with green algae for habitat space. The snail's presence is in turn dependent on whether the snail's primary predator, the green crab, lives in the area. By controlling the population of snails, green crabs have an indirect effect on the abundance of algae. These species interactions are typical of the lower intertidal zone where predator-prey relationships, rather than limiting abiotic conditions, are strong influences on organism adaptations.

Diatoms, seaweed, and other marine plants flourish in the lower intertidal zone. Here they convert sunlight into food for hundreds of large to microscopic primary consumers, which, in turn, feed the secondary consumers that prey upon them (Figure 5-28). The breakdown of dead plant material produces very small particles of food that can drift off to other areas of the estuary, providing the energy for ecosystems beyond the lower intertidal zone. When these small pieces of food fall to the bottom of the ocean they become buried in the mud that has also settled to the bottom of protected estuary waters.

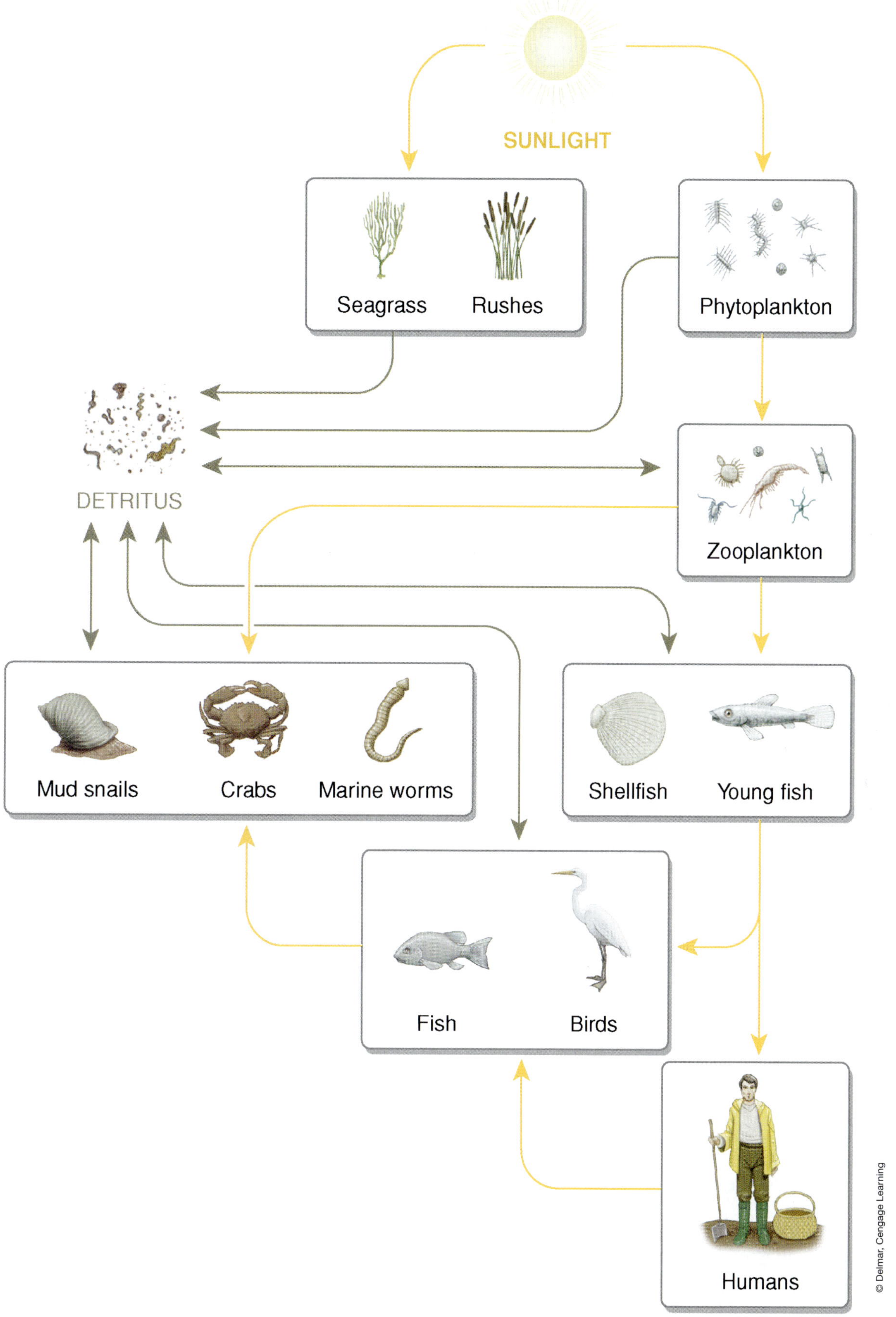

Figure 5-28 Food web and trophic levels in an estuary.

Salt Marshes

Salt marshes—typically surrounding estuaries just above the intertidal zone (Figure 5-29)—are characterized by an assemblage of salt-tolerant grasses. The most tolerant of these grasses, smooth cordgrass (*Spartina alterniflora*), is generally the first species to colonize a salt marsh as it emerges from the upper intertidal zone. Smooth cordgrass has evolved at least three adaptations that allow it to survive in these very salty conditions. First, cordgrass concentrates salt in its roots—this helps the roots gain water as water generally moves from areas of low to high salt concentration. Second, cordgrass can secrete salts through special glands that leave a visible film of salt crystals on the tough exterior of the stems and leaves. Third, specialized cells store water in the cordgrass so that it can be used during times of higher salt stress.

Saltiness, however, is not the only abiotic factor that cordgrass must overcome on the lower marsh. The frequent inundation by high tides means that the root systems are constantly submerged under water where oxygen, needed by the roots, is limited. A spongy tissue with enlarged gas-filled spaces running from the leaves to the roots solves this problem by allowing oxygen to move down the "hollow" stem of the cordgrass to the roots. Smooth cordgrass has also developed symbiotic relationships with two animal species to help with soil oxygenation and the availability of nutrients. The burrowing activity of fiddler crabs (e.g., *Uca pugnax*) increases the amount of water that drains out of the soil at low tide and provides a pathway for oxygen to reach cordgrass roots. The feces from dense colonies of ribbed mussels provide nitrogen to the soil around the roots of the cordgrass.

Because of these specialized adaptations for salt tolerance and oxygen depletion, smooth cordgrass serves as the colonizer that allows the rest of the salt marsh ecosystem to develop and flourish. Once a few seeds are able to germinate and take hold on the upper intertidal zone, generally after a heavy rain when salt concentrations are a bit lower, the cordgrass spreads by sending out horizontal stems, or rhizomes, that grow just beneath the ground surface. In this way, very dense patches of cordgrass can grow rapidly with the tight interlocking system of roots and stems trapping organic matter and fine particles of mud. The deposition of this material around the base of the stems

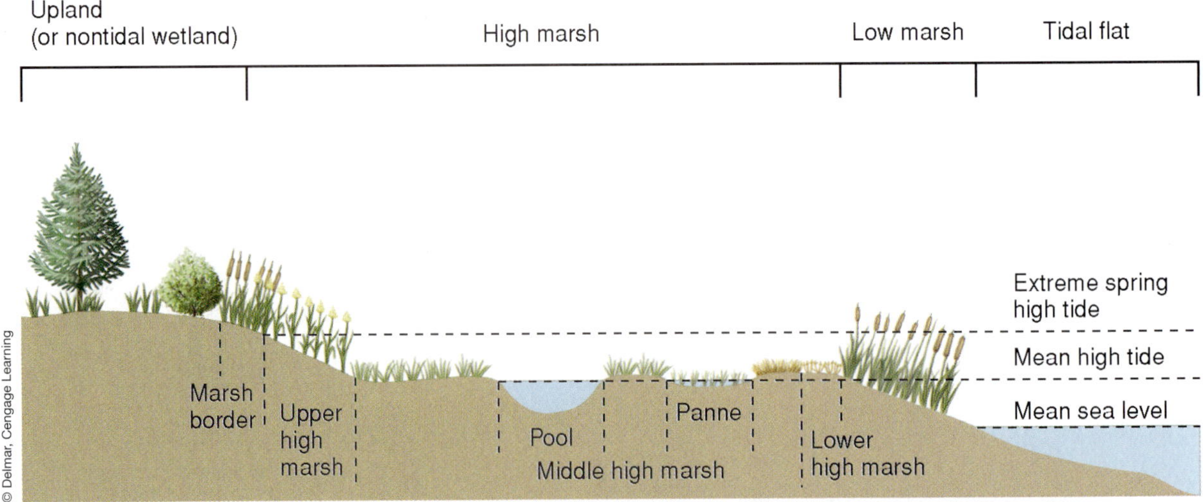

Figure 5-29 Habitats in a salt marsh.

raises the surface slightly—just enough so tidal inundations are less frequent. These are places where less tolerant plant species can grow. Among these plants is another form of cordgrass, short cordgrass (*Spartina patens*), that characterizes the higher salt marsh surface. The ability of the smooth cordgrass to inhabit the harsh intertidal zone leads to a series of successional steps that ultimately result in a complex salt marsh with a zonation dependent on almost imperceptible, yet very significant, elevation differences (see Figure 5-29).

Both cordgrass species are the dominant primary producers on salt marshes through which energy from the sun flows through the food web (Figure 5-30). Despite low species diversity, salt marshes are highly productive. Grasshoppers (*Orchelimum*) and leafhoppers (*Prokelisia*), with their piercing mouthparts, are two of the very few animals able to eat the very tough blades of the smooth cordgrass while it is alive and growing. These insects, in turn, are eaten by a number of other spiders and insects, but less than 5 percent of the energy available in the cordgrass is estimated to move through the food web in this way. Other animals make use of the energy stored in the cordgrass, but only after the grass dies and is decomposed by bacteria, fungi, and small algae. Most of the decomposition occurs in sediments inhabited by worms, shrimp, fish, and crabs that feed on the decomposers. The feces of these animals are further decomposed and broken down by bacteria and other microorganisms until simple nutrients are released in the soil to fertilize the growth of more cordgrass. In this manner, the high productivity of the salt marsh environment is sustained.

Salt marshes are an important part of the larger coastal ecosystem. The dense root systems of the cordgrass provide shelter for the young of many species such as blue crab, white shrimp, and spot tail bass. The protection from predation

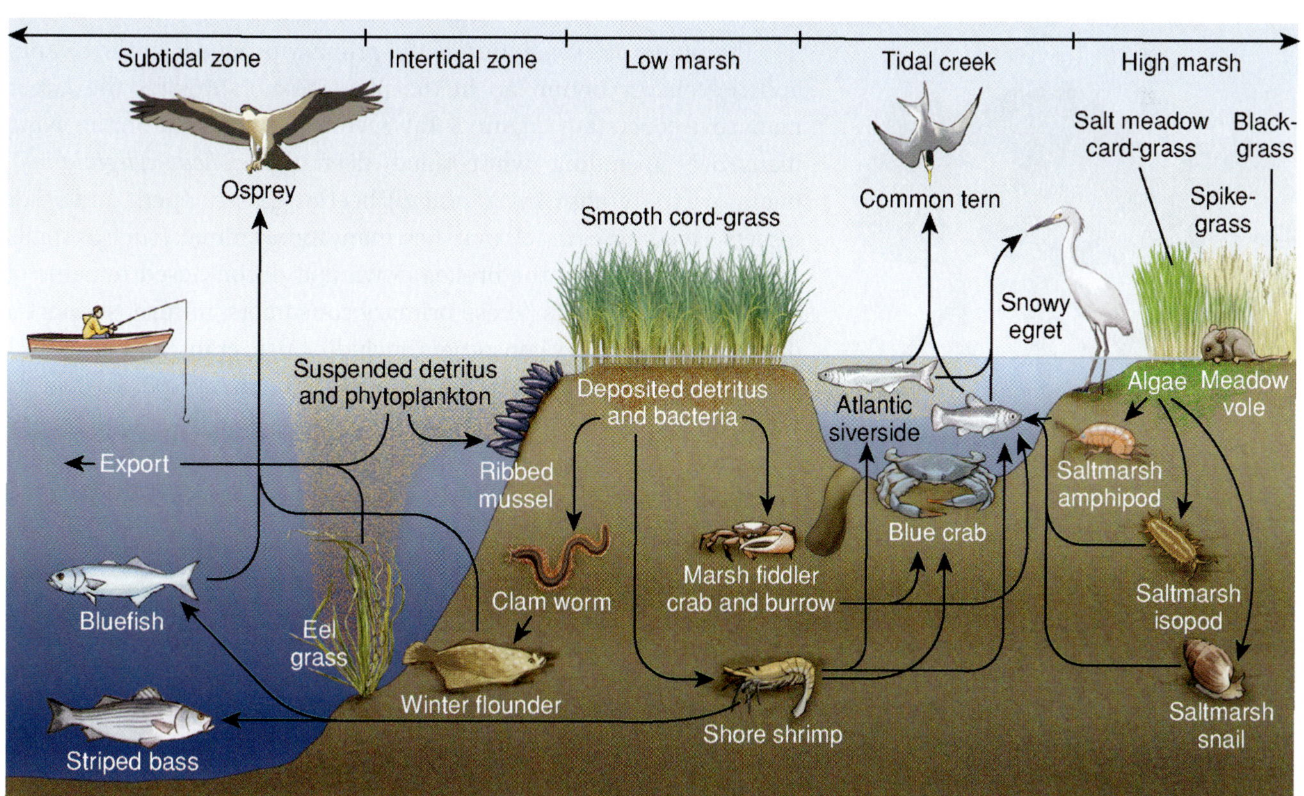

Figure 5-30 Food web and trophic levels in a salt marsh.

and abundant food supply make salt marshes an excellent nursery for juvenile animals that may spend their adult lives in the estuary or even the open ocean. If these animals move from the marsh too soon, they will fall easy prey to other animals living in the tidal and freshwater creeks that lead out of the marsh to the adjacent estuary. Without the protection afforded by the roots of the salt marsh plants, the diversity and health of other coastal environments would suffer.

Mangrove Swamps

In tropical and subtropical areas of the world, including southern Florida, **mangrove swamps** are commonly found on the fringes of estuaries, rather than the salt marshes of more temperate regions. The dominant plants are mangroves—a group of more than 50 species of salt-tolerant trees and shrubs. Each mangrove species has its own environmental niche defined by its tolerance range for two abiotic factors: soil salinity and soil oxygen levels. Red mangroves, like smooth cordgrass in the salt marsh environment, are the most tolerant to salt and low oxygen levels, so they commonly occupy the most seaward position in the swamp. Red mangroves have special roots growing from the tree's trunks that hold the plants up off the ground, in places up to a meter. These "prop" roots send smaller roots down into the ground (Figure 5-31). This allows oxygen to enter the plant through the above-ground roots and then move to the below-ground roots. Black mangroves, in contrast, send out horizontal roots in the soil close to the surface and then grow vertical "hollow" roots up out of the ground to provide oxygen for the plant. If the soil is too saline, the red mangrove will predominate. White mangroves, with no specialized root systems, are the most successful of the mangrove species in more oxygenated, fresher water conditions. All mangrove species have salt-secreting glands, like in salt marsh grasses, to get rid of the salt accumulated from the saline soils.

The mangrove vegetation is the primary producer in the swamp and the energy bound up in the plant tissue supports the larger mangrove ecosystem (Figure 5-32). Several animals feed on the living mangrove, including white-tailed deer (*Odocoileus virginianus*), mangrove tree crabs (*Aratus pisonii*), beetles, grasshoppers, and other insects. However, as in salt marshes, many more animals, such as snails and mussels, feed on the broken down and decomposed remains of dead mangrove plants. These primary consumers, in turn, support a diversity of secondary consumers, including fish, crabs, and birds.

Figure 5-31 Mangroves showing prop roots.

Mangrove swamps serve as a nursery to many species and as a permanent home to others. In Florida, several species living in mangroves are endangered, including the American crocodile and Key deer. The survival of these species depends on the health of the mangrove swamp ecosystem as a whole.

5.3 people and coasts

Whether you are surfing at the beach, digging for clams on a tidal flat, fishing in an estuary, or bird watching around a salt marsh, coastal habitats provide something for everyone to enjoy. In addition, people benefit from the commerce taking place in coastal areas—the ports, the commercial fishing,

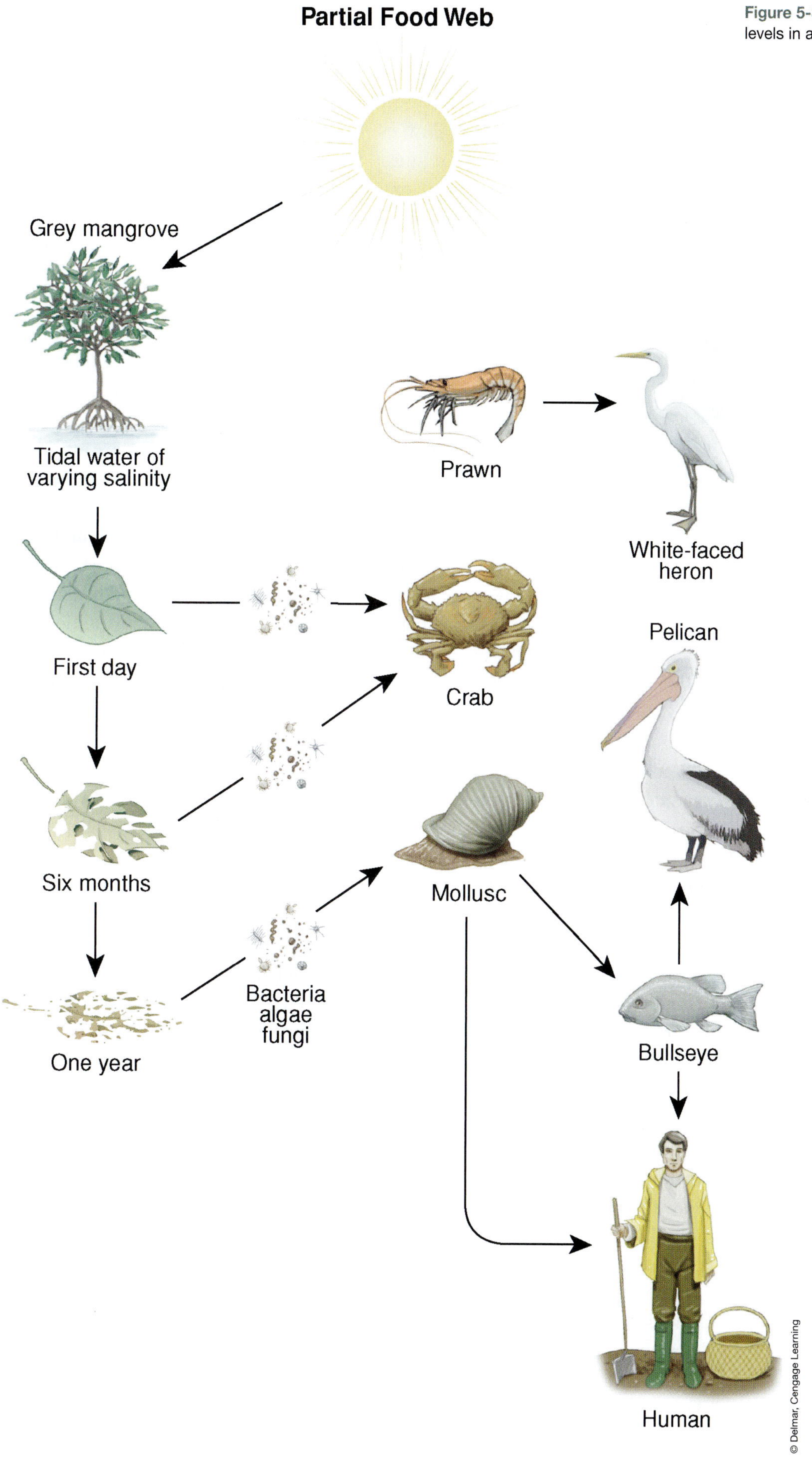
Partial Food Web
Grey mangrove
Tidal water of
varying salinity
Prawn
White-faced
heron
First day
Crab
Pelican
Six months
Mollusc
Bacteria
algae
fungi
One year
Bullseye
Human
© Delmar, Cengage Learning

Figure 5-32 Food web and trophic levels in a mangrove swamp.

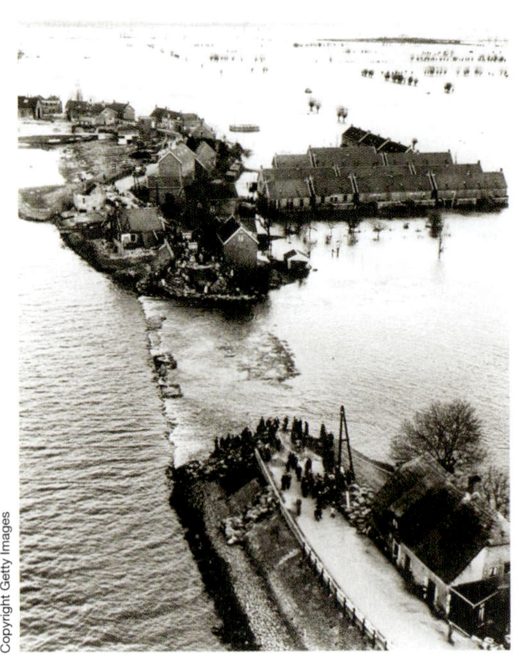

Figure 5-33 Flooding in Holland in 1953.

and the tourism. The coast is a place where people interact with many natural processes and environments. Flooding, erosion, and sedimentation resulting from coastal processes create hazards to human life, property, and commerce. On the other hand, people's facilities such as homes, bridges, and ports, and their efforts to protect them can upset natural ecosystems that are delicately adapted to coastal environmental conditions. How people use their knowledge of coastal processes and ecosystems to soundly manage coastal development while minimizing its effects on the environment is discussed in the sections that follow.

Coastal Flooding

No country in the world is perhaps more aware of the potential effects that coastal flooding can have on people than Holland. With half of their country below sea level, Dutch citizens face the constant threat of flooding from high tides, storm surges, and rising sea level. Any failure in Holland's incredible and expensive system of dikes, sea walls, and surge gates could cause catastrophic coastal flooding. Such flooding occurred on February 1, 1953, and 1,800 people died, 200,000 cattle drowned, and an untold number of homes flooded (Figure 5-33). The current system of flood control works in Holland because sea walls over 13 meters (43 feet) high can easily protect coastal communities from typical storm surges that are 5 meters (16 feet) high. But pumping of ground water to help keep the lowlands dry may continue to cause land subsidence. The potential for catastrophic storms and relative sea level rise has the Dutch government budgeting $25 billion over the next century on additional flood control works.

In the United States, coastal flooding has historically caused much damage and initiated many flood control efforts. A 5-meter-high (16 feet) sea wall surrounds Galveston Island, Texas, built in response to the devastating hurricane in September 1900 that killed over 8,000 of the city's 38,000 residents (Figure 5-34). The risk of flooding due to hurricanes in the eastern United States is so great, and covers such a large area, that engineering defenses alone are inadequate and economically unfeasible for solving the flooding problem. Take New Orleans for example.

During Hurricane Katrina in 2005 (see Chapter 1), the flood waters in New Orleans did not come down the Mississippi River. They came from the sea as the hurricane's storm surge and heavy rains came ashore from the Gulf of Mexico. The result was devastating. Four protective levees failed and 80 percent of the city was flooded, some beneath 6 meters (20 feet) of water (Figure 5-35). Hundreds of thousands were homeless, over 1,500 people died, and the human suffering continued for weeks and months as public services were either destroyed, ineffective, or overwhelmed. Many who evacuated will never return home to New Orleans. This tragedy was in the news for years.

The 2005 flooding and destruction of New Orleans is an example of the dilemma that people face when they choose to live in coastal areas where flooding from storm surges is likely. Did you know that many of the hurricane effects on New Orleans were predicted? The destruction, the health and water quality problems, the homelessness and evacuation needs, even the time it would take to get rid of the flood waters? You see, New Orleans has been trying to live with flooding since it was first settled.

Figure 5-34 Galveston seawall and 1900 hurricane destruction.

The initial settlement of New Orleans was on "high" ground. High ground here is several meters (tens of feet at the most) above sea level and there is not much of it. As the city's population grew, it expanded into lower and lower areas. Canals were built and pumps were installed to drain the low areas and create space for people. Now large parts of New Orleans are below sea level.

People have done more than just spread New Orleans into low areas. Over the years, shipping channels have been dredged through the nearby Mississippi River's delta. These channels have funneled the river's discharge to the Gulf of Mexico and not allowed it to continue building the delta and sustaining its wetlands. As a result, the land has subsided, making New Orleans even more vulnerable to river flooding and hurricane storm surges. The combined effect of canals, pumps, and channeling of the river has made New Orleans a disaster area waiting to happen. And it will likely happen again.

Figure 5-35 2005 flooding of New Orleans. *(Photo by: Commander Mark Moran, of the NOAA Aviation Weather Center, and Lt. Phil Eastman and Lt. Dave Demers, of the NOAA Aircraft Operations Center)*

Coastal Erosion

Another hazard faced by people in the coastal environment is beach erosion. If you were the owner of a beachside resort, your investment would pretty much depend on being next to a sandy beach for your guests to enjoy. Beach erosion is the last thing you would want to deal with as a hotel owner. This is exactly what hotel owners on Miami Beach in Florida have been dealing with for years. If you were the hotel owner, what would you do to protect your investment so tourists from around the world could enjoy the beautiful beaches and warm ocean water? What things could be done to ensure that sand remains on the beach in front of your property? Some of the first attempts to prevent erosion on Miami Beach involved the use of constructed barriers that trap sediment and prevent it from migrating along the beach (Figure 5-36). This might help keep

Figure 5-36 Methods for trapping sediment and preventing beach sand erosion.

A Groin

Groins are structures that extend from the beach into the water. They help counter erosion by trapping sand from the current. Groins accumulate sand on their updrift side, but errosion is worse on the downdrift side, which is deprived of sand.

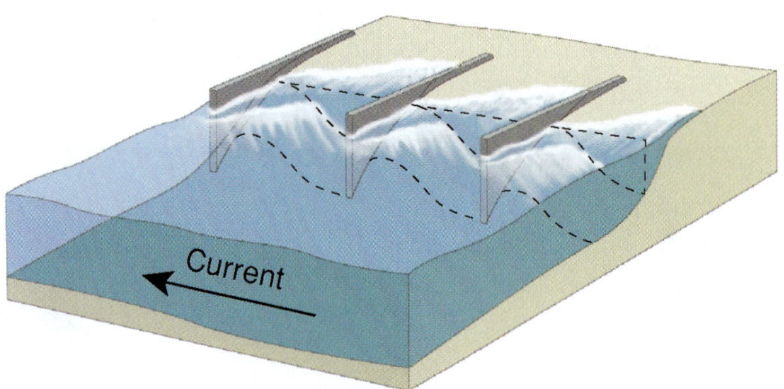

B Seawall

Seawalls protect property temporarily, but they also increase beach erosion by deflecting wave energy onto the sand in front of and beside them. High waves can wash over seawalls and destroy both the seawalls and the protected property.

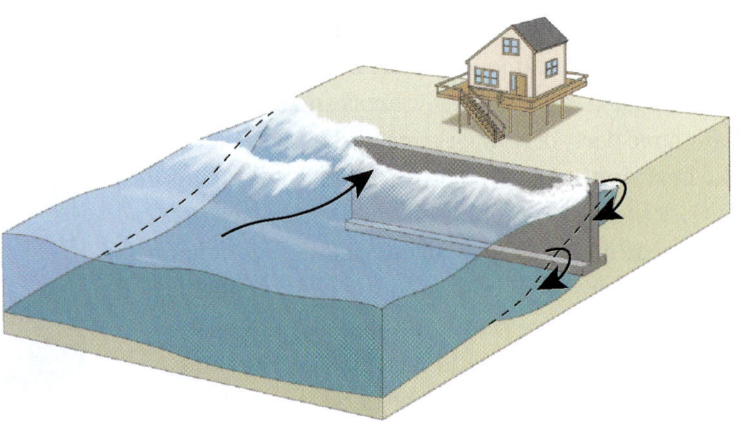

C Importing sand

Importing sand to a beach is considered the best response to erosion. The new sand is often dredged from offshore and can cost tens of millions of dollars. Because it is often finer than beach sand, dredged sand erodes more quickly.

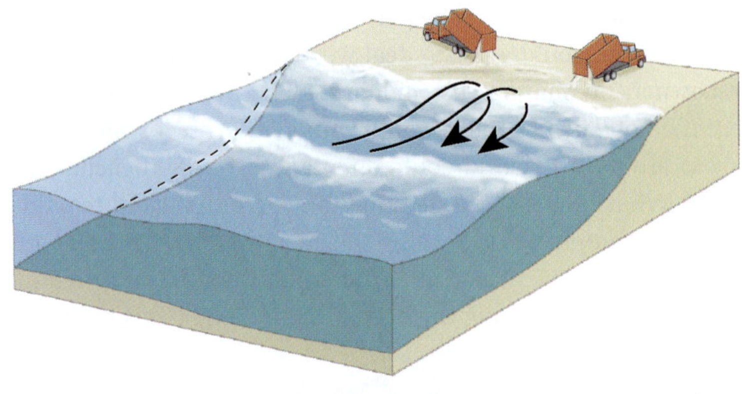

sand in front of a hotel but it also could cause accelerated erosion somewhere down the beach depending on the strength of longshore currents. Because structures that influence sediment transport along coasts can create more problems than they solve, care must be taken in how they are developed and used.

Long-term, persistent beach erosion is largely the result of a limited sediment supply, so in some cases artificially supplying sediment can sustain a beach for a number of years. Beach renourishment projects have pumped slurries of sand from shallow areas offshore to restore and supply sediment to beaches in North Carolina, Florida (including Miami), and elsewhere. These projects are expensive and must be periodically repeated because the sand added to the beach frequently erodes away over time. However, these projects may be worth the expense considering the potential economic benefits that come with a wide sandy beach outside the resort hotels of Miami or other vacation destinations.

Figure 5-37 Cape Hatteras Lighthouse in North Carolina. *(Photo by: Captain Albert E. Theberge, NOAA Corps (ret.), Courtesy NOAA)*

Where beach renourishment has failed, moving buildings and other infrastructure away from impending erosion is a last resort to avoid property damage. In 1999, after years of debate, the famous Cape Hatteras Lighthouse in North Carolina was moved 800 meters (2,600 feet) for $10 million to avoid being washed into the sea (Figure 5-37). When it was originally built in 1870, it was located 500 meters (1,640 feet) from the shore. Slow but persistent erosion and transgression of the shoreline brought it to within 35 meters (115 feet) of the lighthouse. People's structures cannot move with shifting coastal landscapes, so where the encroaching sea is a problem, efforts will need to be undertaken periodically to save structures of value or historical significance.

Where are the greatest threats for coastal erosion problems in the country? What areas should be preparing for action to save historic or valuable beachfront properties? Recent advances in satellite technology are helping with the answers to these questions. Laser infrared detection and ranging (LIDAR) systems mounted on airplanes measure the distance to the surface by sending out pulses of laser energy and measuring the length of time it takes for the energy pulse to be reflected and returned to the airborne instrument. These measurements are very accurate and have a vertical precision of up to 15 centimeters (6 inches). This technology has been used along many U.S. shorelines—such as Texas and North Carolina—to document topographic changes. Once the LIDAR data are obtained, loaded, and processed, they can be viewed in several ways. The data can be used to produce beach profiles (see Figure 5-1) or maps showing different elevations. Producing LIDAR maps and profiles each year allows the shoreline position to be compared and the rate of shoreline retreat or advance to be calculated. Areas of most rapid retreat near resorts, housing developments, and roads can be identified and resources made available to address the erosion problems in those regions with the greatest need. Although this new technology cannot stop the erosion, LIDAR mapping records changes over large areas and enables erosion to be identified long before it becomes a seriously damaging or dangerous problem. Data from this type of coastal monitoring can be a valuable guide for better land use decisions.

Figure 5-38 Dredging operation. (*Courtesy the Earth Science World Image Bank. Copyright EPA*)

Coastal Sedimentation

As discussed earlier, harbors and ports contribute tremendously to the national economy. Ports are located in protected bays safe from the waves of the open ocean. However, it is this same protected setting that accumulates sediment deposited by tides, storm surges, and rivers. Although not a life-threatening hazard, the slow infilling of a harbor or channel by sedimentation can slowly decrease its usefulness, because shallow water prevents ships from using the harbor. The possibility of moving a port with its entire infrastructure of docks, cranes, and warehouses is prohibitively expensive and very few alternative deep water harbors exist. Consequently, the mechanized removal of sediment in order to maintain harbor depth—or dredging—is a common practice where sediment accumulation rates are high (Figure 5-38).

Land Development and Coasts

San Francisco, California, is known as "The City by the Bay." But it is partly on the bay too—the San Francisco Bay. After the famous 1906 San Francisco earthquake, much of the debris and rubble from toppled buildings was disposed of in the most convenient place available—the swampy marshland at the city's edge. A seawall was built around the rubble and additional fill added to create 635 acres of land that was the focal point of the 1915 Panama-Pacific Exhibition. The event showcased ornate temporary buildings and displays designed to prove to the world that San Francisco had fully recovered from the earthquake a decade earlier. Although the site remained idle for 10 years following the exposition, the growing population of the city created a need for more development. To meet the demand, the site of the exposition was developed into what is now the Marina district—a neighborhood of homes, fine restaurants, and a large park. The Marina district has some of the highest housing prices in all of San Francisco. This trendy neighborhood, built on loose rubble and fill from the 1906 earthquake, has completely replaced a coastal salt marsh.

In the early 1900s, development in San Francisco was foremost on the minds of city planners trying to rebuild after the 1906 earthquake. The importance of salt marshes in preserving coastal ecosystems and the dangers of building on poorly consolidated fill were not fully recognized or considered. A growing scientific knowledge of these environments since then has revealed many of the problems associated with filling in marshland for development purposes. The Loma Prieta or "World Series" earthquake of 1989 caused extensive damage in the Marina district 80 km (50 miles) from the earthquake's epicenter. Many locations much closer to the epicenter experienced minimal damage. What caused such an uneven distribution of earthquake damage? The unconsolidated and water-saturated fill underlying the Marina district shook more severely than surrounding areas underlain by bedrock. Little can now be done in the Marina district to resolve the problem because the buildings and facilities are already in place. However, now that the problems of ground stability are better understood, new development on bay fill is severely restricted in San Francisco and other earthquake prone regions of the country.

The history of the salt marsh replaced by the Marina district is similar to the history of many other salt marshes around the country. Between 1950 and 1975, nearly half of all salt marshes in the United States were destroyed when they were drained and surrounded by dikes in order to improve the land for agriculture, homes, and industry. Natural marshes were believed to hold little value for people. However, ecology studies show that salt marshes benefit people as well as the organisms that live in them. Although 50 years ago the many values of salt marshes were not understood, today salt marsh restoration is the major goal of sound land management, not "improvements" that drain, dike, and fill salt marshes. New developments in coastal areas attempt to preserve salt marshes and in places even use them as part of wastewater treatment facilities.

Waste Disposal and Coasts

Treating wastewater using salt marshes is a relatively new technique that emerged with the scientific understanding that filtering pollutants at the land-water interface is a natural function of salt marshes. To this day though, wastewater and sewage is sometimes still dumped into the ocean untreated. Yes, that is right, untreated sewage straight into the ocean—the stuff from your toilet after you flush; the runoff from streets with drippings of oil, gas, and other chemicals; plus the wastes of factories and other industries. The City of Boston discharged sewage into Boston Harbor for over 300 years, completely untreated for most of that time (Figure 5-39). While strong tidal currents in the harbor generally carried the sewage out into the much larger Massachusetts Bay in less than 10 days, the constant additions of sewage kept concentrations of pollutants high

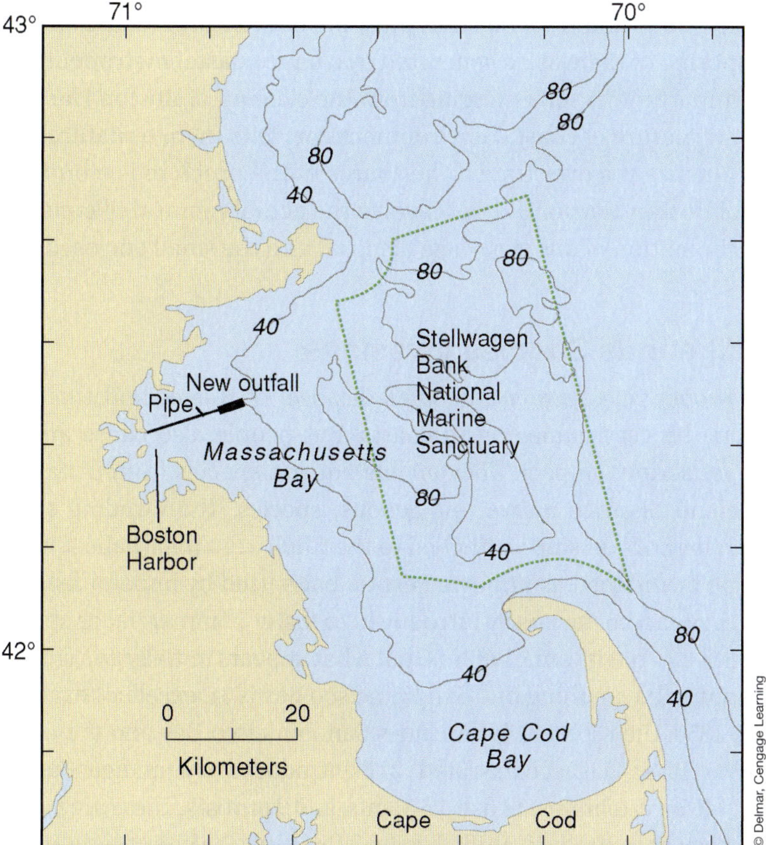

Figure 5-39 Discharge point at the end of sewage tunnel.

and caused much environmental degradation. With the rising tide, the sewage would occasionally wash up on shore. By the early 1980s, Boston Harbor was known as the "filthiest harbor in the nation" with diseased fish, contaminated shellfish beds, periodically unsafe swimming beaches, and degraded habitat.

In 1986, the Massachusetts Water Resources Authority began a court-ordered $4 billion cleanup of the harbor through improved water treatment, enforced reduction of industrial waste, and construction of a huge discharge tunnel. The 14 km (9 miles) long, 8-meter (26 feet) diameter tunnel carries 1.4 billion liters (400 million gallons) of sewage per day out to the more open Massachusetts Bay. In places the tunnel is over 75 meters (250 feet) below the seafloor. It releases sewage on the ocean bottom at 55 points along the last 2 km (1.2 miles) of the tunnel. Release of the sewage at multiple points and into a larger water body dilutes the sewage more quickly. As a result, sewage concentration in Massachusetts Bay has not greatly increased, and the concentration in the harbor has dramatically decreased since use of the tunnel began in September 2000. Sediment at the bottom of Boston Harbor still contains lead and other heavy metals at concentrations above toxicity guidelines for animals living on the ocean bottom, but the current levels are more than 50 percent below levels recorded in the late 1970s. Visible signs of the cleanup's success are evidenced by the return of porpoises, seals, and native fish such as smelt, herring, and striped bass to the harbor's waters. Despite the apparent success of the cleanup, however, the relocation of sewage discharge to Massachusetts Bay through the tunnel was not without controversy. The new discharge points are close to the Stellwagen National Marine Sanctuary (see Figure 5-39). This sanctuary is home to endangered species such as the northern right whale. The controversy surrounding construction and opening of the sewage tunnel highlights the difficult decisions and tradeoffs that must be made to protect the environment while disposing of the wastes generated by a large coastal metropolitan area.

One important lesson emerging from the cleanup of Boston Harbor is that the dynamic nature of coastal environments permits them to naturally cleanse or repair themselves over time if the source of the problem is eliminated. The cleanup of Boston Harbor did not involve the actual removal of contaminants, just the elimination of direct sewage disposal into the small enclosed harbor.

Nonindigenous Species Invasions

Without proper management, people's disposal of wastes and their land use choices can be very damaging to coasts. But people also cause another less obvious, yet serious impact: They enable aquatic species from distant regions to invade and displace native (indigenous) species. The invading species are brought to harbors on ships, attached to their hull or in their ballast water. They can also be introduced as imported exotic baits used by hopeful fisherman. In addition they can be purposely introduced to foster commercial or sport fishing opportunities. What is remarkable is that what appears initially as a small change can turn out to be anything but. San Francisco Bay is an excellent example.

Since 1850, the estuary that includes San Francisco Bay and the delta of the Sacramento River has accumulated 212 introduced (nonindigenous) species, including 146 invertebrates, 32 fish, 25 plants, and 9 protists. These are just the ones that have been specifically identified; many others may be hiding here and there.

The impact of these species on the ecology of the estuary—particularly in the northern embayment of San Francisco Bay—is alarming. For example, invading clams—such as the Asian clam, *Corbula amurensis*—can now filter (to get food) the entire volume of the south and north parts of the estuary in one day! In places, 2,000 nonindigenous clams occupy 1 square meter (10.8 square feet) of the estuary bottom. These filter feeders can be the primary control on the amount of phytoplankton—the important energy producer at the base of the food web—available in the area. As a result, they can control the local energy market. What are the native species to do that need this food resource?

Other changes that have evolved from species invasion include:

Larger and more rigid and rooty Atlantic cordgrass (*Spartina alterniflora*) is displacing the native cordgrass in salt marshes. Expanding areas of Atlantic cordgrass could cause changes in sedimentation patterns, decrease habitat for wetland animals, and decrease the amount of light available to algae.

The Atlantic green crab (*Carcinus maenas*), which invaded the bay about 1990, lives just about anywhere and eats just about anything. It can compete very effectively with other prey species and could change the structure of ecological communities in the estuary.

The 32 or more species of invading fish—all carnivores such as carp, catfish, and bass—are significant predators throughout the brackish and freshwater reaches of the estuary. Some of these fish have contributed to the regional decline or elimination of native species such as the Sacramento perch.

Nonindigenous species continue to invade the San Francisco Bay estuary. Since 1970, about one new species has invaded every 24 weeks. It illustrates how inviting the estuary habitat can be to visiting organisms. Not all visitors are welcome, however—especially those that fundamentally disturb the original ecological balances.

conclusion

Coastal environments are not only attractive places to visit, but they are also important areas of commerce. A variety of distinct coastal ecosystems exist, each the result of physical processes that create unique habitats for plants and animals. Changes are always underway along the coast. These create hazards and land use conflicts that require human ingenuity to resolve. Coastal environments, through the processes that operate within them, can heal themselves if the human impacts that created the problem are managed more effectively. Improved management of coastal environments arises out of a better understanding of coastal processes and the environments they create.

IN THIS CHAPTER YOU HAVE LEARNED:

- The coast is a transition zone between land and ocean where all Earth systems interact.
- The geosphere supplies sediment to coasts and creates shorelines that are variably eroded by waves. Land movements affect shorelines and can be up or down relative to sea level. Some coastal land movements are gradual, but earthquake-related land movements are sudden.

- The gravitational interactions of the Earth, moon, and sun create tides. Tides are regular rises and falls of sea level along coasts. Tides vary from high to low levels about every 6.5 hours. There are times of especially high and low tides each month depending on the relative positions of the Earth, moon, and sun.
- The atmosphere's winds are the main cause of waves. Strong storm winds push large bulges of water, called storm surges, onto land.
- Very gradual sea level changes are caused by changes in the hydrosphere. Increased water storage in polar ice caps during periods of colder global climate decreases ocean volumes and lowers sea level along coasts. Melting of polar ice during periods of warmer global climate increases ocean volumes and causes sea level to rise. Sea level has been gradually rising for the last 21,000 years.
- Several physical processes create beaches and estuaries along coasts. Waves, tides, and longshore currents constantly move sediment on beaches, through inlets, and within estuaries. Sediment erosion and deposition constantly change physical settings along coasts.
- Habitats along coasts are characterized by variable and fluctuating abiotic conditions. Changing temperature, wetness, salinity, and ground stability are abiotic factors that coastal species have adapted to. Special species have adapted to beach, estuary, salt marsh, and mangrove habitats along coasts.
- Many people live along coasts and use them for many purposes. Coasts are enjoyed for recreation and harbors are very important to commerce.
- Coasts affect people by eroding land, depositing sediment in unwanted places, and flooding large areas, especially during storms.
- People affect coasts by displacing or destroying habitat through land and port developments, inappropriate waste disposal, and introduction of nonindigenous species that disrupt ecological balances.
- Coastal studies and monitoring enable people to understand coastal processes and environments. People modify the coast with barriers and other constructed features to protect against coastal hazards. However, sound land use decisions based on understanding of the coast and how it changes are needed to best protect people and ecosystems.

KEY TERMS

Beach
Coast
Dunes
Estuaries
Mangrove swamps
Salt marshes
Storm surges
Tides
Waves

REVIEW QUESTIONS

1. How are ocean waves formed?
2. What are the three factors that control wave size and energy?
3. Describe how beaches are formed.
4. What factors affect the tides? What effects do tidal changes have on coasts?
5. What evidence do scientists have that coastlines have changed over time?
6. Explain how the sand crab is adapted to the beach environment.
7. Describe adaptations that organisms must make to survive in an estuarine habitat.
8. In what ways is cordgrass adapted to the salt marsh environment?
9. Draw and label a diagram illustrating the food web that exists in a mangrove swamp.
10. Why do you think that, with all the hazards that exist from living on the coast, humans still choose to live there?

chapter 6

watersheds and rivers

Students at Blaine High School in Blaine, Washington, have been helping save salmon and their habitat from disappearing in nearby Haynie Creek. The annual migration of Coho Salmon to Haynie Creek was in jeopardy because of the dwindling number of fish returning from the ocean each year to lay their eggs (spawn). To help, the students at Blaine High School incubated Coho Salmon eggs in the classroom, fed the small baby salmon in an aquarium until they were a few months old, and then released them in Haynie Creek. Although this increased the numbers of Coho Salmon in the stream, the students, and the wildlife biologist from the Washington Department of Fish and Wildlife working with them, soon recognized that what was really needed was better habitat in the stream.

Rather than stocking the stream with salmon raised in the classroom, the students are now attempting to save the salmon from extinction by improving their habitat.

What are the habitat needs of salmon? The students at Blaine High School needed to answer this question before they could work to improve habitat on Haynie Creek. The students also learned that natural processes and human land use in the area draining into Haynie Creek—the watershed—can destroy habitat by increasing water temperatures; polluting the water; covering gravel with fine sand, silt, and clay; and reducing the amount of fallen trees in and along the river.

To improve habitat on Haynie Creek, the Blaine High School students planted trees on top of the stream bank to provide shade and keep water temperatures cool, placed old Christmas trees at the base of the stream bank to stop erosion from adding fine sediment to the stream, and placed large logs around the deeper pools to provide more cover for the baby salmon. The tree cover also helps to protect the fish from predators, such as birds, and the roots can filter out pollutants entering the stream via the ground water. The students knew that every bit of improved habitat could help the Coho Salmon population in Haynie Creek survive.

Streams like Haynie Creek are part of river systems that link the land they flow across, the geosphere, to the lakes and oceans of the hydrosphere. Rivers are valuable because:

- *They are ecosystems for wildlife and are locations of great natural beauty and recreational opportunity for people.*
- *People use rivers as a main source of drinking and irrigation water in many areas, as transportation systems to carry materials hundreds of miles inland from coastal areas, and to produce needed electricity at power plants where dams have been built.*
- *River* ***floodplains*** *are excellent agricultural lands but flooding here can also cause extensive damage.*
- *People's activities along rivers can cause them to become polluted, lose habitat, and destroy property.*

Rivers are part of a watershed—the area surrounding the river that drains into the river after a rainstorm. The amount of water and the quality of the water entering the river is controlled by the characteristics of its watershed. The entire watershed must be studied to gain understanding of a river and how it works. An understanding of watersheds and river helps people live more safely and harmoniously with river systems and their ecosystems.

AFTER COMPLETING THIS CHAPTER, YOU WILL UNDERSTAND:
What watersheds are and how they control a river system's characteristics, including erosion, **sedimentation**, and flooding.
The many ways that rivers benefit people.
How people affect rivers.
The challenges of living along rivers, especially recurring flooding.
How people can better protect watersheds and river systems.

 6.1 watersheds and rivers

No matter where you are, you are in a **watershed**—the area (or basin) drained by a stream or river system. You can think of a watershed like a giant funnel collecting all of the water within a drainage area and channeling it into a stream, a river, a lake, or an ocean. Imagine rain falling on the land. Where does this water go? Some soaks into the ground and some flows downhill—or "runs off" —forming a small stream. This stream may join with others to form larger and larger streams and rivers. The water in a watershed eventually merges and flows to a lake or ocean.

River systems developed by the water draining a watershed appear much like a branching tree. The largest river flowing in the watershed represents the trunk and is called a **trunk stream**. Smaller streams flowing into the trunk stream are referred to as **tributaries** and look like branches on a tree.

Each watershed is separated from adjacent watersheds by its divide. Watershed divides typically follow ridges and mountain tops (topographic highs) where the rain falling on one side of the divide flows into one watershed, and rain falling on the other side of the watershed flows into the adjacent watershed (Figure 6-1).

The Mississippi River has the largest watershed in the United States with 41 percent of the total area of the United States draining to the mouth of the Mississippi River, where it enters the Gulf of Mexico (Figure 6-2). The highest

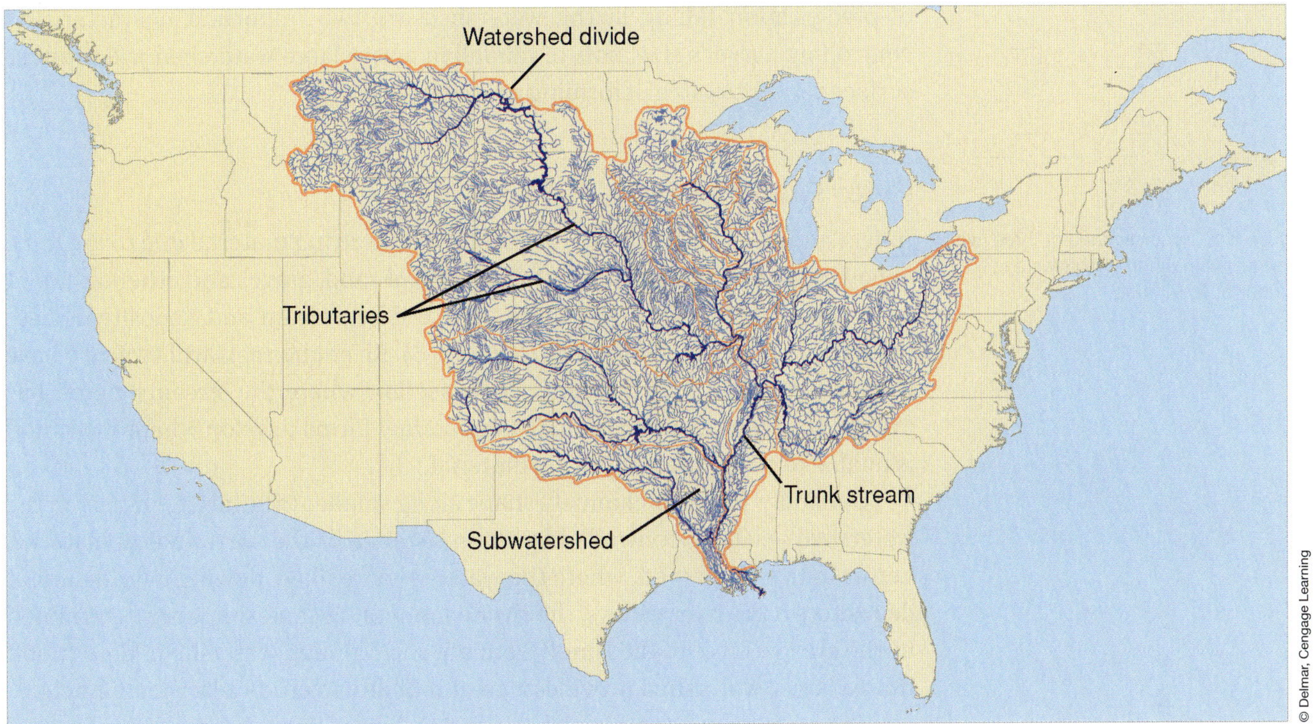

Figure 6-1 The Mississippi River watershed and drainage network.

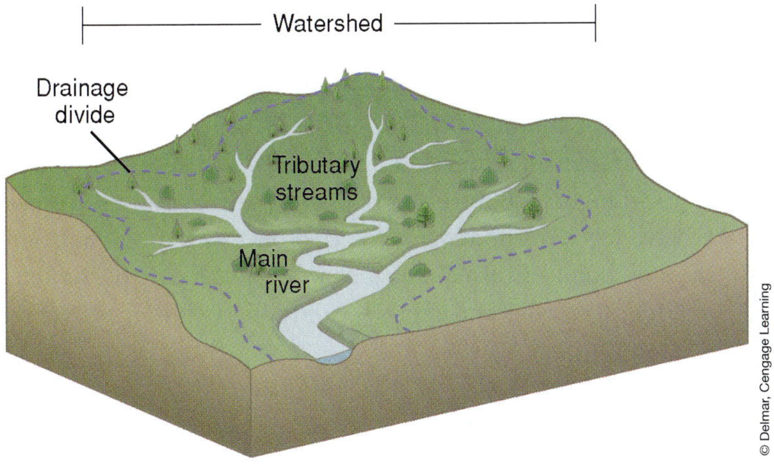

Figure 6-2 Watershed and its components.

peaks in the Rocky Mountains to the west and the Appalachian Mountains to the east are along the watershed divide of the Mississippi River. The Mississippi River has several large tributaries flowing directly into it, such as the Missouri River, Ohio River, and Illinois River, each with its own subwatersheds. These large tributaries serve as trunk streams to a network of other tributaries that become smaller and smaller as the contributing watershed area decreases.

Not only does the amount of water draining from the watershed determine the river morphology—the size, shape, and drainage pattern of the river—but many other characteristics of the watershed are important controls on river morphology. Watershed characteristics ultimately control the amount of runoff

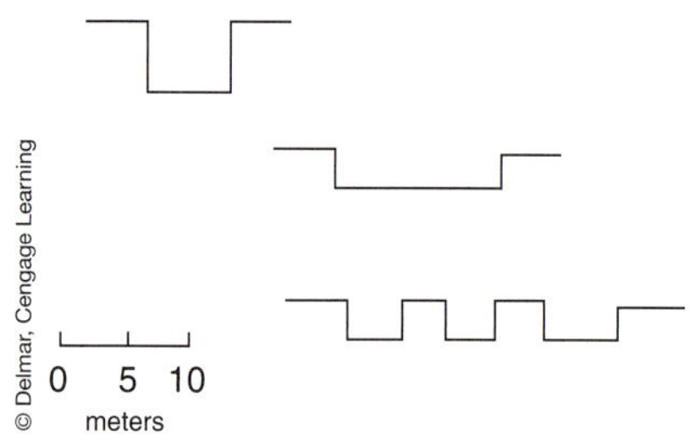

Figure 6-3 Cross sections of three river channels with differing shapes and number of flow paths.

and sediment that enters the river system. Watershed area and the amount of precipitation falling in the watershed are two important characteristics controlling a river's size. Lots of rainfall in a very large watershed will produce surface runoff capable of forming a large river.

River Channels

A river **channel** can form by erosion of the underlying material due to the force of flowing water. This results in the removal of sand, gravel, and other sediment in the channel that in turn can be carried downstream and deposited where the force of the river diminishes (Figure 6-3). River channels will typically have rounded banks with a U-shaped cross section where the stream's runoff has formed the channel by erosion. Other channel forms develop where the runoff is unable to erode the underlying material.

Generally, the river channel's shape along smaller tributaries—higher in the watershed—will be controlled by the resistance of the surrounding material. Lower in the watershed, where the water is more likely flowing over sediment deposited in the river valleys, the river channel's size and shape are determined by the erosive force of the runoff from the surrounding watershed. More runoff from a larger watershed provides a greater ability to erode a larger channel.

The amount of runoff is just one factor controlling the character of a river or stream. Streams with the same cross-sectional area can look very different. Similarly, a stream with multiple interconnected channels (a **braided stream**) can have the same overall cross-sectional area as a single **meandering** channel.

The width of the river will increase when its banks are unstable, such as when vegetation is removed. When they are unstable, they can be easily eroded by the force of the river's flow. Sandy soils erode easily and, because of this, river channels whose banks are composed of sand tend to be braided and wider than channels composed of less easily eroded clay-rich sediments. Because clay banks are fairly resistant to erosion, the erosive force of the river is focused on the channel bottom rather than its banks. Consequently, channels with clay banks tend to be deeper, narrower, and more meandering (winding) than channels with sandy banks (Figure 6-4).

Figure 6-4 Stream velocity is increased (a) when the flow is constricted into a smaller area and (b) when channel roughness is decreased on the channel bottom and banks as accompanies the removal of vegetation. Removing vegetation from the banks increases the erosive force of the stream while simultaneously reducing the resistance of the banks to erosion.

Vegetation is also a very important factor in bank stability. Vegetation that grows on riverbanks helps them resist erosion. River channels with lots of vegetation on their banks, just like rivers with banks composed of clay, tend to be narrower and deeper than river channels without vegetation on their banks. Vegetation growing on sandy riverbanks is critical for maintaining channel stability. If the vegetation is removed, the channel would be very susceptible to widening because the sandy banks would be exposed to the full erosive force of the river's flow.

Erosion and Deposition of Sediment

Rivers do not only have water moving downstream in their channels, but sediment is in motion as well. The amount of sediment that can be moved by the river is closely related to the force of the flowing water in the channel.

Understanding the factors that control the amount of sediment that can be moved by the river is essential for predicting where erosion and deposition will occur. If damages due to flooding are to be avoided, and river habitat protected, then it is important to know where erosion and deposition are most likely to occur.

The amount of sediment a river can move—the river's **sediment carrying capacity**—is related to the velocity of the flow (how fast the water is moving). Factors that increase the flow's velocity also increase the river's sediment carrying capacity. Four factors control a stream's velocity and, therefore, its sediment carrying capacity: gradient, discharge, channel area, and channel roughness. A steeper **gradient** (slope) and greater **discharge**—the total volume of water flowing past a point on the river in a given amount of time—will cause an increase in a stream's velocity. On the other hand, increases in channel area and channel **roughness**—the friction between the water and the bed, banks, wood and large rocks in the channel—will decrease the stream's velocity. During high flows, a river's velocity increases slightly in a downstream direction despite a loss in slope. The increase in velocity downstream results from increases in discharge and decreases in channel roughness that occur as the watershed becomes larger.

Rivers will erode only if the force of the flow is greater than the resisting force of the channel banks and bottom. A certain *incipient motion force that corresponds with threshold velocity* must be reached that exceeds the river channel's resistance before erosion and sediment movement begins. If you throw sand in a stream, the small particles of sand are likely to be washed downstream. If you tossed some gravel into the stream at the same spot, the gravel would probably fall to the bottom of the stream and not move. This is because the threshold velocity needed to transport the gravel is exceeded only during larger flows when the channel is filled with more water. Although the stream is flowing and, therefore, has the capacity to transport sediment, only particles with a low threshold velocity can be moved.

If the threshold velocity necessary to move the sediment is exceeded, the river will erode sediment from the banks and river bottom until the sediment load—the amount of sediment the river is moving—equals the river's sediment carrying capacity. Once the carrying capacity is reached, the stream will no longer erode because it is unable to carry more sediment. To keep sediment moving in a river system from the upper watershed to the mouth of the river, the flowing water must maintain its velocity. When the velocity and, therefore, the sediment carrying capacity decreases, the river will deposit sediment. This commonly occurs when a river's discharge decreases, for example, after a flood. Much flood damage to agricultural crops, homes and other buildings, and aquatic habitat comes from complete or partial burial by the deposited sediment (Figure 6-5).

Photo by Lyn Topka Courtesy USGS

Figure 6-5 Woody debris around a house after a flood.

A River's Equilibrium State

Flowing water generated by runoff in the watershed will carve its own channel as it passes over an erodible material. As the channel size increases with continued erosion, the increasing channel area will cause a decrease in stream velocity (Figure 6-6). At some point the sediment carrying capacity of the slower

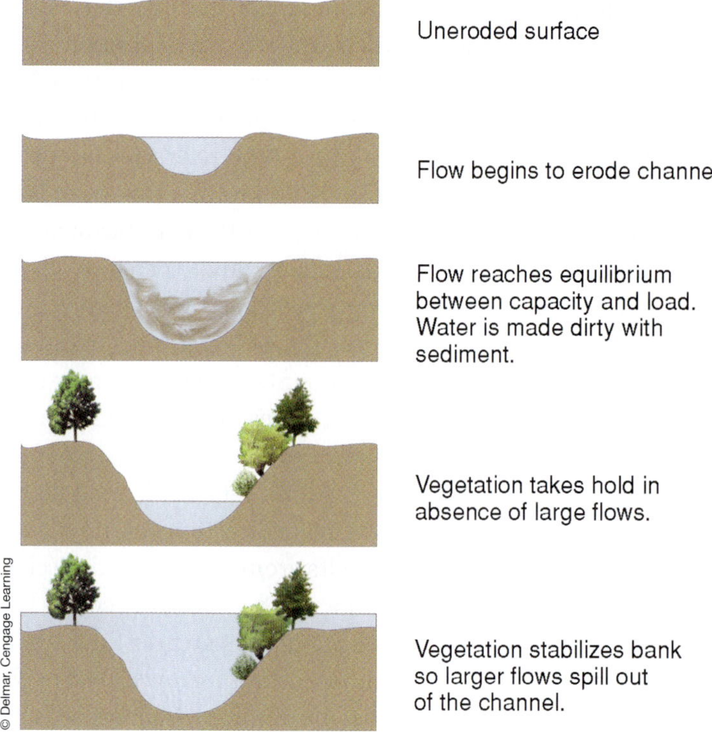
Uneroded surface
Flow begins to erode channel.
Flow reaches equilibrium
between capacity and load.
Water is made dirty with
sediment.
Vegetation takes hold in
absence of large flows.
Vegetation stabilizes bank
so larger flows spill out
of the channel.
© Delmar, Cengage Learning

Figure 6-6 The creation of a channel in equilibrium and the development of overbank flooding.

moving water will match the sediment load already moving in the newly eroded channel. The channel will ultimately reach an equilibrium state where the water flowing through the channel is just able to move the sediment already in transport. The flow passing through this equilibrium channel is unable to increase the channel's size through continued erosion because it does not have the capacity to carry more sediment.

If a much larger flow enters the newly formed channel, the channel will tend to grow in size to reach a new equilibrium with the greater erosive force of the higher discharge see. Several years may pass before a flow is big enough to increase the channel's size. In the interim, vegetation may begin to grow that increases the resistance of the banks to erosion. Consequently, even the largest flows, those that occur once every 100 years or more may not reach the threshold velocity necessary to uproot the vegetation and expose the more erodible soil below. Therefore, channels tend to reach and maintain a certain size without much change as long as the banks remain stabilized by vegetation.

The concept of equilibrium helps us to understand how a river might respond to human activities along the river and in the watershed. Human activity in the watershed that changes the amount of water or sediment coming into the river will cause the river's morphology to change, including changes in channel width, depth, and pattern. Also, rivers will respond to human activities that alter the force of the flow. For example, when a river is filled in and made smaller in certain areas, to build a bridge, for example, the river's velocity will increase as it is constricted into the smaller area. This will increase the force of the flow and possibly result in erosion if the banks and bottom are not protected.

Another example of river response to human activities occurs frequently in urban areas. Greater amounts of runoff are generated in heavily developed watersheds because less water can soak into the ground when the rain falls on roads, parking lots, and rooftops. Consequently, unlined channels in urban areas frequently show signs of erosion as they become larger due to the erosive force of the greater runoff.

River Flooding

River channels in the United States that are formed in erodible sediments are generally in equilibrium with the largest flows that occur every year or two perhaps because of a spring rainstorm with snow melt or a heavy summer thunderstorm. These flows will completely fill the channel with water. Flows larger than this, if they are unable to further erode the channel because of vegetation growth, will overtop the channel banks and spread out over the adjacent surface, causing a flood.

This **floodwater**—the water flowing outside the channel—will decrease in velocity as it spreads out beyond the confines of the channel banks. The decrease in flow velocity causes the sediment in the floodwater to be deposited. This flood sediment generally consists of fine sand, silt, and clay and is deposited when the water overtops the river channel's banks. Repeated floods over hundreds and thousands of years build up layers of fine-grained flood sediment and create a floodplain adjacent to the river. Because floodplain sediments are typically very fertile, floodplains have historically been used for farming and were the center of many ancient civilizations. Natural floodplains are typically inundated with floodwaters every year or two, the frequency of flows large enough to completely fill and overtop the channel. Floodplains are more typical along rivers lower in the watershed where channels are less restraining. Higher in the watershed, the smaller tributaries are more confined between high banks with little opportunity to spread out and deposit the sediment necessary to create a floodplain.

The Nile River of Africa is the longest river in the world (6650 km or 4132 miles). The watershed of the Nile is huge and drains about one-tenth the area of Africa (2,850,000 km^2 or 1,100,000 mi^2). For thousands of years, the Nile River floodplain in Egypt was a fertile and successful farming area. The Nile River would flood every year, bringing with it rich sediment from throughout the watershed. Today, flooding is controlled by dams on the Nile River.

Two dams straddle the Nile River in southern Egypt: the newer Aswan High Dam (completed in 1970), and the older Aswan Dam or Aswan Low Dam (finished in 1902). Although the Aswan High Dam (Figure 6-7) has provided several benefits—it regulates river levels, controls water releases for irrigation, and generates hydroelectric power (an inexpensive source of electricity)—it has also created many problems.

Because rich sediments no longer overflow onto the floodplain, artificial fertilizers now have to be used for farming. The water level downstream of the dam is lower now too, so the banks of the river are exposed and experience steady erosion due to the death of vegetation and the loss of roots that retained the soil. What is happening to all of the sediment that used to be deposited

Figure 6-7 Aswan High Dam on the Nile River. (*Photo by Benjamin Franck, Nov 2005*)

on the floodplain? It is settling out and slowly filling in the reservoirs behind the dams. Consequently, there is less sediment in the river water downstream of the dam. Less sediment means clearer water and more photosynthetic phytoplankton. The increase in phytoplankton has thrown off the balance of the food web and had a negative impact on fishing in the river, which is an important economic resource.

 watershed and river ecology

The water and sediment moving down a river system from throughout the watershed create a range of different habitats from the headwaters to the mouth of the river. Where certain plants and animals are found in or along a river are very much dependent on the force of the flow and the amount and size of sediment in the water. What types of plants and animals do you see living on riverbanks near your home? How do the types of plants and animals vary from headwaters to the mouth of the river? These are the types of questions an ecologist asks to determine the range of habitats that exist throughout a watershed. This information can be used to locate habitat or identify areas where habitat is absent and may need to be restored.

The primary abiotic factors controlling the types and distribution of species in river systems are: discharge, flow velocity, channel shape, substrate particle size, and water quality (including water temperature). Each species has a given tolerance level for each of these factors so species distributions are dictated by how the abiotic conditions vary throughout a watershed and between different watersheds across the country. Within a single watershed, the range of habitats from headwater stream to river mouth may be used by a single animal species at different times of the year or at different times of the animal's life. In addition to habitat changes from the headwaters to the mouth of a river, habitat also varies across a river from the middle of the channel to the top of the adjacent banks. The health of a river ecosystem depends on the existence of varied habitat throughout the watershed and across the channel.

Three distinct zones of habitat typically develop from the center of a river channel to the banks adjacent the floodplain: midchannel, bankside, and banktop (Figure 6-8). The bankside and banktop environments, taken together, are referred to as the **riparian zone**—a transition zone at the interface of land and water.

Midchannel Environment

Within the channel zone, animals live both in the water and in the channel bottom sediment. The force of the flow in the midchannel area is too great for plants to take hold permanently, although some may grow temporarily on exposed sand and gravel bars during low flow periods in the summer. Aquatic plants, mosses, and algae can grow on rocks on the channel bottom if currents are not too strong. These primary producers serve as a source of food for many macroinvertebrates living within the gravel.

The term *macroinvertebrate* refers to any animal without a backbone that can be seen without a magnifying glass or microscope. Macroinvertebrates include crayfish, mussels, snails, spiders, leeches, and the larval stages of many insects. The presence and abundance of particular species vary from region

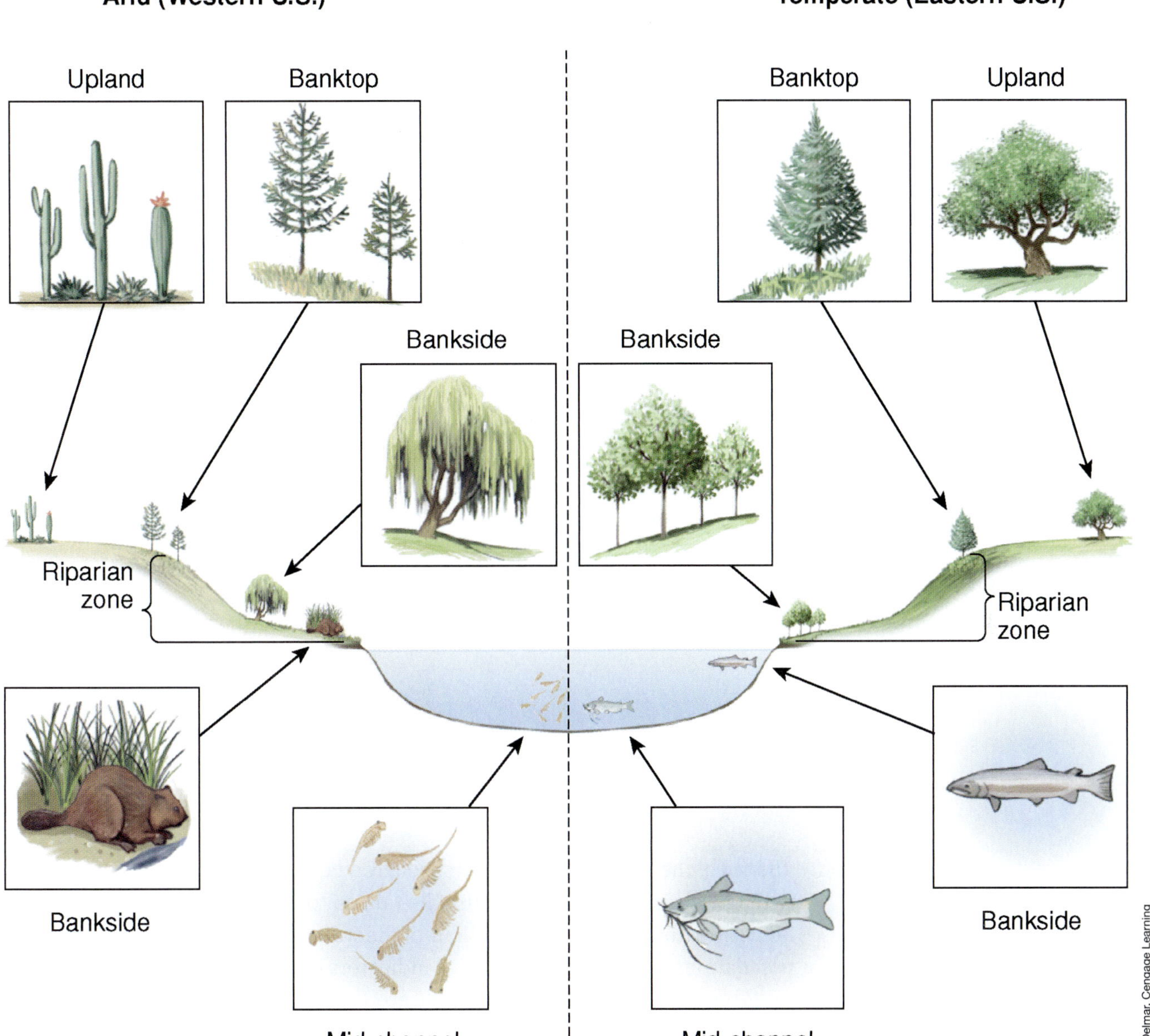

Figure 6-8 Ecological zonation in a river.

to region and within the same watershed due to differences in water quality. Stream macroinvertebrates living in the midchannel environment commonly display adaptations for withstanding strong water currents. Stream dwelling mollusks typically have thicker shells than their still water counterparts to protect them from gravel moving along the stream bottom during high flows. Black fly larvae, common in very fast moving waters, attach to rocks on the channel bottom with hooks. If they are swept away by the currents, the larvae can return to their rock by eating a silk lifeline that it released when initially detached.

Many macroinvertebrates, such as stoneflies, caddisflies, and mayflies, are sensitive to water **pollution** (Figure 6-9). Their absence in a stream where they would typically be found indicates poor water quality. Other macroinvertebrates, such as midge larvae, mosquito larvae, and rat-tailed maggots, are more resistant to water pollution and are sometimes found in very polluted waters.

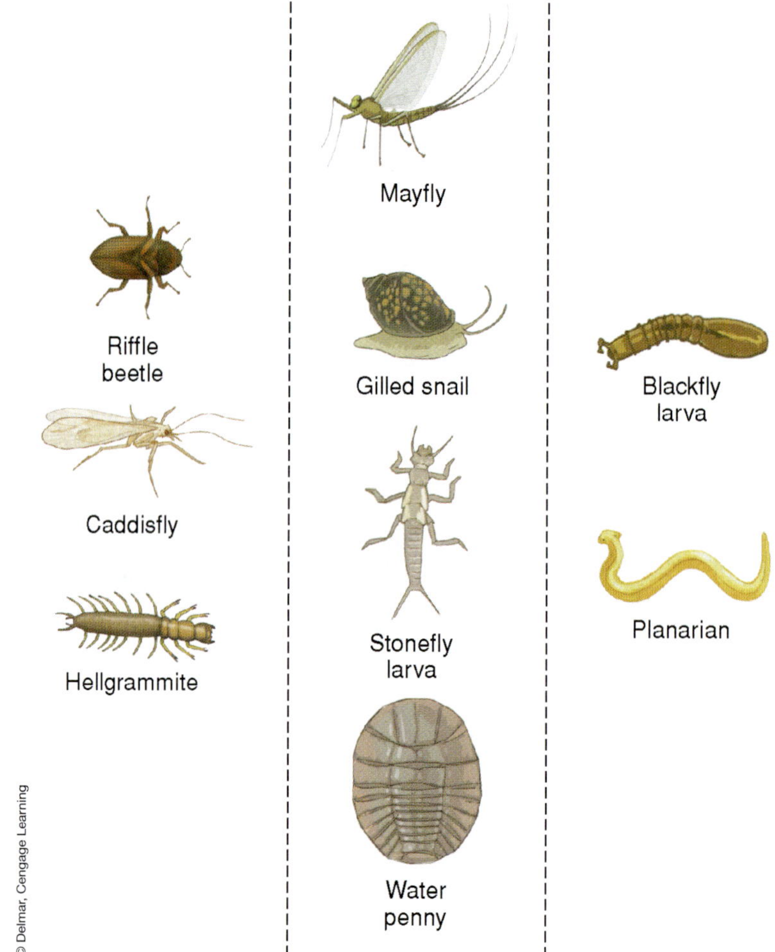

Figure 6-9 Macroinvertebrates as indicators of varying water quality.

The tail of the rat-tailed maggot is specially adapted for polluted stagnant water with low dissolved oxygen because the tail, up to 15 cm (5.9 inches) in length, is an air-breathing tube that pokes out of the water. Studies of macroinvertebrate populations in streams can serve as indicators of water quality.

Above the channel bottom, fish are the most noticeable inhabitants in the water itself. Many fish species feed on the macroinvertebrates that are found on the channel bottom while other fish live more in the water itself. The shape of the channel is a very important control on the types, abundance, and size of fish seen in a particular locality. Narrow channels with well-vegetated silt and clay banks will tend to develop shallow riffles alternating down the stream with deep pools. This provides excellent habitat for trout that rest in the slower moving water of the pools but feed on the adjacent faster moving riffles where more food, generally insects, is likely to float downstream. Trout are cold water fish and are commonly found in streams throughout the northern United States and in higher elevation mountain streams where water temperatures stay below 75°F (23.9°C).

Narrow channels with well-vegetated silt and clay banks also tend to be deeper than wide channels with unvegetated sandy banks. Consequently, water temperatures remain lower during the hot summer months and warmer in the winter. Water in a wide channel spreads out over the channel bottom and warms up more easily. Trout and other cold water fish will be searching for deep pools to stay cool during the warmest time of the year. Pool habitat is also important in winter in the northern United States because fish will tend to retreat to the deeper parts of pools where ice is less likely to form. Fish can survive the winter underneath the ice but not where the water is too shallow and the stream completely freezes.

Several mammals and birds, like bears and eagles, respectively, do not live in the midchannel environment but are an important component of the food web as they prey upon fish and other organisms that do live in the midchannel. These mammals are secondary consumers. The fish they eat are also secondary consumers because they feed on insects and macroinvertebrates. Some macroinvertebrates feed on aquatic plants and are primary consumers, whereas others are decomposers that feed on the remains of dead organisms and their waste products.

Fish seek cover to hide from their predators such as mammals and even other fish. Different species and ages of fish prefer different types of cover; therefore, the types and abundance of fish species depends on the amounts and variety of cover present. Deep pools, overhanging banks, and logs are used by the adults of larger fish species, whereas individual rocks can serve as cover for juveniles or adults of smaller species. The health of the riparian corridor—the bankside and banktop environments—is directly related to the amount of cover it provides for species from the midchannel environment.

Bankside Environment

Because most of the riverbank is above water except during short periods of high flow, terrestrial plants can grow here. Plants growing on the bank display several adaptations to withstand the force and scouring of water, ice, and debris that pass by. Bankside plants such as willows and cottonwoods are fast rooting and able to easily colonize bare sites with abundant sunlight. In just a year or two, new root systems are able to withstand the force of high flow and they begin to capture sediment that promotes further plant growth on the bank. Another adaptation that is particularly important in deserts is that bankside trees are able to grow vertical roots that can extend down to the water table. Once the water table is reached, the roots expand laterally to capture moisture.

Lower on the bank, flow velocities are higher and the length of time that the bank is under water increases. This creates zones with different plant species from the bottom to the top of the bank (see Figure 6-10). Along the lower part of the bank, a number of seasonal herbaceous plants, such as grasses, grow. Slightly higher on the bank, shrubs such as dogwoods and willows take hold. The seedlings are not washed away and the shrub's root systems grow fast enough to help them withstand the high flows reaching this level. At the highest levels of the bank, large trees are able to grow as their roots remain well-aerated above the water table. The types of trees that grow on the upper banks vary with climate and soil type. Red maple and sycamore typify the moisture-rich eastern United States, whereas cottonwoods are more characteristic of the arid western states.

Beavers are common residents of the bankside environment along many rivers, feeding on willow, alders, and cottonwoods. They also use the bankside trees to build dams and dens on rivers and streams. Beaver dens are also built by tunneling directly into the stream bank. Beavers will live in a particular area until their preferred vegetation is depleted. They then move on to another part of the stream, allowing a succession of plants to reemerge after the biological

Figure 6-10 Riverine system with natural environments and varying land uses.

pressures of the beaver have been removed. Herbaceous species will give way to shrubs and trees if the banks are high enough. The total succession though may never become fully completed before the site is disturbed again by returning beaver or scoured by floods. As a result, bankside environments typically have a patchwork of habitats at various stages of succession found in close proximity to each other. This helps maintain ecological diversity and provides the complexity of habitat required for many species.

Banktop Environment

Banktop environments are where trees and other plants grow. A more diverse understory of herbaceous plants and shrubs may be present here because of the greater stability and, therefore, age of the banktop forests compared to the banksides. Further from the bank edge, a transition to more upland vegetation occurs. This transition is most pronounced in arid regions where the tall trees supported along the river give way to shorter drought-resistant desert shrubs and cacti further away.

Riparian environments, the banktop and bankside zones together, provide critical habitat for many animals. This is especially true in desert regions where the contrast between the riparian zone and dry upland environments is striking. Over 80 percent of all amphibian species and 40 percent of all mammal species in California reside in riparian forests. But less than 10 percent of the original riparian forests in California remain after decades of farming, mining, dam construction, and other human activities. Preservation of the remaining riparian forests in California and elsewhere around the United States is an essential first step in sustaining the ecological diversity supported by these riverside environments.

A healthy banktop environment is critical for maintaining the ecological health of other riverine environments as well. The vegetation on the banktop serves as a buffer that can trap and absorb contaminants before they can enter a river. Concentrations of nitrates, a pollutant when found at high levels, are up to 90 percent lower in forested riparian soils compared to adjacent cropland soils. The roots of riparian trees also help to anchor the soil and prevent erosion. Undercutting of the banks, which results when scour removes vegetation on the bankside but not the banktop, creates overhanging banks that provide cover for many fish species in the midchannel. Tall trees on the banktop shade the stream, keeping water temperatures lower in the hot summer—a major benefit for cold water fish like trout. Finally, as the banktop trees mature, die, or are completely undercut by bank erosion, they fall into the stream. This provides cover for fish and diverts flow into secondary channels that serve as habitat for juvenile fish. Even adult fish will use the side channels during high flows to seek refuge from the fast moving floodwaters.

Although the three environmental zones across a river or stream contain distinct species assemblages, the health of all three environments depends on the presence of the others. Beavers build dens in the midchannel area out of trees harvested from the banksides and banktop. Eagles nest in tall trees found at the banktop and feed on fish in the midchannel area. Fish thrive in the cool, clear waters of the midchannel area, shaded by trees of the bankside and banktop environments, and hiding under fallen trees from these areas.

Habitat Variations along Rivers

Many larger animals in a river, such as salmon and trout where they are present, not only depend on the habitats across a river or stream channel but also on the existence of varied habitats developed throughout a watershed. Different habitats may be used by a species during different stages of their lives. Salmon and trout lay their eggs in clean gravel at the bottom of the stream. Too much sand and other fine sediment in the gravel can suffocate the incubating eggs. Consequently, salmon and trout do not usually lay eggs near the mouth of rivers where fine sediment may be deposited as the river slows before entering a lake or ocean. These fish prefer to lay eggs in faster moving tributary streams in the upper watershed where the flow is just fast enough to wash away fine sediment but leave clean gravel on the channel bottom.

After hatching, the juvenile fish tend to remain in smaller streams or side channels on bigger rivers because the force of the flow is generally less and there is more protection. As the fish grow to adulthood they will move down the watershed to occupy deep pools. When multiple trout species are present in a stream, such as Brown trout and Brook trout, the larger species (Brown trout in this case) will typically be found lower in the watershed. The smaller Brook trout are forced into less ideal habitat higher in the river system due to the competition from Brown trout.

River systems and the adjacent riparian vegetation provide corridors of migration for animals that move between the upper watershed to lowlands in search of food and habitat during different seasons of the year. It is along rivers where continuous strips of habitat are less likely to have been disturbed by people. Bears typically are found in mountainous areas in much of the United States but during times of drought they may migrate long distances in search of food. This migration will often occur along undeveloped rivers because of the unbroken riparian forest growing along the riverbanks and the proximity of water and food within the river itself.

6.3 people and rivers

Just as many plants and animals find river environments an excellent place to reside, people also like to live near rivers. Most large cities in the interior of the country are able to thrive because they are located on large rivers—St. Louis and Minneapolis on the Mississippi River, Pittsburgh and Cincinnati on the Ohio River, and Kansas City and Omaha on the Missouri River are examples. Even Chicago is connected to the St. Lawrence River through the Great Lakes and by a canal to the Illinois River. Although there are many advantages to living near rivers, rivers can also be dangerous and pose many hazards for those who choose to live near them (Figure 6-10).

Benefits of Rivers

Historically, many towns and cities in the United States were built where rivers leave the confines of bedrock canyons and flow onto much broader floodplains. These are places where stream gradients change dramatically—waterfalls are commonly found at these locations. Waterfalls were used to power water mills for grinding flour, corn, and other grains or to operate large saws for cutting lumber. The confluences of steep tributaries where they enter the flatter valley bottoms of larger rivers were favored locations for settlement throughout New England in the 18th and 19th centuries (Figure 6-11).

Figure 6-11 Old mill at the fall line in the Eastern United States.

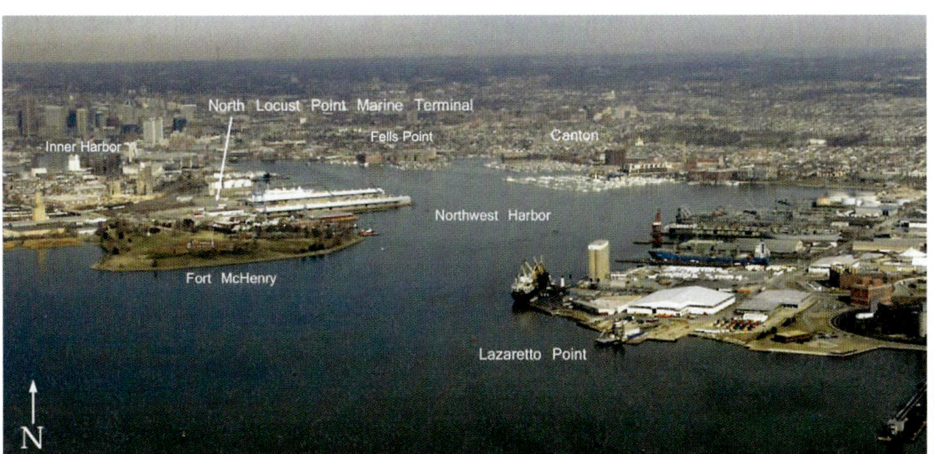

Figure 6-12 A bustling port at a tributary river mouth.

Mills were built on the steep tributaries and crops grown on the fertile soils of the nearby flat valley bottoms. Although the economies of many cities are no longer intimately connected to a river, the historical connection to rivers explains their current location.

People have also tended to settle at river mouths. These are transportation centers where ships from around the world deliver or receive goods. Moving upstream, barges carry products from foreign countries that are unloaded off large ships at river mouths (Figure 6-12). Barges moving downstream carry products to the river mouth for export to other parts of the world. The Mississippi River is the largest portion of a 40,233.6 km (25,000-mile) network of inland rivers on which 1.2 billion tons of cargo valued at $150 billion is transported annually in the United States. Large items such as cars, refrigerators, crates of wheat, and lumber are constantly in transit on the Mississippi River between New Orleans, Louisiana, and Minneapolis, Minnesota for a fraction of the cost it would take to transport the same items by air or land.

River mouths are also important agriculturally. The nutrient-rich sediments deposited on deltas where the river enters the sea create extremely fertile agricultural land. Over 55 percent of the Mississippi River delta (Figure 6-13) is used to grow cotton, soybeans, corn, rice, and many other crops. The delta has been able to sustain agricultural communities for more than 200 years, because the periodic flooding by the river replenishes soils by depositing new sediments on the farm fields. This sediment helps maintain soil fertility by returning nutrients to the soil with each flood.

Floodplains are also productive agricultural areas. In mountainous regions, floodplains along the river are sometimes the only good place to farm. A floodplain's proximity to the river also allows crops to be irrigated. Where are farms or ranches located in your area? How many are located on floodplains next to the river compared to other areas?

Rivers are used by people in many other ways as well. Where does your water that you drink or use for cleaning and cooking come from? Most of the water that people use comes from rivers and lakes (see Chapter 14). In some cases, drinking water travels several hundred miles via canals to reach cities in need of a reliable source of water. The Central Arizona Project carries water 336 miles from the Colorado

Figure 6-13 Farmland in the Mississippi River delta. *(Courtesy Earth Science World Image Bank © Bruce Molnia, Terra Photographics)*

River to Phoenix and Tucson, Arizona. Although these are two cities that are not located on large rivers, they are still dependent on a large river over 200 hundred miles away for their water supply.

Misusing Rivers

According to recent estimates, the average American family uses over 100 liters (105.7 quarts) of water each day for drinking, bathing, washing clothes and dishes, watering lawns, and carrying away wastes. Even more water is required by industry for uses such as cooling equipment, manufacturing products, and creating steam to generate electrical power. In the United States, the greatest amount of water is used to irrigate crops where irrigated crop acreage is now more than 20 million hectares (50 million acres). As populations increase, so, too, does the demand for water.

After this enormous amount of water is used, where does it go? On average, 90 percent of the water used by cities and industry is returned to river systems or to the oceans as wastewater. Some of this wastewater contains pollutants—harmful substances such as disease-causing organisms, toxic chemicals, and metals that have a negative effect on water or the living things that depend on the water. Because pollutants can be hard to trace and remove, water pollution has become one of the most critical environmental problems today.

There are two different ways in which water pollution can occur. If pollution comes from a well-defined location, such as a pipe through which a factory discharges waste into a stream, it is known as **point source pollution**. Although such pollution may be very serious, it usually can be reduced with current technology. A source of pollution that cannot be tied to a specific point of origin is called **nonpoint source pollution**. This type of pollution typically occurs as pollutants are carried along in runoff and flow into a river or seep into groundwater and are carried far away. Nonpoint source pollution is much more difficult to control because it is hard to trace its exact source. Examples of nonpoint source pollution include: chemicals, animal wastes, and sediment from farmland and grazing lands; acid drainage from abandoned coal mines; sediment from construction sites; and chemicals and sediments from roads and highways.

There are two main ways of detecting water pollution in a water body such as a stream or river. The first is to take a sample of the water and measure the concentrations of different chemicals that it contains. If dangerous levels of pollutants are detected, the water is considered polluted. The second way is to examine the fish, insects, and other invertebrates that the water will support. If many different types of living things are found, there is likely no water pollution. If there is no life at all, the water most likely contains pollutants.

One of the most important water pollution considerations is sewage waste. With over 8 billion people on the planet, a lot of human waste is produced every day. Microorganisms live in human waste and water polluted with this waste can lead to some very serious water-related illnesses such as diarrhea, hepatitis, typhoid, and cholera. Most cities and towns in the United States have wastewater treatment systems that effectively kill microorganism in the water before it is returned to the environment. Some viruses and parasites, however, are able to resist water treatment processes.

Animal and livestock wastes are also a concern and are much more difficult to control than human waste. There are more than 100 million cattle in the United States. Each produces 10 times as much waste as a human being. This means that, all together, these cattle produce the equivalent waste of 1 billion people! On top of this, waste is produced by other types of livestock, including horses, sheep, pigs, goats, chickens, ducks, and turkeys. This waste contributes to disease-causing parasites, such as *Giardia* and *Cryptosporidium,* when it is carried by runoff from pastures and barnyards and into river systems. People who drink water polluted with these microorganisms may experience chills, fever, stomach cramps abdominal pains, nausea, vomiting, and diarrhea.

Of greater concern than sewage is water pollution by factories and mines. Many factory and mining processes produce toxic chemicals; heavy metals such as lead, cadmium, and mercury; and strong acids. These substances need to be disposed of and although laws now limit and control the disposal of toxic substances into the environment, some factories still release these pollutants directly into nearby rivers. In the United States alone, around 400,000 factories take clean water from rivers. Many pump polluted water back into these same rivers.

The danger of factory pollutants is that they can build up in the bodies of living things. For example, a pollutant that dissolves in water is released into a river. The pollutant is absorbed by microscopic algae living in the river. The algae, which contain only low levels of the pollutant, are consumed by small water animals. When frogs or fish eat these smaller animals, they also consume the pollutant from the algae that these animals had eaten. The frogs and fish are in turn eaten by birds or other animals. In this way, each larger organism consumes a greater number of the smaller organisms and, therefore, more of the pollutant. When humans eat fish from the river, the pollutant builds up in their bodies in the same way. Over a long time, even tiny amounts of certain pollutants can build up to levels that can cause birth defects or illnesses such as cancer.

Another source of water pollution is runoff from sidewalks, roads, and highways. Roads and highways are typically covered with a mix of toxic chemicals. These include everything from gasoline, motor oil, and brake fluids leaking from cars to bits of worn tires and exhaust emissions. In winter, roads and sidewalks are sprinkled with salts as well to melt ice. When it rains, these substances are washed off along with runoff and into rivers that the runoff enters. There have been cases where heavy rainstorms have washed toxic chemicals into rivers in such concentrations that they killed large numbers of fish overnight.

Toxic chemicals can also enter river systems through fertilizers and pesticides used by farmers. Fertilizers are applied to fields to produce better crops, and pesticides are applied to kill insects that damage crops. When rain falls on the fields, some of these chemicals are washed away with runoff and into nearby streams and rivers. Fertilizers are a problem because they can cause a massive increase in the growth of algae in water. As this algae dies and decays, it removes oxygen from the water, changing the living conditions for other organisms, such as fish, often killing them.

Severe sediment pollution of rivers and streams is another water pollution problem. Much extra sediment is washed into lakes and streams by runoff that flows over fields on which farmers have failed to protect the soil. On construction sites, where the soil is exposed, up to 10 times more sediment can wash away as from an equal area of cropland. Increased sediment also comes from logging areas where large blocks of trees have been removed from steeply sloping land. Another source is strip-mining operations in which soil is removed by giant machines in order to expose the rock below.

The extra sediment that enters river and stream systems results in a number of harmful effects. Sediment damages hydroelectric plants as well as clogging irrigation canals, river channels, and harbors. Suspended sediment also blocks sunlight needed by algae for photosynthesis. Algae are an important food source for many aquatic organisms. Extra sediment can also block the gills of fish, effectively suffocating them.

Cleaning Up Polluted Water

There is no easy way to clean polluted water; most pollutants are very difficult to remove. Point source pollution can be treated in a wastewater treatment plant. A typical plant handles the wastewater in several steps. First, the water is passed through filters that catch solid objects. The water is then placed in a large tank, where heavy particles sink and are filtered out. At this stage, any floating oils are skimmed off the surface. The water is then sent to an aeration tank where it is mixed with oxygen and bacteria. The bacteria feed on wastes in the water and use the oxygen. The water is then sent to another tank where chlorine is added to kill microorganisms.

Nonpoint source pollutants can sometimes be removed through natural cleaning processes. Wetlands are especially effective at removing pollutants. Wetlands, such as those found on floodplains, deltas, and stream banks act as a natural sponge, soaking up runoff and then slowly releasing it. The roots of wetland plants trap sediment, helping to keep rivers and streams clear. These plants also use nutrients, such as fertilizers, in the runoff for growth. Some wetland plants can absorb metals and chemicals as well. Unfortunately, wetlands have historically been seen as areas that were better drained or filled and alarming amounts of wetland habitats have been lost in the United States. Now that the benefits of wetlands are being realized, especially in removing pollutants, more efforts are being made at preserving and restoring these important habitats.

Legislation and Control

With so many people in the United States choosing to live near rivers, a large amount of municipal and industrial waste is generated that needs to be disposed. Historically, convenience predominated over thoughtful planning, and wastes including raw sewage, oil, and toxic chemicals were discarded directly into rivers. Many rivers became so contaminated by the 1960s that animals and fish could not live in them. Some even burned (see Chapter 14)! Many became unsafe for fishing, swimming, and drinking. The U.S. Clean Water Act was passed in 1972 to change this and protect rivers.

The stated objective of the Clean Water Act is to "restore and maintain the chemical, physical, and biological integrity of the Nation's waters." Passage of the Clean Water Act resulted in the creation of the U.S. Environmental Protection Agency, which administers funds for wastewater treatment projects and enforces regulations prohibiting the discharge of toxic pollutants into rivers, streams, and all other water bodies in the country. In the first 30 years following the passage of the Clean Water Act, $80 billion were spent on wastewater treatment facilities. This resulted in an additional 120 million citizens being served by modern sewage treatment.

On the enforcement end, in 2003, the Environmental Protection Agency fined one company $34 million, the largest single fine yet, for spilling nearly 5,678,117.7 liters (1.5 million gallons) of oil during seven separate leaks from a 8,851.4 km (5,500-mile) pipeline running through five states. The largest of these spills, almost 3,785,411.8 liters (1 million gallons), released diesel fuel into the Reedy River, South Carolina, in 1996. The spill spread 54.7 km (34 miles) downstream, killing 35,000 fish and other wildlife.

During the early years of the Environmental Protection Agency, most efforts were aimed at eliminating easily identifiable sources of surface water pollution, such as point or pipe discharges. Although these actions vastly improved the water quality of the nation's rivers, the scientific community has become increasingly aware that a river's health is directly linked to the health of its watershed. Because of this increasing scientific understanding, the Environmental Protection Agency's efforts now include educational projects designed to inform all residents of a watershed how their actions can impact a river's water quality. As more homeowners and farmers stop using chemical fertilizers and pesticides on their lawns and crops, fewer harmful toxins run off into rivers.

After more than 30 years, signs of the Clean Water Act's success abound. The Cuyahoga River, a river that has actually caught on fire (see Chapter 14), now has pleasure boats crowding the river in summertime. Restaurants in Downtown Cleveland line the riverbank with customers sitting out on pleasant evenings—eating, listening to music, and hearing birds signaling the return of life to the river. Twenty-seven species of fish are now found in the Cuyahoga River, including pike, bass, and blue gill.

The many success stories should not make people complacent. Many problems, some of the most intractable ones, still remain. One example is on the Hudson River in New York where 589,670.1 kg (1.3 million pounds) of PCBs were discharged into the river from 1947 to 1977. PCBs, or polychlorinated biphenyls, were a petroleum-based chemical used in the manufacturing of electrical parts before their use was banned worldwide by the early 1980s. PCBs were used in large quantities at two manufacturing plants north of Albany.

Why were these chemicals banned? The chemicals were banned because PCBs can cause cancer in animals and other illnesses related to the nervous system, reproduction system, and immune system. The same ill health effects are thought to be possible in humans as well. The global unity in banning PCBs is testimony to how devastatingly quick this single chemical moves through the food chain. People fishing on the Hudson River are still urged not to eat the fish they catch downstream of the two manufacturing plants that used PCBs.

Floods

River flooding is a problem that affects nearly every community in the country because everyone lives in a watershed. Our close association with rivers means that property damage and the loss of life from flooding can be extreme. The National Weather Service estimates that flooding caused $50 billion in damage in the United States during the 1990s with at least minor losses reported in all 50 states. In addition, over 1,000 people died in the United States due to river flooding during the 1900s. Floods are naturally reoccurring events that result

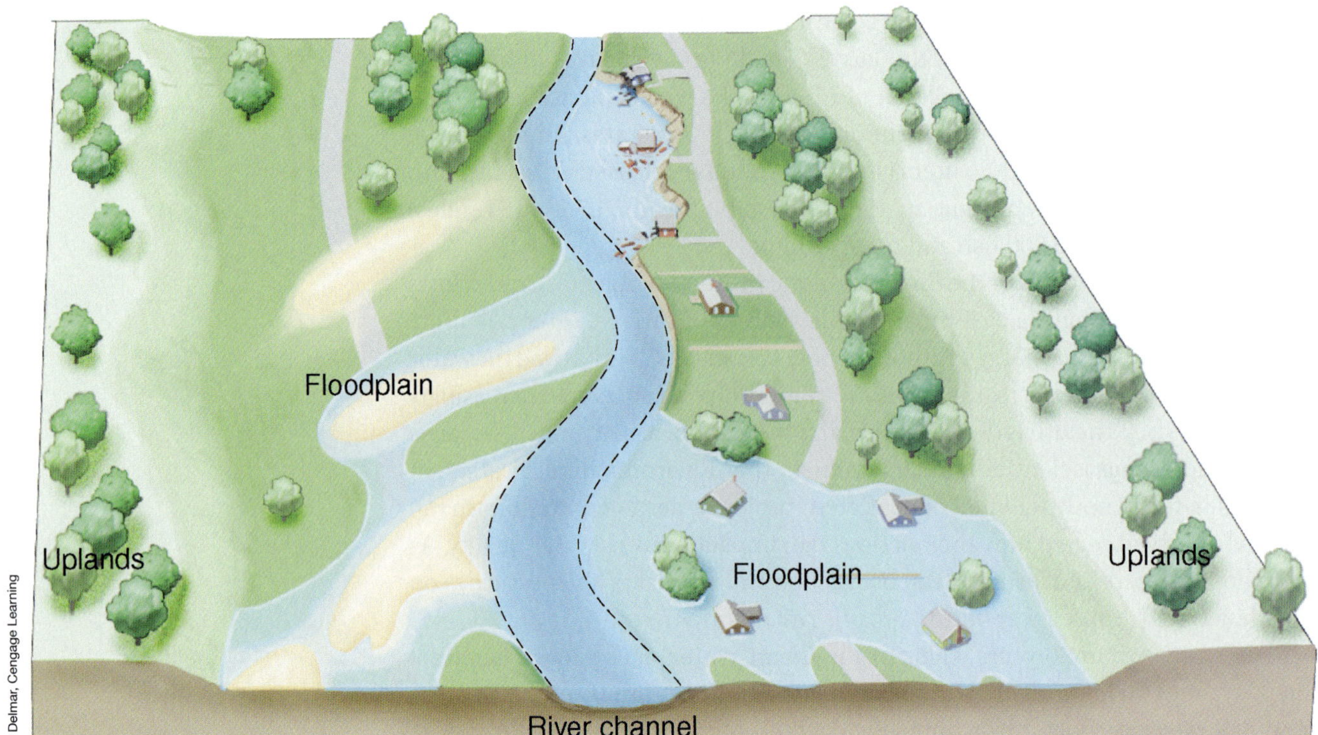

Figure 6-14 Results of overbank flooding, erosion, and deposition.

from unusually heavy rains or, in cold climates, from rapid snowmelt. Damages and deaths result from three different types of hazards: overbank flooding, erosion, and deposition (Figure 6-14).

Overbank Flooding

The most extensive damage due to flooding is caused by floods that overtop the banks of the river channel and spread across the floodplain. On the lower Mississippi River, floodwaters can spread over very large areas because the floodplain is up to 48.3 km (30 miles) wide. The floodwaters can get into people's homes, wash out crops, cover roads, bring in debris, and drown animals and people.

Because overbank flooding spreads over such a large area, flow velocities are generally low and people's lives are not generally in great jeopardy from the force of the water. The Federal Emergency Management Agency, or FEMA, requires at-risk communities to map where overbank floods are in excessive velocities. These are typically greater than 0.3 meter (1 foot) per second. FEMA then requires active (flood walls put up at the start of the flood) and/or passive (house on stilts) flood controls in these areas.

However, the sometimes contaminated floodwaters can cause serious outbreaks of infectious diseases such as tetanus, cholera, and typhoid fever. This is generally not the case in the United States, but in developing countries where cleanup after a flood is slow, the spread of diseases is a very real threat. Pools of standing and stagnant water that might remain for months after a flood serve as breeding grounds for mosquitoes, increasing the risk of malaria and other mosquito-borne diseases. Unsafe drinking water and poor sanitation, problems magnified by floods, cause an estimated 80 percent of all diseases in the developing world. When left untreated, many of these diseases result in death.

Erosion

Although erosion is a natural process, erosion hazards are most severe where riverbanks are poorly vegetated and composed of sand. Overbank flooding is less of a threat in these settings, because the erosion increases the channel's size. The larger channel is able to contain more of the floodwater, but damage results from the erosion. Bridges, power lines, roads, and homes built near the edge of the river can fall into the river as the banks erode.

Erosion damage can occur on any river in the United States but is more characteristic of flooding problems in desert climates where vegetation on riverbanks is limited. Floods in southern Arizona during 1983 caused 13 deaths and over $200 million in damages. At least 3,000 buildings were destroyed or damaged. More than 10,000 people were displaced and cut off from help because roads, bridges, and phone/electric lines crossing rivers were damaged by erosion. The damage would have been far less if the riverbanks in southern Arizona were more resistant to erosion.

Deposition

Deposition of sediment is usually the biggest headache awaiting homeowners returning from a flood. The damage caused by water flowing over the banks is bad enough, but the mud deposited on floors, carpets, and the lower walls of homes makes the situation much worse. However, the most severe damage caused by deposition occurs in the river channel itself. As the channel fills in with sediment, floodwaters are diverted toward the banks, accelerating bank erosion. Furthermore, as sediment fills in the channel, it can hold less water. This causes more water to flow out onto the floodplain where homes, factories, and roads are located. Channel deposition is commonly a direct cause of the damage from overbank flooding and bank erosion.

living with rivers

Despite the hazards associated with flooding, people continue to live by rivers in large numbers. The economic and other benefits of living near rivers outweigh the costs of damages caused by overbank flooding, erosion, and deposition. How can the benefits of rivers be enjoyed while reducing the hazards associated with flooding? Protecting human investments is extremely important, but doing so in a way that does not cause harm to our neighbors or the environment is an essential part of being a responsible resident of the watershed.

Many techniques have been used in the past to try and control floods and limit the damage caused by them. Some are expensive, result in the loss of river and riparian habitat, and can cause damage elsewhere on the river. In recent years, a number of new stream restoration techniques have been employed on rivers that simultaneously provide some flood protection while re-creating lost habitat. Some traditional flood control methods and new stream restoration techniques are described next.

Flood Forecasting

The damages resulting from floods can be minimized if people have ample warning and residents in flood-prone areas can prepare for the impending rise in water levels. Even a couple hours' advance warning of when a river will

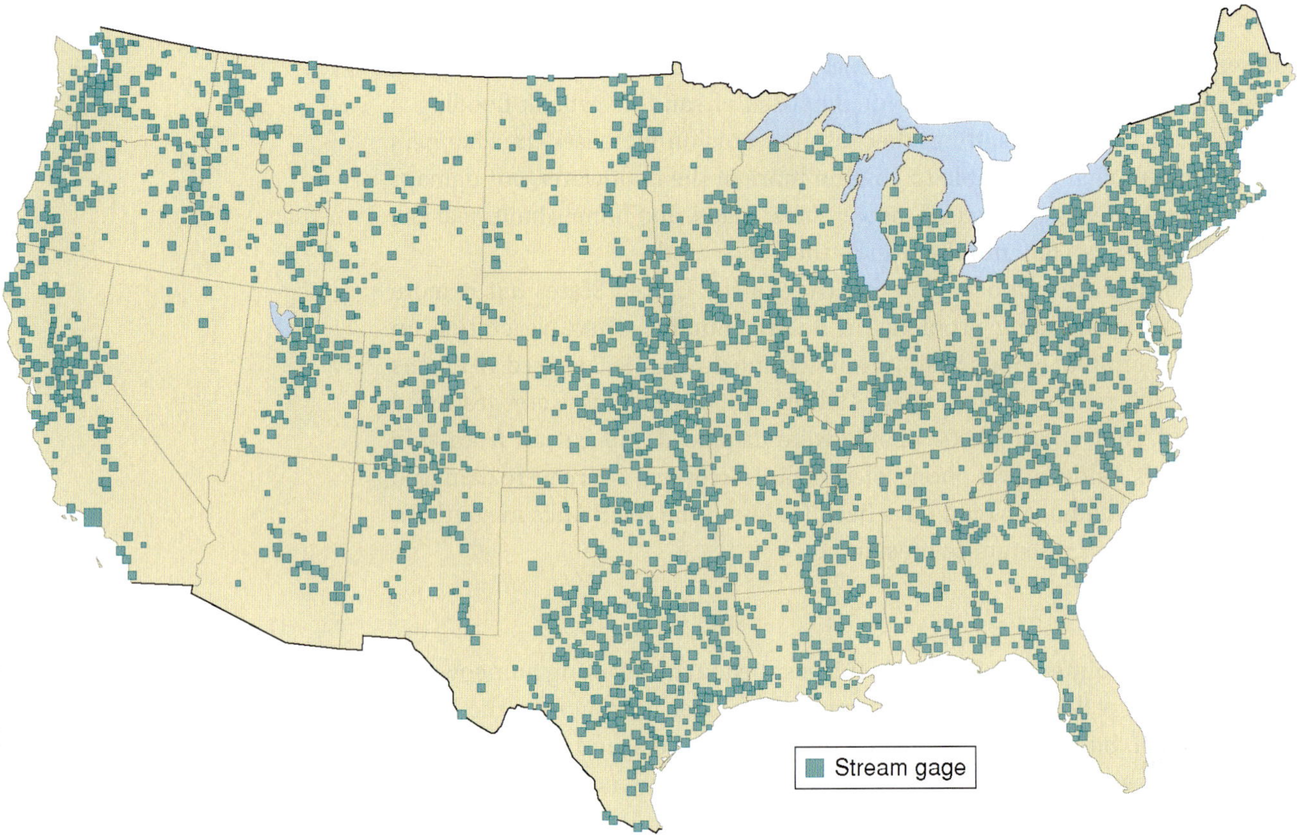

Stream gage

Figure 6-15 Locations of stream gages in the United States.

overtop its banks provides enough time to evacuate residents, move cars, and collect valuables from homes before floodwaters inundate an area. Water, food, and other emergency supplies can be gathered by those whose homes might potentially be cut off by floodwaters for several days. The potential savings to life and property are enormous if the peak height of the flood can be predicted in advance.

How can these predictions be made? The River Forecast Centers, which are part of the National Weather Service (NWS), forecast floods and provide flood warnings in the United States. Although the River Forecast Centers use many sources of data when developing flood forecasts, the United States Geological Survey (USGS) is the principal source of data on river depth and flow. The USGS operates and maintains more than 85 percent of the nation's stream-gaging stations, which includes 98 percent of those that are used for real-time river forecasting (Figure 6-15). Currently, this network comprises over 7,000 stations dispersed throughout the nation, approximately 4,200 of which are equipped with earth satellite radios that provide real-time communications. The NWS uses data from approximately 3,500 of these stations to forecast river depth and flow conditions on major rivers and small streams in urban areas.

Figure 6-16 Stream gage station.

Very simple stream gages are only a ruler that begins near the bottom of the channel and rises to near the top of the riverbanks (Figure 6-16). These simple gages can be seen bolted to the concrete abutments underneath a bridge. They are used to observe how high

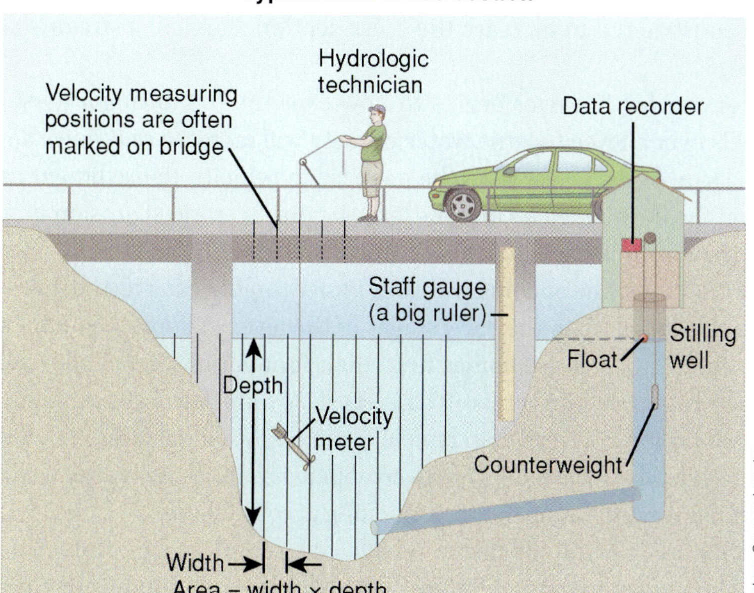

Figure 6-17 Establishing a stage-discharge curve for a stream gaging station.

the river is at any particular time and, during a heavy rain, can be visited often to see how quickly the river level is rising. Look carefully as you approach a bridge near your home and see if you can spot one of these gages.

Many United States Geological Survey stream gages can electronically relay data every 15 minutes on the height, or stage, of the river at the gage. Information on river stage can be used to estimate the discharge by establishing a relationship between the stage and the discharge (Figure 6-17). These stage-discharge relationships are established during previous visits to the gage site when water velocities, river stage, and the channel area occupied by the flow are measured together to determine the discharge associated with various river stages.

If the discharge is known from an upstream river gage, the potential height and arrival time of the flood peak can be calculated for downstream sites where the channel's cross-sectional area is known from previous surveys. During large floods on the Mississippi River, you may hear news reports stating that the river's stage will crest in St. Louis at 11.9 meters (39 feet) in 2 days, for example. This is possible because of the many gages monitoring flow further upstream. With the advanced warning, city officials can determine the likelihood of flooding in various parts of the city and issue warnings to those in jeopardy of being inundated by overbank flow. For such flood forecasting to be accurate, though, a lot of research is required months, and even years, in advance.

Flood Control

Forecasting flood heights in advance can limit damages but by no means completely eliminates them. Levees, dams, flood control channels, and other techniques are used to control overbank flooding and reduce damages. The benefits and problems with levees are discussed next.

Levees—long, high earthen mounds built on riverbanks—are common along large gently sloping trunk streams where the greatest hazard is from overbank flooding. Thousands of miles of levees have been built on U.S. rivers with over 2,575 km (1,600 miles) of levees found on the Mississippi River alone. They are constructed to increase the cross-sectional area of a stream's channel (Figure 6-18).

Levees can fail if a river begins to flow over the top during a flood. When water spills over a levee top, the water velocity will increase as it flows down the steep backside of the levee. With the increase in velocity, the sediment carrying capacity of the flow increases and the levee begins to erode. If erosion progresses through the entire levee, floodwaters will rush through the narrow opening at very high velocities. The opening will grow in size rapidly, causing the flow velocity to eventually decrease. In the early stages of the levee collapse, though, the flow is strong enough to remove homes from their foundations, wash out roads, and scour deep pits over 10 meters (32.8 feet) deep into fertile farm fields. These hazards are far more severe than typical overbank flooding. Overbank flows will move much slower if the banks are overtopped simultaneously over a wide area rather than forced out at one narrow opening across a levee.

In 2005, as a result of heavy winds and storm surges from Hurricane Katrina, there were extensive failures of the levees and flood walls protecting New Orleans, Louisiana, and surrounding communities. The Mississippi River Gulf Outlet ("MR-GO") breached its levees in approximately 20 places, leaving approximately 80 percent of the city flooded with up to 6.1 meters (20 feet) of water. Flooding from the breaches put the majority of the city under water for days, in many places for weeks. The Army Corps of Engineers made emergency repairs to levee breaches and pumps worked at draining the city. By mid-September, the inundation of the city had been reduced from 80 to 40 percent. In the aftermath of Hurricane Katrina, engineers investigated the possibility that a failure in the levee design, construction, and/or or maintenance caused much of the flooding. At least 1,836 people lost their lives in Hurricane Katrina and the subsequent floods. The storm is estimated to have been responsible for $81.2 billion (2005 U.S. dollars) in damage, making it the costliest natural disaster in U.S. history.

Although building levees very high to keep them from being overtopped would seem to make sense, the cost is commonly too expensive. Levees protecting large cities like St. Louis are built very high to contain the largest floods, but most levees in the country are built only high enough to contain moderate-sized floods. Building very high levees costs over $1 million per mile and is cost-effective only in very large cities where the value of the protected property far exceeds the cost of levee construction and maintenance.

In addition to the hazards of levee failures, levee construction also has long-term negative consequences on aquatic and riparian habitat. Because levees contain larger discharges within the river channel than would naturally occur, the resulting higher flow velocities scour the channel bottom. Gravel used by fish to lay eggs can be swept away and other midchannel habitat destroyed.

The construction of levees also tends to block access side channels found on the floodplain because levees are typically built across the entrances to the side channels. Cutting off access to side channels results in the loss of prime habitat for juvenile fish. This forces them to compete for limited habitat in the main channel with bigger and stronger adults. The inability of juveniles to survive will ultimately affect the species' population because fewer juveniles will reach maturity and reproduce.

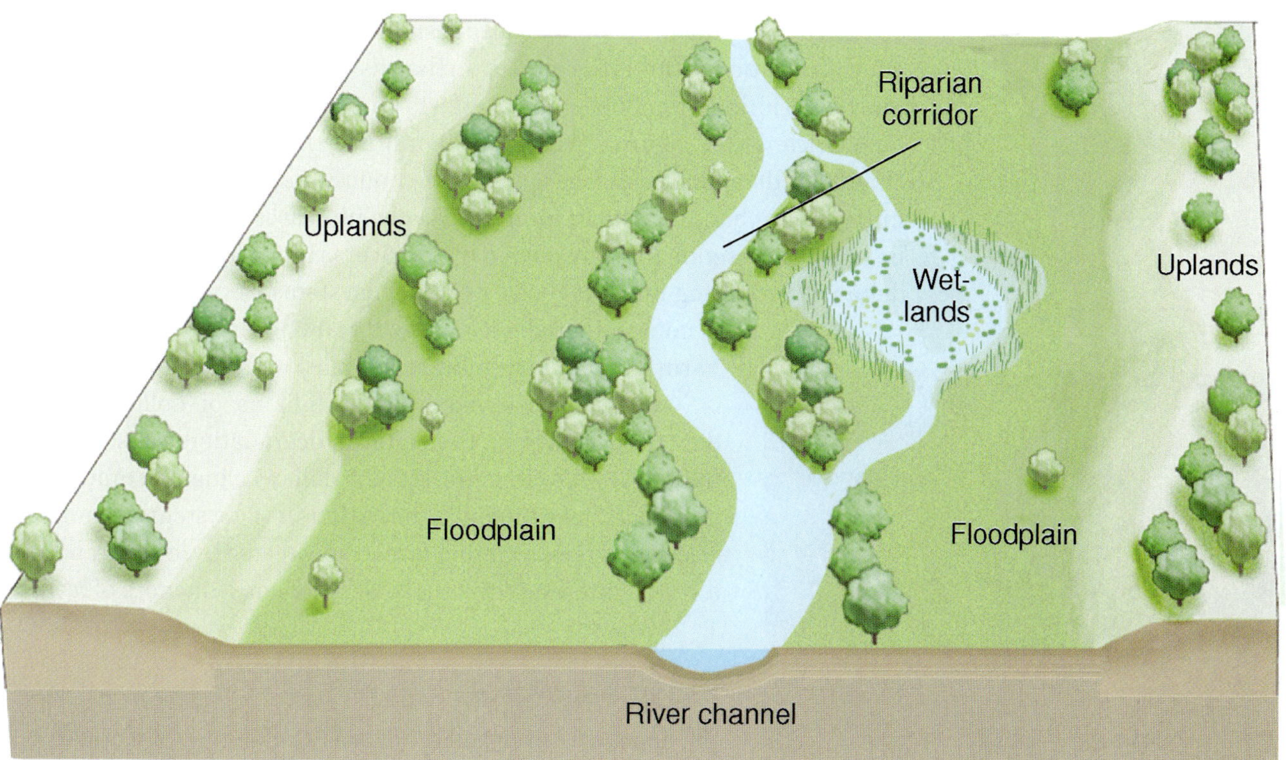

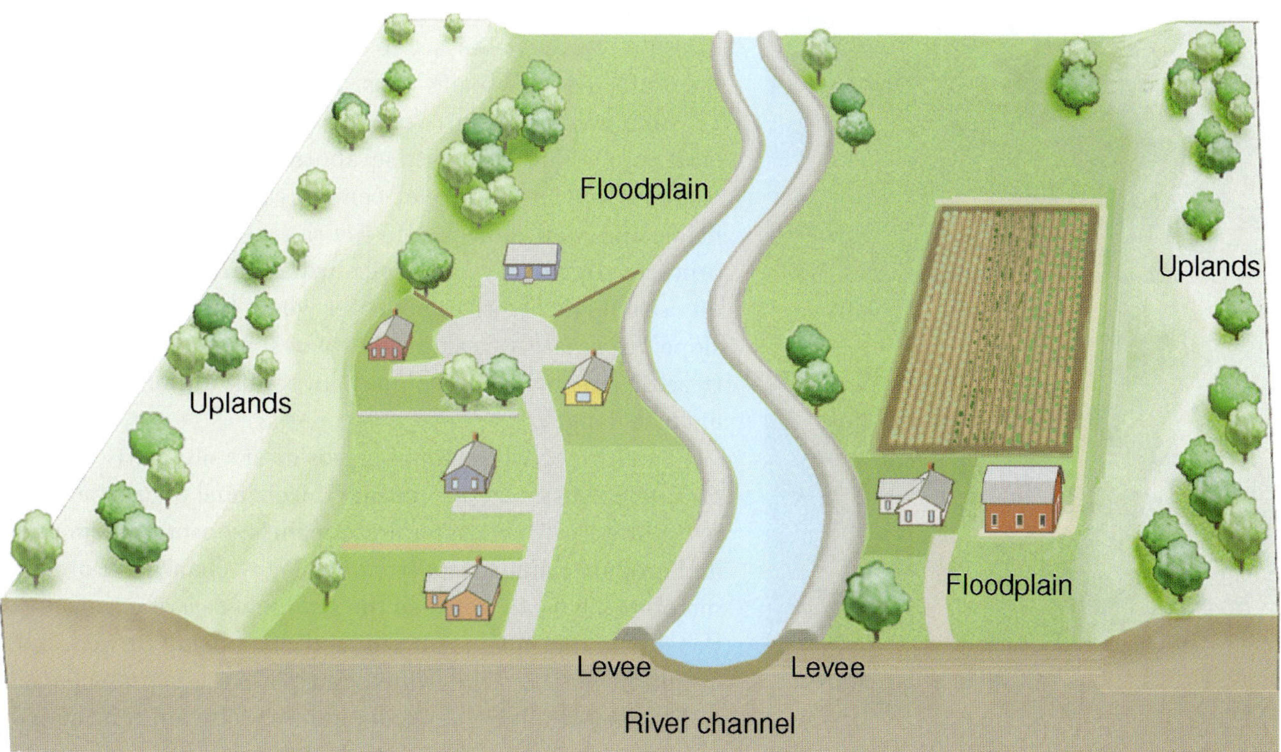

Figure 6-18 Changes in rivers due to levee construction.

Environmental problems can also result miles downstream of the levees. A good illustration of this is the impact on the Mississippi River delta in the Gulf of Mexico resulting from flood control efforts on the lower Mississippi River near New Orleans (see Figure 6-18). Levees confining the floodwaters to the river channel reduce flood damages as intended but increase the velocity of water in the river channel, which, unintentionally, increases scour of the channel bed. The extra sediment transported by the river is ultimately deposited on the delta with resulting impacts to the delta's ecosystems. Some of this sediment is contaminated with agricultural and industrial wastes that harm fish and invertebrate populations.

Image courtesy Colin Mably

(a)

Image courtesy Colin Mably

(b)

Image courtesy Earth Science World Image Bank copyright United States Geological Survey

(c)

Figure 6-19 (a) Unprotected stream bank. (b) Stream bank with wooden armor. (c) Protected bridge base.

Levees have saved humans billions of dollars in flood damages. However, the habitat degradation caused by the nation's levee systems offsets the dollar savings to some degree. The federal investment in the nation's levee systems is enormous so every person, whether directly protected by levees or not, has a vested interest in the successes and consequences of levees.

Erosion Control

Erosion of riverbanks causes millions of dollars of damage each year in the United States. Erosion hazards in the United States have traditionally been treated by armoring riverbanks with large rocks or cement (Figure 6-19). The threshold force that the river must reach before it is capable of transporting the large rocks or scouring the cement is much greater than the previously eroding material underneath. Because the river is unlikely to reach the threshold forces necessary to transport the armor, erosion at the protected site usually stops.

Channel armoring is usually completed where property is at immediate risk. Bridge abutments are almost always protected by large rocks or cement structures. Homes, farm fields, and roads adjacent to the river will also frequently be protected by bank armor. Typically, bank armoring will be seen on the outside of a river meander where the water depth and slope are greatest, making the forces stronger. Nearly every river or stream in the country has bank armor somewhere along its length.

Bank armor, while stopping erosion at a site, can cause bank erosion downstream or scouring the channel bottom adjacent to the protected bank. A natural stream is constantly depositing and eroding sediment as stream velocity changes in order to maintain a balance between the sediment carrying capacity of the stream and the actual sediment load being carried. Bank armor prevents erosion from taking place at the protected site so the river simply erodes sediment from elsewhere if the sediment load being carried is less than the carrying capacity. The erosion commonly occurs on the riverbanks just downstream of where the armor ends. Unable to erode at the site of the bank armor, the stream will be prone to accelerated erosion at the first spot where the bank is unprotected. The new location of erosion could, in turn, be protected by additional armoring.

In this manner, long sections of rivers have become armored over time. Although this may limit property damage caused by erosion, ecological consequences also result. The protected banks prevent undercutting that provides cover for fish. Additionally, not allowing the river to erode at the banks means that fewer trees are undermined and, therefore, fewer fall into the stream. Although excessive erosion is bad for fish habitat because the stream becomes too shallow as the width increases, some erosion is necessary to ensure that wood in the stream is slowly replaced over time. Bank armor has a negative impact on this process by eliminating erosion completely.

Instead of armoring banks with large rocks, eroding riverbanks can sometimes be protected using bioengineering techniques that are designed to provide bank protection while creating habitat for river organisms. These techniques use vegetation to stabilize riverbanks and re-create a more natural condition where protective vegetation was earlier removed by human land use. Log deflectors can turn water away from the riverbank during the first few years of a project while vegetation takes hold on the banksides (Figure 6-20). Eventually, as the log deflectors slowly rot away, the roots of the emerging bankside vegetation will provide the necessary stability to prevent rapid bank erosion. The vegetation will allow some undermining of the bank that will provide some overhanging bank cover and will cause some trees to fall into the river where they can serve as fish habitat in the midchannel environment. These bioengineering methods work well as long as there is some room for erosion. If roads, homes, and other human investments are directly on top of the riverbank, rock armoring may still be needed.

Figure 6-20 Log deflectors turning water away from the riverbank. (*Photo by RG Johnsson, 1966, Courtesy NPS*)

Land Use Planning

Bioengineering techniques are an example of only one stream restoration technique used to reduce flood damages while protecting and restoring aquatic habitat. Perhaps the best approach for the long-term reduction of flood damages and protection of river ecosystems is to not live on floodplains. Government buyout programs accomplish this in some areas where they have compensated landowners, and even entire communities, for moving out of flood-prone areas. After the devastating 1993 flood on the Mississippi River, FEMA used nearly $60 million to buy out over 7500 flood-prone properties in Missouri and Illinois. Nationwide, the federal government has spent $1 billion to move approximately 25,000 homes.

Once the properties are purchased, the land is set aside as open space that can be flooded with little resulting damage. Allowing rivers to spread out in these open spaces creates habitat for river and riparian organisms while simultaneously reducing the pressure on levees protecting property elsewhere along the river. Slowly, as more and more individuals and communities accept financial assistance to move out of flood-prone areas, the damages caused by flooding can be reduced, even for those still living on floodplains throughout the country.

6.5 conclusion

River systems play an integral role in people's lives around the world, even those not living directly adjacent to a river or stream. Everyone's presence in a watershed means that their actions can potentially affect river stability and river environments. Flood control efforts in the United States have prevented billions of dollars in damages but often harm midchannel and riparian habitats and can sometimes create new hazards elsewhere on the river. Newly emerging stream restoration techniques attempt to improve river habitat while limiting flood damages, but they are not appropriate in all settings. Wherever you live, rivers and watersheds are important to you.

IN THIS CHAPTER YOU HAVE LEARNED:

- Watersheds are the land areas that contribute surface runoff to a specific, connected set of streams and rivers—a river system.
- Watersheds can be small or large. The largest in the United States is the Mississippi River watershed, which covers 41 percent of the continental United States.
- Habitat for plants and animals changes across watersheds—from the higher elevations and smaller, cooler, faster streams of headwater regions to the larger, warmer, slower moving rivers at lower elevations.
- The area and amount of surface runoff in a watershed are important controls on stream and river morphology. Channel gradients, depths, and widths are controlled by watershed characteristics.
- Riverbanks and stream banks are also important controls on channel character. Well-vegetated banks resist erosion and help stabilize channel width and depth.
- Habitat varies across a stream or river. Midchannel environments are where many fish and macroinvertebrates live. The bankside environment is where grasses grow at the waterline, brush such as willows grow on the bank, and trees such as cottonwoods grow on top. The banktop environment includes the floodplain where forest trees and animals live.
- A river is said to be in a state of equilibrium when its sediment carrying capacity matches the sediment load in transport. People change a river's equilibrium by doing things that change its flow or the amount of sediment it is carrying.
- Flooding occurs when a river's flow exceeds the volume of its channel. Floodwaters that overtop a riverbank deposit sediment on the adjacent floodplain. Floodplains are productive farmlands and many people live on them.
- Flooding causes billions of dollars of damages and many deaths, both in the United States and throughout the world. Engineered structures such as levees have been used to control flooding, but these also change river morphology and habitat. Flood forecasting, stream restoration, and moving people from floodplains have helped protect lives and property.

- People use rivers for transportation, power generation, and water resources (see Chapter 14). They misuse rivers where they have used them as waste disposal systems. Polluted rivers are less common since passage of the 1972 Clean Water Act, but many are still negatively affected by improper waste disposal.
- People can live better with rivers by protecting their environments and by accepting and respecting their recurring ability to flood.

KEY TERMS

Braided stream
Channel
Discharge
Floodplains
Floodwater
Gradient
Meandering
Nonpoint source pollution
Point source pollution
Pollution
Riparian zone
River
Roughness
Sedimentation
Sediment carrying capacity
Tributaries
Trunk stream
Watershed

REVIEW QUESTIONS

1. What is a watershed?
2. What are two important factors in controlling a river's size?
3. Of what value are rivers to humans?
4. Explain how the area and amount of surface runoff in a watershed affect the morphology of a river.
5. Living beside rivers has both advantages and disadvantages. Describe one of each.
6. How does the presence or absence of vegetation affect stream banks?
7. List the three types of stream or river environments and describe the types of organisms that live in each.
8. When is a river in a state of equilibrium?
9. What is the difference between point source pollution and nonpoint source pollution?
10. In what ways do people manage life on a floodplain?

chapter 7

weather and storms

Students at George Washington High School in Guam, an island in the South Pacific, leave the classroom each day to check a rain gauge (Figure 7-1). Even if they know it has not rained, they have a good excuse for going outside—they must check to see that leaves or twigs have not fallen and blocked the small opening at the top of the gauge. The students received the rain gauge from researchers at the University of Oklahoma who are interested in differences in the amount of rain falling throughout the South Pacific. These students are not the only ones working with University of Oklahoma researchers. Over 160 schools in 22 countries are collecting rainfall data for the schools of the Pacific Rainfall Climate Experiment (Figure 7-2).

Courtesy SPaRCE

Figure 7-1 Students in the Schools of the Pacific Rainfall Climate Experiment checking their school's rain gauge.

Studying **weather** *patterns with student-collected data throughout the Pacific region will help shed new light on the El Niño weather pattern. The El Niño weather pattern is named for years when warm ocean water, which is normally located in the southwestern Pacific Ocean, moves to the eastern Pacific Ocean off the coast of South America. This change in ocean water temperatures can affect weather patterns worldwide and create storms that cause loss of life and billions of dollars in damage. Lots of rain usually falls where the students in the South Pacific are collecting data, but is there less rain in El Niño years when the warm ocean water is on the other side of the Pacific Ocean? Answers to this question will help researchers at the University of Oklahoma determine the impact of El Niño and improve forecasting of future weather disasters in the South Pacific and around the world. Imagine, the data collected by students living on remote islands in the South Pacific may one day allow researchers to predict the weather in parts of the United States, perhaps in your community!*

When was the last time you had great plans for a big day outside, only to have the day washed out by rain, blown away in a strong wind, or covered in ice and snow? What are a couple of recent examples where weather conditions changed your plans? Weather reports can be heard almost constantly on the radio and local television news. The weather report helps us make our daily and weekly plans. "Will I need an umbrella when walking to school today?" "Will my weekend birthday party be held outside?" More importantly though,

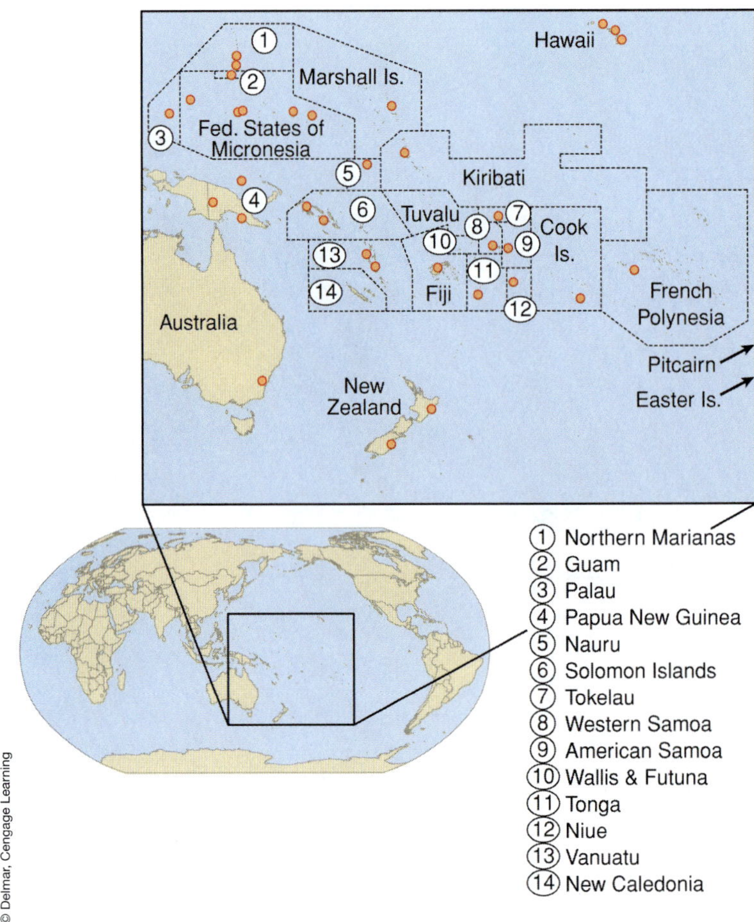

Figure 7-2 Countries in the South Pacific participating in the Schools of the Pacific Rainfall Climate Experiment.

weather reports can save people's lives and billions of dollars in damage by warning people of approaching hurricanes, tornadoes, and other severe weather conditions.

It is amazing what a strong influence the air in the atmosphere above us has on what we do on the Earth's surface. On a warm, calm day we scarcely think of the air even being present. On other days though, the biting cold, the blustery wind, or the dense fog remind us that the air is made of molecules. Moving molecules that make up the air can give almost anyone a bad hair day!

What controls the weather and how can it be predicted? How do plants, animals, and people adapt to extreme weather conditions such as searing heat, bitter cold, strong winds, or torrential rains?

AFTER COMPLETING THIS CHAPTER, YOU WILL UNDERSTAND:
Why the atmosphere is dynamic and constantly moving and changing.
How global atmospheric circulation patterns are developed.
How plants and animals have adapted to severe weather and climate conditions.
What causes severe weather conditions, including monsoons, hurricanes, tornadoes, and droughts.
How weather and severe storms are predicted and how these predictions help save lives and protect property.

 the dynamic atmosphere

In Chapter 1 you learned that the atmosphere is Earth's most dynamic system. The lower part of the atmosphere, the 7 to 17 km (4.35 to 10.6 miles) thick troposphere, is where most of the action is—it is where weather happens. The air's temperature, pressure, moisture level, and movement in the form of wind are all important characteristics in describing the weather. The climate, on the other hand, reflects the long-term average weather conditions that predominate in a given area. The southwestern United States, for example, is characterized by a desert climate that is generally hot and dry. This means that the weather on most days is hot and dry, but there are days, even in the Sonoran Desert of Arizona, when the weather is rainy and cold. The weather, and its constant daily changes, is largely driven by three things: energy from the sun, the Earth's rotation around its axis, and density differences between hot and cold air.

Energy from the Sun

The earth is unevenly heated by the sun. This difference in input of the sun's energy causes differences in surface temperature, weather, and climate. The part of the Earth's surface that receives the most energy, the equatorial regions, is the warmest on average. This is because the sun's energy strikes the Earth's surface most directly at low latitudes (Figure 7-3). The areas of coldest surface temperatures are at the North and South Poles, where the sun's energy strikes the Earth's surface most obliquely. Here energy from the sun is spread out over a larger area

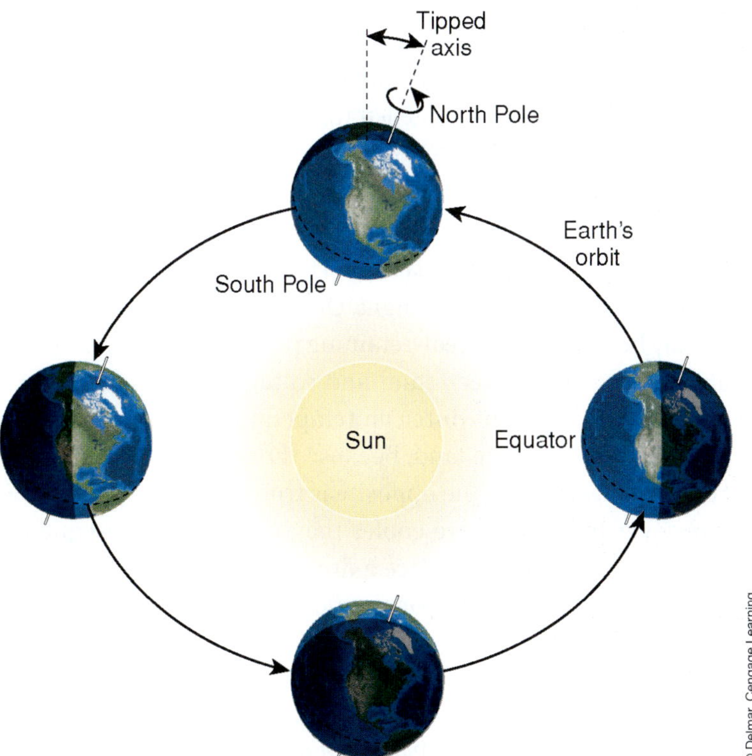

Figure 7-3 The tilt of Earth's axis and its relationship to the sun at different times of the year.

than at the equator and, therefore, less warming occurs. Furthermore, the sun's energy at the poles travels a greater distance through the atmosphere, so more energy is absorbed before reaching and warming the Earth's surface. Because of these factors, equatorial areas receive three times more energy from the sun than the polar areas.

The amount of the sun's energy reaching the equator is about the same throughout the year. Between the equator and the poles, however, the amount of the sun's energy that reaches the Earth's surface varies with the seasons. The cause for these seasonal differences is the 23.5-degree tilt of the Earth's axis relative to the plane of its orbit around the sun. The warmer summer months in the northern hemisphere occur when the North Pole is tilted toward the sun. The colder winter months occur when the axis is tilted away from the sun.

Just the reverse happens in the southern atmosphere where the seasons are opposite in time to those of the northern hemisphere. When the axis at the North Pole is tilted toward the sun, the axis at the South Pole is tilted away from the sun. Therefore, December is the beginning of summer in Australia with people more likely to be heading to the beach for the December holidays than the ski slopes!

Uneven heating of the Earth's surface by the sun's energy not only occurs between the equator and poles but also between the oceans and land. Land surfaces heat up much faster than the ocean. Why do you think water takes longer to heat up than the land surface? First, sunlight can penetrate into the water more deeply so the energy is warming up a greater volume of material. If you were swimming under the water on a sunny day, you could look up and see the sun's rays penetrating beneath the surface. However, if you were buried in a shallow hole in soil, all you would see looking upward is darkness. On land, almost all the sun's energy is being absorbed at the surface. Another reason the ground heats up more quickly than water is that water has a specific heat five times greater than land. Specific heat is the amount of heat required to raise the temperature of 1 gram (0.04 oz) of a given material by 1°C (33.8°F). Therefore, five times more energy is needed to heat 1 gram of water to the same temperature as 1 gram of land.

For the same reasons that water bodies heat up more slowly than land surfaces, they also cool down more slowly if the source of heat is removed. After a summer of energy delivery from the sun, the ocean will be much warmer and it will take several months for it to cool down to its lowest winter temperatures. Land surfaces, on the other hand, can experience dramatic temperature differences in a single day with the sun heating up the land during the daylight hours, only to have it cool down rapidly at night. Ocean water temperatures vary little from day to day because of the heat-retaining properties of water.

The uneven heating between land and water creates temperature differences in the air above them. In winter, air temperatures over the ocean will tend to be warmer than air over the land, because the ocean water retains the heat absorbed during the summer. The opposite is true during the summer when air temperatures over the oceans are cooler than over land. Consequently, cities that are far from the ocean experience a dramatic difference between summer and winter temperatures. The seasonal differences in temperature are far less extreme for coastal cities.

At the equator, the ocean water temperatures are higher than anywhere else on Earth. This makes the air above equatorial ocean waters also persistently warmer than elsewhere. The Pacific Ocean, with the largest body of water straddling the equator, creates a very large area of warm, moisture-laden air above its surface. A large body of air such as this, where weather conditions are similar throughout its extent, is known as an **air mass**. Moist air masses tend to form over oceans and dry air masses over the continents. Cold air masses form closer to the poles and warm air masses form closer to the equator. These temperature differences create density differences that cause air masses to circulate in the troposphere.

Atmospheric Circulation Due to Density Differences

The uneven heating of the surface between the equator and the poles creates density differences that drive vertical air movements (Figure 7-4). Warm air tends to rise because it is less dense than cold air. This is why hot air balloons rise into the sky when the air in the balloon is heated (Figure 7-5). Like air in a hot air balloon, warm air near the surface rises through the atmosphere.

With the warmth at the equator, a lot of water evaporates from the oceans. This is because warm air can absorb more moisture so as it is heated, surface water evaporates into the air. Energy is added to air to warm it, and that energy can move water molecules more quickly, which allows them to pass from liquid into gas and evaporate. As air rises, it expands into the less dense atmosphere. Expansion is work and this uses energy, which cools the air.

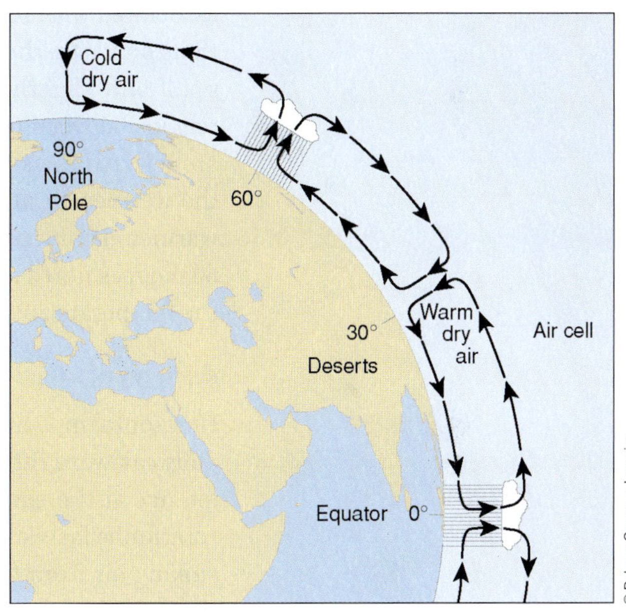

Figure 7-4 Global atmospheric circulation patterns with areas of low and high pressure.

The moisture in the rising and cooling air condenses into tiny water droplets on dust particles suspended in the atmosphere. When enough condensation occurs, the drops of water will become heavy enough to fall as rain. A lot of rain falls near the equator because much moisture-laden air rises here. Because the equator is the warmest area on Earth throughout the year, the areas of highest average rainfall are in the tropics, a climate zone straddling the equator between 15 degrees north and south latitude.

Areas of **low pressure** develop where air rises. This is because the rising air moves atmospheric molecules away from each other and Earth's surface. Low-pressure areas, no matter where they occur, are associated with rainfall as the rising air cools and the water vapor it contains condenses. Air, which is a fluid, moves from high to low pressure. When you run water into your sink, it flows from the higher edges to the lower drain. However, air cannot drain out the bottom. It gets pushed up the top into the upper atmosphere where it cools and rains.

When air is warmed at Earth's surface, it expands and is at lower pressure than surrounding air. The surrounding air moves into this low-pressure area. As air accumulates, it gathers moisture at the low and thus is forced to rise upward, expanding, cooling aloft, and precipitating.

The constant warmth at the tropics creates a steady flow of rising air that pushes the cooler air above it away. Once beyond the influence of the rising tropical air, the cooler air high in the atmosphere

Figure 7-5 Balloons rise and fall due to the density differences of the air filling them.

begins to sink back to the surface. The sinking air warms and becomes denser (the molecules in the air move more closely together) as it descends lower into the atmosphere. With more air pushing from above, the molecules that make up air move more closely together. The warmer air can evaporate moisture and absorbs more water vapor, so these areas of sinking air tend to have lower amounts of rainfall. This is the reason there is a zone of dry tropical deserts adjacent to the tropics between 15 and 30 degrees north and south latitude. Why do you think some deserts occur outside this latitude range?

Deserts are characterized by areas of **high pressure**. Imagine the sinking air pushing down on your shoulders, increasing the **air pressure** on your body. High-pressure areas, regardless of where they occur, tend to be associated with clear skies. Pay attention to your local weather—when you have a blue sky morning, it is likely a high-pressure system. The sinking air warmed enough to absorb the moisture and keep away clouds. As the day progresses, however, the sun will warm the surface of the Earth, and warm parcels of air will rise, cool, and condense, and "fair weather" cumulus clouds will form.

Figure 7-4 shows that the polar regions are also characterized by sinking air associated with high-pressure systems. Because the poles are the coldest locations on Earth, the cold air is dense and it sinks. This sinking air tends to warm up, if only slightly, at the poles, and evaporate moisture. Consequently, polar regions between 60 and 90 degrees north and south latitude are quite dry and are referred to as polar deserts. The constantly sinking air at the poles pushes the warmer air at the surface away to lower latitudes. When these masses of warmer air move beyond the influence of the cold sinking polar air, at around 60 degrees north and south latitude, they begin to rise. This rising air occurs in the temperate zone dominated by low pressure and rainfall.

Atmospheric Circulation Due to the Earth's Rotation

Horizontal movements of air across the Earth's surface result from the Earth's daily eastward rotation around its axis. Because the Earth has a greater circumference at the equator than anywhere else on Earth, the speed of the Earth's rotation is fastest at the equator. When the warm air rises at the equator, the sinking air from the tropical deserts is drawn toward the equator to replace the air rising there. However, because the air at the equator is rotating more rapidly than the air being drawn toward it, the incoming air moves westward in its path to the equator. This westward deflection of air movement is known as the **Coriolis Effect** and produces the **trade winds** that blow from east to west at latitudes between 5 and 30 degrees north and south of the equator (Figure 7-6).

The upward rising air at the equator is so strong that the Coriolis Effect has a minor impact here and surface winds are minimal. This region between 5 degrees north and south of the equator is known as the intertropical convergence zone because the air rushing in from the tropical deserts is converging on this tropical region. Sailors used to refer to this region as the doldrums because of the lack of strong surface winds to move their ships. On the other hand, sailors commonly took advantage of the trade winds between 5 and 30 degrees latitude to cross the oceans.

While air moves toward the equator at latitudes less than 30 degrees, the air between 30 and 60 degrees north and south latitude tends to move toward the poles as it is drawn in beneath the rising air characteristic of temperate

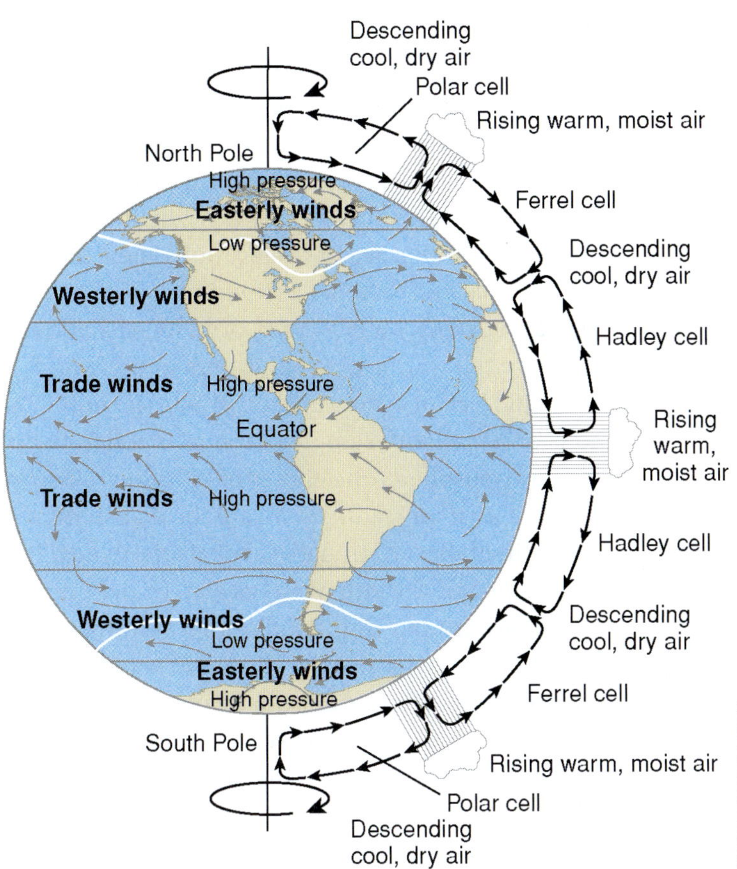

Figure 7-6 Global wind belts.

regions. Because the rotation of the Earth is slower at higher latitudes, air moving from the tropical deserts is rotating more rapidly than air in the temperate regions. This causes it to take an eastward path as it moves poleward and the prevailing deflected winds blow from west to east. These characteristic winds of temperate regions are known as the **westerlies**. The winds forcing the westerly jet stream are aloft, closer to 60 degrees north, and are created by pressure differences. Much of the United States is under the influence of the westerlies.

Trade winds do not only move air from east to west, but they move ocean water as well. In the Pacific Ocean, the warm water at these low latitudes is blown to the western Pacific Ocean where ocean water temperatures are typically higher than anywhere else in the Pacific. But during some years, for still unknown reasons, the trade winds are not quite as strong, or even have become reversed, and warm ocean water collects in the eastern Pacific Ocean off the coast of South America instead. These atypical years, when warm ocean water is in the eastern Pacific Ocean, give rise to what is called the **El Niño** weather pattern.

El Niño—An Example of Ocean-Atmosphere Interaction

El Niño is the interaction of the oceans and the atmosphere every few years that moves warm surface water from the west Pacific to the east Pacific Ocean. People fishing along the coasts of Ecuador and Peru noticed this change over 100 years ago and called it "El Niño," which means "the child" in Spanish. This was named after the Christ Child because the phenomenon commonly occurs around Christmastime.

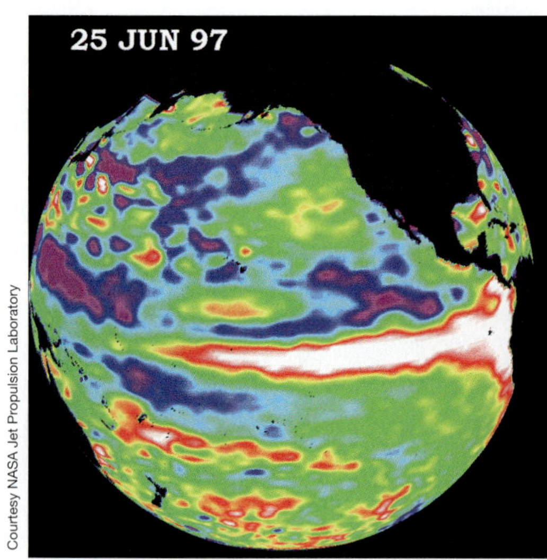

Courtesy NASA Jet Propulsion Laboratory

Figure 7-7 Satellite images of sea level elevation in the Pacific Ocean.

The movement of warm water from the west to the east Pacific Ocean can be measured and monitored in several ways, including wind velocities and directions, sea surface temperatures, surface air pressure, and sea level elevation. Satellite observation of sea level elevation is an excellent El Niño monitoring tool. Because water expands as it warms, warm ocean water has a higher sea level elevation than cold ocean water. Sea level elevations show that the difference between normal and El Niño distributions of warm water in the Pacific Ocean are striking. Much of the warm, west Pacific equatorial waters, which are the warmest ocean waters in the world, move to the eastern Pacific Ocean during El Niño events (Figure 7-7).

El Niño is an excellent example of the interaction between the oceans and the atmosphere. The causes of El Niño are only partially understood, but two important factors are the distribution of warm waters and the direction and strength of equatorial winds. The Pacific has the widest equatorial width of any ocean. Under normal conditions strong westward-blowing winds—the easterlies—help concentrate the warmest water in the west Pacific. The warm water contributes to evaporation and the development of low atmospheric pressures. Under El Niño conditions, the westward-blowing winds decrease, or actually reverse direction, and blow eastward due to changes in air pressure. This atmosphere change is called the **Southern Oscillation**. The Southern Oscillation is so strongly connected to El Niño water changes that the two are commonly linked by name: They are known as the El Niño–Southern Oscillation, or just "ENSO." The shift in wind strength and direction are important reasons why warm water from the west moves to the eastern Pacific Ocean.

El Niño is important to people fishing in the east Pacific because the intruding warm, nutrient-poor water decreases the normal upwelling of cool, nutrient-rich water in this region. The normal nutrient-rich water conditions support abundant sea life, including plankton, fish (like anchovies), and sea birds that eat the fish. When it is replaced by warm water from the west Pacific, the populations of plankton, fish, and sea birds are drastically reduced. The birds starve and cannot feed their young. Many eventually die or move elsewhere (Figure 7-8).

El Niño conditions do much more than affect sea life in the east Pacific—they also have a major impact on global weather. The shift of warm water to the east Pacific brings greater evaporation and lower air pressure systems that generate heavy rains and flooding along western parts of North and South America (Figure 7-9).

The loss of warm water in the west Pacific during El Niño conditions disrupts the normal evaporation and storm systems that bring rain to nearby regions. As a result, droughts develop in places like Australia and Indonesia; at the same time heavy rains and flooding affect usually dry areas in the eastern Pacific. The drought conditions seriously affect agricultural productivity and food shortages; famines can even develop.

The changing weather conditions that accompany El Niño events affect people in many ways. Dangerous storms and food shortages from droughts are two examples. El Niño conditions can also affect human health. Changing precipitation, temperature, humidity, and storm patterns are commonly associated

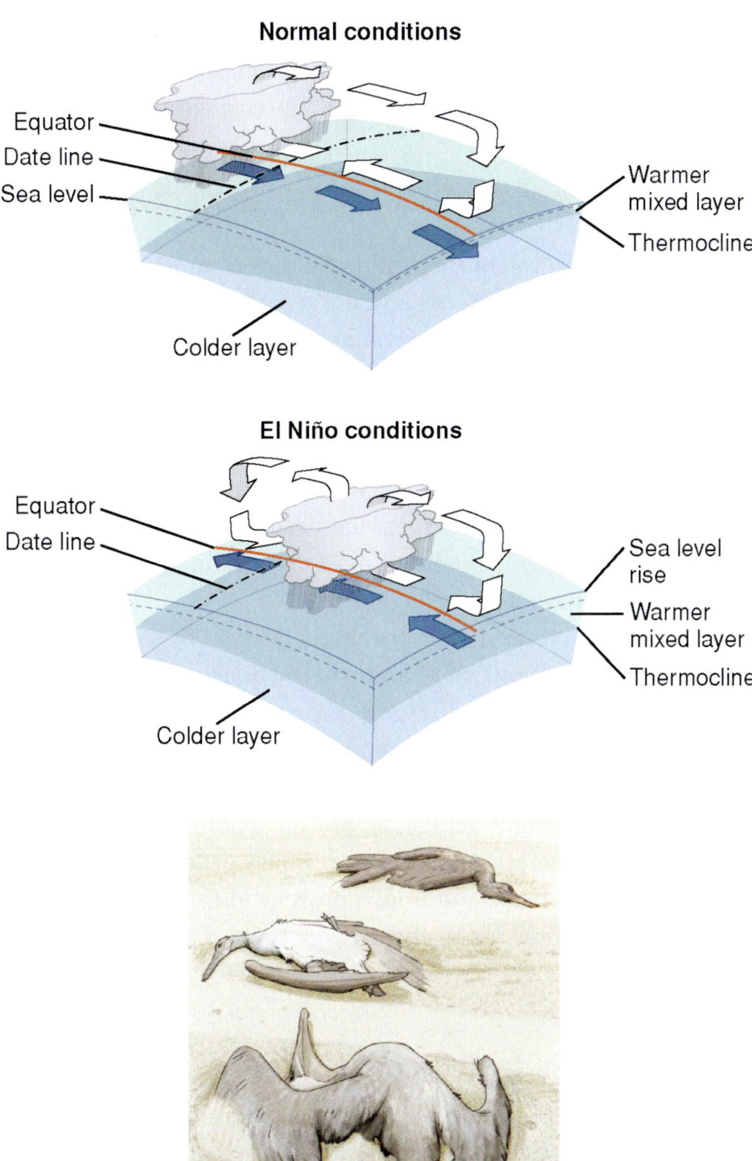

Figure 7-8 El Niño effects.

with increases in infectious diseases such as hepatitis, malaria, dengue, and yellow fever. For example, the density of mosquitoes is much greater in tropical areas during El Niño events. The incidence of dengue, a leading cause of childhood deaths in Southeast Asia for decades, is directly related to mosquito density (Figure 7-10).

Strong El Niño conditions, such as in 1997 and 1998, move tremendous amounts of warm water eastward. It may take years for this water to be completely dispersed and mixed with other water. The remnants of this warm water can influence regional weather for years after El Niño conditions. This may be a reason that many recent storms in the northeast Pacific have brought heavy rains and increased coastal erosion.

Scientists are actively working to better understand El Niño and its causes. As El Niño's impacts on global and regional weather are becoming better understood, scientists are beginning to search for

Figure 7-9 Gulf of Alaska storm. (*Courtesy Earth Science World Image Bank © Bruce Molnia, Terra Photographics*)

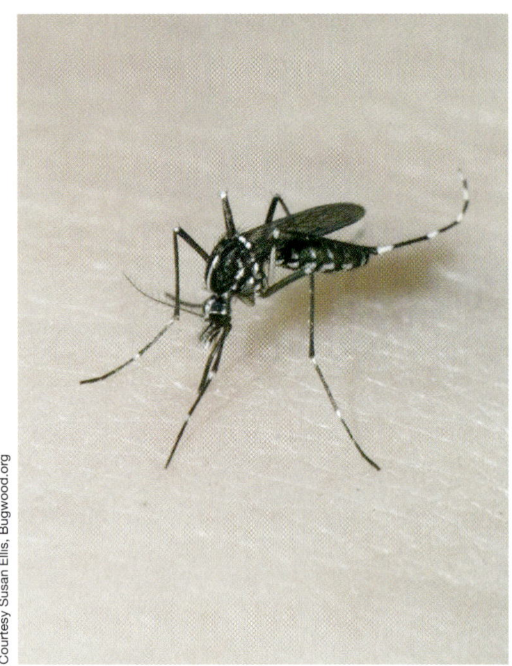

Figure 7-10 A mosquito on a child's arm.

coupled ocean and atmosphere circulation patterns in other oceans. These combined systems, called teleconnections, may be the key to understanding and predicting long-term weather.

7.2 weather, climate, and the biosphere

The distribution of the world's biomes is largely controlled by global air circulation patterns and the resulting climate belts that encircle the Earth parallel to the equator. Biomes, as described in Chapter 2, are communities of plants and animals extending over a large geographical area. Although the weather on any given day might vary, the tropics are generally characterized by hot, moist conditions; the tropical deserts by hot, dry conditions; the temperate zone by cool, moist conditions; and the polar deserts by cold, dry conditions. How do these differences in climate determine the types of plants and animals found in a particular area?

Each climate belt is characterized by one or two distinct biomes. The tropics are home to the dense tropical rain forests near the equator. Barren deserts are commonly found between 15 and 30 degrees latitude where the development of high-pressure systems limits rainfall. Temperate zones are characterized by extensive forests and grasslands. Polar regions are famous for the presence of nearly permanently frozen ground and tundra vegetation. Plants, as primary producers, must be specially adapted to the atmospheric conditions present in each climate zone because they depend most directly on the energy from the sun and moisture falling from the sky.

Plant Adaptations to Tropical Climates

The high rainfall and year-round warm temperatures typical of tropical climates near the equator leads to the development of tropical rain forests with dense vegetation and high species diversity (Figure 7-11). Plants have many adaptations for surviving the heavy rains and warm temperatures found in the tropics. Most plant species have pointed leaves to help drain water off the plant and prevent molds that grow on any moist surface in the warm environment. With few abiotic factors limiting plant growth, plants must compete for the sun's energy. Faster growing tree species are more successful at capturing the sun's energy. These species also spread out at the top to form a broad canopy that captures a maximum amount of sunlight. The canopy also shades the forest floor and prevents other plants from growing tall to compete for sunlight.

Figure 7-11 Dense growth in a tropical rain forest. *(Courtesy NBII Photographer: Randolph Femmer)*

The rain forest canopy, where the sun's energy is focused, supports a diverse ecosystem of other plants and animals. A number of vines, such as lianas (Figure 7-12), begin by growing directly on the canopy over 45 meters (147.6 feet) above the forest floor. The vine really begins to thrive once a root grows down the tree and into the ground where it accesses nutrients and water in the soil. In this way, vines capture the sun's energy on the upper canopy while being rooted in the ground. The tree provides structural support for the vine to reach so high.

Perhaps surprisingly, the soils in tropical rain forests are not very fertile. The high rainfall permeates through the soil, dissolving and washing away the nutrients. Leaves, twigs, and other plant parts that fall to the forest floor are the main source of nutrients in the soil. Tree roots, therefore, tend to be very shallow, so they are close to the nutrients released from the decaying plant matter on the forest floor. The shallow roots growing in the moist soil do not provide much support for the tall trees. In order to stand tall, trees in the tropical rain forest commonly have 3- to 5-meter-wide (9.8- to 16.4-foot-wide) buttresses at their base that provide the support they need so they do not topple over.

Courtesy Guillaume Blanchard

Figure 7-12 Liana vine hanging from a tree with a buttressed base.

Plant Adaptations to Desert Climates

Whereas plants in tropical rain forests must rid themselves of excess moisture, desert plants are especially adapted to prevent moisture loss and access scarce water supplies. Desert climates, typically located between 15 and 30 degrees north and south latitude, are characterized by low annual rainfall and hot summer temperatures. Cacti are known as succulents for their ability to store water for use through the long, hot, dry summer. Cacti with ribs, such as the saguaro (Figure 7-13), are particularly suited for this because the ribs allow the plant to swell and contract like an accordion without tearing the plant apart. Ribs also help prevent the dry, warm desert wind from stripping moisture from the cactus by keeping a cushion of still air surrounding the plant. Cacti do not have typical photosynthetic leaves because the water loss through the leaves would be too great. Instead, photosynthesis occurs in the stems that tend to be green like leaves on plants in other climates.

Courtesy Jörn Napp, Hamburg, Germany

Figure 7-13 Saguaro cactus in the Sonoran Desert of Arizona.

The sharp spines on cacti are actually highly modified leaves. The spines are not photosynthetic but do provide a prickly defense against animals in search of water, which is a scarce commodity in the desert. The spines also help to absorb water and shade the cacti from the intense sun. Dew that forms on the spines during cool nights slips to the base of the spine, where it is absorbed by the plant through small holes located at the spine's base.

Many cacti have shallow root systems that spread out over a large area just below the ground surface. This enables the plant to absorb as much water as possible during the infrequent rains. The moisture is then stored for use during drier periods. Other desert plants, like mesquite trees that do not have the water storage ability of cacti, must have a more regular and reliable source of water. To do this, mesquite trees grow long tap roots that extend up to 80 meters (262.5 feet) beneath the surface until reaching ground water located far below the surface.

Plant Adaptations to Temperate Climates

Temperate climates, typically between 30 and 60 degrees north and south latitude, are characterized by high annual rainfall like the tropics but much cooler temperatures. Very cold winters occur in inland areas far from the ocean. Plants in temperate climates must be adapted to the cold temperatures that accompany the loss of the sun's energy in the winter months. The winter sunlight is insufficient for significant photosynthesis. The potential damage to the tree, if its leaves were to freeze, is great. Shedding the leaves in the fall avoids freezing problems, and the growth of new ones in the spring permits photosynthesis during the time of year when there is the most sunlight. Trees that shed their leaves during one season of the year are known as deciduous trees. Leaves of deciduous trees are generally flat and large in order to maximize the amount of sunlight captured during the summer growing season. Trees in tropical rain forests are not deciduous. Their leaves stay on year-round because there is no cold time of year when having leaves would be harmful to the plant.

Inland temperate regions with less rainfall and greater seasonal temperature variations are conducive to the development of grasslands. The drier and colder winter conditions are too harsh for trees to survive, so various species of grass characterize the plant community. Although mostly farmland today, the Great Plains of the midwestern United States were originally natural grasslands (Figure 7-14). Grasses are specially adapted to open, bright, sunny conditions. The long slender green blades that we associate with grasses are actually leaves that grow from the base of the plant at ground level. Animals that eat grass do not eat the base of the plant; this way the grass leaves can regrow despite constant grazing. Grasses spread through rhizomes, which are stems that travel along just underground and send up new leaves at regular intervals. This allows grasses to colonize the open areas typical of grasslands very rapidly.

Figure 7-14 Natural prairie grassland. (*Courtesy Earth Science World Image Bank © Bruce Molnia, Terra Photographics*)

At the northern edge of the temperate climate zone in the Northern Hemisphere, extensive forests of evergreen trees replace the deciduous forests and grasslands found further to the south. Evergreen trees, such as fir and spruce, are specially adapted to the cold temperatures, limited sunlight, and high snowfall that typify this region (Figure 7-15). As their name implies, evergreen trees retain their leaves, or needles, throughout the entire year, because growing new needles each spring would consume too much energy. The adaptation to evergreen needles permits photosynthesis to begin as soon as temperatures are warm enough in the spring, without expending time and energy during the short summer growing new leaves. Unlike the flat leaves of deciduous trees that are prone to high water loss and freezing, evergreen needles are narrow and covered with a waxy coating to help prevent water loss and freezing in the cold, dry winter months.

At the northern limits of the temperate climate zone in the Northern Hemisphere, much of the annual precipitation falls as snow. The conical shape of evergreen trees living in these northern forests promotes the shedding of snow off the tree branches. This is an adaptation that has developed because trees that do not accumulate snow on their branches are less likely to break under the weight of the snow and are, therefore, more likely to survive. Furthermore, evergreens have long fiber cells in their wood that make the branches more flexible so they can support greater loads of snow.

Courtesy National Park System

Figure 7-15 Evergreen trees in snow.

Plant Adaptations to Polar Climates

Figure 7-16 Tundra vegetation. (*Courtesy Earth Science World Image Bank © Bruce Molnia, Terra Photographics*)

The colder temperatures and shorter growing season in polar climates eventually become too harsh for even evergreen trees to survive, despite their cold weather adaptations. Polar climates, located between 60 and 90 degrees latitude, are not only characterized by cold temperatures, but also very dry conditions. In these cold, dry climates tundra vegetation predominates. Tundra is characterized by lichens, mosses, grasses, sedges, and shrubs (Figure 7-16). All of these plants grow low to the ground so they can be insulated by snow during the bitterly cold winters. Yes, surviving under a cover of 0°C (32°F) snow is quite a bit easier than staying alive while fully exposed to air temperatures that fall below −50°C (−58°F).

Located close to the poles, polar climates receive very little, if any, sunlight in the middle of the winter. Even in the summer when polar regions may receive sunlight for 24 hours of the day, the sun's energy is weak because of the low incidence angle of the sun's rays at these high latitudes (refer to Figure 7-3). Consequently, polar vegetation is specially adapted to carry on photosynthesis at low temperatures and under low light intensities. Many plants also have dark red leaves that allow the plant to absorb more heat from the sun compared to green leaves.

The extremely cold temperatures in polar climates cause the ground to be permanently frozen in permafrost. Warm summertime temperatures last only long enough to thaw the upper few centimeters of the permafrost, leaving the deeper ground permanently frozen year after year. Roots are unable to penetrate into the frozen ground, so tundra vegetation has shallow root systems. This, in turn, is another factor limiting the height of tundra plants, because tall plants generally require a deeper root system for support.

The permafrost helps support the plant communities that are present in the tundra. Although many areas of the polar desert receive less precipitation annually than parts of the Sahara Desert, the permafrost prevents the melting snow at the surface from percolating deep into the soil. Consequently, the water ponds up at the surface, forming bogs and marshy areas that provide a source of moisture for the plants during the short summer growing season.

7.3 severe weather conditions

Severe weather, although commonly detrimental to individuals of a given species, is sometimes critical for maintaining the diversity of habitat that a healthy ecosystem requires. Damage by strong winds to a patch of temperate forest, for example, will create open areas within which grasses and shrubs can flourish. Some animals require both closed forest areas and open grasslands for survival. Without severe weather conditions to maintain habitat diversity, entire ecosystems would suffer. To people, however, severe weather's harm to individuals and their property is a serious concern. The severe weather conditions that are most harmful to people include monsoons, hurricanes, tornadoes, and droughts.

Table 7-1 *Annual average rainfall and temperature patterns in New Delhi, India*

NEW DELHI, INDIA						
Month	Average high	Average low	Warmest ever	Coldest ever	Average dew point	Average precipitation
JAN.	68	48	85	32	44	0.9
FEB.	73	53	91	32	46	0.8
MARCH	83	62	100	43	50	0.6
APRIL	95	72	109	55	52	0.4
MAY	101	80	113	62	57	0.6
JUNE	101	83	112	70	67	2.7
JULY	93	81	111	70	75	7.9
AUG.	91	80	108	68	76	7.9
SEPT.	92	78	102	68	71	4.8
OCT.	90	68	100	57	59	0.7
NOV.	81	57	95	45	51	0.1
DEC.	71	49	91	36	45	0.4

Latitude: 28 degrees, 35 minutes north
Longitude: 77 degrees, 12 minutes east

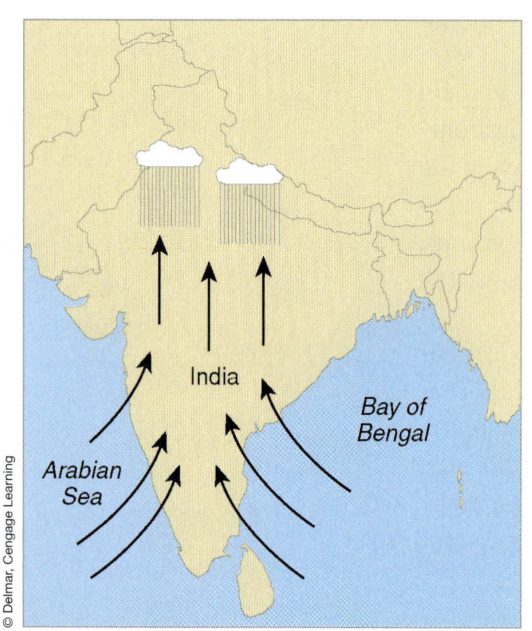

Figure 7-17 Monsoon rains over India.

Monsoons

Monsoons are very heavy rains that occur where very hot landmasses in the summer are surrounded by much cooler ocean waters. The differences in air temperature above these surfaces create differences in air densities that influence regional weather patterns. India, for example, is a large landmass that heats up dramatically in the summer. The average high temperature in New Delhi reaches 38.3°C (101°F) in both May and June. The hot air above this land surface rises and, as it does so, the surrounding cooler air from over the Indian Ocean flows inland to replace the rising air (Figure 7-17). The air coming off the ocean is full of moisture and as it heats up over the hot continent, it also begins to rise. As the moist air rises, it cools, condenses, and falls as rain. Very heavy rain usually results because the temperature differences between the land and ocean water are great. For example, the average monthly rainfall in New Delhi for July and August is 20 cm (7.9 inches; Table 7-1). India is the site of the strongest monsoons in the world. Where in the United States do you think monsoons occur?

Hurricanes

Areas of very warm water develop in the world's oceans between the equator and 30 degrees north and south latitude (Figure 7-18). The air above these warm ocean waters begins to rise and a low-pressure area develops. These "pools" of warm water within the ocean are quite large, so the resulting low-pressure areas can be over 600 km (370 miles) in diameter. These low-pressure areas develop between June and November in the Northern Hemisphere when the ocean water temperature is at its warmest.

The June to November 6-month period of the warmest ocean water is known as hurricane season. A **hurricane** is an organized ring of convective storms with circulating air due to the conservation of angular momentum. Warm surface

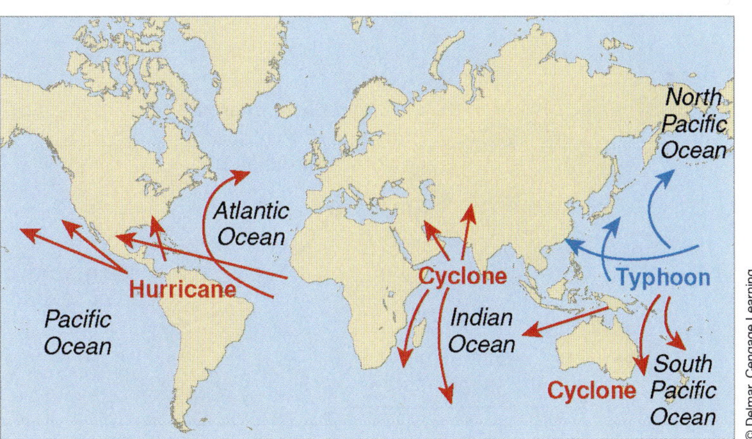

Figure 7-18 Hurricane formations and the general direction in which they move.

waters need to extend more than 200 meters (656.2 feet) deep to fuel this storm. In a hurricane, the rising air in a low-pressure system lifts off the ocean very rapidly and draws in air from the surrounding area. The rapid lift, in combination with the Coriolis Effect produced by the Earth's rotation, generates strong winds. Low-pressure areas with sustained winds over 120 km (74.6 miles) per hour are considered hurricanes. Storms with lower wind speeds are referred to as tropical storms or tropical depressions.

Hurricanes are not stationary locations of rising air. The location of rising air in the center, or eye, of the hurricane moves to the west with the trade winds (refer to Figure 7-6) at about 30 to 50 km (18.6 to 31.1 miles) per hour. Hurricanes will continue to move as long as the ocean water below remains warm enough to keep the air rising into the eye of the storm. Hurricanes forming in the Atlantic Ocean off the coast of Africa typically move across the ocean for 1 to 2 weeks until they reach land in the southeastern United States. Once on land, the hurricane weakens because the ocean's warmth and moisture are no longer providing a source of energy for the storm. Hurricanes also become weaker when they drift into higher latitudes and move over cooler ocean water.

The earth's rotation also creates a counterclockwise spin of the air around the eye of hurricanes in the Northern Hemisphere at the surface. Aloft, the spin of the air is clockwise as it exits the storm (Figure 7-19). The spinning motion around the center of the hurricane is very strong, with the strongest winds and heaviest rain near the storm's eye. The strongest hurricanes, such as Hurricane Ivan in 2004, are able to sustain winds of over 240 km (150 miles) per hour near the eye. The eye itself, where the air is rising vertically, is eerily clear and calm with very little wind—just some weak downdrafts. If it could be converted to power, the amount of energy swirling around the eye of a single hurricane could meet U.S. electricity needs for several years.

Figure 7-19 Hurricane several hundred kilometers east of the Florida coastline.

Tornadoes

The low-pressure systems typifying temperate regions move from west to east with the westerlies (refer to Figure 7-6). As with hurricanes, the Earth's rotation also creates a counterclockwise spinning motion around these low-pressure systems. As a storm moves west

Figure 7-20 (a) United States air circulation patterns that lead to the formation of tornadoes in the Great Plains. (b) How cold air wedges underneath warm air.

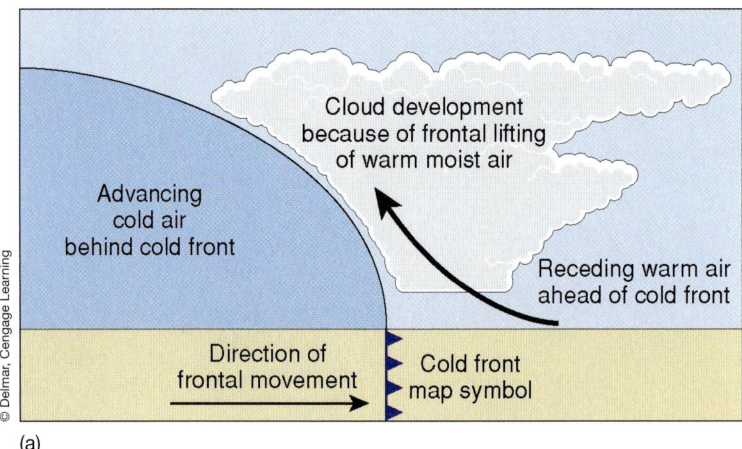

(a)

(b)

to east across the United States, the counterclockwise rotation draws down cold, dry air masses that form in the western interior of the United States and Canada (Figure 7-20). During the winter, this cold air can drop temperatures well below freezing over much of the Great Plains and eastern United States.

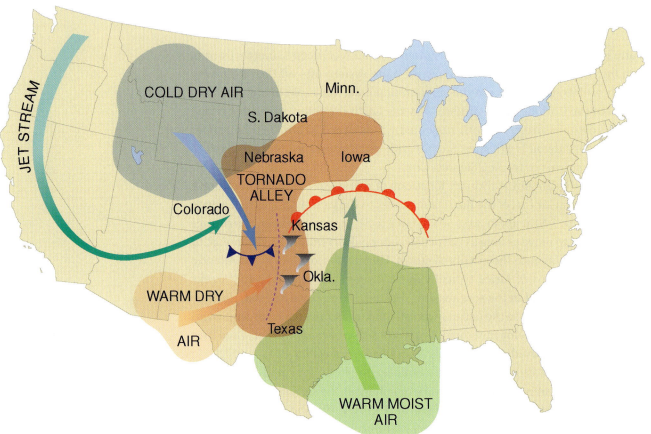

In the spring, these cold, dry air masses may collide with warm, moist air masses being drawn into the Great Plains from the Gulf of Mexico by clockwise circulation around a high-pressure area over the Atlantic Ocean near the island of Bermuda. Given its position over Bermuda nearly every spring and summer, this high-pressure area is known as the Bermuda High. When the cold and warm air masses collide over the Great Plains, the denser cold air wedges underneath the warm air and the warm, moist air begins to rise. As the warm air rises, it cools, water vapor condenses, and rain develops. When the temperature and moisture differences between the colliding air masses are great, the lifting of the warm, moist air is more rapid and creates very unstable conditions in the atmosphere that result in thunderstorms. Strong rising air in a thunderstorm draws surface air into the updraft. Wind has rotation that is parallel to the surface due to such things as different wind speed at different heights above the surface. This is expected due to surface obstructions and resistance at the surface and fewer obstructions with height. When drawn toward the rising updraft in the storm, it becomes vertical rotation. This spinning motion will sometimes be focused into a small area and form visible funnels at the base of the clouds that sometimes reach the ground surface (Figure 7-21)—these are **tornadoes**.

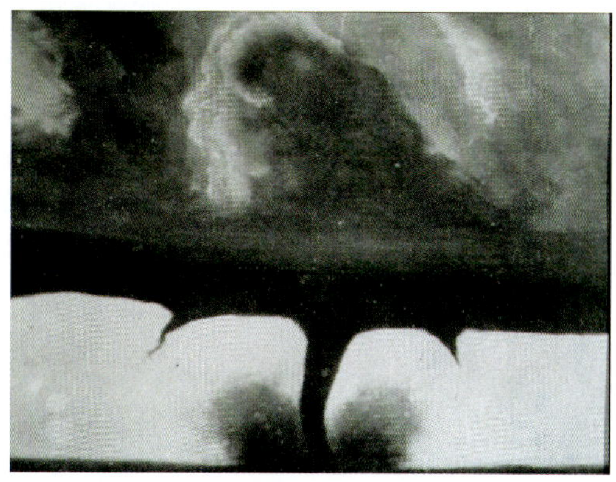

Figure 7-21 Oldest known photograph of a tornado with multiple funnel clouds. (*Courtesy NOAA, National Weather Service*)

Tornadoes typically average only 150 meters (492.1 feet) across and remain on the ground for less than 10 km (6.2 miles). Although smaller in size, wind speeds created by the air spinning around the center of a tornado are much stronger than those in hurricanes. Top wind speeds in a tornado can be over 500 km (310.7 miles) per hour. Tornadoes also move forward faster than hurricanes with some moving across the Earth's surface at over 100 km (62.1 miles) per hour. Tornadoes are short lived compared to hurricanes, with the longest tornado having lasted only 7 hours 20 minutes on March 13, 1917. This record tornado traveled for 754.8 km (469 miles) through Missouri, Illinois, and Indiana.

Tornadoes have occurred in all 50 states but are most common in the midwestern United States during the spring and fall when the temperature differences between the colliding cold and warm air masses are greatest. Because the air in polar regions is colder in the spring after the long winter, tornadoes are typically much stronger and widespread during this time of year. Tornadoes are more common to the south in the early spring and become more common further north later in the spring. This is because the air coming from the north is not as cold as summer approaches.

Although tornado development is most dramatic and common in the Midwest where the cold, dry air and warm, moist air collide, tornadoes can develop in any setting where intense warming of the land surface can cause the air to rise and generate a strong updraft. Because hurricanes and monsoons are also areas of warm rising air, small tornadoes are frequently associated with these much larger weather systems.

Droughts

Stormy weather and rainfall during monsoons, hurricanes, and tornadoes are associated with low-pressure areas where warm air is rising. High-pressure areas, in contrast, are associated with dry weather. Generally, dry weather is considered to be a good thing—the perfect day for a picnic or working outside. Too much of a good thing, however, begins to create problems. Long periods of high pressure centered over the same area can lead to a **drought**, which is an extended period of below-normal rainfall.

Like low-pressure systems, high-pressure areas also move with the global wind patterns (refer to Figure 7-6). This means that high-pressure areas usually move west to east across the United States with the westerlies. In some cases, these high-pressure areas become stationary, remaining in the same area for weeks or months, and prevent rain from falling.

When a high-pressure area becomes stationary, the air continues sinking at the same place, compressing the air below. This compression heats the air up more and more and further prevents rainfall from occurring. Stationary high-pressure systems can occur anywhere but are more common in the tropical deserts between 15 and 30 degrees latitude.

 people and severe weather

People must cope with the weather every day. In most instances, people are insulated against poor weather conditions by staying indoors in heated or air-conditioned homes, bringing an umbrella if going outside on a rainy day, or wearing a windbreaker if the breeze is a bit strong. Long-term climate conditions, in fact, control where people decide to live. Population densities are generally very low in areas with extreme temperatures and very low rainfall. Severe weather, however, can occur in areas with pleasant climates and high population densities. These severe weather conditions threaten people's lives, cause extensive property damage, and place resources such as water supplies at risk.

The severity and extent of the problems vary with the weather conditions involved. Understanding the problems associated with each severe weather condition is essential for limiting the damage and loss of life caused by them. What types of severe weather occurs in your community? What actions can you take to help save lives and prevent damages associated with these weather hazards?

Impacts of Hurricanes

The westerly movement of hurricanes that form in the Atlantic Ocean means that they commonly make landfall in the southeastern United States along both the Atlantic and Gulf of Mexico coastlines (Figure 7-22). Hurricanes also affect California and Arizona if they originate in the eastern Pacific Ocean far enough north to be caught in strong westerlies that carry the hurricanes eastward. Damage, injury, and death during a hurricane, wherever they make landfall, can result from heavy rains, strong winds, and storm surges. Damage from storm surges is restricted to coastal areas, but strong winds and heavy rain can move tens to hundreds of kilometers inland.

Homes, if not properly protected, can lose roofs, siding, and windows due to strong winds, flying debris, and/or fallen trees. Trees blown over by the wind can fall on power lines and across roads (blocking emergency crews from helping injured people) and crush cars that are, unfortunately, in the wrong place.

Rainfall totals in excess of 38.1 cm (15 inches) are not uncommon during a single hurricane. Rains falling in 2 days or less from a hurricane can sometimes exceed more than half of the normal annual precipitation. Because of the heavy rainfall, flooding is a major problem during hurricanes. In coastal areas, the flooding can be worsened by storm surges where seawater is pushed inland by the hurricane's strong winds (see Chapter 5). Over 90 percent of deaths from hurricanes are caused by drowning due to coastal and river flooding. Such was the case on Galveston Island, Texas, in 1900 when a hurricane storm surge flooded the island and killed more than 8,000 residents (see Chapter 5). How many students are in your school? Eight thousand is a big and, unfortunately in this case, tragic number.

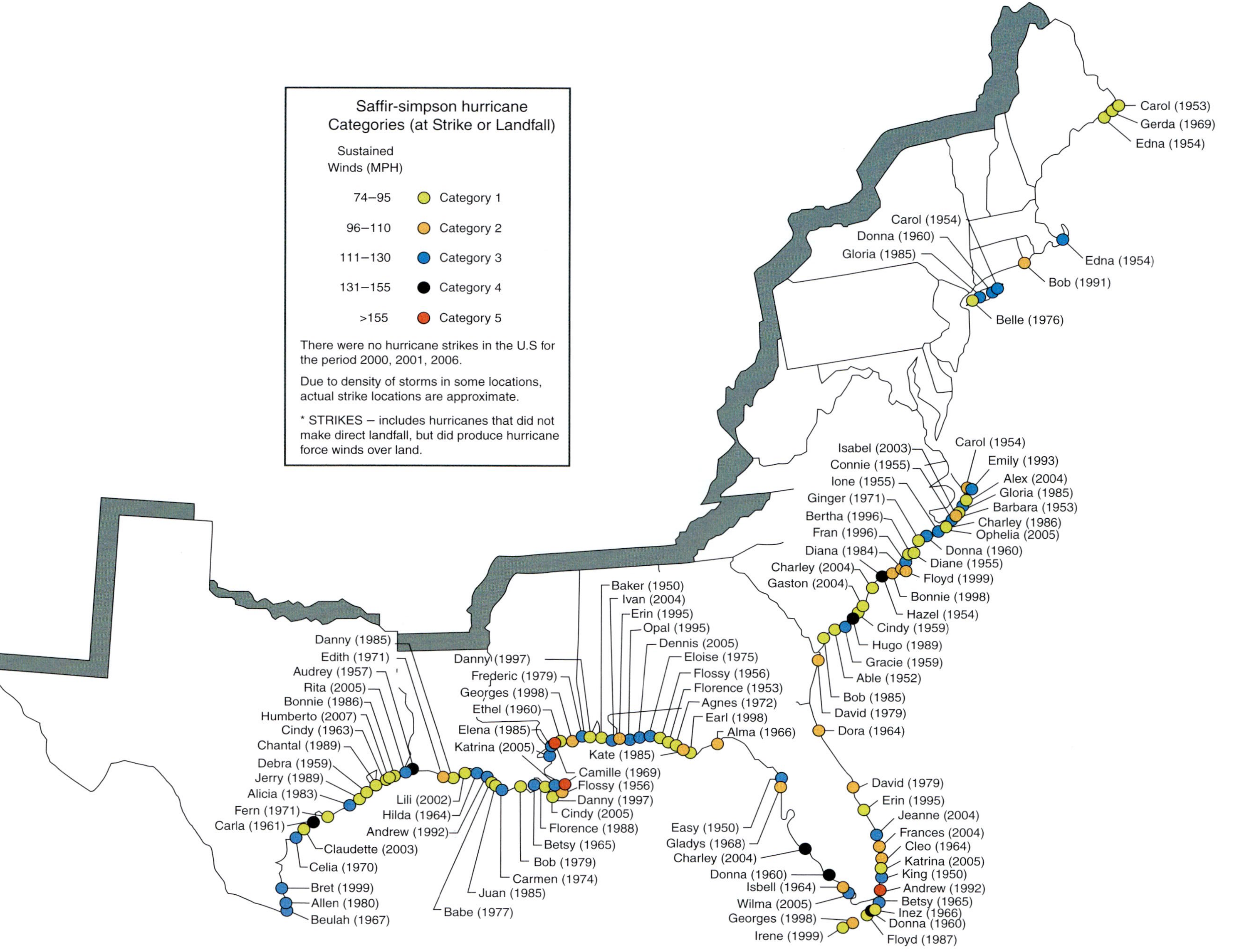

Figure 7-22 Places where hurricanes made landfall in the eastern United States between 1950 and 2007.

Courtesy NOAA, National Climatic Data Center

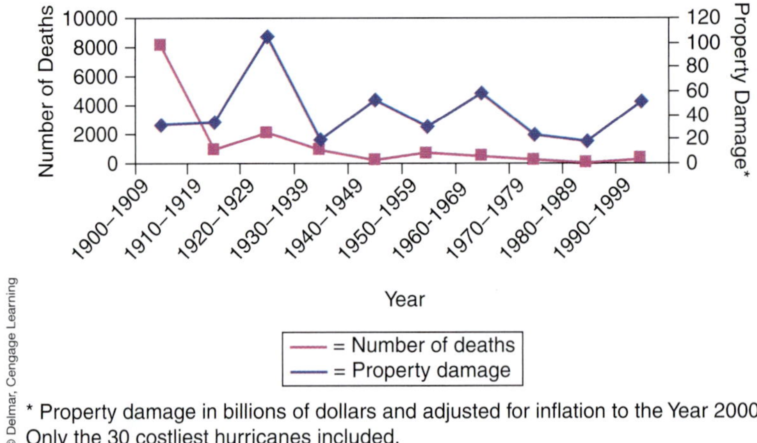

Figure 7-23 20th century tally of deaths and damage costs from hurricanes.

Deaths related to hurricanes have decreased in the past 100 years (Figure 7-23). However, the amount of hurricane-caused property damage has not decreased. Damage and deaths due to hurricanes can be minimized through accurate storm forecasts. An ideal hurricane forecast predicts where the hurricane will strike land, when the hurricane will strike, and how strong the hurricane will be when it makes landfall. Improvements in the science of hurricane forecasting are largely responsible for the decreasing numbers of deaths resulting from hurricanes. Before the advent of satellite photography, hurricanes went largely undetected until they approached land. Coastal residents and sailors were familiar with the old adage, "Red sky at morning, sailors take warning, red sky at night, sailor's delight." A red sky in the eastern morning sky would mean the sun was rising through clouds. The only cloudy stormy weather that moves from west to east in the southeastern United States is associated with tropical storms and hurricanes.

Ships out at sea provided some additional warning time, but these observations did little to predict where, when, and how strongly the storm would hit land. Ship reports would indicate the storm's location but the reports from several ships sometimes proved quite confusing when more than one storm was active at the same time. To avoid this confusion, a system was established in the early 1900s to name hurricanes by giving the first storm of the season a name beginning with the letter "A," the second storm a name beginning with the letter "B," and so forth through the alphabet. This practice continues to this day and, on average, about 10 storms are named in a single season.

Beginning in 1943, planes were also used to fly directly into hurricanes to take measurements of wind speeds and air pressure. These "hurricane hunters" from the U.S. Air Force Reserve still provide valuable information on the strength of hurricanes several days before they make landfall. However, not until 1966, when satellites became available, was the position of a hurricane able to be tracked carefully enough to understand how hurricanes moved and provide more accurate warnings of when and where their landfall would occur. With a "bird's eye" view from space, the position of each hurricane can be tracked across the ocean (Figure 7-24). What do you notice about the tropical storm paths in the Atlantic Ocean during 1998? First, not all hurricanes make landfall. Second, almost all hurricanes in the Atlantic Ocean move to the north as they move eastward. Because the amount of northward movement varies with each hurricane, the location of landfall can vary dramatically.

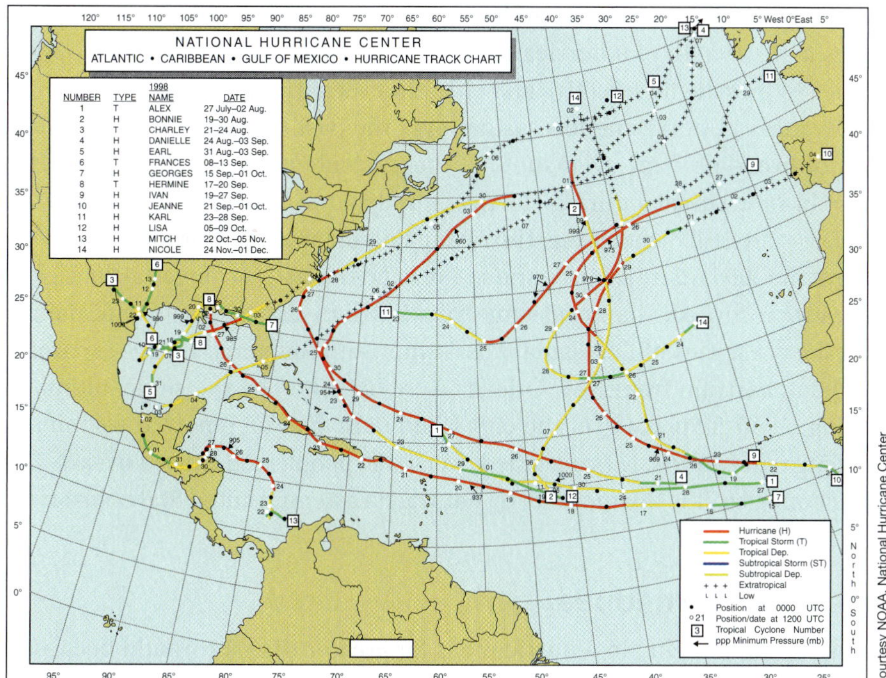

Figure 7-24 Tracks of hurricanes and tropical storms during the 1998 hurricane season in the Atlantic Ocean.

Hurricane warnings given 3 days in advance can predict the location of landfall only within 320 km (200 miles) of either side of the hurricane. In other words, if the prediction states that the hurricane will hit the coast at Charleston, South Carolina, the actual place that the hurricane hits land 3 days later may be 320 km (200 miles) to the north or south of Charleston. Evacuating a 640 km (400 miles) stretch of coastline to save people from the potentially life-threatening hurricane is extremely expensive and, ultimately, would be unnecessary for most of this area. Extremely dangerous conditions are only likely within a 160 km (100 miles) stretch of a hurricane's landfall. Predictions made 24 hours in advance are far more accurate and predict the location of landfall within 130 to 160 km (80 to 100 miles). Evacuating this shorter stretch of coastline is more cost-effective; however, emergency officials like at least 36 hours to warn residents and coordinate the evacuation.

Predictions of where the hurricane will make landfall do not provide information on how strong the hurricane will be when it reaches the coastline. Hurricanes become much stronger very quickly if they pass over areas of warmer ocean water. Measurements from hurricane hunters flying through a hurricane 2 days prior to landfall may greatly underestimate the strength of the hurricane if it later passes over an area of much warmer water. Recently, though, an approach has been developed to identify these areas of warm water and, therefore, enable predictions of when and by how much the hurricanes will increase their strength.

Newly developed satellite technology can measure minute differences in the height of the ocean. Just as warm air is less dense and rises, warm ocean water is also less dense than cooler water and also rises slightly above the cooler ocean water surrounding it. Detection of slightly higher levels on the ocean surface is an indication of warmer ocean water. If a hurricane passes over one of these warmer zones near the coastline, the hurricane will strengthen rapidly just before making landfall. This latest most sophisticated piece of technology has now advanced the science of hurricane forecasting to where the location, timing, and strength of a hurricane can be reasonably predicted far enough in advance to provide for effective and efficient evacuation.

Advances in hurricane forecasting over the past century have saved thousands of lives—perhaps even some of the students at Archbishop Shaw High School in New Orleans during Hurricane Katrina in 2005 (see Chapter 1). But hurricanes are still devastating. Hurricane Katrina was the third most powerful storm of the 2005 Atlantic hurricane season, and the sixth-strongest Atlantic hurricane ever recorded. The storm surge from Katrina caused catastrophic damage along the coastlines of Louisiana, Mississippi, and Alabama. Levees separating Lake Pontchartrain from New Orleans were breached by the surge, ultimately flooding about 80 percent of the city. Wind damage was reported well inland, impeding relief efforts. Katrina killed over 1,400 people and caused over $80 billion in damages. It is the most costly natural disaster in United States history.

Impacts of Tornadoes

The likelihood of being in a hurricane is far greater than a tornado because of a hurricane's immense size. However, for those caught in a tornado, the experience is far more dangerous because of a tornado's much stronger winds. Lives are frequently lost, with an average of 50 deaths occurring each year during the approximately 1,000 tornadoes that develop annually in the United States. The single deadliest tornado was the tristate tornado on March 18, 1925, when 695 people died in southeastern Missouri, southern Illinois, and southern Indiana.

Courtesy NOAA, National Weather Service

Figure 7-25 Damage to a home from a tornado.

Damage in tornadoes is primarily the result of very powerful winds spinning around the center of the tornado. A tornado can completely level all buildings in its path or lift cars and homes from their foundations (Figure 7-25). A tornado in Minnesota on May 27, 1931, is reported to have lifted a train with five coach cars, each weighing about 70 tons, off its track. Hurricanes cover a large area, but the winds are never strong enough to lift a five-car train off its tracks!

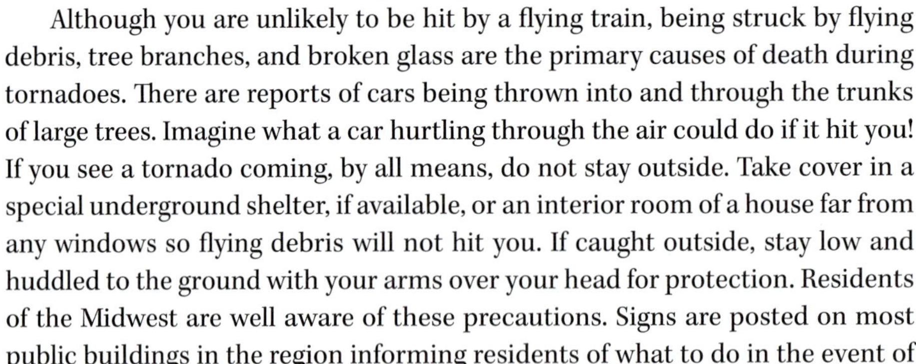

Although you are unlikely to be hit by a flying train, being struck by flying debris, tree branches, and broken glass are the primary causes of death during tornadoes. There are reports of cars being thrown into and through the trunks of large trees. Imagine what a car hurtling through the air could do if it hit you! If you see a tornado coming, by all means, do not stay outside. Take cover in a special underground shelter, if available, or an interior room of a house far from any windows so flying debris will not hit you. If caught outside, stay low and huddled to the ground with your arms over your head for protection. Residents of the Midwest are well aware of these precautions. Signs are posted on most public buildings in the region informing residents of what to do in the event of a tornado.

More than twice as many people have died in tornadoes than in hurricanes since 1960. How can this be true if tornadoes are rarely more than a mile wide but hurricanes frequently are more than 482.8 km (300 miles) in diameter? First of all, far more tornadoes occur each year than do hurricanes. Also, when and where a tornado will form is very difficult to predict. General conditions conducive to tornado formation are known but pinpointing exactly when and where they will occur is still elusive. Consequently, evacuation of residents before a tornado hits is commonly not possible. Once a tornado is spotted, it is not safe to try and evacuate the area because of the risk of being hit with flying debris from the strong winds.

The best way to remain safe and limit the damage to your home is to reside and remain in a specially engineered house meant to withstand strong winds. How can a tornado-resistant home be constructed? The use of foundation anchors spaced 1.2 to 2.4 meters (4 to 8 feet) along the foundation can bolt the house safely to a concrete foundation so that it is not blown away and left swirling in the wind like Dorothy's house in the *Wizard of Oz*.

Tornadoes can also lift roofs off houses. To strengthen the attachment of the roof to the rest of the house, many homes are equipped with hurricane clips. As the name implies, these work well in hurricane-prone areas too. The clips are made of galvanized steel and work best if they attach the roof to the exterior walls every 1.2 meters (4 feet) around the house.

A final piece of tornado engineering included in many new homes built in tornado-prone areas is the construction of in-residence shelters. Small rooms like a bathroom in the interior of a home are usually the safest and least damaged. If these interior rooms are reinforced they can withstand 400 km (250 miles) per hour winds. The walls and roofs of the in-residence shelters must be able to withstand flying debris hitting the structure at 150 km (90 miles) per hour or more. These flying missiles can puncture through conventional walls and roofs, so in-residence shelters are constructed with 15 to 20 cm (6 to 8 inches) thick concrete masonry walls reinforced with steel bars. Other possibilities include walls constructed of multiple layers of alternating plywood and sheet metal. All of the engineering designs to make homes safe in tornadoes also work to withstand the winds of a hurricane that are generally not as strong but often last longer than those in a tornado.

Very few homes, even the best engineered ones, can withstand a direct hit by a powerful tornado. To protect against the low likelihood of damage by a tornado, many people purchase tornado insurance. Insurance companies can afford to pay the cost of rebuilding damaged homes, because a large number of homeowners pay a small amount to insure themselves against the unlikely event that their home will be damaged in a given year. An insurance company may lose money during a year with lots of tornadoes and associated damage, but those years are offset in other years when very few tornadoes occur.

Impacts of Droughts

Tornadoes, hurricanes, and other severe low-pressure system storms create damage and serious problems on time scales of minutes, hours, or perhaps several days. Droughts, on the other hand, do little damage at first. Weeks of dry weather must persist before the damage associated with a drought appears; the worst impacts result from droughts that last several years.

Droughts can cover large regions of the United States at any one time—the midwest, the southeast, the northeast, for example—but usually the conditions that lead to drought in one region result in higher than average rainfall in another. The persistent development of high-pressure regions in the drought-stricken area tends to steer moisture-laden air around it. Typically, the

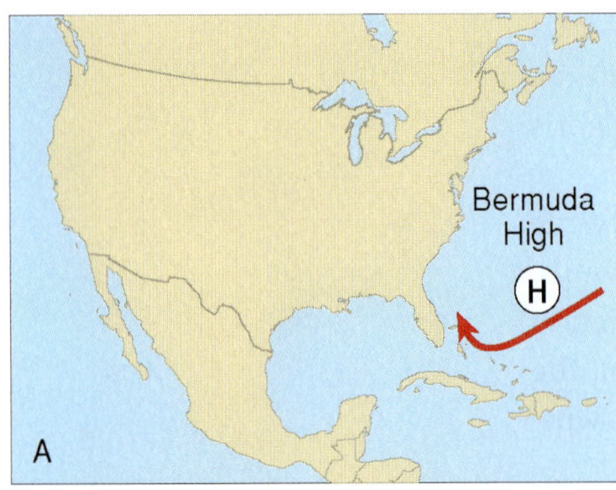

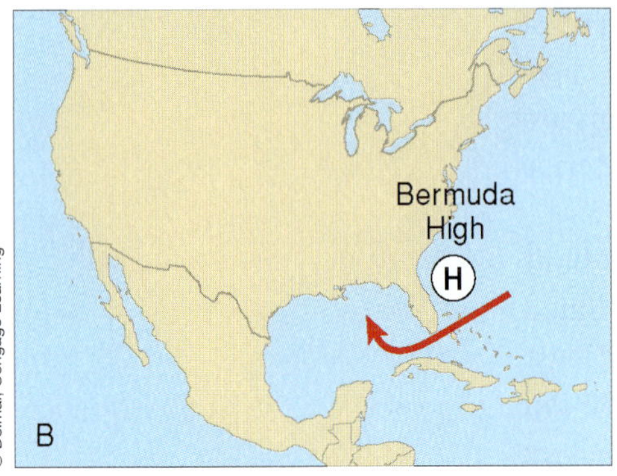

Figure 7-26 (a) Normal position of the Bermuda High, leading to summer rainfall in the southeastern United States. (b) Displaced position of the Bermuda High closer to the coastline, leading to droughts in the southeastern United States.

Bermuda High forms well off the east coast of the United States, forcing moist ocean air toward the southeastern United States. However, during years when the Bermuda High is closer to the American coastline, the southeastern United States is more directly under the influence of a high-pressure system (Figure 7-26). This pushes the moisture-laden air from over the ocean further west than normal and gives rise to wetter than normal conditions in the midwest while simultaneously creating drought conditions in the southeastern United States.

Crop losses can be severe during droughts. A drought in the central and eastern United States between June and September 1980 led to $20 billion in agricultural losses. Severe, persistent, and lengthy droughts can also cause deaths due to the accompanying high temperatures. The 1980 drought that caused so much crop damage is also believed to have caused over 10,000 deaths over a 3-month period—not in a few hours or in a small area, like the hurricane-induced storm surge on Galveston Island, Texas, in 1900, but no less tragic in its human toll. Heat-related deaths average 385 per year. This is more deaths on average than are caused by hurricanes, tornadoes, floods, and lightning combined!

Drought-related tragedies are even worse in poor countries unable to cope with the resulting crop losses and water shortages. The worst area of drought in the past 30 years has been the African Sahel just south of the Sahara Desert. Rainfall has been down 20 to 50 percent across this portion of Africa, with particularly dry years in 1972, 1975, 1984, and 1985. The resulting crop failures have led to the starvation of perhaps more than 1 million people in Ethiopia alone. Droughts do not have the immediate punch of a tornado or hurricane, but their slow accumulation of problems can be devastating to the people that must endure them.

Recognizing the conditions that might lead to drought a year or two in advance of its occurrence can allow for plans to cope with reduced agricultural crop yields, potential water shortages, decreased stream flows that might reduce power generation, increased forest fires, and increased dust from wind-blown soil. The creation of water resource control plans can help alleviate conflicts between different water users when supplies are low during droughts.

Several states have enacted laws to prepare for water shortages during droughts. The Water Resources Planning Act was passed in Pennsylvania in 1992 after several years of repeated droughts. The law provides funds for determining how much water is available for use, how much is needed by various users, and where within the state shortages might develop if water use exceeds the amount available. Perhaps, most importantly, the law creates a committee whose members represent all of the various water-using stakeholders. The committee is charged with deciding how much water each stakeholder can use during periods of drought. These committee meetings are no doubt filled with much discussion, but water allocation decisions made beforehand can help avoid damages and economic losses associated with droughts.

Areas of drought in some parts of the world appear to coincide with the El Niño weather pattern. Droughts in India, southern Africa, and Central America seem to occur when ocean water temperatures are warmer than normal in the eastern equatorial Pacific Ocean. A better understanding of when and how El Niño forms will help alleviate problems related to droughts because several months, perhaps in the future several years, will be available to plan for expected droughts. The rainfall data being collected by students at the over 160 schools involved in the Schools of the Pacific Rainfall Climate Experiment may one day help that goal be realized.

 human impact on the atmosphere

An **air pollutant** is any gas or particle released into the atmosphere at a concentration that has a negative effect on human health or the health of the environment. Some air pollution has natural origins, such as volcanic eruptions, lightning-triggered forest and grass fires, and the decay of marshland organisms. Although there are natural causes, it is human activity that produces the most significant amount of the pollutants that are of greatest concern today. On a global level, these air pollutants contribute to several environmental problems, including acid rain, photochemical smog, and the depletion of the ozone layer.

Precipitation that is more acidic than normal is called **acid rain**. Acid rain forms when sulfur dioxide and nitrogen oxides react with water vapor in the air, forming nitric acid and sulfuric acid. The acids fall to the ground dissolved in precipitation in the form of snow, sleet, or fog as well as rain.

The sulfur dioxide that contributes to acid rain forms whenever sulfur-containing fossil fuels, such as coal and oil, are burned. These fuels are commonly burned in power plants that produce electricity, but certain industries such as oil refineries and paper mills burn them as well. Of particular concern in the United States are coal-burning power plants in the midwest. These plants burn coal that contains significant amounts of the mineral pyrite (FeS_2) and other sulfur-bearing compounds. The sulfur dioxides that these plants generate are carried by upper-level winds blowing from a generally westerly direction toward the eastern coast of the United States and Canada. As a result, the acids that form in the atmosphere return to the Earth's surface as acid rain in areas far from their source.

Most of the nitrogen oxides that contribute to acid rain are generated when atmospheric nitrogen combines with oxygen at high temperatures in the engines of motor vehicles. A significant amount of nitrogen oxides are also formed by the burning of fossils fuels in power plants.

Acid rain can harm both plant life and animal life. It can change the conditions of aquatic ecosystems and vegetation in lakes, streams, ponds, and rivers. For example, many fish, and particularly their eggs, cannot survive in acidic water. Acid rain also damages plants by stripping away nutrients in soil needed by plants to grow. This loss of nutrients limits the growth of trees as well, making them weak and more susceptible to damage from pests and disease.

Photochemical smog is a thick, brownish haze that shrouds many urban areas. It is formed when nitrogen oxides and hydrocarbons in the air react with sunlight. Like acid rain, the nitrogen oxides that contribute to photochemical smog are generated by motor vehicles. Hydrocarbon pollution comes from motor vehicles as well, mainly from the evaporation of gasoline from carburetors, crankcases, and gas tanks. In addition, unburned hydrocarbons are released from motor vehicle exhaust.

The major component of photochemical smog is **ozone** (O_3), a gas molecule made up of three oxygen atoms. It is produced by the reaction of nitrogen oxide and hydrocarbon gases in sunlight. Because these reactions are powered by the sun, the greater the intensity of sunlight and the warmer the day, the larger the amount of ozone produced. On cloudy days, ozone production is reduced and after the sun sets, it stops altogether. Ozone levels also follow traffic density patterns, gradually rising in the morning and peaking between noon and 4 P.M.

A number of harmful effects result from exposure to photochemical smog and ozone, its main pollutant. Ozone irritates the eyes, nose, and throats of humans. It is also a powerful lung irritant that can cause respiratory problems and permanent lung damage, not just to humans but animals as well. Ozone is harmful to plants because it interferes with photosynthesis. As a result, plant growth is stunted. This effect can reduce crop yields and hurt the agricultural industry.

The severity of photochemical smog depends on weather conditions. As you have learned, warm air heated by the Earth's surface normally rises into cooler air above. The rising air carries air pollutants, such as nitrogen oxides and hydrocarbons, higher into the atmosphere where they are blown away. Sometimes, however, a **temperature inversion** occurs where a warm air mass moves into a region of cold air. The warm air rises above the colder air near the Earth's surface. The warm layer prevents the colder air from rising higher into the atmosphere and dispersing its pollutants. In these instances, smog can build up to dangerous levels in the trapped air. A thermal inversion can last several days and only wind or rain can break up the top layer of warm air.

Although ozone close to the Earth's surface is a pollutant, ozone about 30 km (18.6 miles) above the Earth serves as a protective shield as it absorbs and filters out some of the sun's harmful ultraviolet radiation. This ozone exists in a layer of the upper atmosphere called the ozone layer. Exposure to ultraviolet radiation is dangerous to humans because it can cause certain types of skin cancer, cataracts, premature skin aging, and weakening of the immune system. Ultraviolet radiation can also harm crops and destroy sensitive marine life such as phytoplankton.

When sunlight strikes an ozone molecule in the ozone layer, ultraviolet radiation is partly absorbed as the molecule breaks apart into an oxygen molecule and an oxygen atom. When the oxygen atom collides with an oxygen molecule, they react to form a new ozone molecule. Ozone is constantly being made and destroyed through this process. Each time this cycle occurs, some ultraviolet radiation is absorbed and prevented from reaching the Earth's surface.

In the mid-1980s, scientists discovered a thinning of the ozone layer, including an extremely thin area over Antarctica. The destruction of the ozone layer was linked to the human manufacturing of gases containing chlorine and fluorine, called **chlorofluorocarbons**, or "CFCs." These gases were widely used in products such as aerosol spray cans, air conditioners, refrigerators, fire extinguishers, and cleaning agents. CFCs react with ozone molecules, blocking the cycle that absorbs ultraviolet radiation, thereby allowing the radiation to pass through the atmosphere and to the Earth's surface, increasing exposure to humans and the environment.

Controlling Air Pollutants

Air pollutants, such as nitrogen oxide, sulfur dioxide, hydrocarbons, and CFCs are difficult to control because they are dispersed by the wind. Pollution produced in one area travels to neighboring regions where no pollution is generated. For this reason, solving air pollution problems in the United States can only occur with the cooperation of state and local governments. On a global level, the collaboration of national governments is required.

Over the last 20 years, the governments of many nations have met several times to come up with plans for reducing global air pollution. For example, in 1987, more than 170 nations signed the Montreal Protocol on Substances that Deplete the Ozone Layer. These governments agreed to restrict the use of CFCs gradually. By 1996, all industrialized nations halted production. Measurements taken in the upper atmosphere have shown that the level of CFCs is beginning to decrease. However, scientists do not expect the ozone layer to return to normal until about 2050.

In the United States, Congress has passed numerous laws to reduce air pollution. For example, the Clean Air Act was originally passed in 1963. This act dealt with reducing air pollution by setting limits on the amounts of air pollutants that power plants and factories could emit into the atmosphere. It did not, however, take into account air pollution by motor vehicles, which had become the largest source of many dangerous pollutants. Amendments to the Clean Air Act through the 1960s and 1970s set standards for the air pollutants emitted by motor vehicles, as well as authorizing research on alternative fuels and motor vehicles that emit fewer air pollutants. After a lengthy period of inactivity, the Clean Air Act was again amended in 1990. It strengthened and improved previous amendments with further reductions in air pollutants emitted by power plants, factories, and motor vehicles. It also set dates and enforcement policies for achieving reductions.

In the United States, there are several ways that air pollutants are monitored so that the regulations set by the Clean Air Act can be enforced. The first is the use of small plastic tubes, open at one end to the atmosphere and with a chemical absorbent at the other end. Air is exposed to the absorbent over a period between a week and a month, after which the tubes are collected and analyzed in a laboratory. This method is inexpensive and more commonly used to measure air pollution at specific spots such as major roads or close to a factory. The second method involves collecting air samples and bringing them back to a laboratory for examination. The samples of air are typically pumped through a filter or chemical solution, which is then analyzed. The final method is the most expensive because it involves sophisticated electronic equipment in which air samples are exposed to electromagnetic radiation. Air pollutants are detected based on the extent to which they absorb or reflect the radiation.

There are two main ways that air pollution can be controlled. The first is by using devices that capture air pollutants already created. For example, today, motor vehicles are equipped with catalytic converters; these are devices that change harmful gases to less harmful ones. Hydrocarbon and nitrogen oxide gases from the engine pass over small beads coated with metals inside the catalytic converter. The metals cause chemical reactions that convert these gases into nitrogen, carbon dioxide, and water. Power plants and factories use devices called scrubbers on their smokestacks to control pollution. Scrubbers are most widely used to remove sulfur dioxide from smokestack gases. They work by spraying a mist consisting of ground limestone and water into the gases flowing from the smokestacks. The mist reacts with sulfur dioxide to produce calcium sulfate sludge, a solid that can be removed and disposed of.

The second method for controlling pollution is by limiting the amount of pollutants produced in the first place. For example, the air pollutants produced by cars would be less if there were fewer cars on the road. Another option is to increase the fuel efficiency of automobiles, or even use alternatives to the internal combustion engine, such as electric cars. Power plants and factories can reduce air pollutants by shifting from high- to low-sulfur coal. Sulfur can also be removed from high-sulfur coal before burning it. The public can help by reducing the amount of electricity they use, which means less fuel is burned at the power plant and less pollution is released.

conclusion

Weather affects our lives every day and we must be constantly aware of the changes in the weather because of the potential for severe storms to threaten lives and property. The long-term consistency of weather patterns determines climates and the impact of those climates on the distribution of plants, animals, and people around the world. Many people live where severe weather is possible. Are you one of them?

IN THIS CHAPTER YOU HAVE LEARNED:

- Weather consists of the daily changes in air temperature, pressure, moisture content, and movement. Climate is the long-term average weather conditions in an area.
- Weather and climate are mostly due to changes in the very dynamic lower part of the atmosphere—the troposphere.
- Most of the air movements and changes in the troposphere are caused by energy from the sun. The sun's heating of lands and oceans through the seasons causes differences in overlying air temperature and density. Hot air rises and colder air sinks.
- Low-pressure areas in the atmosphere are developed where air rises, and high-pressure areas are developed where it sinks. Air moves from high- to low-pressure areas. As the Earth rotates on its axis, air masses move regularly westward at low latitudes (trade winds) and eastward at high latitudes (westerlies).
- Organisms are adapted for successfully living in the Earth's diverse climates—from warm, humid equatorial rain forests to cold and largely frozen polar regions.
- Severe weather, including monsoons, hurricanes, tornadoes, and droughts, causes many problems for people. Severe weather can destroy property and cause deaths every year. Severe weather effects are very expensive for society.
- Accurately predicting severe weather is increasingly possible, especially the location and strength of hurricanes. Accurate prediction can guide evacuations and give people time to find appropriate shelter from severe weather. Even predicting droughts can help people to better manage valuable water resources.
- People cannot control weather or severe storms, but they can anticipate them, prepare for them, and make individual choices that minimize their negative impacts.

KEY TERMS

Acid rain
Air mass
Air pollutant
Air pressure
Chlorofluorocarbons
Coriolis Effect
Drought
El Niño
High pressure
Hurricane
Low pressure
Monsoons
Ozone
Photochemical smog
Southern Oscillation
Temperature inversion
Tornadoes
Trade winds
Weather
Westerlies

REVIEW QUESTIONS

1. List the important characteristics in describing the weather.
2. Beginning at sea level, what are the layers of the atmosphere? What characterizes each layer?
3. What causes the different seasons?
4. What are the respective roles of density and Earth's rotation in atmospheric circulation?
5. In what ways are plants adapted to polar climates? Desert climates?
6. How does a hurricane form and travel?
7. How does a tornado form, and why is it so destructive?
8. What conditions can lead to droughts?
9. What are some of the effects of droughts on the biosphere?
10. Describe three ways in which humans have controlled air pollution.

chapter 8

urban environments

Biology classes at Desert Vista High School in Phoenix, Arizona, are determining how bird populations are affected by the growth and development of their city. As part of the Ecology Explorers Program, these students are helping Arizona State University researchers answer questions such as: Where do birds live in the Phoenix metropolitan area? What landscape designs attract birds? How do bird populations differ among commercial, residential, and industrial areas of the city? Initially, students investigate their school using binoculars and a detailed map to record the number and type of bird species they see at various places around the school. Once comfortable with collecting ecological data, students conducted field studies around their

homes or favorite location within the city. Some students investigated how having a pet or different types of vegetation around their homes affected the types and number of birds seen.

The results of the research conducted by over 2,000 students at 50 schools in Phoenix each year are shared with university researchers at a poster session held at Arizona State University's Center for Environmental Studies. The Ecology Explorers Program not only allows students to present their research to scientists, but they also begin to appreciate the complex natural ecosystem present right under their noses: along their streets, where they play sports, or outside the local mall.

What do you notice in a big city—large skyscrapers, the fancy old buildings, or statues of famous people (Figure 8-1)? Surrounding these structures, even in the largest cities, is a mostly unnoticed assortment of plants, animals, and insects that have adapted to, and even thrive in, the built environment. Students participating in the Ecology Explorers Program noticed far more life than they thought existed throughout the Phoenix Metropolitan area. What types of birds or other types of animals and plants do you see in your neighborhood? How do the types of plants and animals you see vary between the most developed part of your community and the more natural areas?

The variations you see in your community are similar to what occurs in communities around the world. The creation of an urban environment alters natural ecosystems. Those species best adapted to the ecological niches created by the urban setting can survive or even thrive. For example, ledges on

Courtesy: Laslovarga

(a)

Courtesy: Gerold Kogle

(b)

Courtesy: Ludovic Lepeltier and Damien Leblois

(c)

Figure 8-1 (a) Skyscraper in Bangkok, Thailand; (b) Ornamentation on the roof of the Helmsley Building in New York City; (c) Statue of Heroes in Budapest, Hungary.

Courtesy USFW Service, Photo by Craig

Figure 8-2 Peregrine falcon nesting on ledge of New York City skyscraper.

Courtesy: Jan van der Crabben

Figure 8-3 Rat in an underground sewer system.

buildings can mimic rocky cliffs that provide nesting for peregrine falcons (Figure 8-2), and rats can thrive in dark, moist sewer systems safe from predators (Figure 8-3).

Some species that adapt to urban settings come into conflict with the human residents. Wealthy residents of Park Avenue in New York City were not too happy when a red tail hawk built a nest on a ledge above the entrance to an apartment building and began to leave droppings on the sidewalk right by the front door. How can these conflicts be resolved? What are some examples of conflicts between the natural ecosystem and the urban environment in your community or the area near you? Programs like Ecology Explorers are a great way to understand the needs of birds and other organisms living in these special human-constructed environments. With these understandings, conflicts between people and the natural ecosystem can be mitigated and perhaps resolved.

In Chapter 3 you learned that most of the world's future population growth will reside in cities. Currently, 77 percent of Americans live in **urban areas***. These are defined by the U.S. Census Bureau as communities with populations of 2,500 or more, or if there are at least 390 people per square kilometer. This is far higher than in 1790, when only 5 percent of Americans lived in cities and towns (Figure 8-4). The migration of Americans into urban areas has sometimes occurred very rapidly. The population of the*

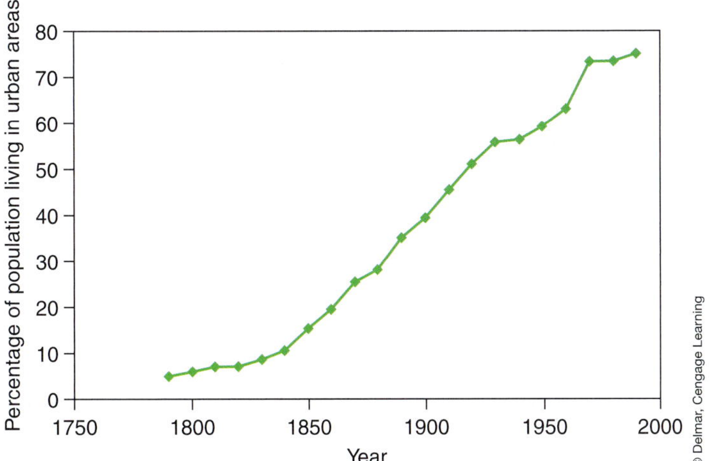

Figure 8-4 Percentage of Americans living in urban areas during each decade between 1790 and 1990.

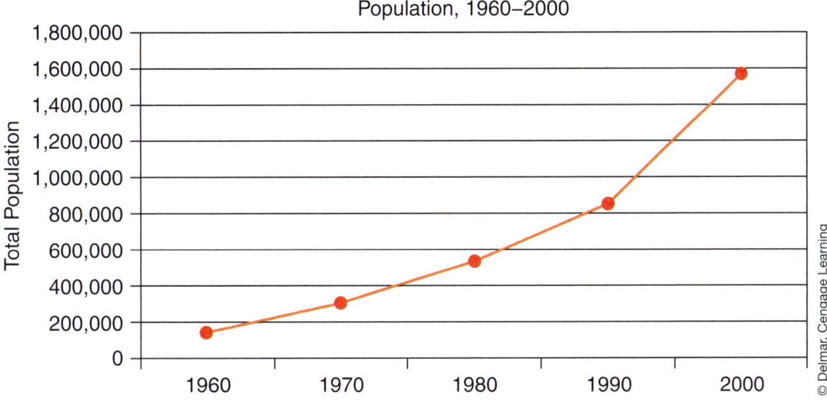

Figure 8-5 Population growth in Las Vegas between 1960 and 2000.

Las Vegas area, for example, grew more than 10-fold between 1960 and 2000 (Figure 8-5). Urban planners in Las Vegas and other rapidly developing areas of the world, whose job it is to decide where buildings and roads should be built, must ensure human safety and comfort while following laws that protect the environment. Whether as a city planner or just a city resident, it is important to understand the changes and interactions created by human development.

AFTER COMPLETING THIS CHAPTER, YOU WILL UNDERSTAND:
What urbanization is and why it happens.
How urban development affects Earth systems.
How urban settings affect the people that live within them.
How urbanization can be managed to protect Earth systems.

 8.1 urbanization

Urban environments have many impacts and influences. Because so many people now live in urban areas, and most of the increases in world population will come to live in very large cities (see Chapter 3), understanding **urbanization** is very important. To start, it is helpful to learn what urbanization is.

What Is Urbanization?

When you think of a city, what images come to mind? Tall buildings and crowded streets? People living in towns and cities are said to live in urban areas. How big is your community? Would you classify it as a town or city? Would another designation better describe your community?

City planners, geographers, and others who study where and how people live use three general classifications to describe the extent to which an area is developed: urban, suburban, and rural. Although a precise definition for these categories is difficult to agree upon, urban areas are typically more densely populated and have a greater number of buildings. Cities are urban areas, but so are smaller towns. Suburban areas are those places surrounding larger cities where single-family residential homes are common. Population densities are much lower in suburban areas, but much of the landscape is still affected by human development with crisscrossing road networks connecting homes and shopping areas. Rural areas are the remaining places in the "country" with large open areas, commonly farmland or forest, between homes. Population densities in rural areas are very low. With these loose definitions in mind, how would you classify the area you live in?

Differences between cities complicate defining what the urban environment is. Some cities are densely packed with tall skyscrapers, whereas others are spread out over great distances. New York City has 16 skyscrapers over 60 stories high in a total area of 750 square kilometers (289.6 square miles). In contrast, Mexico City's tallest building is 55 stories high but the city spreads out over 2,400 square kilometers (926.6 square miles). Use Table 8-1

Table 8-1 *Populations of the world's 20 largest cities as of 2004.*

Rank	City	Population	Year
1.	Mumbai (Bombay), India	11,914,398	2001
2.	Shanghai, China	10,996,500	2003
3.	São Paulo, Brazil	10,677,019	2003
4.	Seoul, South Korea	10,207,296	2002
5.	Moscow, Russia	10,101,500	2001
6.	Delhi, India	9,817,439	2001
7.	Karachi, Pakistan	9,339,023	1998
8.	Istanbul, Turkey	8,831,805	2000
9.	Beijing, China	8,689,000	2001
10.	Mexico City, Mexico	8,591,309	2000
11.	Jakarta, Indonesia	8,389,443	2000
12.	Tokayo, Japan	8,340,000	2003
13.	New York City, U.S.	8,085,742	2003
14.	Teheran, Iran	7,796,257	2004
15.	Cairo, Egypt	7,629,866	2004
16.	London, U.K.	7,712,036	2001
17.	Lima, Peru	7,029,928	2004
18.	Bogotá, Colombia	6,712,247	2001
19.	Bangkok, Thailand	6,320,174	2000
20.	Rio de Janeiro, Brazil	5,974,081	2003

to determine the densities of these two cities with similar populations. How well would a definition for an urban environment based on population density work in these cases?

In the absence of a clear method for defining what is meant by urban, most countries establish an arbitrary definition based on population. Whereas the U.S. Census Bureau defines an urban area as any town or city with more than 2,500 residents, other countries have chosen a different population figure (Table 8-2). Whatever the population chosen to distinguish between urban and rural areas, most people would agree that population centers with more than 1 million people represent an urban environment. Nearly 40 percent of the world's population, or 2 billion people, live in densely populated urban areas with a million or more people. It is truly an amazing feat of human ingenuity that water, food, shelter, electricity, and other services can be made available to millions of people in a single urban area such as Tokyo, New York City, Mexico City, or Shanghai. How many of the other 20 largest cities in the world can you name (see Table 8-1)?

The number of large cities has grown rapidly in the past century. In 1900, only 12 urban areas around the world had a million inhabitants or more. Now 400 cities have a population of a million or more. The number of people living in urban environments worldwide is currently under 50 percent but is expected to grow to over 60 percent of the world's population by 2030. With this dramatic expansion of urban settings, the surrounding Earth systems are becoming increasingly stressed. Soundly managing the growth of urban areas is one of the great challenges of the next century.

Urban Origins

People living in a city are much like organisms living in a natural ecosystem. Everybody performs a different task to keep a city going, just as all organisms have their own special niche in an ecosystem. By sharing responsibilities, the residents

Table 8-2 *The minimum population for towns in selected countries to be considered urban areas.*

Country	Minimum Population	Urban Population
Sweden	200	83%
Denmark	200	85%
South Africa	500	57%
Australia	1000	85%
Canada	1000	77%
Israel	2000	90%
France	2000	74%
United States	2500	75%
Mexico	2500	71%
Belgium	5000	97%
Iran	5000	58%
Nigeria	5000	16%
Spain	10,000	64%
Turkey	10,000	63%
Japan	30,000	78%

Courtesy Realman208

Figure 8-6 Gold mask found with mummy of King Tut in Egypt.

Courtesy Jastrow (2006)

Figure 8-7 Cuneiform tablet.

of a city can greatly improve their health and standard of living. But people did not always live in cities and divide up the work as they do today. Before the first cities emerged about 6,000 years ago, most people lived in small groups engaged in hunting and gathering their food. Little time was left to do much of anything else when people were spending their entire day collecting food. A reliable supply of food and water is needed in order for cities to develop and flourish. Today planes, trucks, and trains deliver these necessities to just about anywhere in the world; however, the first cities emerged in large river valleys where nearby water and soil resources combined to make productive farming possible (also see Chapter 14). The first city in the world, Uruk, developed about 6,000 years ago along the Euphrates River, in what is now Iraq.

The development of cities marked a major shift away from small groups of hunters and gatherers. As agriculture provided a way to provide large numbers of people with a reliable food supply, people, for the first time, had the freedom to spend time on activities other than food collection. Of course some people had to grow the food, as farmers do today, but because they could grow more than they needed, it could be shared with others. This allowed for innovations in technology, education, and government. People developed methods for using metals such as gold, silver, copper, and iron for creating tools, weapons, and jewelry (Figure 8-6). Writing emerged as a way of communication and documenting historic events (Figure 8-7). With these advances came the development of more complex societies.

The economic, technological, and educational opportunities that emerged with the world's first cities are still a major reason why people live in them today. Basic services such as electricity, public transportation, and public education are so widespread in modern cities that we scarcely remember that the governmental institutions that make them possible are at least partly modeled after the world's earliest urban centers. With their basic needs met, people can focus on a range of other activities. Cities enable people to do more. If someone wants to be a painter, businesswomen, surgeon, professional basketball player, or famous chef, he or she will likely find the best opportunities in a city.

 8.2 urbanization and earth systems

The emergence of cities created new environments never before seen on Earth. As discussed in Chapter 2, plants and animals occupy specific niches within a habitat. Urban environments place stress on organisms adapted to the preexisting conditions as new abiotic and biotic factors arise that organisms must adapt to. In some ways the homes, buildings, and paved surfaces provide the physical framework for the environment that creates new habitats for plants, animals, and insects. The physical structure, or skeleton, of the city that the organisms occupy is much like how coral reefs provide specific ecological niches for a diverse suite of species (Figure 8-8). The coral reef is built of a community of living organisms dependent on other organisms for their survival to protect them from predators, to provide food, and to remove waste. Urban areas provide this type of complex interaction

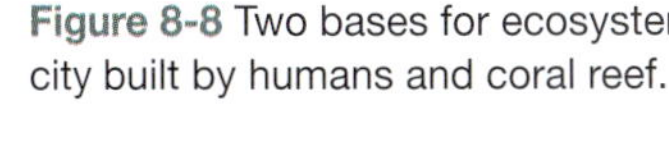

Figure 8-8 Two bases for ecosystems: city built by humans and coral reef.

(a)

(b)

in support of people and their needs. However, it must be noted that urbanization has generally led to the fundamental loss of habitat for species and loss of species' biological diversity as well as a decrease in species' richness and health. Animals thriving in urban areas are typically highly tolerant of pollution and degraded habitats and are not indicative of complex, sustainable, natural ecosystems.

The intensity of development in urban areas controls the amount and types of changes to Earth systems. A town like Decatur, Illinois, with a population of around 80,000 is considered urban but has a much different environmental impact than a large metropolitan area of nearly 7 million people like Chicago, Illinois. Atmospheric pollution from factories and millions of cars in large cities creates occasional health hazards, especially for residents with breathing problems. But, as summarized later, there are some common impacts on Earth systems from urban developments.

Biosphere Impacts

Urban developments have big impacts on the biosphere. They destroy, displace, or otherwise significantly modify natural ecosystems; develop new habitat that organisms adapt to; and create conflicts between people and other species trying to use the same space. Urbanization's impacts on natural ecosystems are the most obvious.

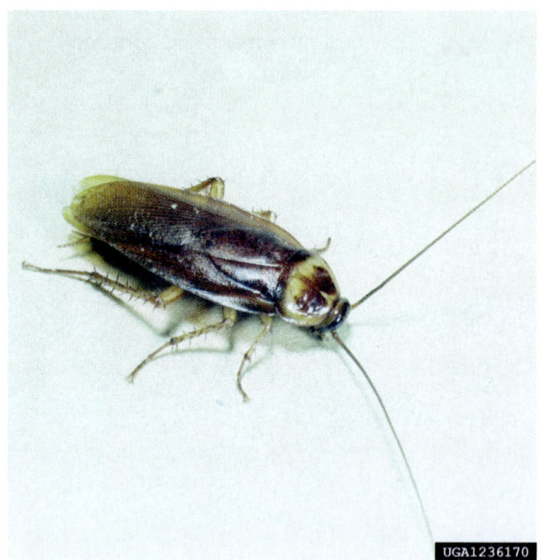

Figure 8-9 A cockroach (*Courtesy Clemson University - USDA Cooperative Extension Slide Series, Bugwood.org*)

Construction of the buildings, roads, and utility and communications systems that are needed in cities requires the removal of previous ecosystems. Wetlands get filled in, forests get cut down, and river channels become controlled around cities. The organisms once at home in these natural ecosystems either die out or move on; they become displaced.

But urban areas are not without plants, animals, and insects. Many species thrive in urban environments. Coyotes that live in Boston and other U. S. cities are an example (see Chapter 2). Falcons that use skyscrapers for nesting areas and city pigeons for food are another. Do you know of a city that does not have rats scurrying around in dark, dingy underground habitats? Perhaps some of the most successful adaptations are those of insects—spiders, yellow jackets, and termites, for example. The cockroach may be the most successful urban species of all (Figure 8-9).

Cockroaches can live just about everywhere people do. Though commonly out of sight during the day, they emerge from cracks and other dark places at night to scrounge for food. Cockroaches eat just about everything—from regular "people" food to wallpaper. There are probably many more cockroaches in cities than people.

Where species try to use the same urban habitats as people, conflicts come about. The red tail hawk living over the entrance to a Park Avenue apartment building in New York City is one example. However, birds interfere with people's urban activities in other ways, too. Dallas-Fort Worth Airport in Texas, the third busiest airport in the United States, has spent millions of dollars minimizing the likelihood that birds will strike an airplane as it takes off or lands. Of particular concern is the possibility that a bird will be sucked into a jet's engine, causing a deadly accident. In addition to setting up "birdar," a bird detecting radar system that warns pilots of large flocks of starlings and grackles in the vicinity, airport authorities try to alter the habitat around the airport to make it less attractive to birds. Some techniques that have been employed to discourage birds from feeding and flying around the airport include mowing grassy areas, removing trees, and limiting access to water. Because habitat alteration has not been entirely successful, the airport has also tried setting off propane cannons and playing taped recordings of birds in distress to try and scare the flocks away from the area. What other examples can you think of where animals come into conflict with urban people?

People, as part of the biosphere, are also affected by urban developments. Some people thrive in the city and would not live anywhere else. Others only like the country. Over time, the distribution tends to work out and most people probably live where they want. The people in cities do experience some differences however. For example, newborn infant survival rates are higher in cities because health care and good sanitation can be more available. However, densely populated areas are also where infectious diseases can spread more rapidly.

In November 2002, severe acute respiratory syndrome (SARS), a new respiratory disease, was identified in southern China. In less than a few months, more than two dozen countries on four continents had reported over 8,000 cases of the disease with more than 750 people dying as a result. People infected with the disease had almost always traveled recently to countries where SARS occurred and returned to a new country as a carrier of the disease.

To prevent hundreds of thousands of people from being infected by these returning travelers, health officials in several countries, including the United States and Canada, quarantined people who were either infected with SARS or had come in close contact with infected individuals. These quarantines lasted up to 20 days and prevented people from coming in contact with others who were, or might have been, infected with SARS. People under quarantine could not leave the hospital or their homes, go to work or school, or visit family and friends. Coming into contact with others would risk spreading the disease further, potentially resulting in a widespread outbreak. Without our urban and global lifestyle, the SARS outbreak would have remained within China and not spread around the world.

This example highlights a two-sided characteristic of densely populated areas. On one hand, small problems such as a few cases of infectious disease can quickly become bigger problems. On the other hand, tremendous resources can be brought to bear on addressing and solving problems. Fixing problems in urban environments can go a long way toward fixing them in societies as a whole. The eradication of malaria in the developed world (see Chapter 4) is an example.

Atmosphere Impacts

The Earth's atmosphere within a city is commonly affected by emissions from cars, factories, and power plants (see Chapter 15). Ground-level ozone is a major component of air pollution in most cities. Not to be confused with natural ozone in the upper atmosphere that provides protection against the sun's harmful radiation, ozone at the ground surface is created when sunlight reacts with engine exhaust. Another significant contributor to air pollution is particulate matter: tiny, often microscopic, particles suspended in the air. Finally, sulfur dioxide and nitrogen dioxide are two gases produced by the burning of fossil fuels that further pollute the air in and around cities.

The health effects of severe air pollution can be deadly. In 1880, 2,200 residents of London, England, died as a result of a toxic smog of sulfur dioxide and particulate matter created from the burning of coal used to heat homes and power factories. A similar incident in 1948 killed 50 people in Denora, Pennsylvania. Although killer "fogs" are unusual today in the developed world due to government regulations controlling the release of air pollutants, cities in developing nations in Asia and other parts of the world commonly have air pollution levels three times higher than the World Health Organization's air quality standards. These continuing problems lead to more than 200,000 deaths each year around the world.

Air quality in the United States has greatly improved since passage of the **Clean Air Act** of 1972. However, health problems in U.S. cities still persist. Health effects vary from minor coughing to severe heart disease and lung cancer. Individuals who are particularly sensitive to air pollution include infants; the elderly; and those suffering from asthma, bronchitis, or emphysema. Even if you are not in one of these high-risk groups, you can take steps to reduce your exposure to air pollution. The Environmental Protection Agency (EPA) recommends that all individuals stay indoors when pollution levels are high, or to limit outdoor exposure to the early morning or evening hours because ozone levels increase during the daylight hours. If outdoor exposure is unavoidable, the EPA suggests avoiding overexertion, because a faster breathing rate will mean more pollution is entering the lungs.

The largest cities usually have the worst air pollution problems, but population is not the only factor controlling the severity of pollution. The local geography and climate play a large part as well. Cities in relatively low-lying areas surrounded by mountains in warmer climates can often have severe pollution problems because all of the car exhaust and power plant emissions remain trapped within the encircling mountain ranges. This problem is particularly true in Los Angeles, where air pollution problems are the worst in the country (Table 8-3).

The air in the center of the city, particularly at night, can be as much as −12.2°C (10°F) warmer than surrounding rural areas (Figure 8-10). This **urban heat island effect** is caused by the presence of fewer trees and shrubs

Table 8-3 *U.S. cities with the most polluted and cleanest air.*

METROPOLITAN AREAS MOST POLLUTED BY YEAR-ROUND PARTICLE POLLUTION (ANNUAL $PM_{2.5}$)	
Rank	**Metropolitan Areas**
1	LOS ANGELES-LONG BEACH-RIVERSIDE, CA
2	VISALIA-PORTERVILLE, CA
3	BAKERSFIELD, CA
4	FRESNO-MADERA, CA
5	PITTSBURGH-NEW CASTLE, PA
6	DETROIT-WARREN-FLINT, MI
7	ATLANTA-SANDY SPRINGS-GAINESVILLE, GA
8	CLEVELAND-AKRON-ELYRIA, OH
9	HANFORD-CORCORAN, CA
9	BIRMINGHAM-HOOVER-CULLMAN, AL
11	CINCINNATI-MIDDLETOWN-WILMINGTON, OH-KY-IN
12	KNOXVILLE-SEVIERVILLE-LA FOLLETTE, TN
13	WEIRTON-STEUBENVILLE, WV-OH
14	CHICAGO-NAPERVILLE-MICHIGAN CITY, IL-IN-WI
15	CANTON-MASSILLON, OH
16	CHARLESTON, WV
17	MODESTO, CA
18	NEW YORK-NEWARK-BRIDGEPORT, NY-NJ-CT-PA
18	MERCED, CA
20	ST. LOUIS-ST. CHARLES-FARMINGTON, MO-IL
21	WASHINGTON-BALTIMORE-NORTHERN VIRGINIA, DC-MD-VA-WV
22	LOUISVILLE-ELIZABETHTOWN-SCOTTSBURG, KY-IN
22	HUNTINGTON-ASHLAND, WV-KY-OH
24	YORK-HANOVER-GETTYSBURG, PA
24	LANCASTER, PA
24	COLUMBUS-MARION-CHILLICOTHE, OH

Table 8-3 *Contiuned*

TOP 25 CLEANEST CITIES FOR YEAR-ROUND PARTICLE POLLUTION (ANNUAL $PM_{2.5}$)	
Rank	**Metropolitan Areas**
1	SANTA FE-ESPANOLA, NM
2	HONOLULU, HI
3	CHEYENNE, WY
4	GREAT FALLS, MT
5	FARMINGTON, NM
5	ANCHORAGE, AK
5	ALBUQUERQUE, NM
8	BISMARCK, ND
9	KENNEWICK-RICHLAND-PASCO, WA
10	LUBBOCK-LEVELLAND, TX
10	BILLINGS, MT
12	IDAHO FALLS-BLACKFOOT, ID
13	GRAND JUNCTION, CO
13	COLORADO SPRINGS, CO
15	BELLINGHAM, WA
16	RAPID CITY, SD
16	FARGO-WAHPETON, ND-MN
18	PUEBLO, CA
19	FORT COLLINS-LOVELAND, CO
20	SALEM, OR
21	DULUTH, MN-WI
21	ALBANY-CORVALLIS-LEBANON, OR
23	SALINAS, CA
24	CAPE CORAL-FORT MYERS, FL
25	PORT ST. LUCIE-FORT PIERCE, FL

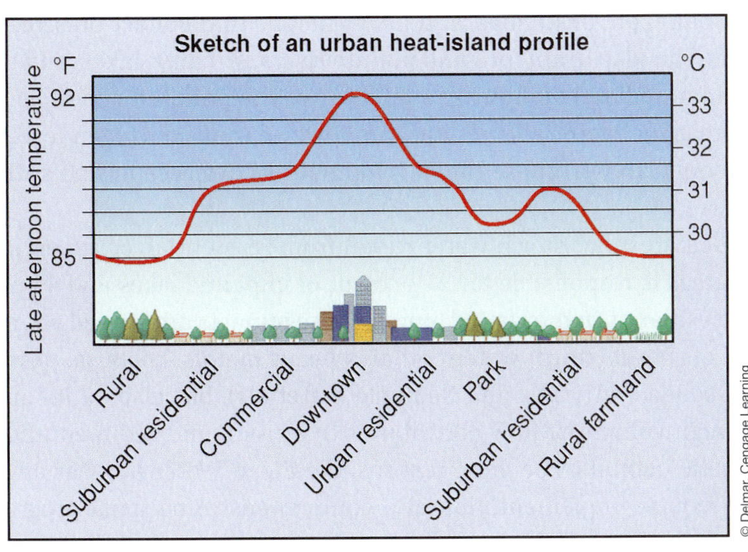

Figure 8-10 Temperature differences between a city and an adjacent rural area.

in the city that not only provide shade, but a cooling effect that results when water is transpired through their leaves. Furthermore, the dark surfaces of buildings absorb the sun's rays instead of reflecting them away, causing the air temperature around the buildings to rise. The maximum temperature in July has increased by up to −17.2 °C (1°F) during the last 30 years or more in cities such as Baltimore, Phoenix, and Los Angeles because these urban areas have grown larger. The urban heat island effect is actually even more noticeable at night when buildings and paved surfaces radiate the heat they have absorbed during the day, slowing the typical drop in air temperatures that occurs after sunset.

The urban heat island effect affects both human health and the health of the ecosystem. Air quality becomes worse because the formation of smog increases in warmer temperatures. Fine airborne particles in smog can cause throat irritation and breathing problems that are particularly bad for those who suffer from asthma or heart and lung diseases. Smog also slows the rate of photosynthesis, so it retards the growth rate of trees. Higher temperatures in cities during a summer heat wave put more people at risk for heat exhaustion and heat stroke. Increased temperatures can also cause physiological stress in other animals, change the types of plants and animals that live in an urban area, and even lead to changes in the distribution of diseases.

Hydrosphere Impacts

Although air pollution generated by urbanization is a serious problem associated with large cities, all urban areas, regardless of size, must ensure high-quality drinking water to their residents while preventing pollution to the surrounding hydrosphere. Water in urban areas can be contaminated in many ways. The roofs and paved surfaces of buildings, parking lots, and other structures prevent water from percolating into the ground after a rainstorm. Consequently, a greater amount of water runs off into streams after a rainstorm. This urban runoff can affect the environmental conditions in the receiving stream in several ways. The greater flow rate can cause erosion that harms aquatic habitat, such as washing away gravels that fish use to lay their eggs. Rain water can be heated up as it drains off hot buildings and roads in the summertime, resulting in thermal pollution problems. Contaminants, such as gasoline, oil, pesticides, and heavy metals that collect on streets and parking lots, can also drain into and pollute rivers, streams, lakes, and oceans during and immediately following a rainstorm. The polluted runoff generally carries increasing bacteria levels and can lower dissolved oxygen levels, lead to algal blooms, and promote the development of diseases in fish and other organisms.

The EPA states that storm water runoff from residential, commercial, and industrial areas is responsible for 21 percent of impaired lakes and 45 percent of impaired estuaries in the United States. These impacts are caused not only by the quality of runoff (storm water contains heavy metals, bacteria, pesticides, suspended solids, nutrients, and floatable materials), but also by its quantity because a high volume of flow contributes to erosion and sedimentation and affects aquatic habitats. For these reasons, the Clean Water Act was amended in 1987 to require implementation of a comprehensive national program for addressing storm water discharges. In response to the 1987 amendments, the EPA developed the National Pollutant Discharge Elimination System (NPDES) Storm Water Program.

As far as urban areas go, people's most important impacts relate to providing needed water and disposing of used (waste) water. Cities have developed tremendous water supply systems. Most depend on surface water resources and use dams, reservoirs, and aqueducts to capture and distribute water. Treatment facilities that ensure that the water is clean are also important parts of these systems. New York City uses a total of 5,180 square kilometers (2,000 square miles) of watershed to capture the 4.9 billion liters (1.3 billion gallons) of water its residents use each day.

Disposing of wastewater, including sewage, is also an important challenge for all urban areas. Historically, much city wastewater was just released into the nearest river, lake, or ocean. This still happens in different places around the world, but both health and environmental protection are causing more and more wastewater treatment before it is released. Treatment can even make city wastewater drinkable again.

The **Clean Water Act** of 1972, similar to the Clean Air Act, has greatly reduced water pollution problems in the United States. However, many problems still exist because the contaminants released into lakes and rivers decades ago still persist in these environments. Even for urban residents, fishing is a major American pastime; but eating lots of fish that was caught in polluted urban waterways is potentially a serious health risk. PCBs, mercury, and other chemicals accumulate in fish over time as they feed on contaminated sediments directly or consume other fish and organisms that have already consumed toxic chemicals. As in many areas of the country, local and state governments in the Great Lakes region of the country have issued health warnings, recommending that no more than one fish from these areas be eaten per month.

Geosphere Impacts

The more obvious impacts of urbanization on the geosphere are the changes that accompany excavations for foundations, roads, and underground facilities such as sewers and subways. Another impact that almost all cities have comes from how they dispose of solid (municipal) waste. Municipal waste includes everything from food scraps to old shoes; it is everything you throw into the garbage and someone hauls away. If it is hauled away, where does it go? In the United States, 15 percent of city garbage is burned, 30 percent is recycled, and 55 percent goes into modern garbage dumps called landfills.

Did you know that Athens, Greece, started the first city garbage dump 2,500 years ago? Even so, garbage was commonly just thrown into the streets or piled outside the city gates. It was not until studies in England in the early 1800s linked disease to unsanitary conditions that cities began to be seriously concerned about garbage disposal. Today's landfills are much more than garbage dumps. They are engineered facilities that isolate wastes from interactions with air and surface or ground water. Liners are used at the base to prevent liquids (called leachate) from migrating out of the landfill and potentially contaminating the environment. Garbage is placed on top of the liner, strongly compacted and finally covered with clean, vegetated topsoil (Figure 8-11). Gases such as

Figure 8-11 Modern landfill design.

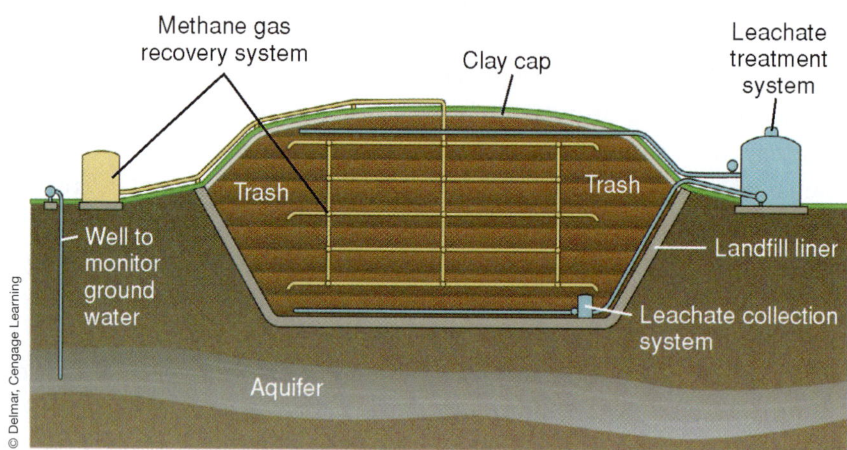

methane, which are generated from decaying organic matter in the garbage, are collected and burned rather than being allowed to seep into the atmosphere or migrate and collect in structures such as basements and parking garages. Providing municipal waste disposal for large cities, where landfills can be huge, is an especially costly and continual challenge. In 1987, a barge transporting garbage from Long Island, New York, was not allowed to unload (Figure 8-12). The barge traveled over 9,656 kilometers (6,000 miles) in 2 months while searching for a place to unload the garbage. It finally ended up back where it started and the garbage was eventually burned.

The need for municipal solid waste disposal may seem obvious because all people generate garbage and need to throw it away. However, another impact on the geosphere in older cities is not so obvious. Even though much of a city's area is covered by buildings and pavement, some soil remains in yards, parks, school grounds, and planters. In a more general sense, the dust on sidewalks and window sills is also part of a city's "soil." This soil can be contaminated with toxic substances; a classic example is lead. A toxic form of lead was once common in paint and gasoline. Health concerns, especially for young children who come into contact with toxic forms of lead, include brain development disorders. Although lead has been banned from paints since 1976 and from gasoline since 1986, older parts of cities still have soils affected by their use. In New Orleans, for example, a large part of the inner city residential area has lead concentrations in the soil that are at least 500 to 1,000 parts per million (ppm) (Figure 8-13). The EPA level of concern for soil lead exposure in places where children play is 400 ppm, which is significantly lower than the soil concentrations reported for inner city residential areas. Not surprisingly, about one-quarter of the children in the inner city of New Orleans are exposed to excessive levels of lead. Protecting against this type of legacy is important in many older cities across the United States.

Figure 8-12 Long Island garbage barge.

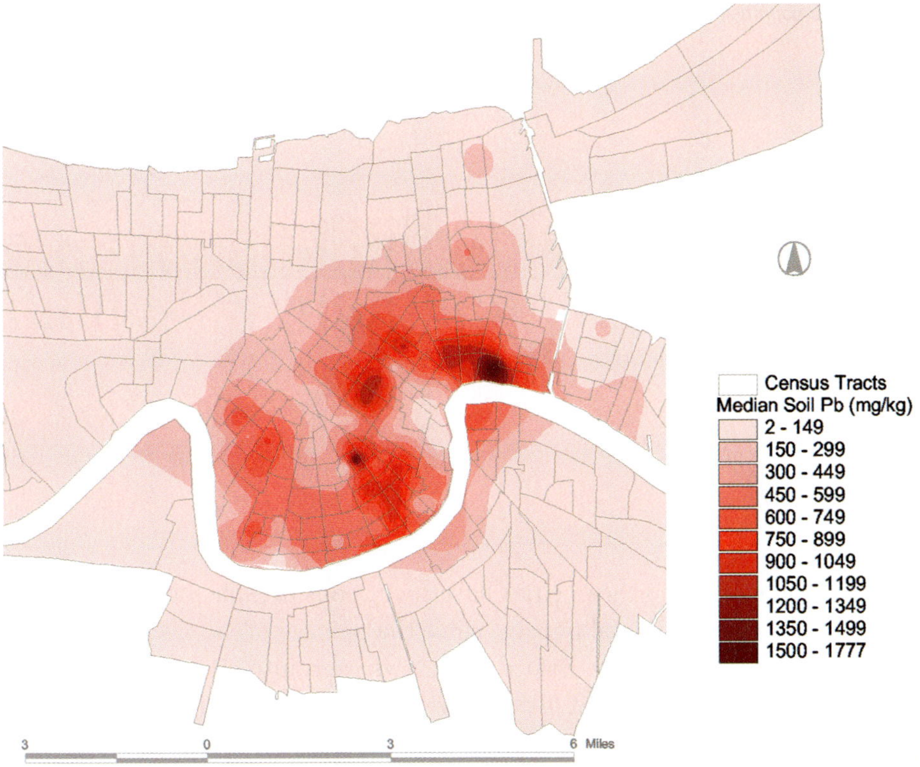

Figure 8-13 Soil lead concentrations in New Orleans.

© Howard Mielke

 assessing urban impacts

The urban environment provides people with their basic survival needs—air, food, water, and shelter. Some people live in only a small area of a city and need not travel more than a few blocks to work, eat, and enjoy a night out. However, the food they eat and the materials used to build their apartment where they sleep come from a much larger area. Your impact on other environments, therefore, potentially extends far beyond the area you immediately occupy. The full sum of the areas you affect is known as an ecological footprint.

You might be surprised how much land you affect, how much energy you use, and how much waste you generate if you consider all the complex linkages between, for example, the can of soda you drink after school and the raw products needed to create it and deliver it to you. The amount of crop and grazing land that each American uses to grow all of their food is estimated to be over 1.5 hectares (3.7 acres). The total amount of land an average American needs to support all of their needs, including food, lumber, energy, and other resources is over 9.0 hectares (22.2 acres). Only about 3 percent of this total footprint, or 0.3 hectare (0.7 acre), is actually in an urban setting where the houses, schools, shopping malls, and roads are built. The average ecological footprint of Americans is greater than that of citizens in most other countries in the world (Table 8-4).

Although the water and energy consumption of the average American is much greater than most other countries, the impact to natural ecosystems surrounding America's urban areas is not necessarily greater than that in other

Table 8-4 *Total ecological footprint of people in selected countries.*

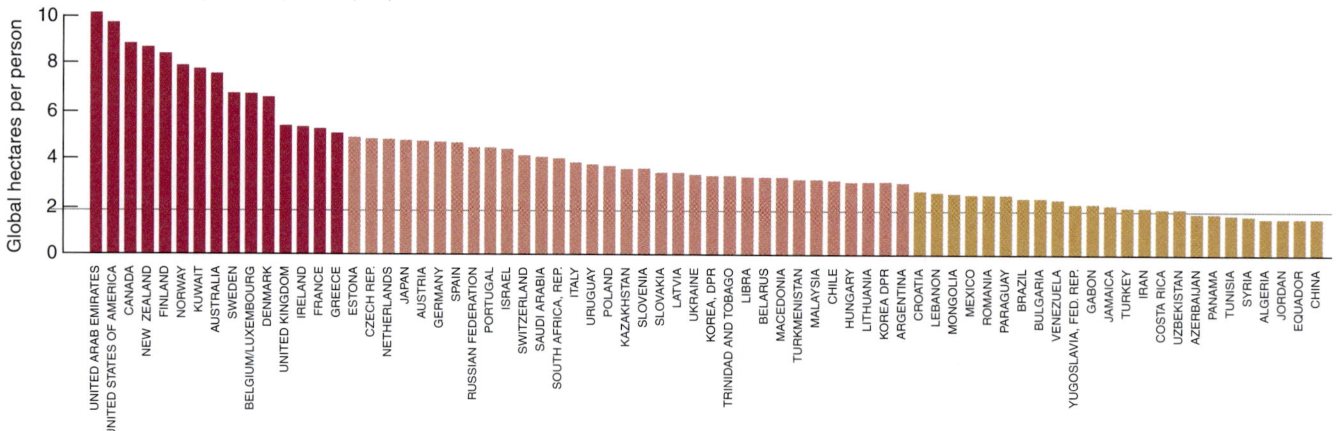

areas of the world. American cities have spent considerable money and effort to minimize environmental impacts since the passage of the Clean Water and Clean Air Acts in the 1970s. Before these laws went into effect, wetland areas were commonly drained and filled during the growth of cities to make more room for buildings. As the scientific understanding of wetlands has increased, it has become apparent that wetlands are a critical component for maintaining ecosystem health and diversity. Rather than filling wetlands, federal law requires that wetlands be protected. Where wetlands must be damaged or destroyed due to the pressures of development in urban areas, other wetlands must be created so that there is no net loss in wetland area.

One important aspect of wetlands in preserving ecosystem health is their ability to filter pollutants and absorb urban runoff such that increased flow into streams is minimized. Recognizing these potential benefits to reduce the environmental impacts of urbanization, the coastal City of Arcata in northern California began to restore its wetland marshes in the 1970s. When the area was first occupied by European settlers in the 1860s, the freshwater and saltwater marshes were soon diked and drained for agriculture. Over the next century, the former wetlands were occupied by railroad tracks, lumber mills, and even a landfill. When the landfill was condemned in 1973 for contaminating Humboldt Bay, the city decided to restore the wetlands.

Arcata now uses wetlands as a natural wastewater treatment facility. The former wastewater treatment plant used chlorination procedures but still left suspended solids and possibly pathogens in the water it released. These problems have been solved by treating the wastes in the restored wetlands. In addition, the 100 acres of wetlands have become a home or rest stop for over 200 species of birds. This has allowed Arcata to simultaneously reduce its pollution problems, improve the surrounding ecosystem, and expand its economy by attracting birdwatchers from around the world.

Cities elsewhere in the United States reduce their environmental impact in other ways. Cars and trucks are a large contributor to air pollution in many urban areas. Buses, trains, and other forms of mass transportation can reduce this impact by having large numbers of people ride together rather than

Table 8-5 *Alternative fuels that can be used in cars and buses to improve air quality.*

Fuel
Liquefied Petroleum Gases (LPG)
Compressed Natural Gas (CNG)
Liquefied Natural Gas (LNG)
Methanol, 85 Percent (M85)
Methanol, Neat (M100)
Ethanol, 85 Percent (E85)
Ethanol, 95 Percent (E95)
Electricity

separately in individual cars. The City of Bloomington, Indiana, has taken an extra step to ensure that its mass transportation system improves air quality. All of the city's buses, and those of the city's public schools and Indiana University in the city, operate on soybeans. Soybean oil, converted into biodiesel, releases far less particulates, carbon monoxide, and hydrocarbons into the air compared to normal petroleum-based gasoline or diesel. The biodiesel used is actually only 20 percent soybean oil mixed with regular diesel fuel, but no modifications to the engine are required before using. Because soybeans are grown by farmers in the surrounding area, the switch to cleaner burning fuels has also improved the local economy. Several other types of cleaner burning alternative fuels are also available that are used in mass transit systems in other parts of the country (Table 8-5).

Recycling and incineration, especially incineration that generates electricity, are being increasingly used to deal with the large volume of municipal waste. But the new buzz word in dealing with city garbage is "prevention." The city of Seattle's goal is now "zero waste." Preventing garbage requires new ways of doing things—like more electronic records and communications rather than paper. Seattle is a national leader in recycling and recycling will continue to be important. If Seattle's residents can decrease the amount of waste going to their landfills by 40 to 60 percent, it will save them $2 million a year.

8.4 conclusion

Cities have improved the quality of life for millions of people worldwide for thousands of years by promoting technological, social, and educational advances. These changes in the life of humans have also resulted in changes to the natural environment. New abiotic and biotic pressures created by the urban landscape result in changes in the ecosystem with organisms that are better capable of adapting to the changing landscape becoming more prolific and others being displaced. But urbanization that soundly manages its wastes, provides efficient transportation, and conserves energy and other resources can help mitigate the negative environmental consequences of the world's rapidly growing human population. Because about 60 percent of the world's people will live in urban centers by 2030, understanding how environmentally sound urbanization can be achieved is important.

IN THIS CHAPTER YOU HAVE LEARNED:

- Urban areas are densely populated by people. In the United States, any community with over 2,500 people is considered urban, but urbanization is most complete in the world's largest cities.
- Urbanization helps people. Because the first cities were developed 6,000 years ago, they have provided economic, technological, and educational opportunities for people.
- Urban impacts on the biosphere include displacement or destruction of natural ecosystems. In their place, new environments are developed where some plants, animals, and insects live—not just people. Although health care for people is generally more available in urban centers, dense populations also raise infectious disease risks.
- Urban impacts on the atmosphere include degradation of air quality and thermal pollution. Large cities become "heat islands" that warm the air around them.
- Urban impacts on the hydrosphere come from increased surface runoff that can pollute surface and ground water. Urban centers also require extensive systems to deliver and treat water that is used by residents.
- Urban impacts on the geosphere include the many excavations needed to build cities and the accompanying destruction, displacement, or contamination of soils. Soundly disposing of solid waste is a key challenge for urban centers. Landfills have been the most common approach. Recycling of solid waste is becoming increasingly viable and important.
- Environmentally sound urbanization is possible and will be important as an increasing number of the world's people come to live in cities.

KEY TERMS

Clean Air Act
Clean Water Act
Urban area
Urban heat island effect
Urbanization

REVIEW QUESTIONS

1. Why did the first cities on Earth develop in large river valleys?
2. In what ways has urbanization both added to and taken away from habitats?
3. Describe three ways in which animals are adapted to urban habitats.
4. "Urbanization is good for the human species." Choose whether or not you agree with this statement and support your stand with evidence from the chapter.
5. Explain the urban heat island effect and how it affects both human health and the health of the ecosystem.

6. Describe three ways in which urbanization affects the hydrosphere.

7. In what ways do urban residents protect the geosphere?

8. Describe three ways in which urbanization has affected the geosphere.

9. Explain the role of wetlands in preserving ecosystem health.

10. Explain the role of the Clean Water Act and Clean Air Act in improving urban environments.

unit three

living with a dynamic earth

chapter 9

restless earth

In anticipation of a wet adventure, students from Sequim High School pack up their raingear and field notebooks for a field trip. They are heading west to the rainy side of the Olympic Mountains where they will identify and record plant species. Another group of students pack much lighter. They are expecting sunshine because they will be recording plant species to the east of the mountains. These two groups of students, separated by less than 80.5 km (50 miles), will be in two very different climate zones within the state of Washington.

The Olympic Mountains rise up along the west coast of Washington. They are a young part of the geosphere that uplifts volcanic and sedimentary rocks originally formed on the ocean floor. The mountains create a barrier to the moisture-laden winds coming ashore from the Pacific Ocean (Figure 9-1). On

Figure 9-1 Rain shadow effect.

the windward (west) side of the mountains the air rises, cools, and precipitates tremendous amounts of rain. The over 250 centimeters (100 inches) of rain that fall here each year create a rainforest. The air that blows over the mountains is depleted in moisture and only 80 km (50 miles) east of the rain forest; the annual precipitation is just 40 cm (16 inches) of rain a year. Sequim High School is located in this rain shadow of the Olympic Mountains.

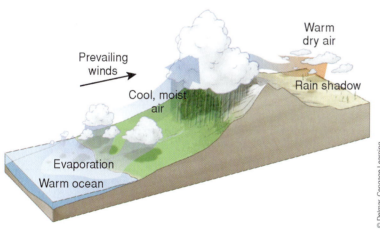

Biology students at Sequim High School have a unique opportunity to integrate earth science, climatology, biology, and ecology. After examining and mapping rainfall data on each side of the Olympic Mountains, students travel to several locations to identify and record the species of plants that they find. Students then correlate rainfall data with specific indicator plant species to establish which ones are the best indicators of rainfall. Finally, students map the distribution of the indicator species and compare it to rainfall data to produce a model of how rainfall in the Olympic Mountains affects biodiversity. Their map shows where interactions of the Olympic Mountains (geosphere), onshore winds (atmosphere), and the Pacific Ocean (hydrosphere) have influenced the distribution of life in the biosphere.

Movements in the Earth's geosphere, called **tectonic movements**, *are responsible for the character, size, and location of mountains around the world. In addition to forming mountains, tectonic movements also form ocean basins and assemble or divide continents. The processes causing these changes can be violent or imperceptible. The landscape can be changed in the blink of an eye by a volcanic eruption, earthquake, or landslide. The land and oceans can be changed over millions of years through gradual erosion or tectonic movement. When the land and oceans change, so do Earth's environments.*

AFTER COMPLETING THIS CHAPTER, YOU WILL UNDERSTAND:
How the rigid geosphere (the **lithosphere**) is broken into pieces called plates.
How and why plates move.
How plate movements cause many changes in the lithosphere, including the formation, growth, or splitting of continents; the formation of mountains; and the formation or destruction of ocean basins.
Plate movements shape Earth's physiography and influence climate and biome development.
The five major physiographic realms in the United States.
How Earth's dynamic geosphere and changing physiography have influenced where people live.

 the moving geosphere—plate tectonics

Most of Earth's dynamic and changing character is directly or indirectly tied to movements in Earth's crust and mantle (see Chapter 1). Recall from Chapter 1 that the rigid crust is about 8 km (5 miles) thick below the oceans and up to 50 km (30 miles) thick beneath the continents. The bottom of the oceanic and continental crust is a distinctive boundary with the upper mantle. However, rigid rocks continue downward beneath the crust into the mantle. The rigid rocks in the mantle only extend downward for a few tens of kilometers (miles) beneath the oceanic crust but may be over 300 km (190 miles) thick beneath the ancient continental crust. Rigid rocks of the crust and upper mantle combine to form the solid lithosphere (Figure 9-2).

The mantle below the lithosphere is not stable—it can slowly flow like very thick molasses and material slowly sinks, rises, and moves from place to place in ways that are still being investigated by scientists. However, the movements in

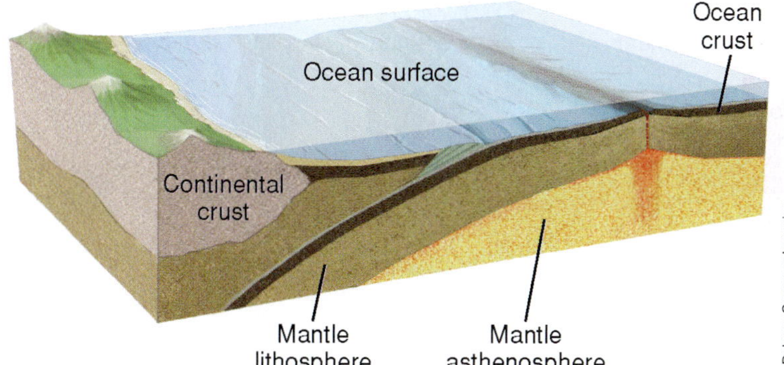

Figure 9-2 Lithosphere thickness.

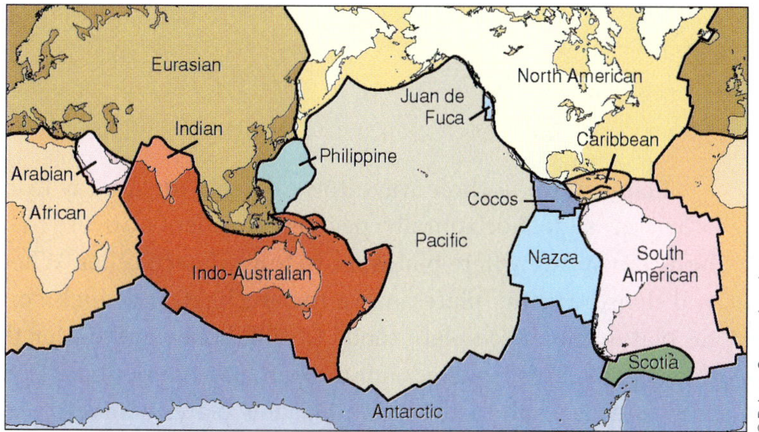

Figure 9-3 Major lithospheric plates.

the mantle have clear influences on other parts of the geosphere, on the hydrosphere, and even on the atmosphere (see Chapter 1). The key thing that movement of mantle material does is break the overlying lithosphere into pieces and move these pieces about like so many floating islands.

The thin and rigid pieces of the broken lithosphere are called **plates** (Figure 9-3). There are oceanic and continental plates and big and small plates but all are moving about on Earth. Both continents and parts of oceans can be included on a single plate. For example, the North American plate includes the entire North American continent and about half of the Atlantic Ocean. Unless you live near a boundary between two plates, such as the San Andreas fault in California where many earthquakes occur, you probably are not reminded very often that the plate you live on is moving. Plates move very slowly—at rates of a few centimeters a year on average or about the same rate that your fingernails grow. The boundaries between the plates are where the lithosphere is broken or faulted; movement between plates is localized along these faulted boundaries.

What causes the plates to move? The earth's upper mantle is hotter in some areas than others. This creates instability where hotter and less dense mantle moves upward relative to cooler and more dense mantle. When hotter mantle rises and cooler mantle sinks, it causes a slow circulation of mantle material (Figure 9-4). The mantle movements are slow—at about the same rate as plate motions or only centimeters per year.

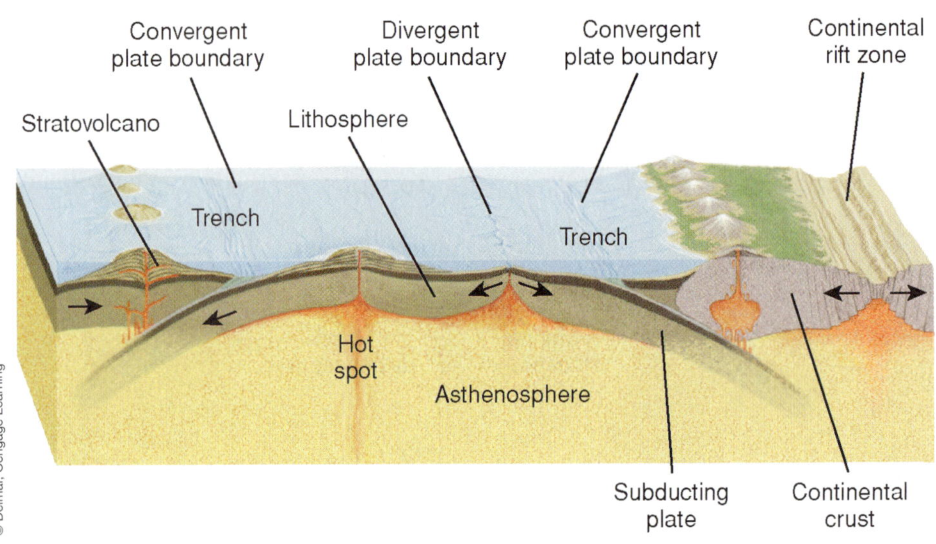

Figure 9-4 Mantle movements and plate boundary types.

Figure 9-5 Surface trace of the San Andreas fault.

Plates move in three ways: toward each other, away from each other, or slide horizontally past each other. **Convergent plate boundaries** are where plates move toward each other. Where thin and dense oceanic plates move toward light and thick continental plates, the dense plate tends to sink below and under the less dense plate. This is called **subduction** and the boundary between the plates is a subduction zone. Oceanic plates can sink below (be subducted) either continental plates or other oceanic plates. Where continental plates move together, they are too light and thick to sink under one or the other so they collide in great collision zones.

Divergent plate boundaries form where plates move away, or diverge, from each other. They can form within continents or oceans. As the plates move apart, melting in the mantle generates magma (molten rock in the geosphere) that migrates up along the boundary (geologists say that the magma is emplaced along the boundary). Eventually continents can be split apart and oceans formed at divergent plate boundaries.

Plates slide horizontally past each other at **transform boundaries**. These boundaries can be developed between any two plates regardless of their lithospheric character (Figure 9-5). These boundaries—large fault zones—create spectacular surface features as they cut across Earth's surface (Figure 9-6). The surfaces along which the plates slide are called transform (or strike slip) faults. Most transform boundaries are on the ocean floor and are relatively short, but there are some very long transform boundaries on land.

 9.2 shaping earth's surface

Rigid lithospheric plates have been in motion since they first formed, at least some 3.5 billion years ago. These movements have brought continents together, split them apart, and opened and closed oceans. They are a major influence on the shape of Earth's surface.

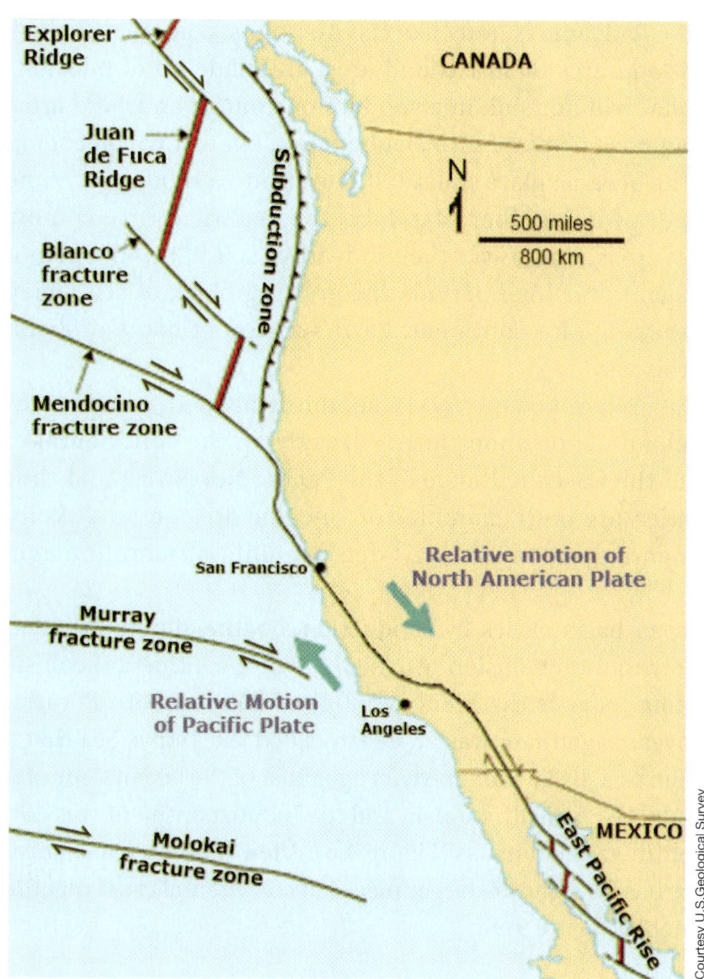

Figure 9-6 San Andreas fault system.

Creating and Growing Continents

Continents such as North America are made up of many pieces of tectonic plates that were assembled gradually over time by converging plates. At convergent boundaries the only material that will sink in the subduction zone is oceanic plate material. Any continental material on a plate will not sink because it is too buoyant. Imagine trying to sink a piece of plastic foam under water. As soon as you let go it will pop back up to the surface. The same is true for continental material—it is too buoyant to sink into the mantle.

Large continents are gradually assembled out of many smaller pieces of lighter crust. But how do these initial small pieces of crust form to begin with? Primitive continental crust is formed by volcanism that is localized above subduction zones. Volcanoes above subduction zones rim most of the eastern and northern Pacific Ocean. If two plates collide that are both made of oceanic lithosphere, the older of the two plates sinks below the other (subducts). The reason is that the older plate is colder and denser than the younger plate. Part of the older plate changes as it sinks into the hot upper mantle and releases water-rich fluids. The fluids released from the subducting older plate rise through the overlying mantle where they cause the mantle to partly melt. Part of the resulting magma will eventually make its way to the surface to form volcanoes. The volcanoes in the Aleutian Islands in Alaska are examples of this type of volcanism (Figure 9-7). The Aleutian volcanoes form by subduction of the Pacific plate beneath the North American Plate.

Courtesy U.S. Geological Survey

Figure 9-7 Kanaga Volcano on Adak Island in the Aleutian Islands stands 2,111.5 km (1,312 miles) high.

Volcanic islands like the Aleutians, commonly called volcanic island arcs or just island arcs, are made up of buoyant material that will not sink in a subduction zone. If an island arc moves on an oceanic plate into a subduction zone, it will get scraped off as the oceanic plate sinks. If it moves into a subduction zone and collides with another island arc, the two will form a combined piece of crust about twice their individual size. If this happens again and again over long periods the growing block of crust may become the size of a continent. Earth's first continents probably formed this way.

When oceanic crust subducts beneath continental crust, a chain of volcanoes forms directly on the continental crust. The volcanoes of the Cascade Range of the Pacific Northwest and the Andes of South America are both examples of volcanic arcs on land. Volcanic arcs help continents grow by adding large amounts of mantle magma to the crust.

If an ocean basin closes by subduction, continents are brought together in a very slow-motion collision. An example of a continental collision that is still happening today is the Indian continent colliding with Eurasia. Roughly 100 million years ago there was an ocean called the Tethys Sea that separated India from Eurasia. By 50 million years ago most of the ocean floor of the Tethys Sea had subducted beneath Eurasia and the Indian continent was added to the Eurasian continental landmass (Figure 9-8). A few times in Earth's history, plate convergence has brought the larger pieces of continental crust together to form supercontinents (Figure 9-9).

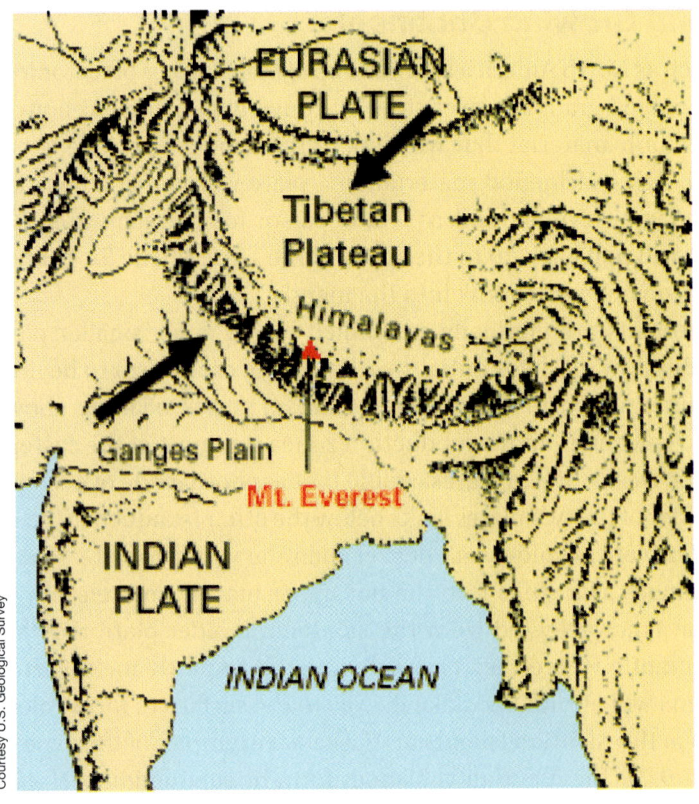

Courtesy U.S. Geological Survey

Figure 9-8 The collision of the Indian and Eurasian Plates is still occurring today.

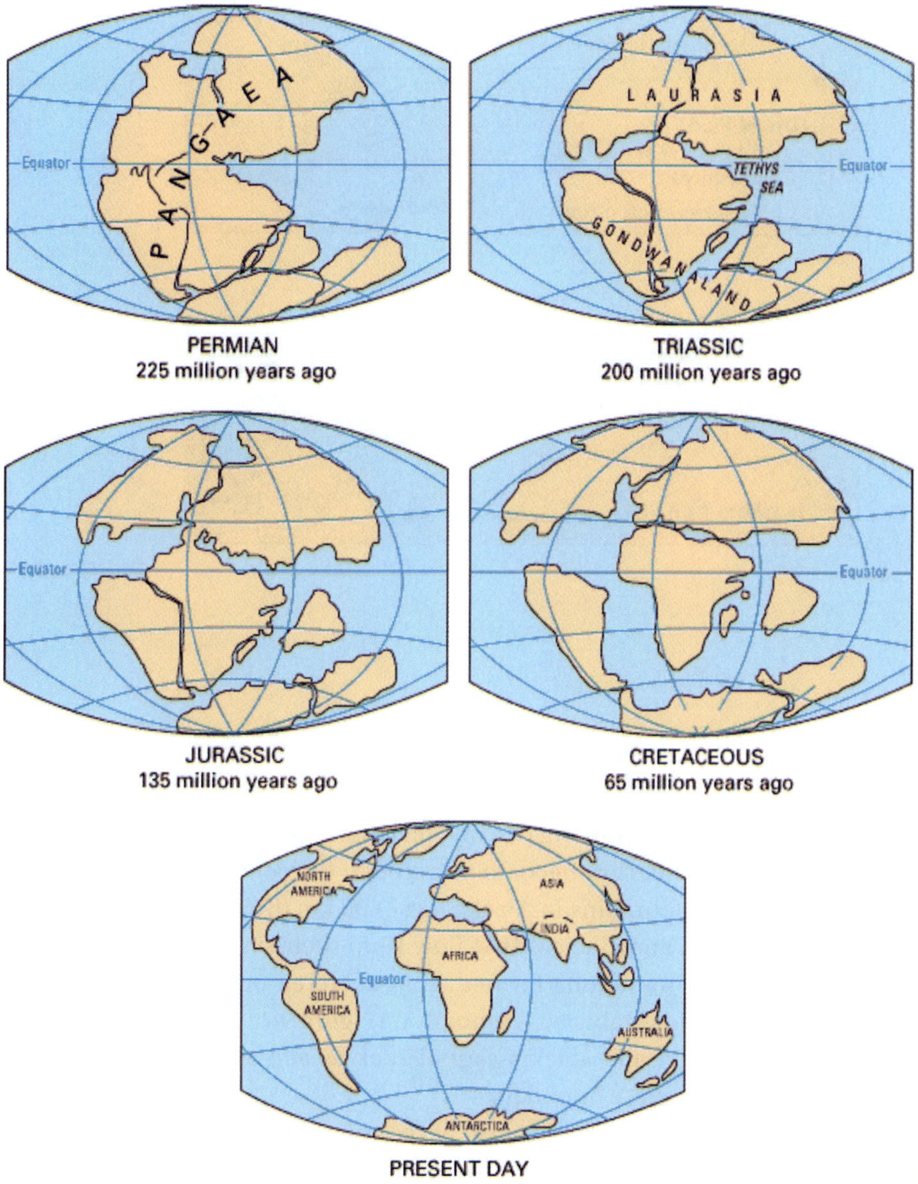
PANGAEA
Equator
PERMIAN
225 million years ago
LAURASIA
TETHYS SEA
GONDWANALAND
Equator
TRIASSIC
200 million years ago
Equator
JURASSIC
135 million years ago
Equator
CRETACEOUS
65 million years ago
NORTH AMERICA
ASIA
INDIA
AFRICA
Equator
SOUTH AMERICA
AUSTRALIA
ANTARCTICA
PRESENT DAY

Courtesy U.S. Geological Survey

Figure 9-9 The reconstruction of Pangaea.

Making Mountains

In addition to creating and growing continents, mountains—really big mountains—form by plate convergence. The world's highest mountains above sea level are where continental plates collide, deform, and stack up huge parts of the Earth's crust. Deformation by folding and faulting is the key process that creates these mountains (Figure 9-10). Because converging plate boundaries can be very long, the mountains that form along them can also extend long distances and become mountain ranges.

Deep sea trenches that form on the seafloor at converging ocean and continent plate boundaries can collect very large amounts of sediment shed from the nearby continent. In these places, the continental plate can be a giant sediment plow. During plate convergence and subduction, these sediments are

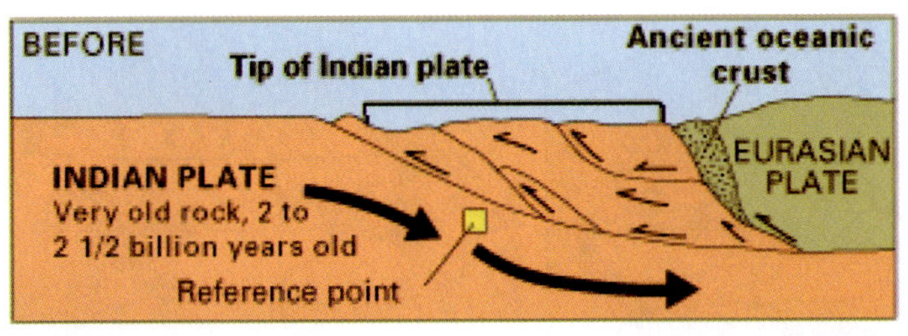

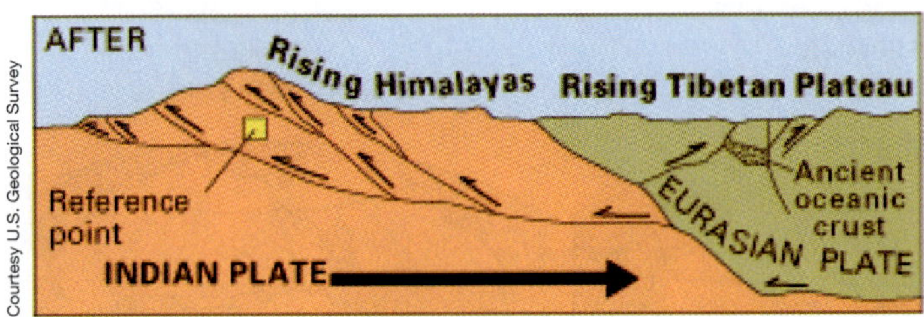

Figure 9-10 Deformation in the India-Eurasia collision zone.

Figure 9-11 The Olympic Mountains in Washington. (*Courtesy Earth Science World Image Bank © Bruce Molnia, Terra Photographics*)

scraped off the sinking oceanic plate and folded and faulted against the continent; they become accreted to the continental margin. The Chugach Mountains in Alaska, over 600 km (375 miles) long and full of icefields and glaciers that flow from their summits to the nearby coast, are mountains formed from accreted ocean sediments. The Olympic Mountains of Washington (Figure 9-11) also formed by the compression and stacking up of rock as accretion took place at a convergent plate boundary.

The compressive forces at convergent plate boundaries can deform the lithosphere and create mountains hundreds of miles away from the plate boundary. The Rocky Mountains are a good example. About 70 million years ago compressive forces from plate convergence along the western edge of North America were transmitted eastward through the continental crust. These forces folded and faulted the upper crust throughout the Rocky Mountain region. This deformation and uplift helped form the Rocky Mountains we see today.

The Earth's highest mountains form where plate convergence causes continents to collide with one another. The collision of the Indian continent with Eurasia has produced the majestic Himalaya Mountains. The Himalayas are home to 96 of the world's 109 mountain peaks whose summits are at 7,300 meters (24,000 feet) or higher. Continental collisions do not have to be as large as that between India and Eurasia to form spectacular mountains. The collision of a small continental block along the Gulf of Alaska coast has produced rugged peaks like Mt. St. Elias. This mountain, 5,489 meters (18,008 feet) high, is only 56 kilometers (35 miles) inland from sea level at the coast. Such spectacular mountains are very young. In fact, Mt. St. Elias is actively growing as the plate collision causing it continues. Such young, steep mountains are rapidly eroded by streams, glaciers, rock falls, and landslides. If they were not growing, they would not be able to maintain their majestic heights and rugged features.

Large mountains and long mountain ranges also form because of magmatic processes where oceanic plates converge and subduct beneath continental margins. The magmas produced by subduction are emplaced into Earths' crust above the subduction zone. Some of these magmas produce volcanoes on the Earth's surface, but many of them get trapped and crystallize within the crust. Over long periods, the rocks crystallized from these magmas significantly increase the volume of the Earth's crust. The increased amount of crustal rocks forms mountains. The Sierra Nevada Mountains in California and the longest mountain range in the world, the Andes in South America, formed this way.

Splitting Continents

At the same time continents are growing or deforming into mountains at convergent plate boundaries, they are being split apart somewhere else at divergent plate boundaries. Divergent plate boundaries are weak zones in oceanic or continental lithosphere where much faulting occurs. During the initial stages of plate divergence, called **rifting**, the lithosphere is pulled apart and great fractures cut through the crust to the Earth's surface. Rigid layers of rock crack from the tensional forces and blocks of rock drop down between the fractures (Figure 9-12). The down-dropped blocks form long linear valleys called rift valleys. The East African rift valley is an example of a divergent plate boundary in its early stages (Figure 9-13). If divergent movement continues in the East African Rift Valley, it may eventually become a narrow ocean much like the Red Sea is today.

Making Ocean Basins

The Red Sea, which separates East Africa from Saudi Arabia, is an example of a narrow ocean basin that has formed at a divergent plate boundary. Along the axis of the Red Sea, two plates are dividing and splitting apart. As the pieces drift apart, the rift zone widens and deepens to form a narrow ocean basin. Magma upwelling from the mantle is reaching the seafloor and forming new oceanic crust along the Red Sea axis. In several millions of years, the Red Sea will become wider and East Africa will be farther away from Saudi Arabia.

This process has created all the Earth's major ocean basins, even the very large Atlantic and Pacific Oceans. The volcanism and magma upwelling that occurs at divergent plate boundaries continually creates new oceanic crust. As divergent movement (spreading) continues, the ocean basins get bigger and bigger. The Atlantic Ocean basin has taken about 200 million years to grow to its present size.

Making Plains

Mountains are only temporary features on Earth's surface. They are inherently unstable landforms as gravity constantly works to lower them. Rocks uplifted in mountains gradually, or in some cases catastrophically as when rock falls or landslides occur, move downslope into valleys and streams. The continued

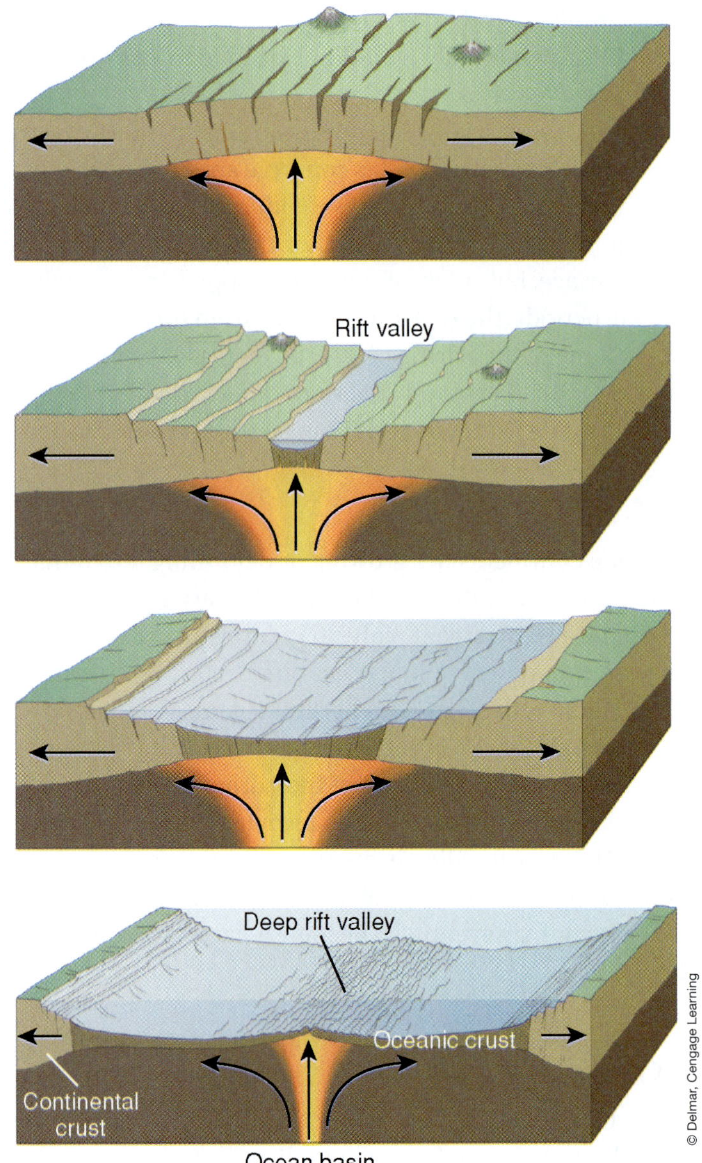

Figure 9-12 Evolution of a rift valley with continent splitting and ocean basin formation.

erosion of mountains and uplands supplies sediment to the major streams. These streams carry the sediment and deposit it on flood plains. Large plains are places where flood plains merge and sediment deposited by streams fills in irregularities in the landscape (Figure 9-14). Some of this sediment is directly from rivers and streams but some large plains are also where rising sea level once flooded continents with ocean water.

The movement of the shoreline across continents during times of sea level rise (transgression) or fall (regression) is very effective at eroding smaller uplands, moving and re-depositing stream sediments, and creating coastal plains (Figure 9-15). The plains that are created this way are commonly the result of many sea level rises and falls.

All plains develop on parts of the geosphere remote from active plate boundaries. They are a clear indication that the forces characteristic of plate boundaries have not been active in moving, deforming, or otherwise directly shaping

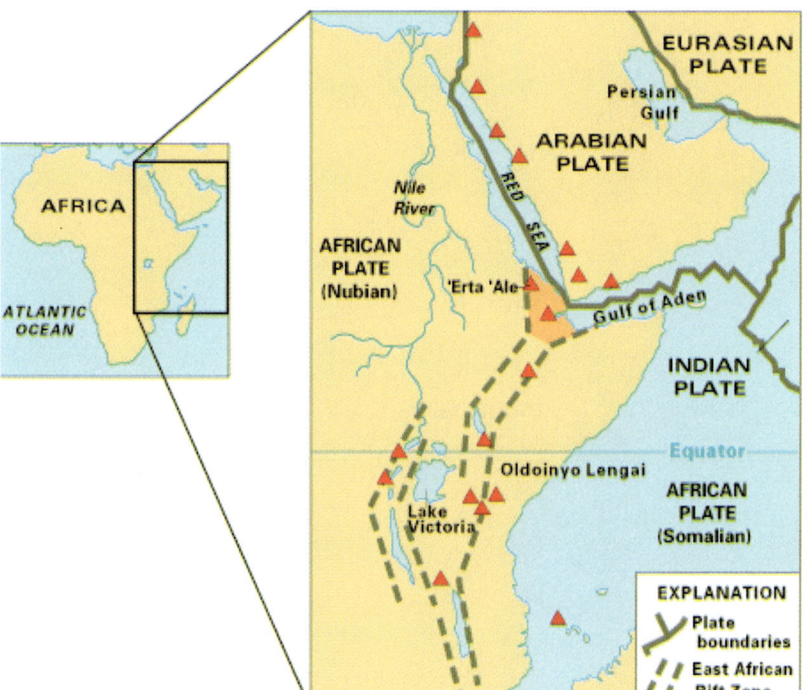

Figure 9-13 East African rift zone and location of volcanoes.

the landscape. On the other hand, the formation of plains is at least indirectly tied to plate movements, especially where the sediments deposited on them are supplied by mountain belts. In addition, many sea level changes that cause shoreline movements across parts of continents may be tied to changes in the size of oceans basins caused by plate movements.

Figure 9-14 Interior plains grassland. (*Courtesy Earth Science World Image Bank © Bruce Molnia, Terra Photographics*)

9.3 physiography, environments, and biomes

You learned above how movements of the geosphere, particularly plate movements, control major features of the landscape. The shapes of the land surface are called its **landforms** and **physiography** describes them and other natural features on Earth's surface. Where landforms are generally similar, areas can be identified that are physiographically distinct. For example, the United States can be divided into five broad physiographic realms (Figure 9-16).

The Appalachian Highlands extend from Alabama in the south to Maine and beyond in the northeast. In general, this is an area of moderate relief mountains and ridges with valleys scattered between. The irregular topographic relief reaches elevations of about 1,830 meters (6,000 feet) in North Carolina but elsewhere the higher elevations are commonly between 880 and 1,200 meters (2,900 and 3,900 feet). The highest elevations are underlain by igneous and metamorphic rocks but folded sedimentary rocks underlie the valley and ridge regions.

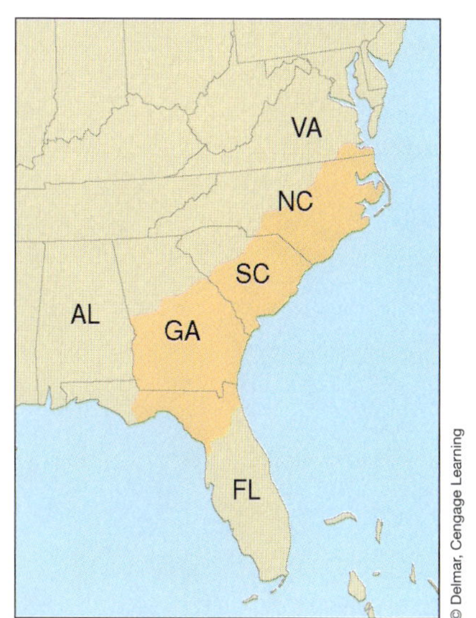

Figure 9-15 The southern coastal plain.

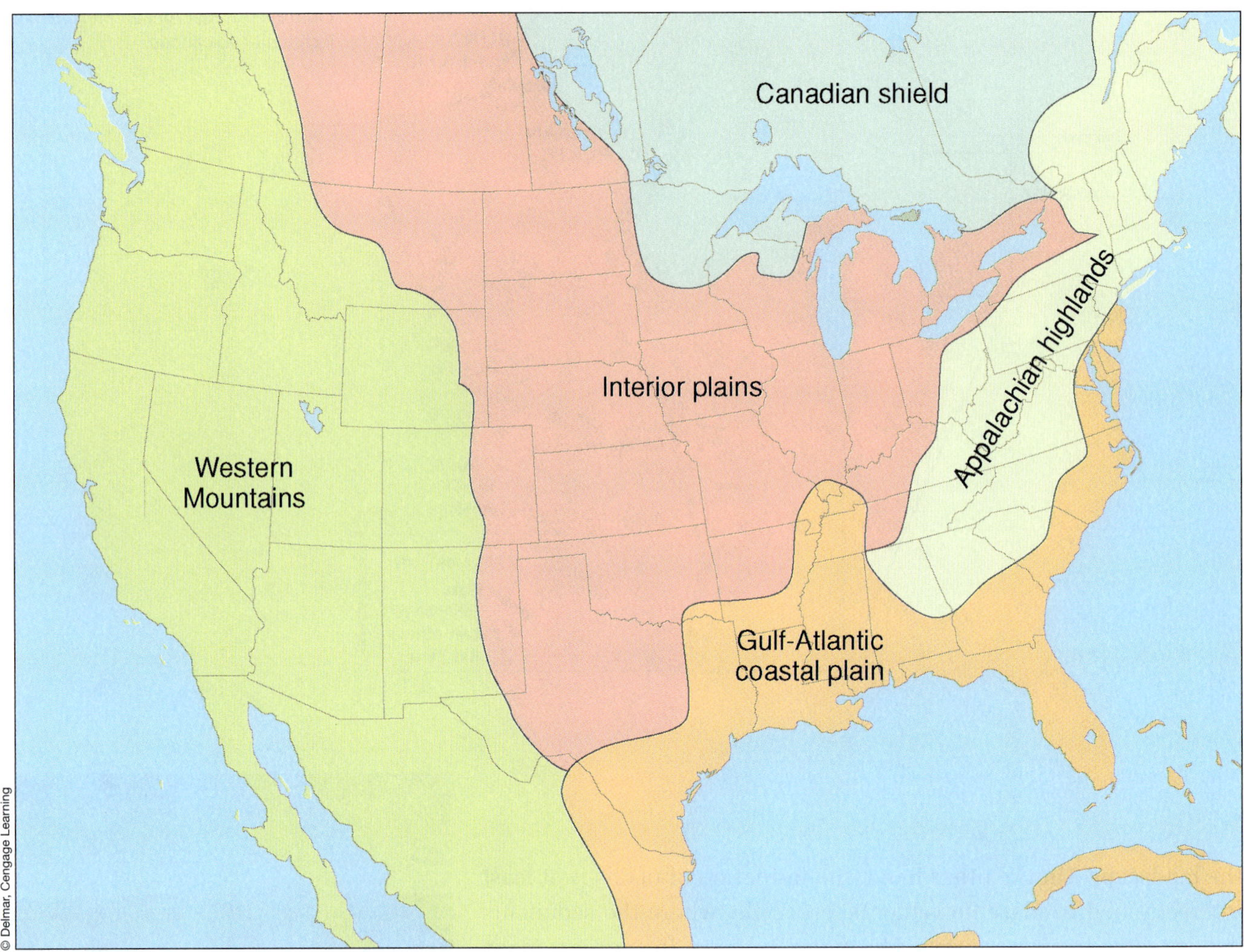

Figure 9-16 Physiographic regions of the United States.

The Coastal Plain physiographic realm extends south from New York along the Atlantic coast and west along the Gulf of Mexico coast to Texas. Elevations on the low relief coastal plain (some say it is flat!) are less than 304.8 meters (1,000 feet). It is underlain by sediments, most deposited at times of higher sea level.

The Western Mountains physiographic realm includes all of the areas from the Rocky Mountains west to the Pacific Ocean. In general, this area is one of high mountains, plateaus, and intervening valleys. Many of the mountains are young and growing. Geologically this area is very complex with many different ages of igneous, metamorphic, and sedimentary rocks. It includes the active volcanoes of the Cascade Range in northern California, Oregon, and Washington.

The Canadian Shield physiographic realm covers part of Minnesota and Wisconsin. It is characterized by small changes in elevation (low topographic relief), thin and poorly developed soils, and abundant lakes. It was extensively covered by glaciers as recently as 12,000 years ago. These glaciers beveled the underlying igneous and metamorphic rocks, which are part of the ancient core of the North American continent. As the glaciers melted and retreated back into Canada, many rivers flowed across the land and many lakes formed.

The Interior Plains physiographic realm extends across most of the central United States. They also have low topographic relief, but here the soils are well developed (see Chapter 13) on the underlying mostly sedimentary rocks.

All of the physiographic realms in the United States can be further subdivided into smaller physiographic regions. But the reason to recognize areas with distinct physiographic character should now be obvious. Identifying these different parts of the landscape identifies different parts of the physical environment that the biosphere has adapted to. The land's physiography influences the biosphere through its effects on regional climate and biome development. The size and location of landmasses, especially their latitude, elevation, and spatial relation to oceans, directly influence physical environments.

Land's Latitude

Recall from Chapter 7 the importance of latitude for climate conditions on land. You learned that the high latitude regions absorb much less solar energy than the low latitude regions. As a result, the climate at the poles is much colder than at the equator. You also learned that land and water have different heating and cooling properties and that this contrast also affects climate. Let us revisit the polar regions and consider how both latitude and the size of landmasses affect the environment.

The polar regions are the areas north of the Arctic Circle (66.5° N), and south of the Antarctic Circle (66.5° S). Both the Arctic and Antarctic are known for their frozen and desert-like conditions. Review your notes from Chapter 7 and recall why the polar regions are so cold. Although the north and south polar regions receive the same amount of sunlight, the two places are very different in terms of climate and life forms. The Arctic climate is known for its long cold winters and short cool summers. The average winter temperature is −34°C (−29.2°F) and the average summer temperature is 3 to 13°C (37 to 54°F). Yet the Arctic climate is mild in comparison to the Antarctic and is home to many land plants and animals that inhabit the tundra. The Arctic is also home to many communities of people. People have lived in the Arctic for at least 4,000 years.

Antarctica is the driest, coldest, and windiest continent on Earth. Winter temperatures average −60°C (−76°F) and summer temperatures average −30°C (−22°F). Most of Antarctica, unlike the Arctic, has no tundra and is buried beneath an ice cap up to 4 km (2.5 miles) thick. Only patches of land on coastal peninsulas are exposed long enough to grow primitive mosses, algae, and lichen. Antarctica does not have native communities and people can only manage to survive in Antarctica for brief visits. The only people who manage to stay for several months at a time are scientists who stay in research stations specially designed for the harsh climate.

Although interior Antarctica supports no life, coastal Antarctica is home to a rich and diverse population of sea animals and birds, including penguins, seals, killer whales, fish, and many migratory birds. The Southern Ocean that surrounds the Antarctic continent provides a rich supply of phytoplankton and krill, the foundations of the marine food web. Despite the harsh climate, the Antarctic environment is among the most fragile on Earth. Similar to the Arctic, the environment is very vulnerable to slight changes in climate.

Why do you think the climate in Antarctica is so much colder than in the Arctic? Both polar regions are at high latitudes, have the same length of seasons, and receive the same amount of solar radiation. Why are the climates not the same? A major difference between Antarctica and the Arctic is the distribution of continents and oceans (Figure 9-17). The North Pole lies in the middle of an ocean and the South Pole sits in the middle of a continent. Recall from

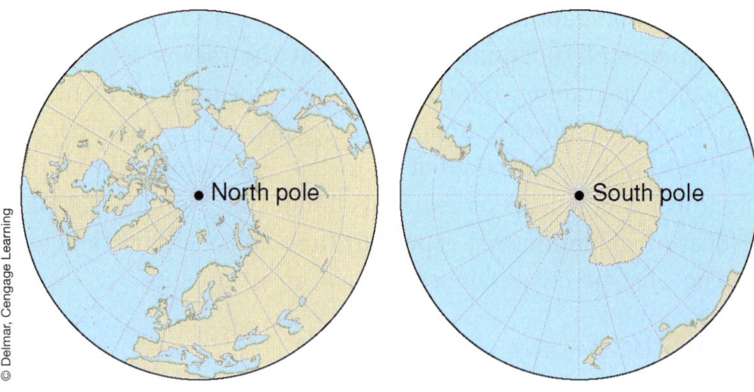

Figure 9-17 Arctic and Antarctic regions.

Chapter 7 that water is more easily heated by the sun's radiation than land. In the Arctic, the ocean stores heat that can later warm the overlying air and nearby land. Because the Antarctic is a large landmass, it is not as easily heated nor does it effectively store heat like the Arctic Ocean does.

Land's Elevation

If you have climbed a mountain, you may have noticed some of the local effects of elevation. As you climb to higher elevations the air is cooler, the sun's rays are more intense, and you get out of breath more easily. What causes these changes? The air is thinner at higher elevations (see Chapter 1) and absorbs less of the sun's radiation. If you camped out overnight on the mountain, you would also notice that the nights are much cooler and the temperature difference between day and night is much greater than at lower elevations. If the mountain you climbed was high enough, you would also notice different plants and animals living at higher elevations. All the major climates can be experienced by moving vertically from low to high elevations. This is why a person can travel through a tropical rain forest, temperate forest, grassy plateau, and eventually to very dry Arctic-like conditions—even with glaciers—as he or she moves from the coast inland to the high Andes Mountains in countries like Bolivia.

A Closer Look at Mountains

Most mountains are part of a range or long belt of mountains. Mountain ranges not only influence local weather patterns, but also influence the weather and climate of large regions. You learned above how the Olympic Mountains in Washington affect the weather and climate of the entire Olympic Peninsula. Mountains like the Olympic, strongly influence precipitation patterns and the distribution of ecosystems. The mountains create a barrier to the moisture-laden winds coming ashore from the Pacific Ocean. The resulting differences in precipitation create completely different biomes that support completely different plants and animals. There are similar effects created by the Andes Mountains in South America, the Himalaya Mountains in Eurasia, and even the coastal mountains of California. Rain shadows vary from small local features as at the Olympic Mountains in Washington to large regions such as Tibet, located inland of the Himalayas.

A Closer Look at Plains

The low relief regions of continents also have a variety of climates and biomes. The climate of plains is in large part dependent on their location on a continent. For example, the Coastal Plains and the Interior Plains of North America are home to different biomes.

The Coastal Plains of the southern United States host an environment distinct from the Interior Plains and from the adjacent Appalachian Highlands. The Coastal Plains are nearly flat and have many river networks and coastal wetlands. They are humid regions that receive much of their precipitation from clouds blown onshore from adjacent parts of the Atlantic Ocean and Gulf of Mexico. The natural vegetation of the Coastal Plains region is mostly temperate forest with a combination of oak, hickory, and pine.

The interior of the United States, west of the Appalachians and east of the Rockies, is a low relief area like the Coastal Plains, but both temperate forests and temperate grasslands are developed here. Temperate grasslands have no trees or shrubs. Grasslands thrive on continents in low relief regions with the appropriate climate conditions. In interior North America, summer temperatures can exceed 38°C (100°F) and winter temperatures can be as low as −40°C (−40°F). Precipitation is usually restricted to late spring and early summer and averages 51 to 89 cm (20 to 53 inches) per year. Seasonal drought and occasional fires affect biodiversity and limit tree growth.

In North America, grassland dominates the Midwestern states between the western headwaters of the Mississippi River and eastern edge of the Rocky Mountains. The rise of the western mountains influenced the development of the North American grasslands by capturing moisture coming with storms from the west. Before the formation of the Rocky Mountains forests dominated the regions now covered by grasslands. The Rockies also influence the movement of air masses between the cold Canadian Arctic and the warm Gulf of Mexico. Alternating warm and cold air masses move across the plains in generally north and south directions parallel to the Rocky Mountains. The winters can be very cold and the summers very warm.

9.4 conclusion

The world population is over 6 billion and is very unevenly distributed around the world (Figure 9-18). Think about what natural factors determine where people live. Imagine that you are one of the first settlers in North America and that you are deciding where to live. What sorts of natural features of the land would you look for? What climate zone would you choose? Where could you best grow food to feed your family? Where would you be least likely to encounter natural disasters such as floods, hurricanes, earthquakes, or volcanic eruptions? All these questions have answers that follow from an understanding of Earth's physiography, climate, and biomes—all influenced by the distribution of lands, mountains, and oceans caused by geosphere movements.

Living close to an active plate boundary is a mixed blessing. Volcanic soils are fertile and feed large populations. The landscapes in active tectonic regions are mountainous and beautiful. Yet volcanoes are common here and their eruptions can be very destructive. Earthquakes also present hazards for those living

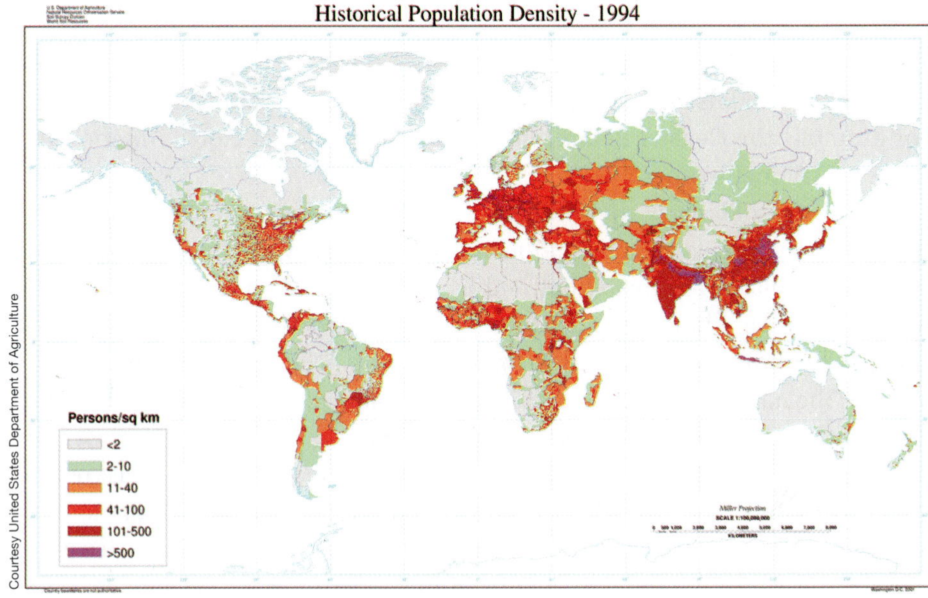

Figure 9-18 Historical population density—1994.

near active plate boundaries. In Chapters 10, 11, and 12 you will learn about the hazards associated with volcanic eruptions, earthquakes, and unstable land—hazards that all reflect a changing and dynamic geosphere.

IN THIS CHAPTER YOU HAVE LEARNED:

- The upper rigid part of the geosphere, called the lithosphere, is broken into thin pieces called plates. Plates include oceanic or continental lithosphere and vary tremendously in size. Some are larger than continents.
- Plates move about on Earth due to instabilities in the underlying and mobile upper mantle. Plate boundaries localize the convergent (moving toward each other), divergent (moving away from each other), or transform (moving horizontally past each other) movements of plates relative to one another.
- Plate movements cause many changes in the lithosphere, including the formation, growth, or splitting of continents; the formation of mountains; and the formation or destruction of ocean basins.
- Plate movements directly or indirectly shape the landscape and are a major control on the physiography of Earth. Earth's physiography, the physical nature of natural features, includes the distribution, size, elevation, and latitude of landmasses and oceans. These Earth characteristics control climate and biome development.
- The United States has five major physiographic realms: the (1) Canadian Shield, (2) Interior Plains, (3) Appalachian Highlands, (4) Coastal Plain, and (5) Western Mountains. All of the physiographic realms can be subdivided into smaller physiographic regions that share similar landforms. The relation of biomes to physiography is well illustrated by the influences of mountains and plains on forest and grassland development.
- Earth's dynamic lithosphere and changing physiography are major influences on where people live and how safe they are from natural hazards such as volcanoes and earthquakes.

KEY TERMS

Convergent plate boundaries
Divergent plate boundaries
Landforms
Lithosphere
Physiography
Plates
Rifting
Subduction
Tectonic movements
Transform boundaries

REVIEW QUESTIONS

1. What causes the plates of Earth's lithosphere to move?
2. What evidence do scientists have that the plates move?
3. In what three ways do plates move relative to one another?
4. Name and describe the five major physiographic realms in the United States.
5. What effect does plate movement have on Earth's climate and biomes?
6. What are factors that affect how mountains form and change over time?
7. Explain what happens in a subduction zone and why it happens.
8. Explain why the climate in Antarctica is so much colder than in the Arctic.
9. What are the positive and negative effects of living close to an active plate boundary?
10. What effects does Earth's physiography have on where humans choose to live?

chapter 10

volcanoes

Students from Mt. Baker High School in Deming, Washington, are very interested in a nearby mountain. It is a volcano, Mt. Baker, in the Cascade Mountain Range. The students have good reason to want to learn about Mt. Baker. It has erupted in the past and could erupt again.

To learn more about Mt. Baker, the students conduct field studies on the flanks of the volcano. They hike across the mountain slopes, examine rock outcrops, collect rock samples, and record their observations.

Back in the laboratory, the students continue their study of Mt. Baker by examining data and images obtained from satellites as well as maps that show the geologic features of the volcano. They discover that Mt. Baker has had

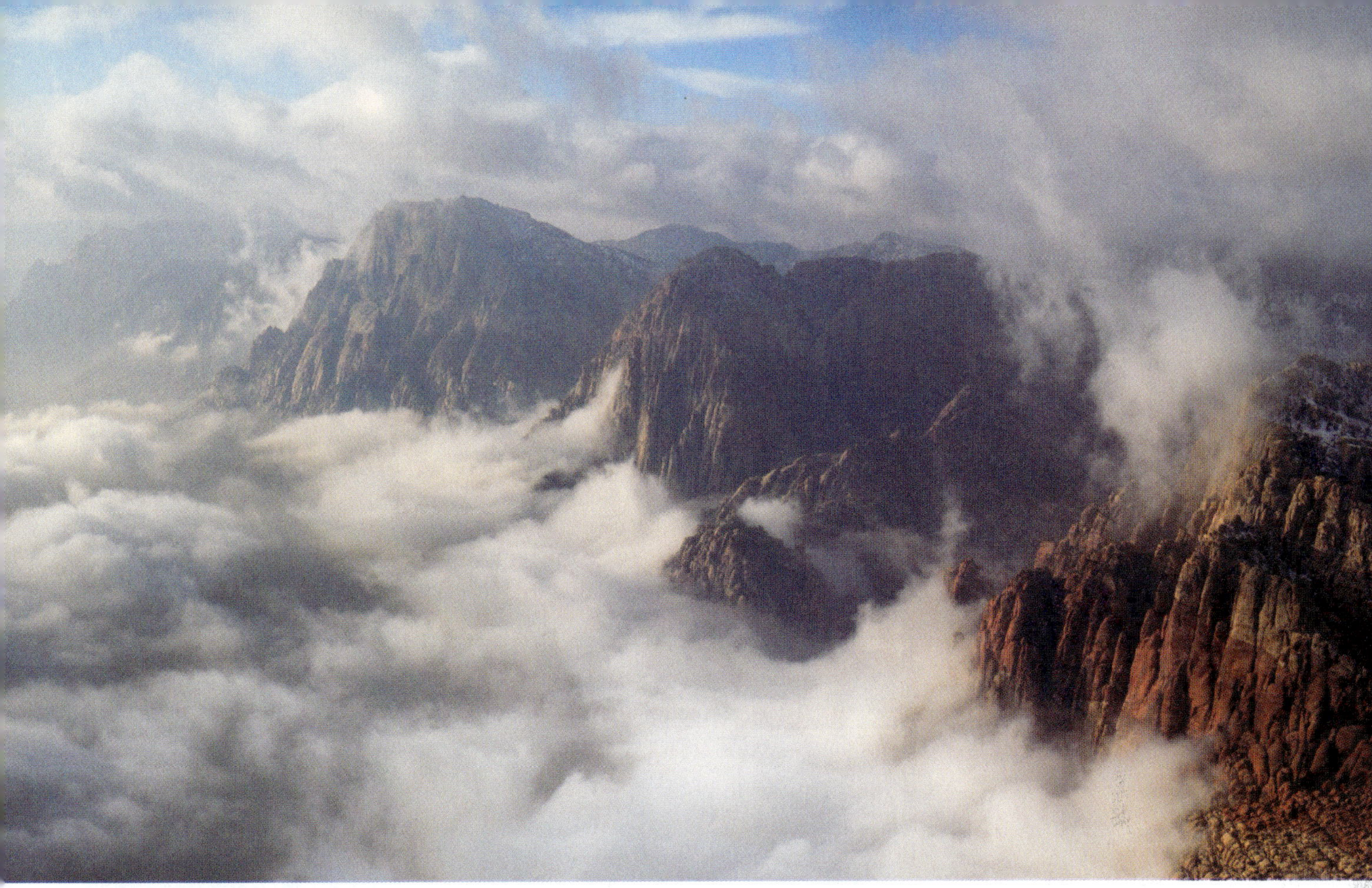

many explosive eruptions in its past. These eruptions have sent hot lava flows and other kinds of volcanic materials far down its flanks and into the nearby valleys.

Mt. Baker is 1 of 12 volcanoes in a chain along the crest of the Cascade Range in the western United States that have been active during the last 2,000 years (Figure 10-1). One of these volcanoes, Mt. St. Helens, erupted in 1980, killing 57 people and causing over a billion dollars of property and resource damage. Since the eruption of Mt. St. Helens, scientists and government officials have been working to improve monitoring of the volcanoes throughout the Cascades to predict and prepare for future eruptions. If you lived near Mt. Baker, you would probably be aware of the risks of living near a majestic but explosive mountain. But even if you do not live near any volcanoes, they can affect you. In some cases, this can happen even if the volcano erupts halfway around the world. As you learned in Chapter 4, volcanic eruptions send materials high into the atmosphere that can change the weather patterns of the entire planet and lead to global cooling.

Volcanic ash deposited thousands of miles from its source helps develop fertile soils used for food production. Processes that accompany volcanism provide geothermal energy (see Chapter 15), metals such as copper in mineral deposits (see Chapter 16), and broken and crushed rock used in construction (see Chapter 16). However, the more local effects of eruptions on people, other organisms, and their habitats are the most obvious. It is these effects that can be hazardous and destructive. An estimated 500 million people live where

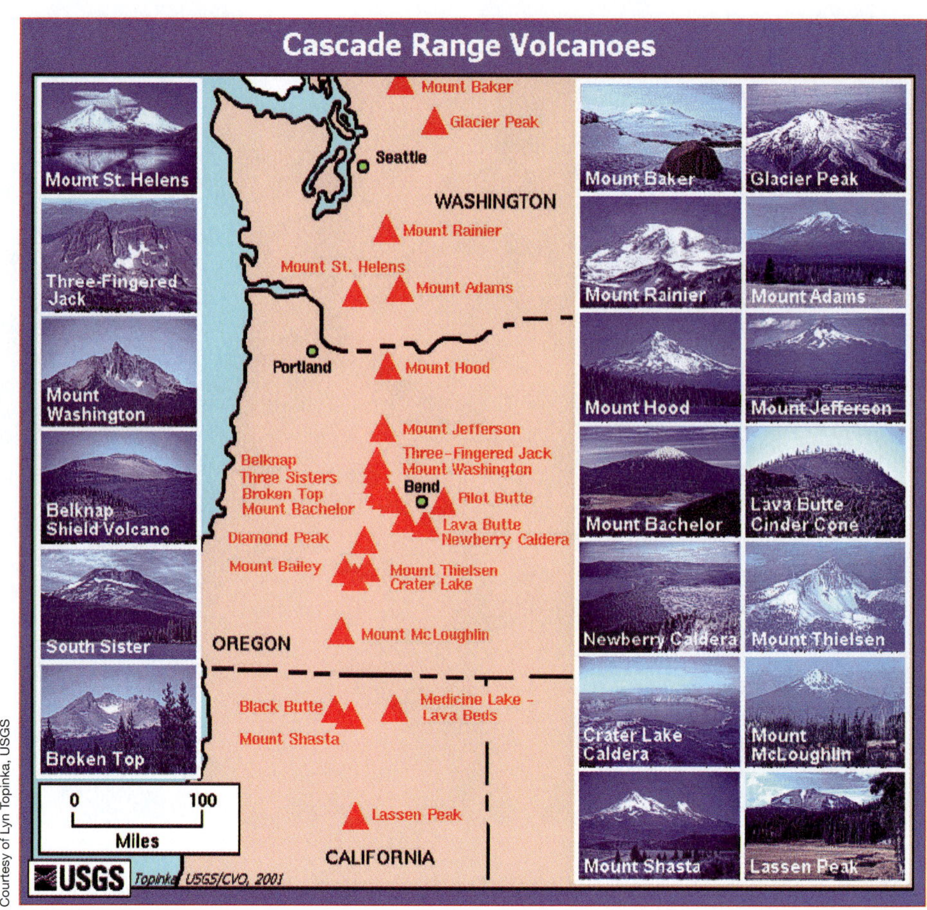

Figure 10-1 Cascade Range volcanoes.

they can be directly affected by volcanic eruptions. This chapter explains how volcanoes affect people and the environment, how people come to understand volcanoes, and how they can live more safely with them.

AFTER COMPLETING THIS CHAPTER, YOU WILL UNDERSTAND:
What volcanoes are and their role in the geosphere.
The two main types of volcanoes and why they erupt differently.
The ways volcanoes have an impact on the hydrosphere, atmosphere, and biosphere—especially people.
How people study volcanoes, evaluate their potential impacts, and predict future eruptions.
What people do to live more safely with volcanoes.

10.1 what volcanoes are

A **volcano** is a place where molten material from deep in the geosphere migrates up and erupts on Earth's surface. The molten material is called *magma* when it is within the geosphere and *lava* when it reaches the surface. Volcanic eruptions also release gases such as carbon dioxide and steam to the atmosphere. The hot lava erupted at volcanoes cools and solidifies to form volcanic rocks. Over a period of many thousands of years, these rocks accumulate on one another from one eruption to the next and gradually construct the mountains we call volcanoes. Two major types of volcanoes are formed: shield and stratovolcanoes.

Figure 10-2 The gentle slopes of a shield volcano, Mauna Loa, Hawaii. (*Courtesy the Earth Science World Image Bank © Bruce Molnia, Terra Photographics*)

Shield Volcanoes

Volcanoes that have gently sloping sides and the curved shape of a warrior's shield are called **shield volcanoes**. The volcanoes of Hawaii are excellent examples (Figure 10-2). Mauna Loa, an active shield volcano on the island of Hawaii, is about 105 km (65 miles) across at its base on the bottom of the ocean. This huge mountain rises almost 5,000 meters (16,400 feet) from the ocean bottom to reach sea level and another 4,168 meters (13,675 feet) above the sea to reach its summit. Mauna Loa is the tallest mountain in the world!

Shield volcanoes erupt very fluid lava. This characteristic of shield volcano lava enables it to flow rapidly out of fissures and vents and down the flanks of the volcano (Figure 10-3). It is red hot when it is erupted and is called basaltic lava. Basaltic lava is very fluid because of its composition: It contains a low amount of **silica**. Silica, a combination of the elements silicon and oxygen, is the most abundant component in Earth's rocks and minerals. The amount of silica in magma is the major control on how fluid, or **viscous**, it is.

Because basaltic lava has only about 50 percent silica, it is fluid enough to flow at high speeds for tens of kilometers before it cools and solidifies to form dark-colored rocks called **basalt**. Each basaltic lava flow is only a few meters thick and it takes many thousands of separate lava flows, accumulating on top of each other over a period of about a million years to build a volcano like Mauna Loa.

Shield volcanoes form wherever large amounts of basaltic lavas are erupted but excellent examples form within oceanic plates. Deep below oceanic shield volcanoes, in regions of higher heat flow called "hot spots," moving mantle material partially melts to form basaltic magma (Figure 10-4). Because these magmas are liquid and less dense than the surrounding mantle material, they are buoyant and rise to Earth's surface to form the basalt in shield volcanoes.

There is more basalt on Earth than any other kind of volcanic rock. Divergent plate boundaries and other regions where Earth's crust is pulled apart, or extended, are other places where basalt magmas are formed in the mantle and erupted onto Earth's surface. As explained in Chapter 9, divergent plate boundaries can split continents and create oceans. These plate boundaries localize the basaltic magmatism and lava eruptions that form new oceanic crust. There is a tremendous amount of basaltic volcanism in these settings, but because so much is on the deep seafloor and relatively gentle in its eruptive style, it is not of special concern to people. However, eruptions on some shield volcanoes can be a direct concern to people.

Stratovolcanoes

The spectacular cone-shaped mountains of the Cascade Range, such as Mt. Baker and Mt. Jefferson (Figure 10-5), are **stratovolcanoes**. These volcanoes are capable of explosive eruptions. They can erupt debris 30 km (20 miles) up into the

Figure 10-3 Eruption styles at a shield volcano.

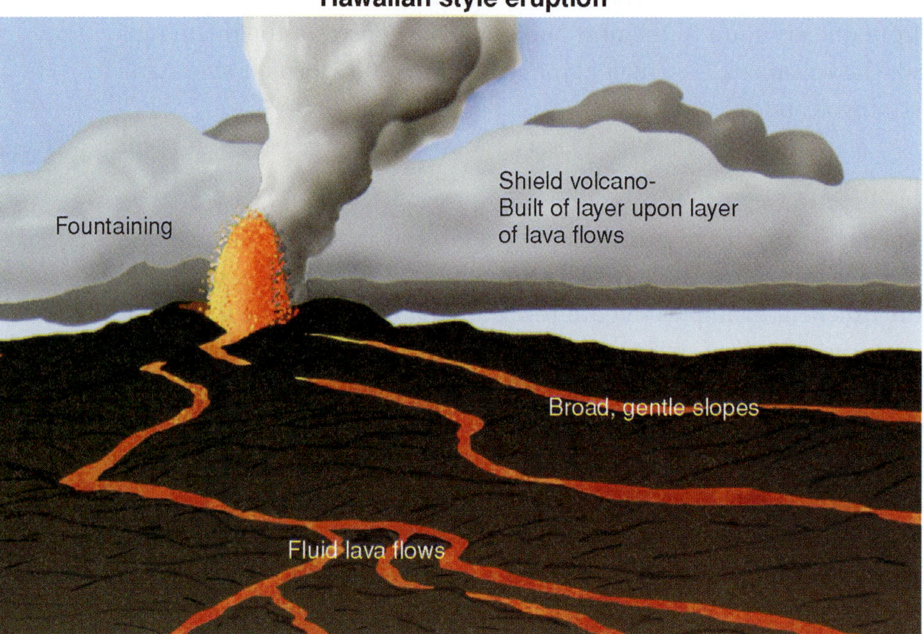

Cascades style eruption

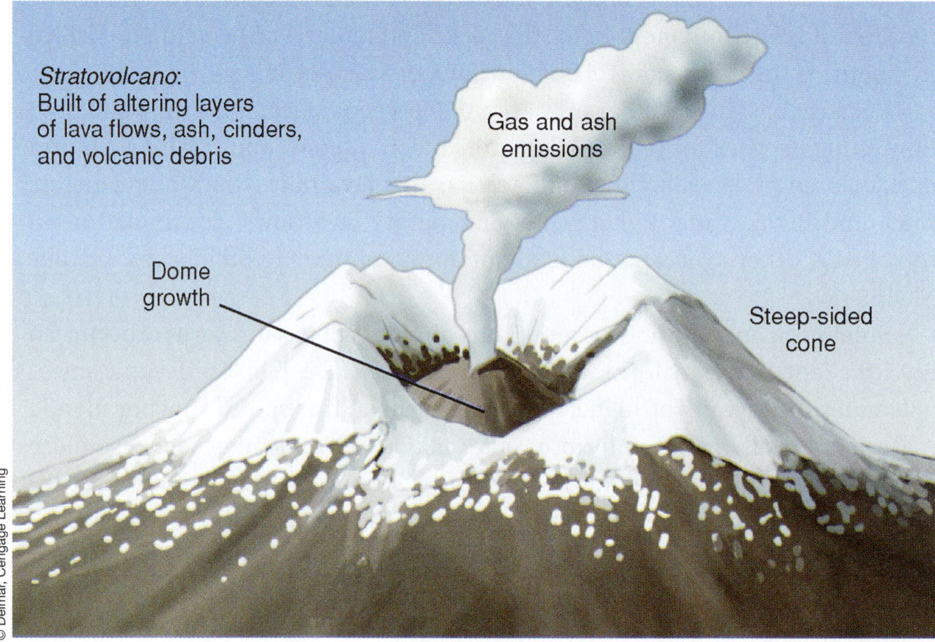

atmosphere, send hot masses speeding down their flanks, and rain volcanic ash down hundreds of kilometers (miles) away from the volcano. Stratovolcano eruptions make those at shield volcanoes look gentle.

Stratovolcanoes erupt many different kinds of materials, varying from lava flows to ash. Volcanic ash is not like the ash in fireplaces or campfires. Volcanic ash is composed of tiny rock fragments, crystals, and frozen lava (glass) ejected high into the atmosphere. The variety of materials that erupt at stratovolcanoes produces their characteristic form. Alternating layers and lenses of lava flows, ash, and other deposits accumulate to produce the steeper slopes of a cone-shaped stratovolcano (Figure 10-6). The variety of these materials is why stratovolcanoes are also called composite volcanoes.

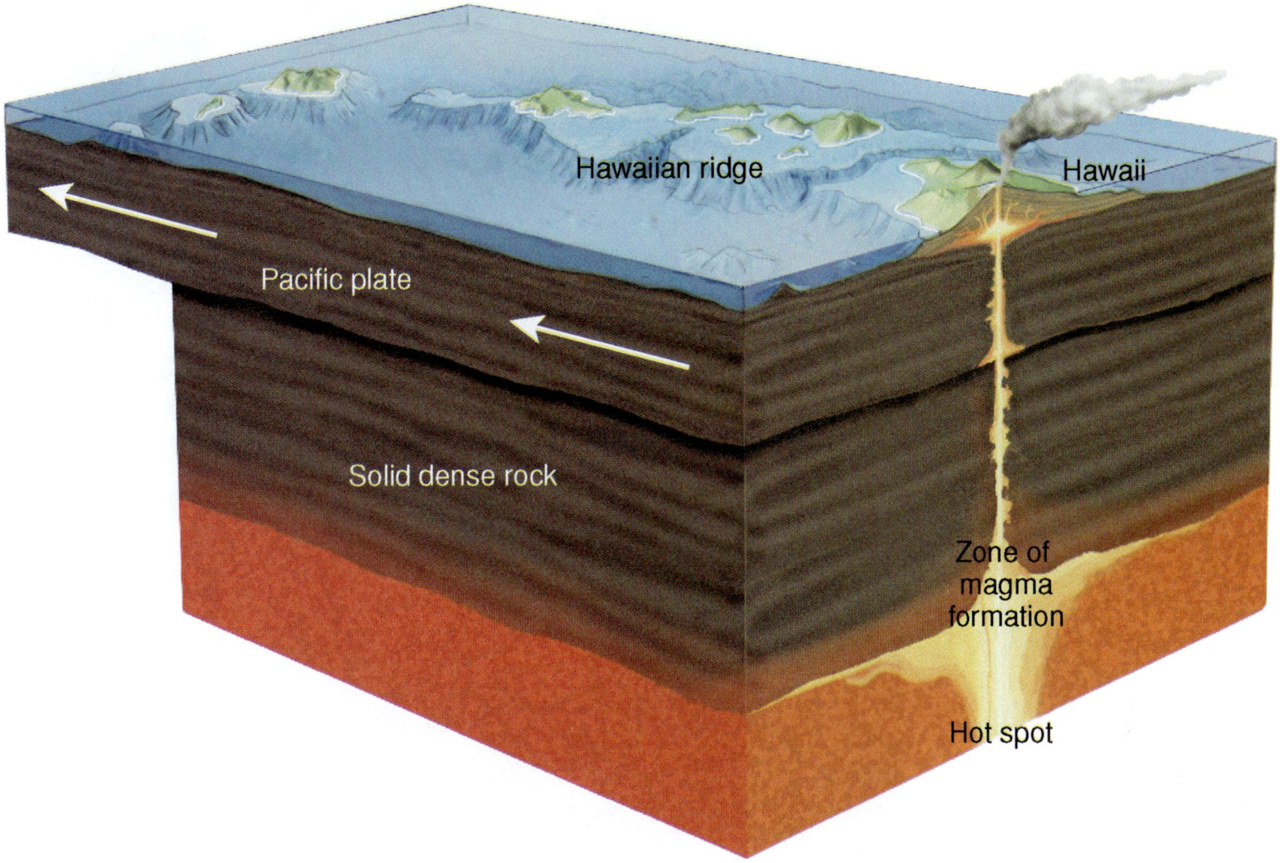

Figure 10-4 Movement of the Pacific Plate over the fixed Hawaiian "Hot Spot" and creation of the shield volcanoes of the Hawaiian Ridge-Emperor Seamount Chain.

Stratovolcanoes are a composite of many different materials because they erupt lavas with variable compositions. In most cases, stratovolcanoes erupt lava with intermediate to high silica contents, about 60 to 70 percent silica compared to the 50 percent in basaltic lava at shield volcanoes. Magma with about 60 percent silica is called **andesitic** and magma with 70 or more percent silica is called **rhyolitic**. Andesite and rhyolite are the volcanic rocks that form when these lavas solidify. Because of their high silica content andesitic to rhyolitic lavas at stratovolcanoes are more sticky (viscous) and less fluid than basaltic lavas at shield volcanoes. They do not flow on the surface as easily as basaltic lava. In addition, magmas at stratovolcanoes commonly contain more dissolved gases, especially water vapor, than shield volcanoes. The amount of dissolved water is especially important because it is a major influence on how explosive eruptions are.

Figure 10-5 Classic Cascade stratovolcano, Mount Jefferson in Oregon.

Magmas below stratovolcanoes can pool in the crust. The pressure deep underground keeps gases dissolved in the magma, but once the magma moves upward to lower pressure regions, the gases separate. When water vapor separates from magma, it dramatically expands in volume and increases pressure, much like what happens in a can of soda when it is shaken. Local increases in water vapor pressure are

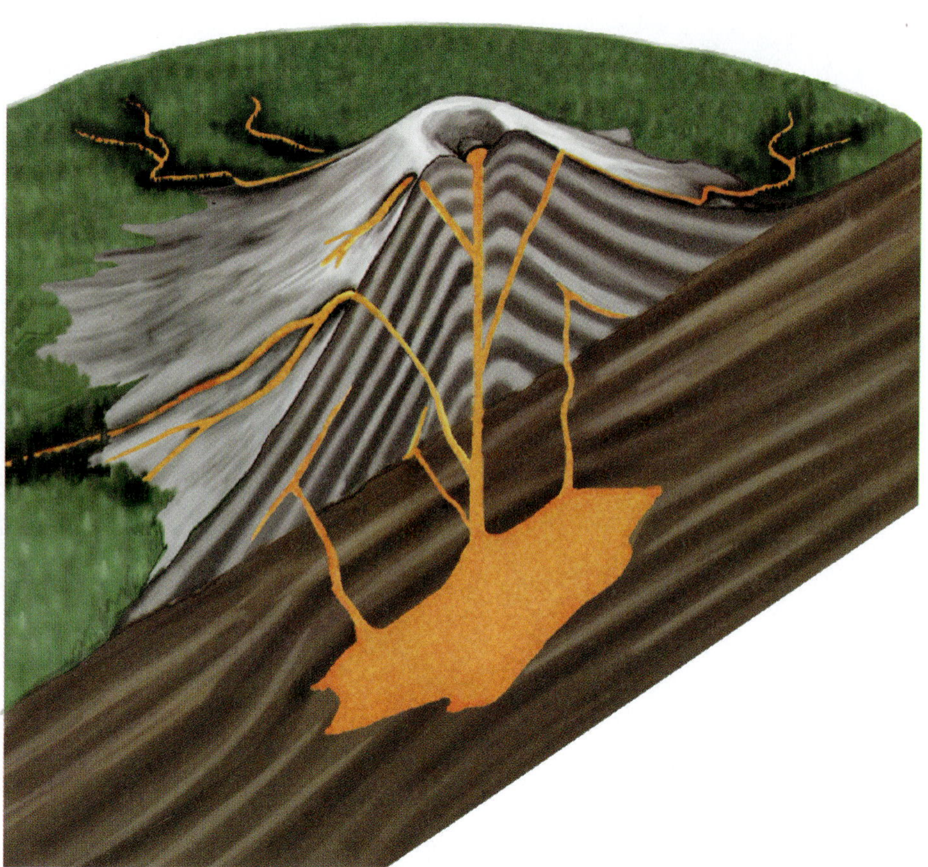

Figure 10-6 Stratovolcano.

also created where a hot magma body within a stratovolcano converts ground water to steam. The high-pressure vapor confined underground is an explosive eruption waiting to happen. If continued upward magma movements release the pressure, an explosive eruption is likely.

Because of the compositional complexity of stratovolcano magmas, they erupt in many different ways and produce many different kinds of volcanic materials. These materials include:

Lava flows: Andesitic lava can flow down slopes to the base of a stratovolcano. Individual flows can be tens of meters thick and several kilometers long. More rhyolitic lavas do not flow well at all and form plugs and domes near the eruptive vents. The more viscous magmas can plug up underground channels (**conduits**) and help trap water vapor.

Pyroclastic flows: The initial blast of a stratovolcano consists of a very hot, dense cloud of volcanic gases, rock debris, ash, and water called a **pyroclastic flow**. The hot debris of a pyroclastic flow can travel down slope from the eruption vent at speeds of over 320 km (200 miles) per hour.

Tephra falls: **Tephra** is the most general term for material blasted into the air by a volcano. It includes ash and larger rock and lava fragments. Larger tephra falls back onto the flanks of the volcano but small tephra—ash—blasted high into the atmosphere can travel thousands of kilometers (miles) before it falls to Earth.

Lahars (mudflows): During and after some eruptions, water from rainfall, melting snow and ice, or the sudden failure of a natural dam mixes with volcanic ash and other eruption debris on the stratovolcano flanks to form giant, fast-moving mudflows called **lahars**.

Debris-avalanches: Volcanic debris-avalanches are giant landslides of volcanic rock and slope debris. They start as a sudden massive rockslide that breaks apart into a jumbled flowing mass of rock fragments ranging in size from grains of sand to blocks hundreds of meters (yards) across.

Stratovolcanoes characteristically form above subduction zones at converging plate boundaries (Figure 10-7). In Chapter 9 you learned that subduction zones form where dense oceanic crust sinks below another oceanic plate or an overriding, less dense continental plate. The down-going oceanic plate carries with it water-rich seafloor sediments and altered basalt. As the plate sinks into the mantle, increasing heat causes the water-bearing minerals to break down (dehydrate) and release their contained water. This water is then free to migrate upward into the mantle above the subduction zone and below the overriding crust.

Water lowers the melting point of the mantle and causes it to partially melt. The resulting magma can then migrate upward into the crust where it pools below stratovolcanoes. In some cases, these magmas heat the crust enough to cause more partial melting and magma formation. Eventually parts of these magmas break through to the surface and erupt.

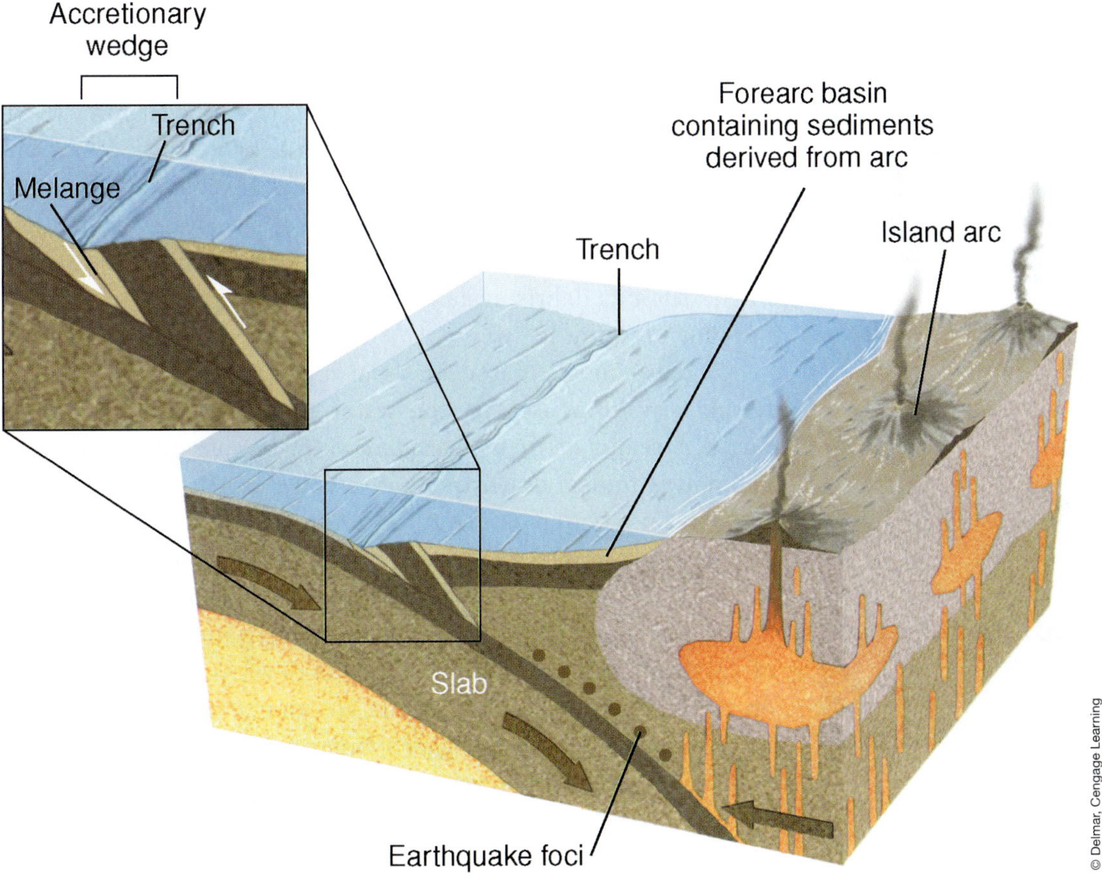

Figure 10-7 Formation of stratovolcanoes at subduction zones.

 volcanoes and earth systems

Many stratovolcano eruptions can have large and seemingly instantaneous effects on the world around them. They are direct evidence of the dynamic and changing character of the geosphere where their key role is the transfer of material and energy from the mantle to the crust. As you learned in Chapter 9, all crust, including continental crust, has an origin ultimately tied to volcanism. Some crust was formed very early in Earth's history but some is forming today wherever volcanoes are active. In this sense, volcanoes are an essential part of Earth's systems and their interactions.

Figure 10-8 Valley and drainage of Upper Muddy Creek filled with volcanic debris from Mt. St. Helens. *(Courtesy: U.S. Department of the Interior, U.S. Geological Survey Cascades Volcano Observatory, Vancouver, WA)*

The materials that volcanoes erupt—lava, tephra, and gases—have significant effects on the hydrosphere, atmosphere, and biosphere. As emphasized in Chapter 4, these effects can be very local (on the flanks of a volcano), regional, (downstream or downwind of a volcano), or global if large amounts of ash are erupted high into the atmosphere. These impacts also vary from almost instantaneous (seconds and minutes), in the case of a violent eruption, to those that take thousands of years to change the environment. Although stratovolcanoes have more obvious environmental consequences, shield volcanoes and other places of extensive basaltic volcanism have significant environmental effects too.

Courtesy Ann Gard, USGS

Figure 10-9 Spirit Lake covered by forest debris. USGS scientists collecting water samples in a boat.

Impacts on the Hydrosphere

Volcanic eruptions send lava flows, pyroclastic flows, lahars, and debris-avalanches into lakes and streams (Figure 10-8). These materials clog the drainages with debris that continues to be eroded into streams for many years. Lakes also collect much debris from eruptions, even the remains of nearby devastated forests (Figure 10-9). In addition, lava flows and other mass movements such as debris-avalanches can dam drainages and create new lakes (Figure 10-10).

Figure 10-10 Debris avalanche blocks drainage and creates a new lake. *(Courtesy: U.S. Department of the Interior, U.S. Geological Survey Cascades Volcano Observatory, Vancouver, WA)*

Because much of the material entering lakes and streams during eruptions are hot, the water temperature is commonly raised. Streams can have elevated water temperatures for many kilometers (miles) downstream of a volcano. Hot rocks within a volcano raise water temperatures underground too. Hot springs, which in some places become geysers, are common around volcanoes (Figure 10-11). Water from hot springs can be a major reason that streams have elevated water temperatures for hundreds of years after an eruption. Hot springs develop in volcanic areas on the seafloor (Figure 10-12), and volcanic eruptions on oceanic islands or coastlines can trigger submarine landslides that initiate large and fast-moving waves called tsunamis. Because earthquakes are a much more common cause of tsunamis than volcanoes, you will learn more about them in Chapter 11.

Impacts on the Atmosphere

Volcanoes have major effects on the atmosphere. The ash and gases they erupt can change atmospheric composition both near and far from volcanoes. The dense, dark clouds of ash and other debris that are blasted high into the atmosphere during stratovolcano eruptions are the most obvious impact (Figure 10-13). Because much ash is composed of very small rock and frozen lava (glass) fragments, it can be blown downwind hundreds to thousands of kilometers (miles).

Everywhere the ash goes, it adds material to the atmosphere that was not previously there. Volcanoes also erupt large amounts of gases, primarily water vapor, carbon dioxide, and sulfur dioxide (Table 10-1). These gases can be released to the atmosphere even when the volcanoes are not erupting. They essentially seep through the ground from hot rocks and underground magma bodies.

Sulfur dioxide is emitted to the atmosphere during volcanic eruptions at rates varying from less than 20 tonnes per day to greater than 10 million tonnes per year. For example, the June 15, 1991, eruption of Mt. Pinatubo in the Philippines released 17 million tonnes of sulfur dioxide. Sulfur dioxide is a colorless gas, but once it is in the atmosphere, it quickly combines with water vapor to form tiny suspended droplets (an aerosol) of sulfuric acid. These aerosols are easily blown around the world (Figure 10-14). They can cause acid rain and volcanic smog (Figure 10-15) and lead to temporary global cooling (see Chapter 4).

Volcanic eruptions around the world emit between 130 and 230 million tonnes of carbon dioxide to the atmosphere each year. Other than water vapor, it is the most abundant gas emitted by volcanoes (see Table 10-1). Carbon dioxide is a colorless and odorless greenhouse gas that is heavier than air. In some cases, it is concentrated enough to accumulate in low areas on Earth's surface (Figure 10-16). More commonly, carbon dioxide is dispersed into the atmosphere. Although people's activities release more than

Courtesy the Earth Science World Image Bank © Larry Fellows

Figure 10-11 Some underground hot springs form into geysers.

Courtesy UCSB, Univ. S. Carolina, NOAA, WHOI

Figure 10-12 Seafloor hot spring.

Table 10-1 *Composition of gases released at volcanoes.*

Volcano Tectonic Style Temperature	Kilauae Summit Hot Spot 1170°C	Erta' Ale Divergent Plate 1130°C	Momotombo Convergent Plate 820°C
H_2O	37.1	77.2	97.1
CO_2	48.9	11.3	1.44
SO_2	11.8	8.34	0.50
H_2	0.49	1.39	0.70
CO	1.51	0.44	0.01
H_2S	0.04	0.68	0.23
HCL	0.08	0.42	2.89
HF	—	—	0.26

(from Symonds et. al., 1994). http://volcanoes.usgs.gov/Hazards/What/VolGas/volgas.html

Figure 10-13 Ash-rich stratovolcano eruption.

150 times the carbon dioxide than do volcanoes, volcanic emissions of carbon dioxide are still important to understanding how greenhouse gases are affecting global climate (see Chapter 4).

Impacts on the Biosphere

As you have probably already recognized, all the changes created by volcanoes on the landscape, hydrosphere, and atmosphere must have an impact on the biosphere. Aquatic and terrestrial habitats and everything living in them can be devastated on the flanks of volcanoes during eruptions (Figure 10-17). However, over time habitat can be reestablished and organisms can reoccupy volcanic areas.

Volcanoes also can create important habitat. The mountains they form provide elevation changes that enable a wide variety of habitats to develop. For example, the tropical jungles of the coast give way upslope to treeless and snowy summits at elevations over 4,000 meters (13,000 feet) on the island of Hawaii (Figure 10-18). A very special habitat is developed where hot springs occur on seafloor volcanoes. This unique ecosystem is home to unusual varieties of bacteria, clams, tube worms, crabs, and other organisms (Figure 10-19). It is here that hydrogen sulfide-eating bacteria flourish and form the base of the food chain. This is one of the few ecosystems on Earth where chemical reactions rather than sunlight and photosynthesis provide the original energy for food chains (see Chapter 2).

Figure 10-14 Distribution aerosols from the 1991 Mt. Pinatubo eruption.

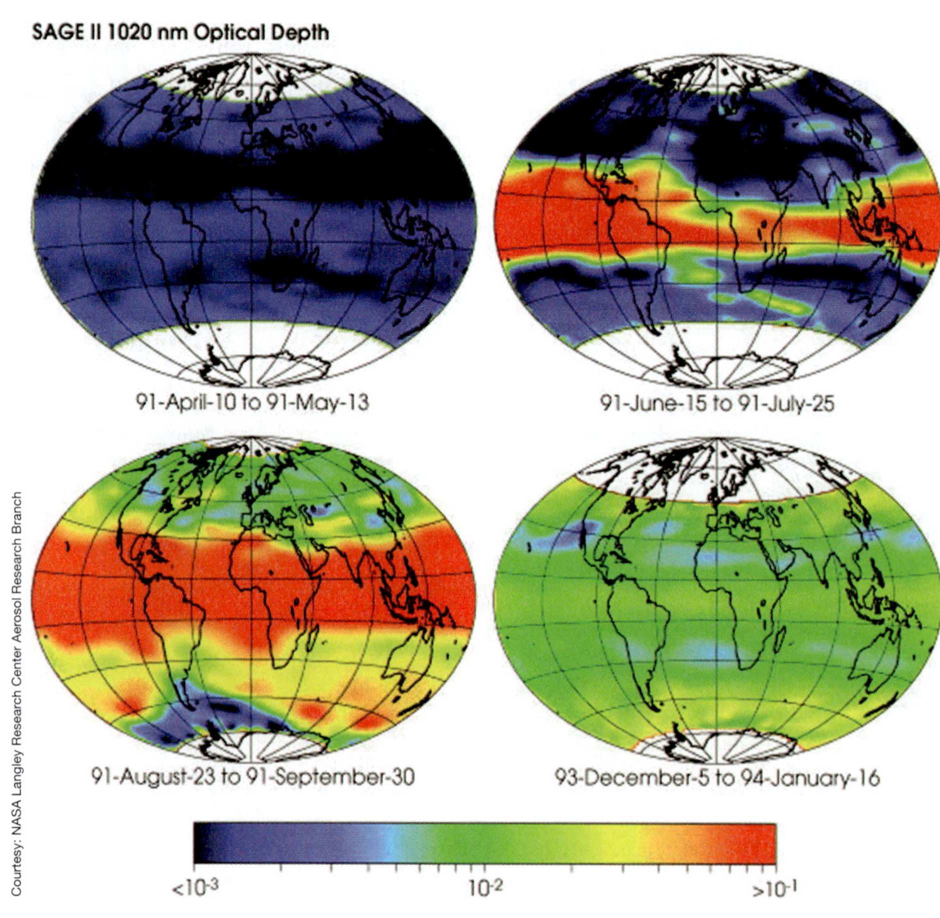

Figure 10-15 Volcanic smog in Hawaii. (*Courtesy U.S. Department of the Interior, U.S. Geological Survey, Menlo Park, California, USA*)

Figure 10-16 Mammal skeleton in carbon dioxide sink. (*Courtesy U.S. Department of the Interior, U.S. Geological Survey, Menlo Park, California, USA*)

Figure 10-17 Trees on the slopes of Smith Creek Valley, east of Mount St. Helens, blown down by the lateral blast. (*Courtesy Lyn Topinka, USGS*)

Figure 10-18 Snowy summit on island of Hawaii. (*Courtesy USGS Library*)

Volcanoes indirectly affect the biosphere by their role in developing soils. As you will see in Chapter 14, soils are where parts of the geosphere, hydrosphere, atmosphere, and biosphere are intimately mixed. Volcanic materials are rich in minerals that help to form fertile soils.

Figure 10-19 Aquatic organisms living by hot springs on the seafloor. (*Courtesy OAR/National Undersea Research Program (NURP)*)

 volcanoes and people

Volcanoes have affected people for all of recorded history. In 79 A.D., Mt. Vesuvius—a stratovolcano in present-day Italy (Figure 10-20)—violently erupted, burying the Roman town of Pompeii in a pyroclastic flow that killed over 3,300 people. A Roman citizen, Pliny the Younger, watched the eruption from a distance of 30 km (19 miles) and described what he saw in a letter. His account includes earthquakes that he felt before the eruption, the shape of the eruption, and the adverse effects on the people around him. But it was 1,700 years later that Pompeii was rediscovered and archeological excavations began to tell Pompeii's story (Figure 10-21). Pompeii preserves a community's final terrifying moments during a volcanic eruption. Since 79 A.D. volcanic eruptions have killed over 200,000 people. Since 1980, about 30,000 people have died in eruption-related disasters.

Two recent examples of eruptions illustrate the great loss of property and habitat possible even if human casualties are low. The 1980 eruption of Mt. St. Helens in Washington claimed 57 human lives and caused over \$1 billion in damages.

Courtesy USAF

Figure 10-20 Mt. Vesuvius eruption.

© Delmar, Cengage Learning

Figure 10-21 Excavation site at Pompeii with mold of people who died in the 79 A.D. eruption of Mt. Vesuvius.

(a)

(b)

Figure 10-22 Impact of the Soufriere Hills volcanic eruption on the island of Montserrat.
((a) Courtesy U.S. Department of the Interior, U.S. Geological Survey, Menlo Park, California, USA (b) Courtesy U.S. Department of the Interior, U.S. Geological Survey, Menlo Park, California, USA)

The first phase of the eruption involved a massive landslide triggered by earthquakes that uncorked a lateral blast of ash and debris-laden hot gas. The ash cloud was as hot as 300°C and traveled at speeds of 100 to 400 kph (60 to 250 mph). The cloud rolled over a 500 km^2 (200 square miles) area and reached as far as 18 km (11 miles) away, laying waste to everything in its path. Every tree was uprooted or knocked over like matchsticks. The loss of trees to the timber industry was the most costly of the damages. Significant damage was also done to streams and rivers that were choked with ash and debris. Millions of dollars have been spent on flood control and dredging of river channels downstream of Mt. St. Helens since the 1980 eruption.

Ten years after the Mt. St. Helens eruption, the Soufriere Hills volcano on the Caribbean island of Montserrat erupted. Before 1995, Montserrat was a bustling island with a population of 11,000. Sheep and cattle roamed the hills, and chartered yachts came into port on weekly tours. In 1995, the Soufriere Hills volcano, which had been dormant for 400 years, started erupting.

Debris-avalanches and hot ash clouds erupted over the southern part of the island (Figure 10-22). A 1997 eruption buried most of the southern island and left the former capitol and several other towns uninhabitable, buried under mountains of gray ash. Since 1995 more than half of the residents have left the island. The remaining people have relocated to parts of the northern end of the island that have been declared safe. In addition to most of the inhabitants losing their homes and most of their belongings, Montserrat's economy has been devastated by losses in tourism and farmland. Constant monitoring keeps the remaining residents informed of volcanic activity and hazards.

Volcanoes and Agriculture

Many people live near volcanoes because fertile soils develop around them. As one volcanologist commented "People are willing to take high-risk gambles for the most basic things of life—especially food." The weathering of volcanic rocks over thousands of years produces some of the most fertile soils on Earth. Volcanic ash is like a time-release abiotic fertilizer. Chemical weathering releases nutrients from the volcanic debris and makes them available to plants.

The land around Mt. Vesuvius in Italy is an excellent example of the importance of volcanic soils for agriculture. The best farming in southern Italy is in the region that was blanketed by volcanic ash and tephra during two large eruptions of Mt. Vesuvius 35,000 and 12,000 years ago. The rest of the region has poor soils developed on limestone and is not good for farming. The ash and tephra from the two ancient eruptions weathered to rich soils.

In tropical rainy regions, the formation of fertile soil and growth of vegetation after an eruption can take only a few hundred years. Examples of such places include the northeastern side of the island of Hawaii, and in many parts of Indonesia, the Philippines, and central and South America. Fertile volcanic soils are helping sustain a large and growing global population.

Although the short-term consequences of volcanic eruptions can be devastating for vegetation, over the long term, they are essential. Whether or not the benefits of volcanoes outweigh the risks, one thing is sure: People will continue to live dangerously close to them.

Volcano Hazards

Volcanoes can endanger people and property in many ways. Stratovolcanoes are of particular concern because many people live around them and their very explosive eruptions can be devastating. Stratovolcanoes are also a place where hazards are present even when eruptions are not in progress. The hazards present at a stratovolcano are illustrated in Figure 10-23. These include lava flows, pyroclastic flows, tephra falls, debris-avalanches, and noxious gases.

Lava Flows

Movement of lava down the flanks of volcanoes can directly engulf homes and communities (Figure 10-24). Although lava flows occur at both shield and stratovolcanoes, they are most commonly a hazard at shield volcanoes because basaltic lavas can travel farther and faster than the lavas common at stratovolcanoes.

Pyroclastic Flows

Because of their speed, volume, density, and scorching hot temperatures, pyroclastic flows are especially dangerous. They are a mixture of hot ash, gas, and rock that behave as a fluid. These flows can quickly travel tens of kilometers (miles) from their source as evidenced by the example of Pompeii, which was completely buried by a pyroclastic flow from Mt. Vesuvius almost 30 km (19 miles) away. A more recent example is the port town of St. Pierre on the

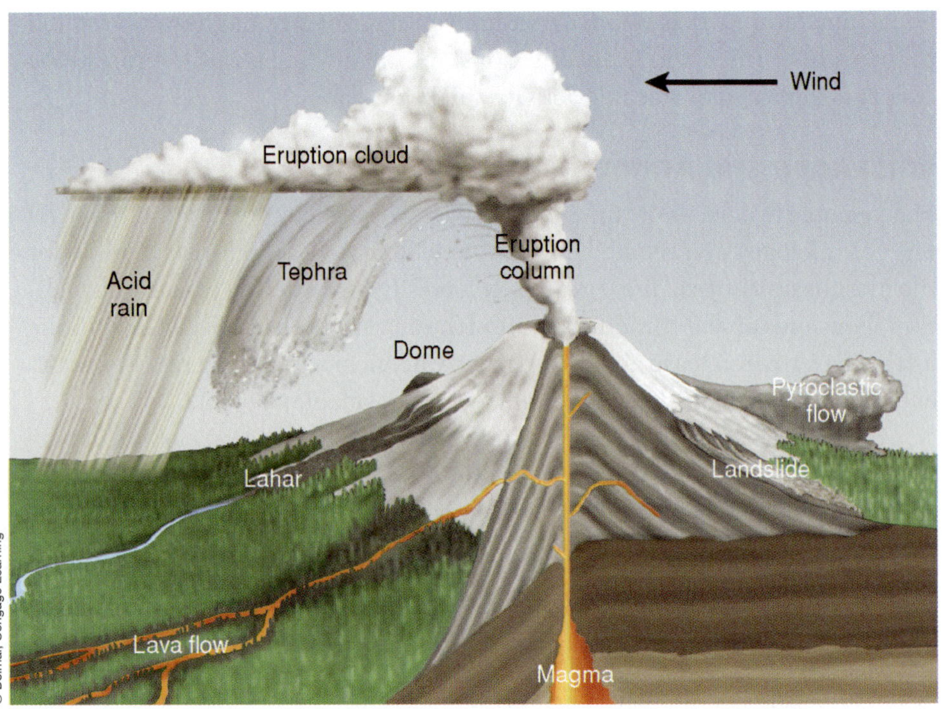

Figure 10-23 Hazards developed around stratovolcanoes.

Figure 10-24 Hawaiian lava flow engulfing a house.

Caribbean island of Martinique. In 1902, the nearby stratovolcano, Mt. Pelee, sent an 800°C (1472°F) pyroclastic flow down its slopes that within seconds devastated everything in its path (Figure 10-25).

Tephra Falls

Tephra varies from house-size blocks to tiny ash particles. Lava spurted from shield volcano vents can be spectacular (Figure 10-26) but these eruptions tend to be small, localized, and reasonably predictable. Tephra ejected by a stratovolcano is another matter. Large blocks of rock and lava can be thrown far down the side of the volcano. The sudden eruption of large tephra has killed even knowledgeable scientists studying stratovolcanoes. But the biggest impacts on people are caused by the dense clouds of erupted ash (Figure 10-27).

Ash from volcanic eruptions causes a host of problems downwind. Although it may look like dirty snow, it is much heavier. If ash fall is accompanied by rain, it will absorb water and become even heavier. A few centimeters (inches) of ash is very difficult to remove and can collapse roofs.

Volcanic ash infiltrates openings and is highly abrasive. It will destroy moving parts in machines such as car and jet engines. More than 80 commercial airplanes have unexpectedly encountered volcanic ash in the past 15 years. Seven of these incidents resulted in loss of jet engine power and crashes were only narrowly avoided. From the vantage point of an airplane, volcanic ash clouds look like typical clouds and unsuspecting pilots steer passenger jets straight into them. The following hair-raising account describes a close call over Alaska.

> *As the crew of KLM Flight 867 struggled to restart the plane's engines, "smoke" and a strong odor of sulfur filled the cockpit and cabin. For five long minutes the powerless 747 jetliner, bound for Anchorage, Alaska, with 231 terrified passengers aboard, fell in silence toward the rugged,*

(a)

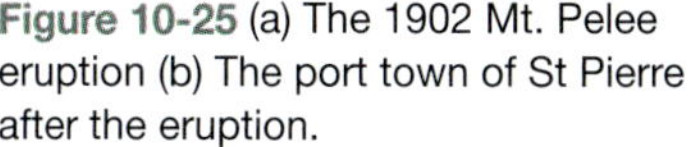

(b)

Images courtesy of: Library of Congress

Figure 10-25 (a) The 1902 Mt. Pelee eruption (b) The port town of St Pierre after the eruption.

snow-covered Talkeetna Mountains (7,000 to 11,000 feet high). All four engines had flamed out when the aircraft inadvertently entered a cloud of ash blown from erupting Redoubt Volcano, 150 miles away. The volcano had begun erupting 10 hours earlier on that morning of December 15, 1989. Only after the crippled jet had dropped from an altitude of 27,900 feet to 13,300 feet (a fall of more than 2 miles) was the crew able to restart all engines and land the plane safely at Anchorage. The plane required $80 million in repairs, including the replacement of all four damaged engines.

Courtesy USGS Library

Figure 10-26 Fountaining magma at a Hawaiian volcano.

An increase in commercial air traffic every year increases the chances that planes will encounter volcanic ash plumes. For example, more than 10,000 passengers and millions of dollars in cargo fly across the North Pacific region each day. Figure 10-28 shows North Pacific and Russian Far East air routes. These routes pass over or near more than a hundred

Courtesy USGS Library

Figure 10-27 Ash particles and a village buried in the ash fallout.

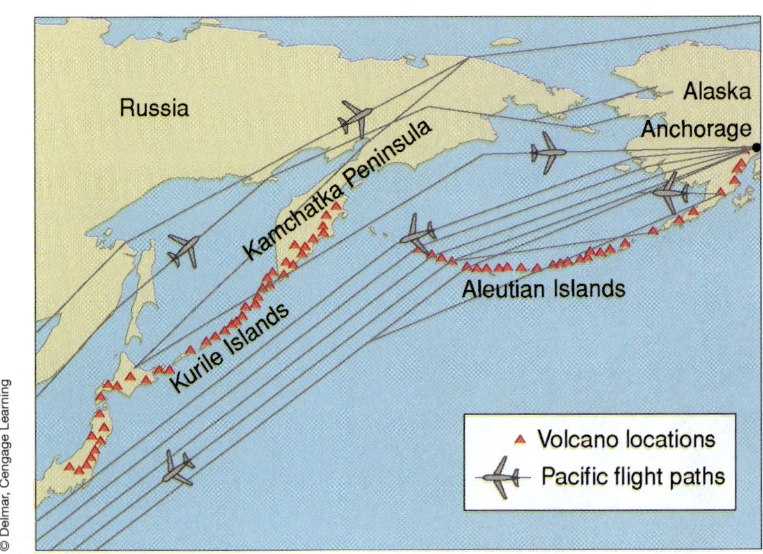

Figure 10-28 Northwest Pacific flight paths and stratovolcano locations.

active volcanoes. In the North Pacific region several explosive eruptions occur every year. Ash from these eruptions is usually blown to the east and northeast, directly across the air routes.

Debris-Avalanches

Debris-avalanches and the landslides that spawn them can be a real problem on the flanks of stratovolcanoes. A very large debris-avalanche occurred at Mt. St. Helens on May 18, 1980. Eyewitness accounts from geologists Keith and Dorothy Stoffel, flying over the volcano in a small plane at that moment, tell the story of the event. These trained observers

> *... noticed landsliding of rock and ice debris in-ward into the crater ... the south-facing wall of the north side of the main crater was especially active. Within a matter of seconds, perhaps 15 seconds, the whole north side of the summit crater began to move instantaneously.... The nature of movement was eerie.... The entire mass began to ripple and churn up, without moving laterally. Then the entire north side of the summit began sliding to the north along a deep-seated slide plane. I (Keith Stoffel) was amazed and excited with the realization that we were watching this landslide of unbelievable proportions.... We took pictures of this slide sequence occurring, but before we could snap off more than a few pictures, a huge explosion blasted out of the detachment plane. We neither felt nor heard a thing, even though we were just east of the summit at this time.*

To escape the eruption, the Stoffels put their plane into a steep dive to gain speed, and luckily were able to outrun the mushrooming eruption cloud. The Stoffels were able to escape harm and their account helped scientists unravel how the eruption progressed.

At 8:32 A.M. on May 18, an earthquake 1.5 km (0.9 mile) beneath the volcano triggered the landslide that led to the Mt. St. Helens debris-avalanche. In 15 seconds the north flank of the volcano collapsed in a series of blocks that merged down slope into a gigantic debris-avalanche. Moving at speeds of 160 to 185 kph (110–115 mph), it traveled over 6 km (3.7 miles) north of the summit with enough momentum to flow up and over a ridge more than

300 meters (980 feet) high! The avalanche covered about 6.2 km² (2.5 square miles), traveled more than 20 km (12 miles) down the North Fork and Toutle River, and filled the valley to a depth of about 45 meters (150 feet; Figure 10-29). The volume of the deposit was almost 4 km³ (1 cubic mile). The removal of almost half of the volcano by the debris-avalanche released pressure and triggered an explosive eruption out the side of the mountain and vertically into the atmosphere (Figure 10-30).

Figure 10-29 Area affected by Mt. St. Helens debris-avalanche. *(Courtesy the Earth Science World Image Bank © USGS Cascade Volcano Observatory)*

Lahars

Lahars can cause more destruction than an explosive eruption. During a volcanic eruption, all the heat brought to the top of the volcano melts snow and ice very quickly and lahars can form within hours. In tropical climates, heavy rainfall also creates lahars. Heavy monsoon rains accompanied the 1991 eruption of Mt. Pinatubo in the Philippines, and within hours lahars reached the lowlands. The Mt. Pinatubo eruption deposited 4 km³ (1 cubic mile) of ash and rock fragments onto the slopes of the volcano. Over the next four rainy seasons, lahars carried half of the eruption material off the volcano, causing widespread destruction in the lowlands. Since that 1991 eruption, lahars have destroyed the homes of more than 100,000 people in the area surrounding Pinatubo (Figure 10-31). Lahars are a danger for as long as there is loose volcanic material on the slopes of a volcano.

Let us examine why lahars are so destructive. Consider how powerful the current is in a large river and how dangerous rivers are when they flood. Imagine if thick mud replaced the water, how would this change the power of the river? Lahars are large rivers of dense mud with a consistency similar to wet concrete. They are powerful enough to destroy most everything in their path, including trees and buildings. They can flow faster than 60 kph (37 mph)

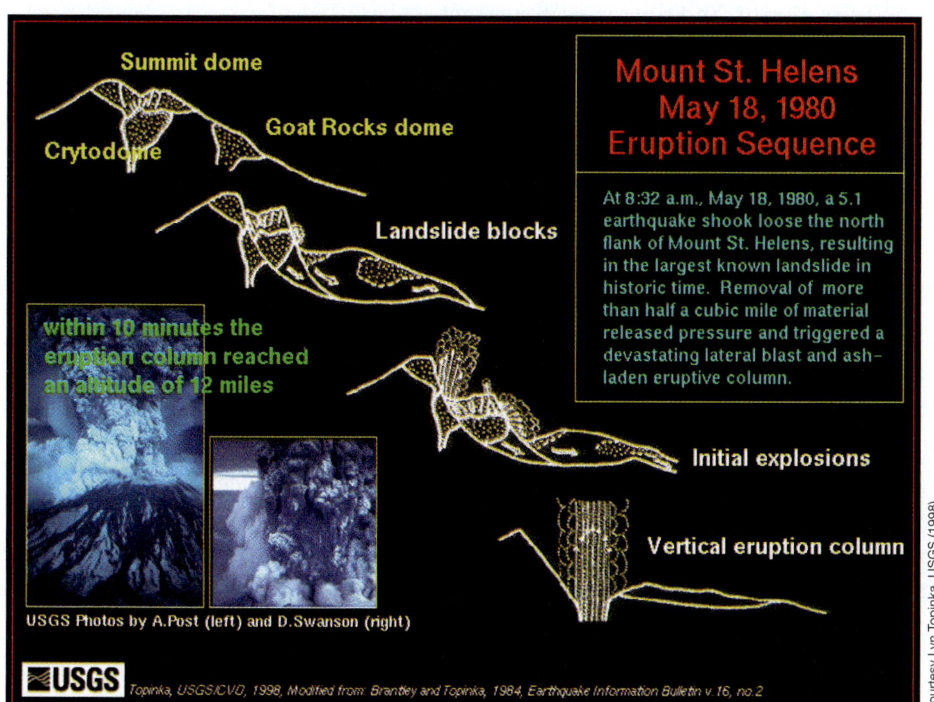
Mount St. Helens
May 18, 1980
Eruption Sequence
At 8:32 a.m., May 18, 1980, a 5.1 earthquake shook loose the north flank of Mount St. Helens, resulting in the largest known landslide in historic time. Removal of more than half a cubic mile of material released pressure and triggered a devastating lateral blast and ash-laden eruptive column.
Summit dome
Goat Rocks dome
Crytodome
Landslide blocks
within 10 minutes the eruption column reached an altitude of 12 miles
Initial explosions
Vertical eruption column
USGS Photos by A.Post (left) and D.Swanson (right)
USGS
Topinka, USGS/CVO, 1998, Modified from: Brantley and Topinka, 1984, Earthquake Information Bulletin v.16, no.2

Courtesy Lyn Topinka, USGS (1998)

Figure 10-30 Mount St. Helens May 18, 1980 eruption sequence.

Figure 10-31 Destruction from the lahars resulting from the 1991 Mt. Pinatubo eruption. *(Courtesy U.S. Department of the Interior, U.S. Geological Survey, Menlo Park, California, USA)*

per hour, build up great momentum, and can travel long distances from the volcano. An ancient lahar that formed during an eruption of Mount Rainier in Washington traveled almost 100 km (62.1 miles) downstream from the volcano!

Volcanic Gases

Volcanic gases include water vapor, carbon dioxide, sulfur dioxide; and lesser amounts of hydrogen sulfide, hydrogen chloride, hydrogen fluoride, and helium. How much of a problem volcanic gases cause depends on the amount erupted, how long the eruption lasts, and the wind direction. For example, on the island of Hawaii, the Kilauea shield volcano releases about 2,000 tons of sulfur dioxide gas each day during eruptions. Since 1986 when Kilauea eruptions became continuous, there has been a steady supply of sulfur dioxide emissions and air pollution and acid rain have become problems.

Volcanic gases emitted in high concentrations can be a problem for people. For example:

- **Sulfur dioxide** irritates skin, eye, nose, and throat membranes and tissues of the upper respiratory tract.
- **Hydrogen sulfide** is colorless and flammable but luckily very smelly. Even small amounts smell strongly of rotten eggs or sewer gas. It irritates the upper respiratory tract, but with enough exposure it causes pulmonary edema and even death.
- **Carbon dioxide** is colorless and odorless. It is not a poison, but higher concentrations decrease the available oxygen in the air and cause people and other organisms to suffocate. Because it is denser than air, it sinks and accumulates in low areas and depressions; high concentrations of carbon dioxide can be deadly.
- **Hydrogen chloride** is a strong acid. It can seriously irritate skin, eyes, nose, throat, and the respiratory tract.

A special type of hydrochloric acid-rich gas forms where lava flows into the sea. The extremely hot lava vaporizes the seawater and chemical reactions produce steamy white plumes rich in hydrochloric acid (Figure 10-32). These acid-rich plumes are dangerous to nearby organisms, but they tend to quickly dissipate into the atmosphere.

Figure 10-32 Acidic plume that developed where lava flowed into the sea on the southeast coast of Kilauea Volcano. *(Courtesy U.S. Department of the Interior, U.S. Geological Survey, Menlo Park, California, USA)*

A particularly scary example of the dangers of volcanic carbon dioxide comes from Cameroon in West Africa. Basaltic volcanoes in Cameroon erupted as recently as 400 years ago. Craters, rimmed basins at the volcano summits, have filled with water to form lakes scattered across the countryside. Carbon dioxide seeping from basalt below the lakes gradually accumulates in their bottom waters. High pressures at the bottom of the lakes, in places over 200 meters (650 feet) deep, enable large amounts of carbon dioxide to be dissolved; perhaps as much as 5 liters (5.3 quarts) of carbon dioxide in 1 liter (1.1 quarts) of water. This is extremely carbonated water!

At 9:30 P.M. on August 12, 1986, something happened to disturb the stratified nature of one of the crater lakes, Lake Nyos. Heavy rains apparently filled the upper part of the lake with colder and denser water. Winds may have concentrated this denser water on one side of the lake where it could sink down to deeper levels and displace the carbonated bottom water upward. As the bottom water moved upward, it came under lower pressures and could no longer keep the carbon dioxide dissolved.

The carbon dioxide rapidly came out of solution and formed a rising cloud of bubbles. The bubbles gathered at the lake's surface into a heavy cloud of carbon dioxide and water droplets. This mist flowed down nearby valleys, killing over 1,700 people, thousands of cattle, and unknown numbers of birds and other animals. Many victims did not know what was happening. They just seemed to quietly suffocate, some while asleep in their beds. To protect against similar disasters in the future, long plastic pipes now continuously siphon bottom water from Lake Nyos before it can become strongly carbonated.

The many ways volcanoes, especially stratovolcanoes, can be hazardous indicates that they deserve people's respect, concern, and study to better understand how to live safely around them.

10.4 living with volcanoes

Living safely with volcanoes requires people to understand three things about them:

- When they will erupt
- How large eruptions will be
- Where hazards will result from an eruption

It is all about predicting and preparing for eruptions as people have no means of controlling volcanoes. Predicting and preparing for eruptions, including evacuations, is difficult, but it can be done. It requires teams of scientists working closely with government authorities and planners. In 1991, the eruption of Mt. Pinatubo, in the Philippines, was predicted to within hours. An international team of geologists worked around the clock in the weeks leading up to the eruption. Civil authorities ordered the evacuation of tens of thousands of people. Many may have perished had they not left the area in time.

Successfully predicting and responding to eruptions requires monitoring volcano activity, measuring and comparing previous eruptions, delineating hazardous areas, and developing warning systems and response plans. A key step to safely living with volcanoes is community planning that guides land use in ways that decrease people's exposure to volcanic hazard risks. In the United States, scientists at volcano observatories in Alaska, Washington, California, Hawaii, and Wyoming are working to better understand volcanoes. Their mission is to (1) detect activity leading to an eruption; (2) provide real-time emergency information about future and ongoing eruptions; (3) identify hazardous areas around active and potentially active volcanoes; and (4) improve understanding of how volcanoes erupt and change the environment.

Monitoring Volcanoes

Monitoring volcanoes detects changes in volcanic activity (Figure 10-33). Events that precede eruptions such as earthquakes, changes in the shape of the volcano, or increases in gas emissions can be monitored (Figure 10-34).

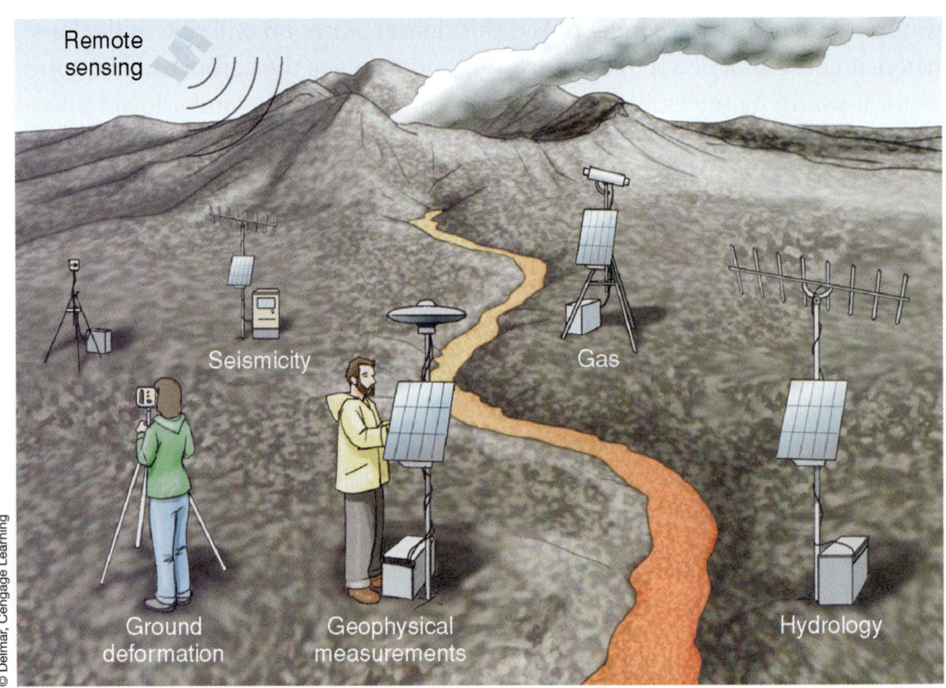

Figure 10-33 A variety of volcano-monitoring tools.

Figure 10-34 Scientists collecting gas samples on an active volcano. *(Courtesy U.S. Department of the Interior, U.S. Geological Survey, Menlo Park, California, USA)*

Worldwide, only a few active volcanoes are closely monitored. Most active volcanoes are not monitored and are potentially very hazardous if located in populated regions.

Volcano-monitoring tools are designed to detect and measure changes in a volcano caused by magma moving underground. The movement of magma will usually trigger earthquakes, cause rising or sinking of a volcano's summit or slopes, and lead to the release of volcanic gases from the ground. If a volcano shows signs of erupting, scientists will install instruments that record activity 24 hours a day. They also make several visits a week to conduct various surveys and to maintain instruments. By monitoring volcanoes, scientists are sometimes able to anticipate an eruption within days to weeks.

Earthquakes are detected by an array of instruments called **seismometers**. These instruments measure how much the ground shakes. Earthquake activity in most cases increases before an eruption. Earthquakes from rising magma are weak and occur less than 10 km (6 miles) beneath the volcano. Before an eruption there are typically swarms of hundreds of weak earthquakes. Because the earthquakes are weak, the seismometers must be placed within a 20 km (12-mile) radius of the volcano in order to detect them. If a swarm of earthquakes is detected beneath a volcano, scientists will work around the clock to detect subtle and significant changes in the earthquake pattern that might indicate an eruption will happen.

Ground deformation is detected with instruments called **tiltmeters**. Tiltmeters detect minute changes in ground slope. Scientists have used tiltmeters to successfully predict eruptions on the island of Hawaii in time to evacuate areas in danger of lava flows. Volcano monitoring also includes measuring variations in gas compositions and changes in local electrical and magnetic fields caused by underground magma movements.

Continuous monitoring of volcanoes on the ground takes teams of scientists and is expensive. There are hundreds of volcanoes that pose a threat to either the immediate area and/or areas downwind. Many volcanoes are in very remote and rugged locations such as Alaska. How can all of these volcanoes be monitored? It is not possible now to provide on-the-ground monitoring of all the volcanoes that pose a threat. However, monitoring volcanoes by satellite (remote sensing) helps scientists keep an eye on them.

Satellites and new remote-sensing techniques have added to the tools scientists use to monitor volcanoes globally. Satellite sensors are useful for three main purposes:

- They can identify and track eruption clouds.
- They can measure sulfur dioxide concentrations in clouds and the atmosphere.
- They can monitor "hot spots" on volcanoes.

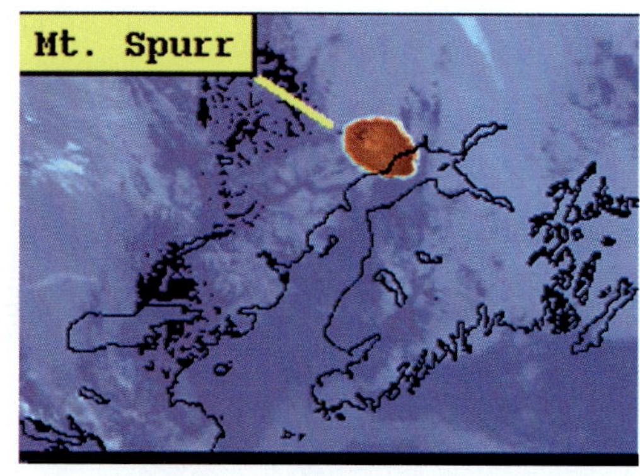

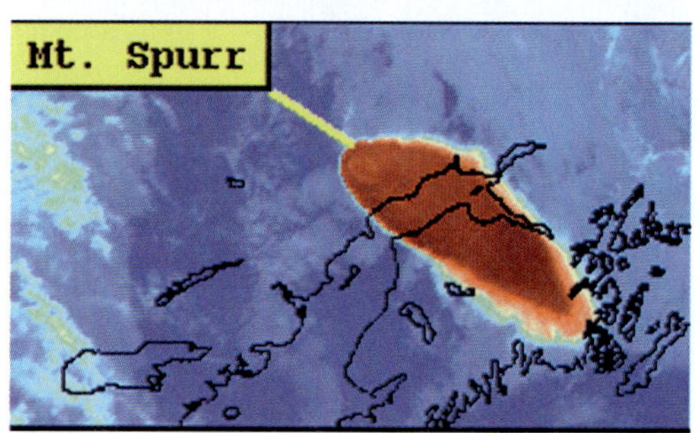

Figure 10-35 Satellite monitoring of a volcanic eruption.

Satellite observations are excellent for tracking eruption clouds (Figure 10-35) but satellite monitoring does not replace the need for ground-based monitoring methods. Ideally, volcanoes are monitored by a combination of ground-based techniques and satellite observations. However, for most of the remote volcanoes of the world, satellite observations of volcanic activity may be all that is available.

In terms of air-flight safety, identifying and tracking eruption clouds by weather satellite is essential. Weather radar on airplanes cannot detect ash. Weather satellites *can* distinguish between weather and ash clouds and have proven very useful in identifying and tracking eruptions. For example, on August 18, 1992, Mt. Spurr in Alaska erupted. The eruption cloud was located and tracked by weather satellite. Airplanes were warned of the location of the eruption cloud and flights were temporarily prevented from landing or departing from Anchorage International Airport.

Determining Eruption Size

Even if scientists can predict the timing of an eruption, it is difficult to predict the size and duration of an eruption. Someday scientists may be able to predict eruption size, but for now hindsight provides the most information. By measuring the size and strength of current eruptions and estimating the size of past eruptions, scientists can try to find eruption patterns for a volcano. These patterns provide clues for the possible size of future eruptions.

Eruption size is expressed using a scale called the **Volcanic Explosivity Index (VEI)**. The VEI is an exponential scale from 1 to 8; each whole number increase on the VEI scale represents a 10-fold increase in the energy released by the eruption. For example, a magnitude 4 on the VEI scale releases 10 times more energy than a magnitude 3 eruption. The VEI takes into account several factors, including the volume of the rock and ash erupted from the volcano, the height of the eruption column, and the length of the eruption in hours (Table 10-2). Prehistoric eruptions are difficult to assign VEIs without written accounts of the height of the eruption cloud.

Table 10-2 *The Volcanic Explosivity Index (VEI).*

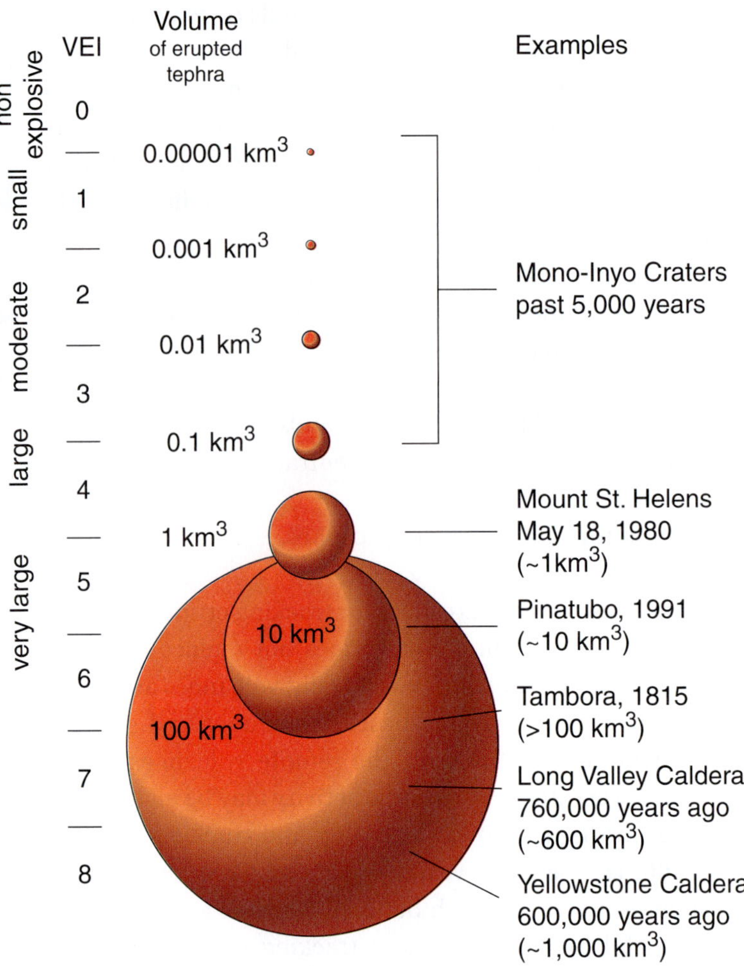

The best we can do to predict eruption size is to find clues about past eruptions. In order to evaluate potential volcanic hazards, scientists have to dig way into the volcano's past. To do this they map out all past eruption deposits. From maps and samples of the previously erupted material, scientists can determine the age and size of previous eruptions. The eruption patterns defined by these data are clues into what to expect from future eruptions.

Warning Systems

Close monitoring of volcanic activities is only one important factor in preventing disaster. If scientists have evidence of an eruption that is about to happen, there also needs to be a warning system in place. In the case of the 1991 eruption of Mount Pinatubo in the Philippines, scientists and government officials worked effectively together in predicting the eruption and evacuating the area. Thousands of lives were saved and millions of dollars in property losses avoided. The total costs of the monitoring and evacuations were far less than the loss of property would have been.

To warn the public of volcano danger in the United States, the volcano observatories use alert code levels that correspond to increasing levels of volcanic activity. As a volcano becomes more active, a corresponding higher alert level is issued. The alert level ranking gives the public and government a framework

Table 10-3 *Volcano alert codes—Alaska.*

Green:	**No eruption anticipated**—Volcano is in quiet, "dormant" state.
Yellow:	**An eruption is possible in the next few weeks and may occur with little or no additional warning**—Small earthquakes detected locally and (or) increased levels of volcanic gas emissions.
Orange:	**Explosive eruption is possible within a few days and may occur with little or no warning. Ash plume(s) not expected to reach 25,000 feet above sea level**—Increased numbers of local earthquakes. Extrusion of a lava dome or lava flows (non-explosive eruption) may be occurring.
Red:	**Major explosive eruption expected within 24 hours. Large ash plume(s) expected to reach at least 25,000 feet above sea level**—Strong earthquake activity detected even at distant monitoring stations. Explosive eruption may be in progress.

from which they can plan their response to a possible volcano emergency. The alert level warning systems are different for the different USGS Volcano Observatories. Table 10-3 lists the warnings used for volcanoes in Alaska.

An example of a well-coordinated volcanic warning system is the Volcanic Ash Advisory Centers (VAACs) that were designed to warn aircraft of eruption clouds. The eruptions of Redoubt and Mt. Spurr in Alaska alerted the international aviation industry to the dangers of ash clouds in areas of active volcanism. In response, collaboration between the Alaska Volcano Observatory and other government agencies (such as the FAA) vastly improved ash cloud-notification procedures. Soon after these eruptions, the VAACs were established in 1995 at a meeting of the International Civil Aviation Organization. The VAACs make sure there is a link between observatories, meteorological agencies, and air traffic control centers.

To meet their goal of an effective warning system, the VAACs are divided into different volcanically active regions around the world. Volcano observatories in each region are in charge of keeping track of the activity in their area by analyzing satellite imagery. The remaining challenge for the international community of volcanologists and government planners is to design a similar global network that will effectively warn and evacuate populations on the ground.

Preparing for Volcanic Eruptions

Monitoring volcanic activity and issuing warnings is the job of scientists. To effectively communicate warnings and manage volcano emergencies, civil authorities and emergency planners need to work with scientists. What do you think is needed in a volcano emergency plan? The following is a list of the basic elements of an emergency plan prepared by the United Nations Disaster Relief Coordinator and United Nations Educational Scientific and Cultural Organization in 1985:

- Identification and mapping of the hazard zones
- Register of valuable movable property (excluding easily portable personal effects)
- Identification of safe refuge zones to which the population will be evacuated in case of a dangerous eruption

Figure 10-36 Mt. Rainier with city in the foreground. *(Courtesy U.S. Department of the Interior, U.S. Geological Survey Cascades Volcano Observatory, Vancouver, WA)*

- Identification of evacuation routes, their maintenance and clearance
- Identification of assembly points for persons awaiting transport for evacuation
- Means of transport, traffic control
- Shelter and accommodation in the refuge zones
- Inventory of personnel and equipment for search and rescue
- Hospital and medical services for treatment of injured persons
- Security in evacuated areas
- Alert procedures
- Formulation and communication of public warnings; procedures for communication in emergencies
- Provisions for updating the plan

The preceding list gives you a feel for the scope of activities that are included in emergency plans. The cities around Mt. Rainier in Washington (Figure 10-36) are developing emergency response plans that meet these standards. Mt. Rainier is one of 16 volcanoes around the world that have been designated to be "Decade Volcanoes" by the United Nations. These volcanoes are the focus of scientific studies and community planning that will map volcanic hazard areas, determine the size of past eruptions, and guide land-use planning so that the risks to people and property around the volcano are reduced. Living with Mt. Rainier is an example of what people need to do to live more safely with volcanoes around the world.

10.5 conclusion

Volcanoes are striking evidence of the dynamic nature of the geosphere. They are commonly beautiful mountains that host vibrant ecosystems and diverse life. They have been beneficial to people, particularly because of the rich soils that volcanic material helps make. But erupting volcanoes are dangerous. By understanding how volcanoes work, people can learn how to live more safely with them.

IN THIS CHAPTER YOU HAVE LEARNED:

- Volcanoes are places where molten magma comes to the surface from deep in the geosphere. Volcanoes transfer matter and energy from Earth's mantle to its crust.
- The form of volcanoes defines two major types. Gentle slopes are developed at shield volcanoes where fluid basaltic lava is erupted. Cone-shaped stratovolcanoes have steeper slopes and explosively erupt a wide range of materials.
- Shield volcanoes are developed within tectonic plates above hot spots in the mantle and about anywhere the Earth's crust is extended. Stratovolcanoes form above subduction zones at converging plate boundaries.

- Volcanoes erupt lava, tephra, and gases that clog and heat parts of the hydrosphere, change the atmosphere's character, and devastate organisms in the biosphere. But volcanoes also create habitat for the biosphere.

- Fertile soils are developed around volcanoes and many people live near them. Since 1980, about 30,000 people have died in eruption-related disasters. Eruption hazards include lava flows, pyroclastic flows, tephra falls, debris-avalanches, and volcanic gases.

- Scientists predict when eruptions will occur by monitoring volcanic activity such as earthquakes, ground movements, and gas emissions. Satellites help monitor remote volcanoes and the progress of eruptions.

- Studies of a volcano's history can help estimate the size of future eruptions and where the most hazardous areas around the volcano are. Maps showing where the risks of volcanic impacts are greatest are important to emergency planning and response.

- People live more safely with volcanoes if they make land use choices that lower the risks of negative impacts, have eruption prediction and warning systems in place, and have emergency plans that help them respond to eruptions.

- People cannot control or prevent volcanic eruptions.

KEY TERMS

Andesitic magma
Basalt
Conduits
Lahar
Pyroclastic flow
Rhyolitic magma
Seismometer
Shield volcano
Silica
Stratovolcano
Tephra
Tiltmeter
Viscous
Volcanic Explosivity Index (VEI)
Volcano

REVIEW QUESTIONS

1. Describe how volcanoes form.
2. Compare and contrast the two types of volcanoes.
3. How does the amount of silica in magma affect the properties of lava?
4. What are the positive and negative aspects of living near volcanoes?
5. What effects can volcanic eruptions have on the geosphere?
6. What effects can volcanic eruptions have on the atmosphere?

7. What effects can volcanic eruptions have on the hydrosphere?

8. What effects can volcanic eruptions have on the biosphere?

9. In what ways can volcanic gases affect humans?

10. How can scientists predict volcanic eruptions?

chapter 11

earthquakes

The largest earthquake ever recorded in North America shook Kodiak and the rest of southern Alaska in 1946. Smaller earthquakes seem to occur here all the time and more large ones are expected. Students from Kodiak High School in Kodiak, Alaska, use global positioning systems (GPS) instruments to precisely measure the latitude, longitude, and elevation of selected landscape features near their school.

The very precise GPS measurements are used to monitor ground movements. Knowing the way the ground very gradually moves is needed to better understand the potential for hazardous earthquakes on Kodiak Island. One of the GPS stations monitored by the Kodiak High students is part of a nationwide network

of GPS stations called EarthScope. The EarthScope program monitors movements of the geosphere that can help us better understand where and how earthquakes occur.

Believe it or not, you do not have to live in classic "Earthquake Country" like Alaska or California to be vulnerable to earthquakes. From 1975 to 1995 only four states did NOT have earthquakes; fortunately, however, most of these earthquakes did not cause significant damage. You may be surprised to learn that parts of the midwestern and eastern United States have experienced strong earthquakes in the past. In 1886, a strong earthquake shook Charleston, South Carolina, toppling buildings and killing many people. Therefore, buildings throughout the United States, including your home, must be built to withstand the ground shaking caused by earthquakes that can cause serious harm and costly damage.

Most building codes now take earthquake hazard potential into account. In Alaskan towns like Kodiak, where communities have a recent history of large earthquakes, scientists (and high school students) are involved in measuring the movements of the Earth's surface. Scientists use these measurements in attempts to predict the next earthquake. South central Alaska is especially prone to earthquakes because it is near a convergent plate boundary (see Chapter 9). In other states that are more distant from tectonic plate boundaries and their related faults, earthquake risks can be relatively low. You can investigate the potential for earthquake hazards in the areas where you live by examining the

Courtesy USGS

Figure 11-1 USGS seismic hazard in the conterminous United States.

map in Figure 11-1. The potential for earthquakes guides many community decisions about building codes and the engineering design of infrastructure such as dams, bridges, and nuclear power plants.

Scientists have studied earthquakes and learned much about the processes that cause them. Earthquakes are a direct result of movement on faults—especially those formed at tectonic plate boundaries—and are direct evidence of the dynamic nature of Earth's geosphere.

AFTER COMPLETING THIS CHAPTER, YOU WILL UNDERSTAND:
What causes earthquakes.
Where and how earthquakes are located.
How earthquake size is measured.
How earthquakes can and cannot be predicted.
How earthquakes affect people.
How people can live more safely with earthquakes.

11.1 what earthquakes are

Earthquakes, like volcanoes, are evidence of an active and changing Earth. How many earthquakes do you think happen around the world annually? Each year there are 500,000 earthquakes detected by instruments, 100,000 that people can feel, and 100 that cause damage. If you live in an earthquake-prone region such as California, you are probably not surprised at the large number of earthquakes recorded around the world.

The destructive power of earthquakes depends not only on the strength of the earthquake, but on a number of other factors such as its location, the design and construction of affected buildings, and the nature of the soil and bedrock it shakes. What do you think happens to the Earth to cause an earthquake?

Courtesy the Earth Science World Image Bank © Michael Collier

Figure 11-2 San Andreas Fault.

Courtesy USGS

Figure 11-3 Fence offset along the San Andreas Fault in 1906.

Fault Rupture Causes Earthquakes

Earthquakes are caused by movements on **faults**. Faults are surfaces along which movement occurs between two large blocks of rock. These blocks and the faults at their boundaries come in all sizes—from those several meters (tens of feet) long to those over a thousand kilometers (miles) long. Movements can be dominantly vertical with one side up relative to the other or horizontal with one side sliding past the other. One of the most famous active fault systems, the San Andreas Fault in California, is the boundary between the Pacific and the North American tectonic plates (Figure 11-2).

The blocks of the Earth's crust usually do not slide easily past one another. Friction makes them stick or get "locked" together. As forces—stress—push or pull the rocks along faults, they gradually bend and deform. Continued and increasing stress can overcome the strength of the rocks that are bending along the fault and cause them to break or rupture. This releases the stress and the rocks "unbend" and quickly move to a new position. The rupturing of rocks along fault zones releases energy, in some cases, tremendous amounts of energy. This energy causes a vibration—an earthquake—which moves outward from the site of fault movement through the Earth and along its surface. Because faults vary so much in size, earthquakes also vary tremendously in size from very weak ones that only the most sensitive instruments can detect to those that demolish entire cities. In general, larger faults cause larger earthquakes.

Earthquakes were measured and observed long before scientists understood their cause. Prior to the powerful earthquake that rocked San Francisco in 1906, scientists were not sure if the shaking during an earthquake created new breaks (faults) in the Earth, or if movement of the Earth along existing faults generated the earthquakes. Through careful observations and field studies of the San Andreas Fault (Figure 11-3), investigators identified a mechanism that explains how fault movement causes earthquakes. Figure 11-4 illustrates observations made of the San Andreas Fault before and after the 1906 earthquake. The diagram illustrates the shape of a fence line at three time intervals: shortly after the fence was built (which was long before the earthquake), years after the fence had been built (shortly before the earthquake), and shortly after the earthquake in 1906. What does this diagram tell you about the changes in the shape of the Earth's surface on either side of the fault before and after the earthquake? Think of the Earth's crust as if it were a rubber band that becomes stretched to the point of breaking.

The change in shape of the Earth's crust after it breaks is called **elastic rebound**. Surveys of the San Andreas Fault observed before and after the 1906 earthquake led to the advance of the elastic rebound theory. The energy released when a fault slips and the Earth's surface changes shape is called **elastic energy**. Earthquakes transmit elastic energy through the Earth and around the Earth's surface.

Seismic Waves

Energy released by fault rupture causes vibrations called seismic waves to pass through the Earth. There are different kinds of seismic waves, and they all move in different ways. The two main types of seismic waves are **body waves** and **surface waves**. Body waves travel through the Earth's inner layers, whereas surface waves move along the surface of the Earth, like ripples on a pond.

There are two different types of body waves: **primary** or "P" waves and **secondary** or "S" waves (Figure 11-5). During an earthquake, P waves cause the first jolt, or shock, that people feel. They are the fastest kind of seismic wave and can move through both solid rock as well as fluids. P waves move through the Earth by alternately compressing and expanding rocks. An S wave is slower than a P wave and is the second wave you feel in an earthquake. S waves can only travel through solid rock and move the rock in a shearing, or side-to-side, motion.

Surface waves travel more slowly than body waves and cause most of the earthquake damage to buildings and other property. Most of the shaking felt from an earthquake is due to surface waves that can cause the ground to move up and down and side to side.

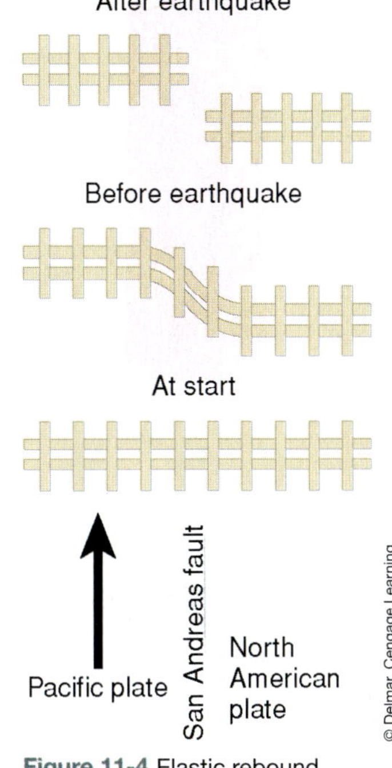

Figure 11-4 Elastic rebound theory.

What People Feel in Earthquakes

People who have experienced earthquakes commonly report that they first hear a noise similar to an approaching train. Suddenly they feel a powerful jolt as if a giant hammer hit the ground, followed immediately by ground shaking both

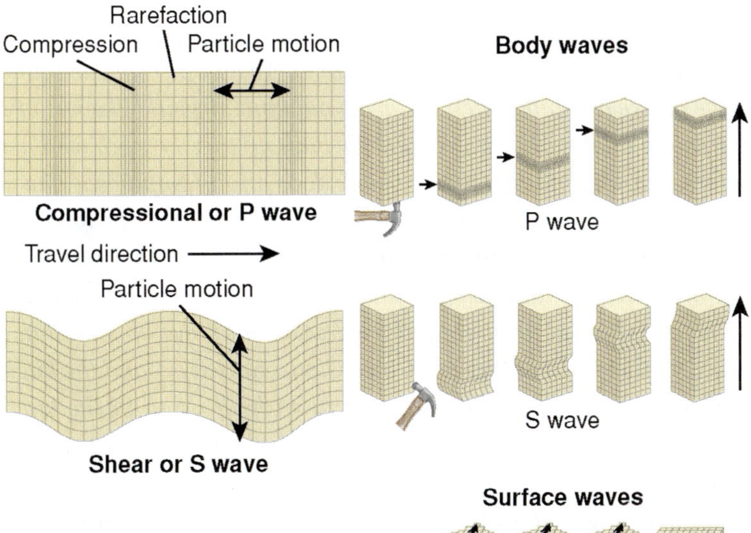

Figure 11-5 Propagation of body and surface waves.

up and down as well as back and forth. Toward the end of the earthquake, the shaking can get violent enough to make the ground appear to roll as if it were acting like waves on the surface of the ocean.

A woman reported her experience after the Loma Prieta earthquake in California in 1989:

> *. . . The hanging lamp begins to sway, ever so slightly, above the table, then the light flickers, and after a few choice words from Dad, everyone heads for the doorways. By this time, the knick-knacks on the shelves are furiously rattling, the floor feels like a huge serpent has decided his back is hurting him and he needs to adjust himself. Then the lights go out . . . pictures begin to fall, windows are rattling, and the wooden structure about you gives off these horrible creaking-crackles . . . What you see for at least a few seconds outdoors is a bizarre swaying of everything—trees, telephone poles, light poles, other houses—that makes you think your vision has gone askew. Almost as quick as it began, it stops!"*

Small earthquakes may not even be felt by people. They may gently sway window blinds and rattle dishes on a shelf. A small earthquake can stealthily arrive and pass by in a few seconds. But large earthquakes leave no doubt. They can last tens of seconds or longer. In 1964, the largest earthquake recorded in North America occurred along the convergent plate boundary between the Pacific and North American plates in Alaska: This was the Great Alaskan Earthquake. This earthquake knocked people to the ground and they were unable to get back on their feet for several minutes. As one person later wrote to his relatives:

> *We are in good shape now. But we had some anxious moments for a while. This was a vicious earthquake, and we were pretty well shaken up. I'm glad we were all home, because I would have hated to have been downtown and watch the streets and buildings dissolve! We had just finished dinner when the quake hit, and at first it started out kind of gentle like one of our earth tremors up here, but it very quickly built up to a strength we'd never felt before. Claudia made it out the back door and promptly sat down in the snow. It was impossible to stand up without some sort of support and she grabbed the earth as though she could hold it still by brute force. I was in the back door and being batted back and forth by the door jambs as the house wobbled around, and tried to keep an eye on Claudia, the kitchen, and Gloria all at the same time. Gloria rattled around the kitchen for a while and finally grabbed something to hold on to in the kitchen doorway, and we just stood and watched our cupboards empty themselves onto the floor.*
>
> *It seemed to go on forever. I guess it lasted five or six minutes, but I was afraid our house would go, it was shaking and creaking and groaning so much and I thought of the car in the garage. I started around the side of the house to get it out in case the house collapsed, but I couldn't stand up. The earth moved so violently that it was impossible to walk, and I got as far as the fence and held on. It was almost enough to make a person seasick.*

11.2 where earthquakes occur

Earthquakes are not randomly distributed around the world but tend to occur in certain locations (Figure 11-6). Earthquakes, like volcanoes (see Chapter 10), are evidence of an active and changing Earth. Also like volcanoes, earthquakes occur mostly at plate boundaries. In fact, 90 percent of all earthquakes in the world occur along plate boundaries (Figure 11-7). The biggest earthquakes are located along convergent and transform plate boundaries because very large faults are developed along them.

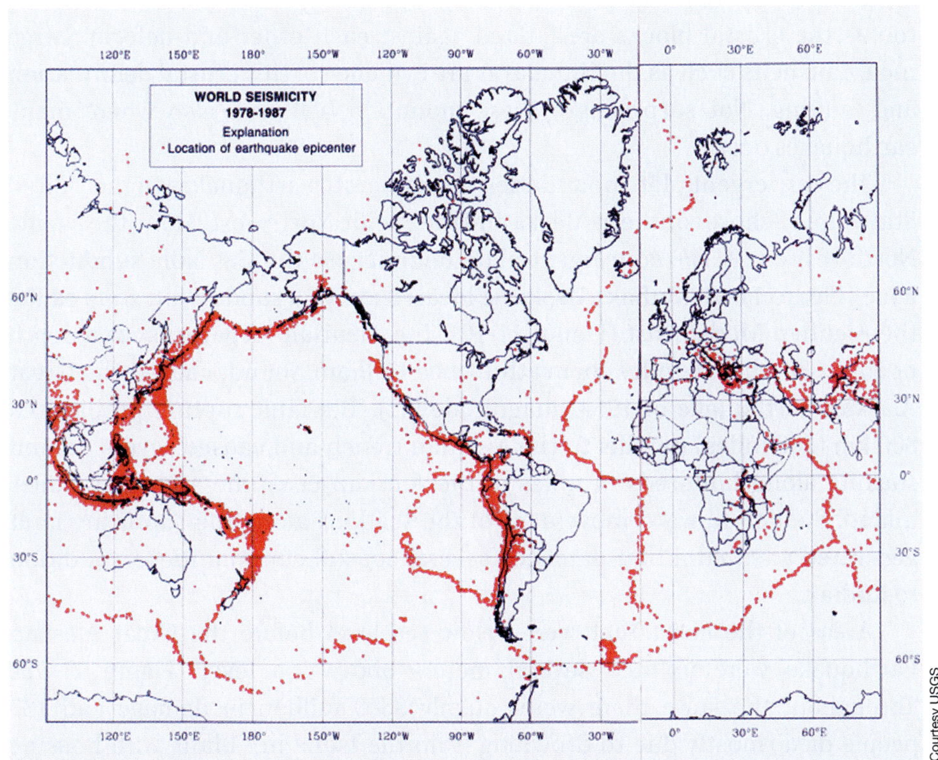

Figure 11-6 World seismicity from 1978 to 1987.

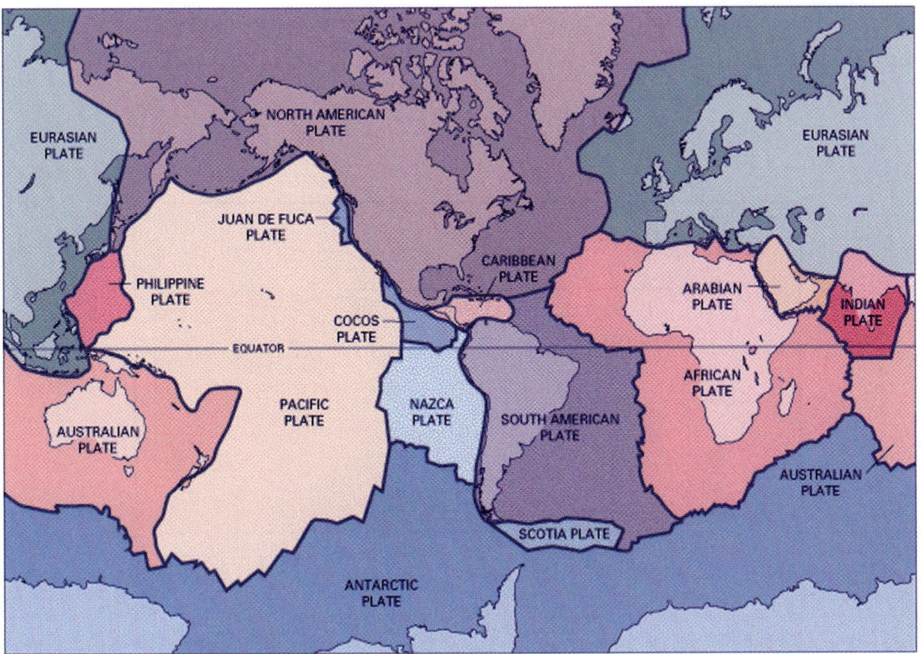

Figure 11-7 Global plate boundaries showing boundary types.

Earthquakes at Convergent Plate Boundaries

The plate boundaries that localize volcanoes around the Pacific Ocean (see Chapter 10) are also major sources of earthquakes. Convergent boundaries, where one plate sinks into the mantle below another, are called subduction zones (Figure 11-8). Subduction zones, where intense faulting occurs, generate earthquakes from shallow crustal levels to deep within the mantle. Subduction zone faults have the potential to generate very powerful earthquakes called "great" earthquakes.

Convergent plate boundaries are also where collisions between blocks of continental crust occur (see Chapter 9). Earthquakes are common here too as the crustal blocks are faulted against each other and deform. Large mountain belts such as the Himalayas are evidence of this crustal deformation and faulting. Not surprisingly, these mountain belts are also where many earthquakes occur.

The convergent plate boundaries that generate earthquakes in the United States are offshore southern Alaska and the Pacific Northwest. Along the Pacific Northwest coast, the convergent plate boundary is the Cascadia subduction zone (Figure 11-9). Kodiak, Alaska, is located above a subduction zone called the Aleutian Megathrust (Figure 11-10). The Aleutian Megathrust at a depth of about 30 km (20 miles) beneath Prince William Sound, caused the Great Alaskan Earthquake of 1964 (magnitude 9.2). The fault rupture propagated 800 km (500 miles) parallel to the Aleutian trench and caused severe ground shaking along hundreds of miles of the Alaskan coastline and many miles inland. Sudden upward movement of the seafloor along the rupturing fault generated a tsunami that damaged several coastal communities in addition to Kodiak.

Areas of the coast that were below sea level before the Great Alaskan Earthquake were uplifted several meters above sea level (Figure 11-11). Throughout the state there were roughly $300 million in damage and 130 people died (mostly due to drowning from the tsunami). Bluffs with housing developments collapsed in landslides near the coast. Rockslides and snow avalanches damaged roads, bridges, railroad tracks, power facilities, and harbor and dock structures.

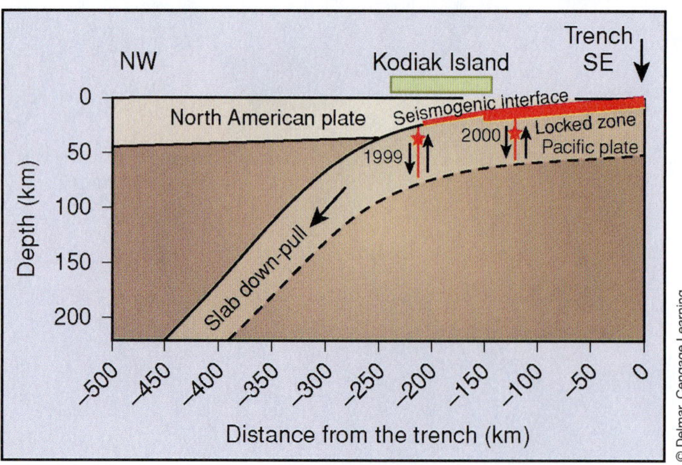

Figure 11-8 Subduction zone showing the location of earthquakes under Kodiak, Alaska.

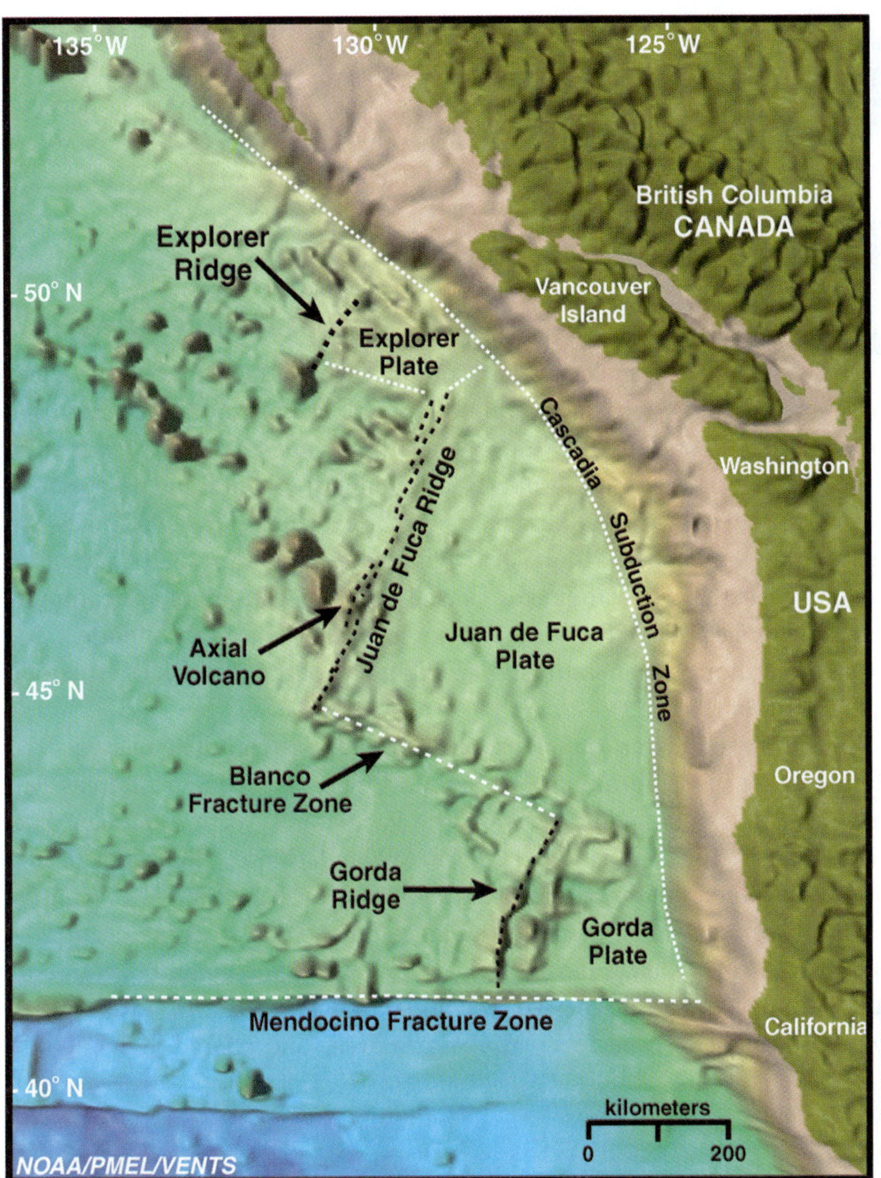

Courtesy: NOAA/PMEL/VENTS

Figure 11-9 Northeast Pacific plate boundaries showing Cascadia subduction zone.

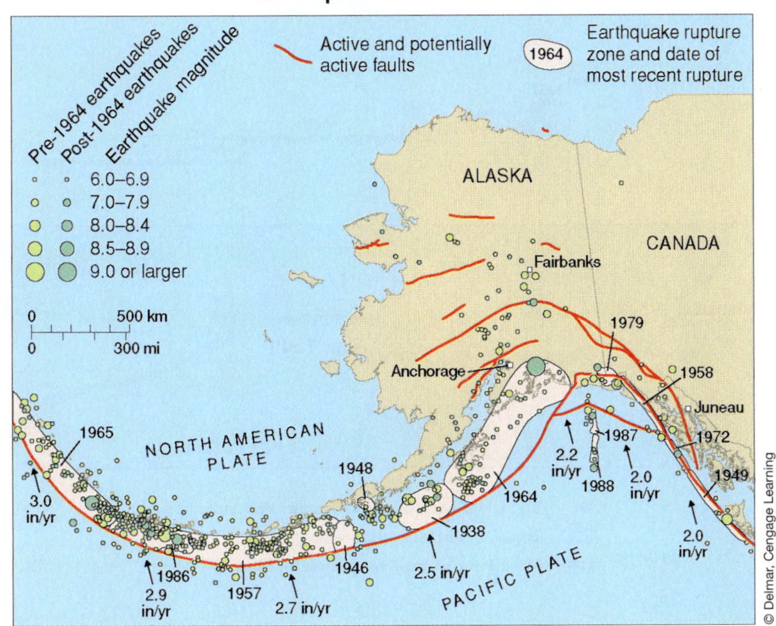

© Delmar, Cengage Learning

Figure 11-10 Earthquakes in Alaska and their relation to the Aleutian Megathrust subduction zone.

Courtesy: USGS

Figure 11-11 Uplifted barnacle covered rocks in the Gulf of Alaska several meters above sea level.

Instruments have not recorded great earthquakes from the subduction zone beneath the Pacific Northwest, but scientists have found evidence along ancient shorelines that they have occurred. Scientists are continuing to study this region as large cities and many people could be affected if a great earthquake were to occur in this area.

Earthquakes at Transform Plate Boundaries

Large earthquakes are also common at **transform plate boundaries**. These are the plate boundaries where large blocks of the Earth's crust slide horizontally past one another. The faults here are called *transform faults* or just *"strike-slip" faults* because the ground movement is horizontal and parallel to the trace (or strike) of the fault. These faults can be over 1,600 km (1,000 miles) long and generate very large earthquakes. Unlike subduction zone faults, such as the Aleutian Megathrust, earthquakes caused by strike-slip faults are commonly generated in the crust of the earth and not from deep within the mantle. One of the great earthquake-generating strike-slip faults of the world is the Anatolian Fault in Turkey (Figure 11-12).

In the United States, transform plate boundaries and their related strike-slip faults are in California and Alaska. In California, the San Andreas Fault system marks the boundary between the Pacific and North American plates (Figure 11- 13). This fault is over 1,200 km (800 miles) long. Thousands of small earthquakes occur on the San Andreas Fault system each year.

The two biggest historical earthquakes on the San Andreas occurred in 1857 and 1906. The latter destroyed parts of San Francisco and is commonly called the 1906 San Francisco earthquake (Figure 11-14). Intermediate-size earthquakes also occur on the San Andreas and related faults, like the

Figure 11-12 North Anatolian Fault with historic earthquakes plotted.

Location of August 17, 1999 Turkish Earthquake

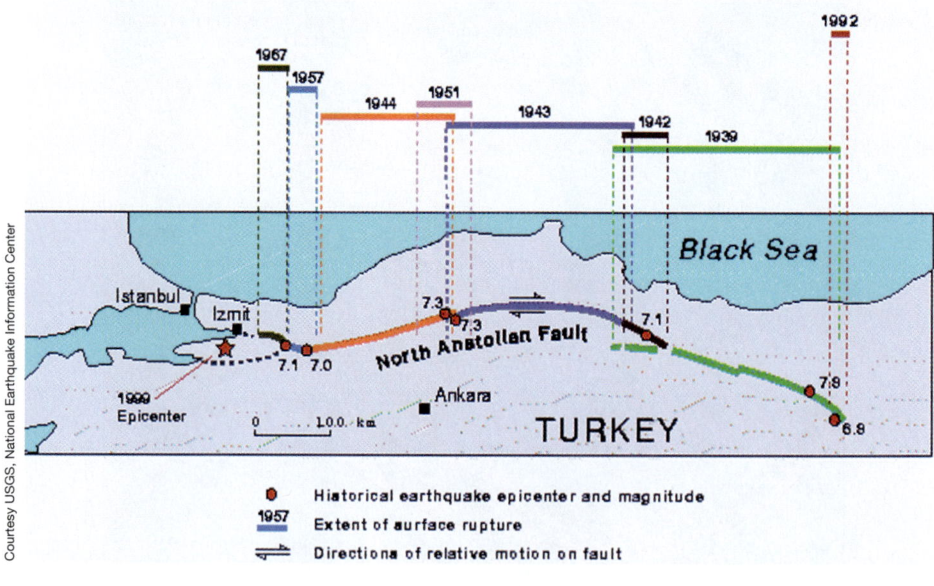

Courtesy USGS, National Earthquake Information Center

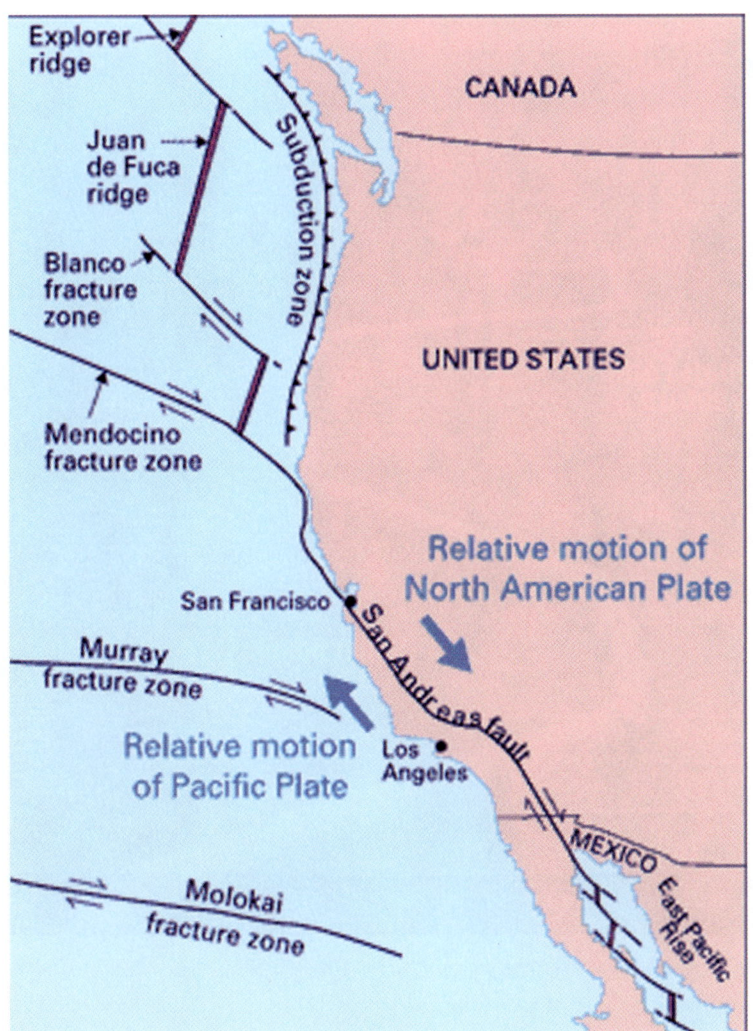

Figure 11-13 Pacific-North American plate boundaries.

Loma Prieta earthquake in 1989. Because these earthquakes can cause significant damage, scientists and engineers intensely study California faults to better understand how to prepare for and predict earthquakes.

In Alaska the big strike-slip faults are part of the Fairweather and Denali fault systems. Like the San Andreas, the Alaska strike-slip faults also localize movements between the Pacific and North American plates. Very large earthquakes have occurred along the Fairweather and Denali faults but they are not close to densely populated areas and have not caused as much damage as the San Andreas in California. An excellent example of preparing for an earthquake is provided by the construction of the Trans-Alaska Oil Pipeline—it passes right over the Denali Fault where the Earth's surface was severely ruptured during an earthquake in 2002, but it was not damaged (Figure 11-15).

Figure 11-14 Destruction caused by the 1906 San Francisco earthquake. *(Courtesy the Earth Science World Image Bank © USGS)*

Earthquakes at Divergent Plate Boundaries

As you can tell from examining Figure 11-7, earthquakes are also concentrated along divergent plate boundaries at the **mid-ocean ridges**. Here the tectonic plates move away from each other and the earthquakes are commonly not large. These earthquakes also tend to be remote from people and do not cause damage.

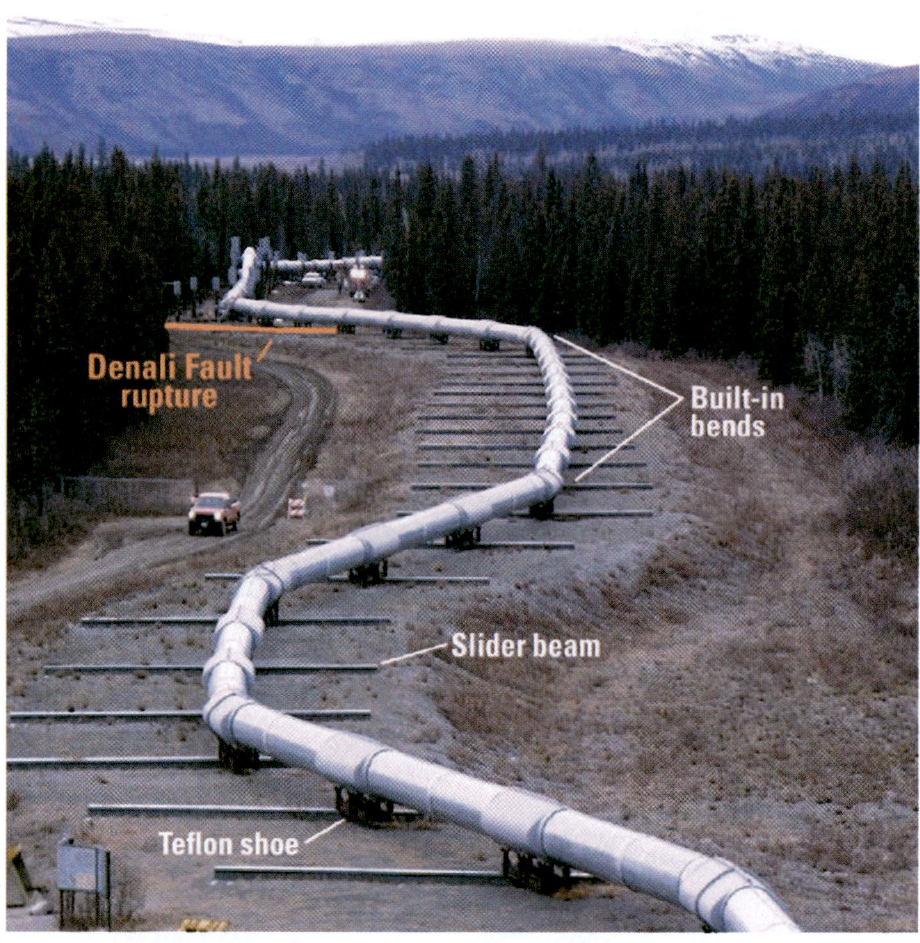

Courtesy USGS

Figure 11-15 Trans-Alaska Pipeline where it crosses the Denali Fault.

Other Earthquakes

Not all large earthquakes occur at tectonic plate boundaries. Some historic great earthquakes have occurred in unexpected places. In 1811, near New Madrid, Missouri, a great earthquake shook vast areas in Missouri and surrounding states. Buildings were destroyed, large areas subsided, lakes were formed, the Mississippi River was diverted, and over 60,702.8 hectares (150,000 acres) of forests were destroyed. Check the location of Missouri on Figure 11-7. Is there an active plate boundary in Missouri? The 1811 New Madrid earthquake shows that being prepared for earthquakes can be helpful throughout the United States.

Figure 11-16 Seismometer. (*Courtesy the Earth Science World Image Bank © Bruce Molnia, Terra Photographics*)

11.3 measuring earthquakes

Scientists can learn many things about earthquakes by measuring seismic waves and determining how strong they are. Scientists use instruments called *seismometers* (or **seismographs**) to measure ground movements caused by seismic waves (Figure 11-16). Seismometers record motion in three directions: up-and-down and horizontal in both east-west and north-south directions. No matter

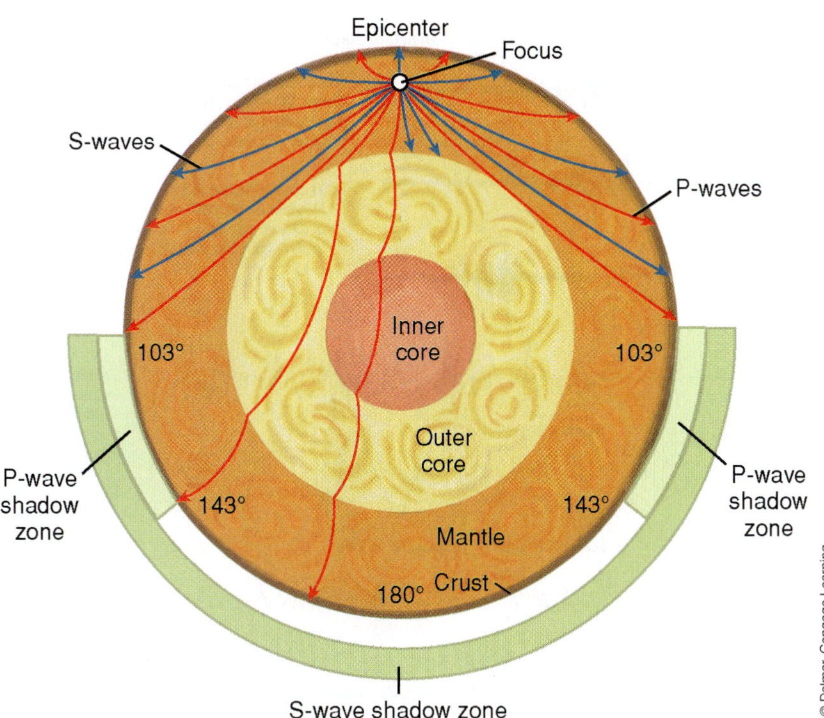

Figure 11-17 P, S, and surface waves traveling through the Earth.

where you set up a typical modern-day seismometer, it will take only a few hours to record seismic waves generated by an earthquake somewhere on the Earth.

Using seismometers, scientists can measure the three different types of seismic waves associated with almost every earthquake (Figure 11-17). These different wave types generally arrive at seismometers in three distinct groups; the seismometers then create a graphical output called a **seismogram** (Figure 11-18). From examining the seismogram, can you infer how the three different groups of seismic waves feel in an earthquake?

Primary or "P" waves are the first waves to arrive and are compressional waves. They can travel through rock at about 5 km (3 miles) per second, or about 15 times faster than the speed of sound through air. The second group of waves to arrive at a location are the secondary or "S" waves. S waves are shear waves that move through material by distorting it from side to side, at right angles to the direction the wave is moving. S-waves travel through solid rock (not liquid) at a speed of about 3 km (2 miles) per second.

The last waves to be recorded by a seismometer during an earthquake are surface waves, which (unlike P and S waves) travel only along the Earth's surface. There are two types of surface waves: those that move the ground from side to side and those that move the ground up and down in a rolling motion. Surface waves cause the most movement of the Earth's surface and, therefore, cause the most damage.

Accurately Locating Earthquakes

Using measurements made by seismometers at different locations, scientists can determine where an earthquake originated in the Earth. The location within the Earth where an earthquake originates is called the **focus**. The location on the Earth's surface directly above the focus is called the **epicenter**

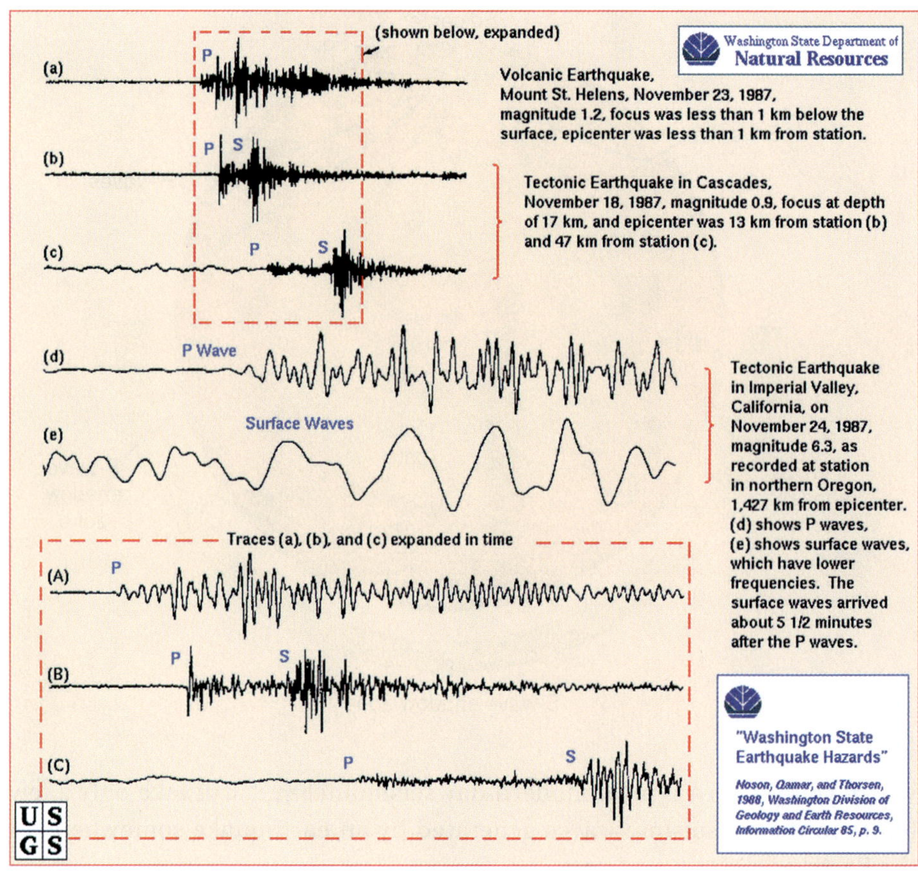

Figure 11-18 Seismograms showing P, S, and surface wave traces. (*Courtesy: U.S. Department of the Interior, U.S. Geological Survey Cascades Volcano Observatory, Vancouver, WA*)

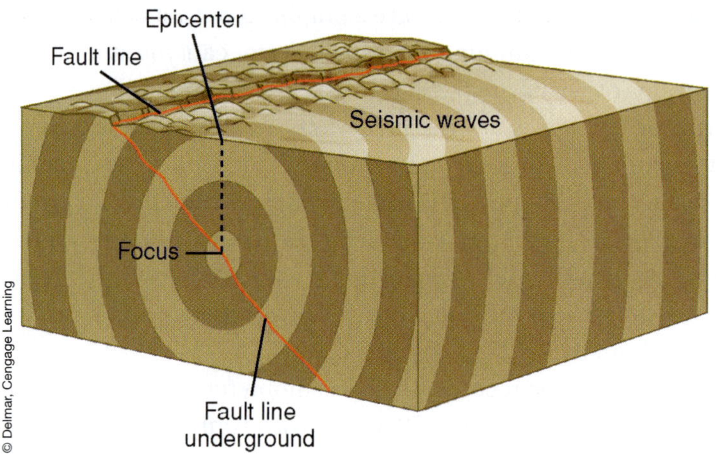

Figure 11-19 Focus and epicenter of an earthquake.

(Figure 11-19). The epicenter is the closest that people can get to where an earthquake originates. Knowing where the focus and epicenter are located, and how strong an earthquake is, can be very helpful in determining where damage is likely.

It is necessary to have three readings from seismographs to triangulate an earthquake's location. The three seismograms in Figure 11-20 show the arrival times of P and S waves at three different locations from the same earthquake.

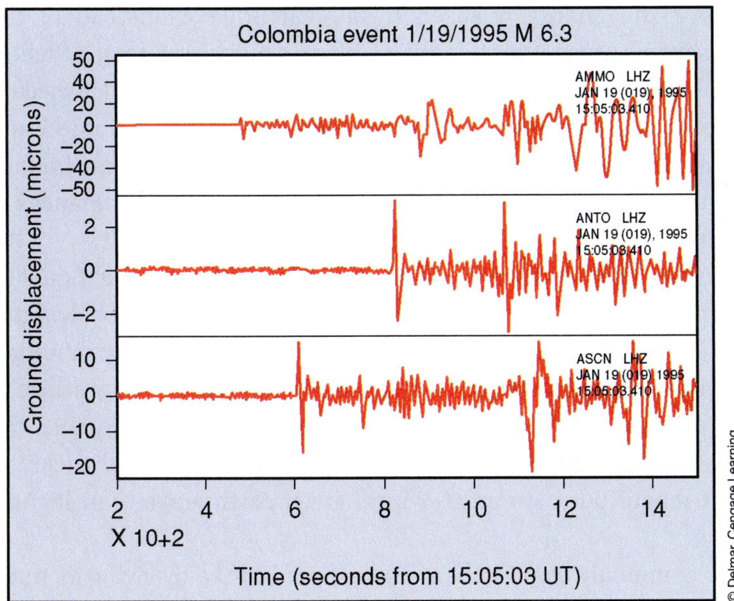

Figure 11-20 Three seismogram recordings from actual locations.

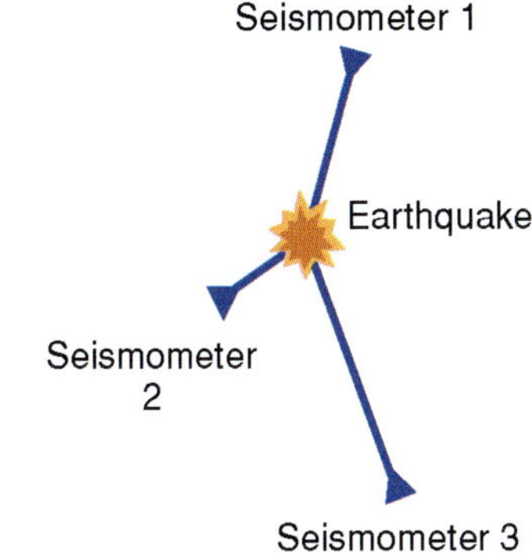

Based on the difference in arrival times of the P and S waves, can you guess which seismic station was closest to the epicenter and which was farthest?

Underground explosions such as nuclear blasts can also create seismic waves that travel through the Earth. In the 1960s, a global network of seismometers was established with the objective of precisely locating and estimating the size of nuclear explosions to make sure that countries were complying with the nuclear test ban treaty. After the nuclear test ban treaties were signed, the global network of seismic stations grew to over 3,500 by the early 1990s. This network now provides very precise measurement of earthquakes around the world.

Determining Earthquake Size

The two key measures of earthquake size are its strength (**magnitude**) and the amount of damage it causes (**intensity**). You may have heard earthquake magnitude reported as a number on the **Richter scale**. Charles Richter was the scientist who first developed a method for estimating and comparing the energy released by earthquakes. The Richter method estimates the magnitude of earthquakes based on the strongest seismic wave measured by seismometers, taking into account the distance of the seismometer from the epicenter (Figure 11-21).

Because earthquake size varies greatly, the Richter scale measures earthquake size with an exponential scale that ranges from 1 to 10. In other words, the ground motion of an earthquake with a magnitude of 6 is 10 times stronger than that of an earthquake with a magnitude of 5; a magnitude 6 earthquake is 100 times stronger than a magnitude 4 earthquake.

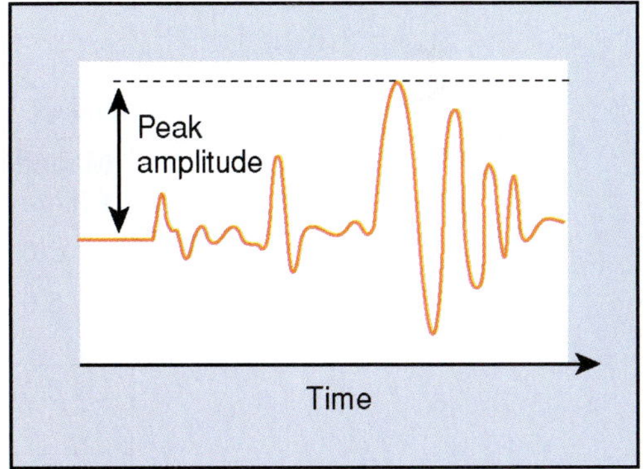

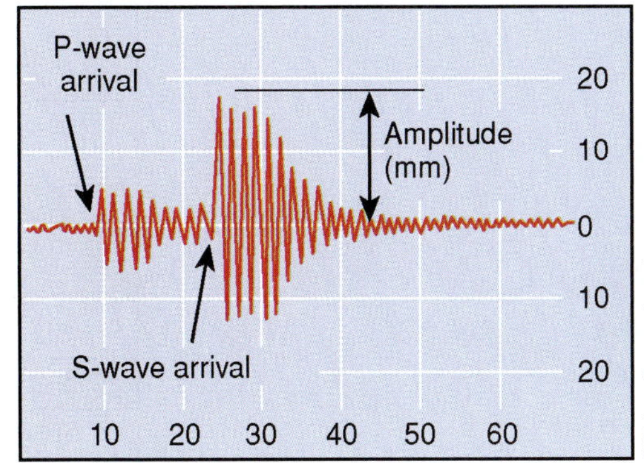

Figure 11-21 Seismogram with amplitude of body wave shown.

If we measure the energy released by an earthquake, instead of the amplitude of the seismic waves, then a 1 unit increase in magnitude on the Richter scale corresponds to a difference of 32 times as much energy released. Based on this, a magnitude 6 earthquake releases 32 times the energy as a magnitude 5, and a magnitude 7 earthquake releases 1,024 times as much energy as an earthquake with a magnitude of 5 ($32 \times 32 = 1024$). To get a sense of the vast differences in energy released in earthquakes of different magnitudes, see Table 11-1.

Today scientists use another method for estimating an earthquake's magnitude that takes more information into account. This method calculates an earthquake's "moment" magnitude and is a more accurate determination of the amount of energy an earthquake releases. The **moment magnitude** is calculated from the area of the fault rupture generating the earthquake, the amount of movement on the fault, and the strength of the faulted rocks. In general, moment magnitudes are greater for a given earthquake than Richter magnitudes.

News reports commonly identify the size of an earthquake by giving its magnitude without clarifying whether it is a Richter magnitude or moment magnitude. In fact, some news reports refer to all magnitudes as Richter magnitudes and locate an earthquake on the Richter scale. This is the case even though scientists have been using (and reporting) moment magnitudes, not Richter magnitudes, for several years. The largest earthquake ever recorded had a moment magnitude of 9.5; it occurred in Chile in 1960. The 1964 Great Alaskan earthquake had a moment magnitude of 9.2 and the earthquake offshore Sumatra, Indonesia, that caused the devastating tsunami in December 2004 had a moment magnitude of 9.0. These were really big earthquakes because they released about a thousand times more energy than "intermediate" size earthquakes like the one in Bam, Iran, in 2002, with a moment magnitude of 6.6.

Table 11-1 *Richter scale magnitude and released energy.*

Magnitude on Richter Scale	Energy Released in Joules	Comments
2.0	1.3×10^{8}	Smallest earthquake detectable by people.
5.0	2.8×10^{12}	Energy released by the Hiroshima atomic bomb.
6.0–6.9	7.6×10^{13} to 1.5×10^{15}	About 120 shallow earthquakes of this magnitude occur each year on the Earth.
6.7	7.7×10^{14}	Northridge, California earthquake January 17, 1994.
7.0	2.1×10^{15}	Major earthquake threshold.
7.4	7.9×10^{15}	Turkey earthquake August 17, 1999. More than 12,000 people killed.
7.6	1.5×10^{16}	Deadliest earthquake in the last 100 years. Tangshan, China, July 28, 1976. Approximately 255,000 people perished.
8.3	1.6×10^{17}	San Francisco earthquake of April 18, 1906.
9.3	4.3×10^{18}	December 26, 2004 Sumatra earthquake.
9.5	8.3×10^{18}	Most powerful earthquake recorded in the last 100 years. Southern Chile on May 22, 1960. Claimed 3,000 lives.

Before seismometers were widely used, locating the source of an earthquake depended on eyewitness accounts of ground shaking and observations of damage to buildings and the Earth's surface. In the late 1800s, an Italian geologist created an earthquake intensity scale, the **Mercalli scale** (Table 11-2), based on people's observations of earthquakes. The Mercalli scale is still a widely used measure of earthquake effects.

Intensity differs from magnitude, which measures the energy released by an earthquake. Energy and intensity are not the same. An energetic earthquake in one region might cause much less damage than an equally energetic earthquake in another region. The intensity of earthquake effects also depends on factors that may have no direct relation to the energy released by earthquakes.

The Mercalli scale depends on people's descriptions rather than actual measurements of ground motion by seismometers. It is an especially useful tool for measuring earthquakes in remote regions of the world where there are not many seismometers. Indeed, the Mercalli scale provides the only

Table 11-2 *Modified Mercalli scale.*

Mercalli Intensity at Epicenter	Magnitude	Witness Observations
I	1 to 2	Felt by very few people; barely noticeable.
II	2 to 3	Felt by a few people, especially on upper floors.
III	3 to 4	Noticeable indoors, especially on upper-floors, but may not be recognized as an earthquake.
IV	4	Felt by many indoors, few outdoors. May feel like heavy truck passing by.
V	4 to 5	Felt by almost everyone, some people awakened. Small objects moved. Trees and poles may shake.
VI	5 to 6	Felt by everyone. Difficult to stand. Some heavy furniture moved, some plaster falls. Chimneys may be slightly damaged.
VII	6	Slight to moderate damage in well built, ordinary structures. Considerable damage to poorly built structures. Some walls may fall.
VIII	6 to 7	Little damage in specially built structures. Considerable damage to ordinary buildings, severe damage to poorly built structures. Some walls collapse.
IX	7	Considerable damage to specially built structures, buildings shifted off foundations. Ground cracked noticeably. Wholesale destruction. Landslides.
X	7 to 8	Most masonry and frame structures and their foundations destroyed. Ground badly cracked. Landslides. Wholesale destruction.
XI	8	Total damage. Few, if any, structures standing. Bridges destroyed. Wide cracks in ground. Waves seen on ground.
XII	8 or greater	Total damage. Waves seen on ground. Objects thrown up into air.

way of estimating earthquake effects over the long historical record, because seismometers and the concept of earthquake energy are relatively recently developed scientific tools. By connecting points of equal intensity on a map, people can approximately locate the epicenter of an earthquake and estimate how far away from the source the earthquake was felt.

Maps that show the distribution of modified Mercalli intensities using contours of equal intensity are called earthquake intensity maps (or **isoseismal maps**). These maps are generated from direct observations made by people and are helpful for characterizing historic earthquakes, such as the New Madrid earthquakes of 1811 to 1812 (Figure 11-22). Based on the effects of the New Madrid earthquakes, they are estimated to have had a magnitude of at least 8.0 on the not-yet-invented Richter scale.

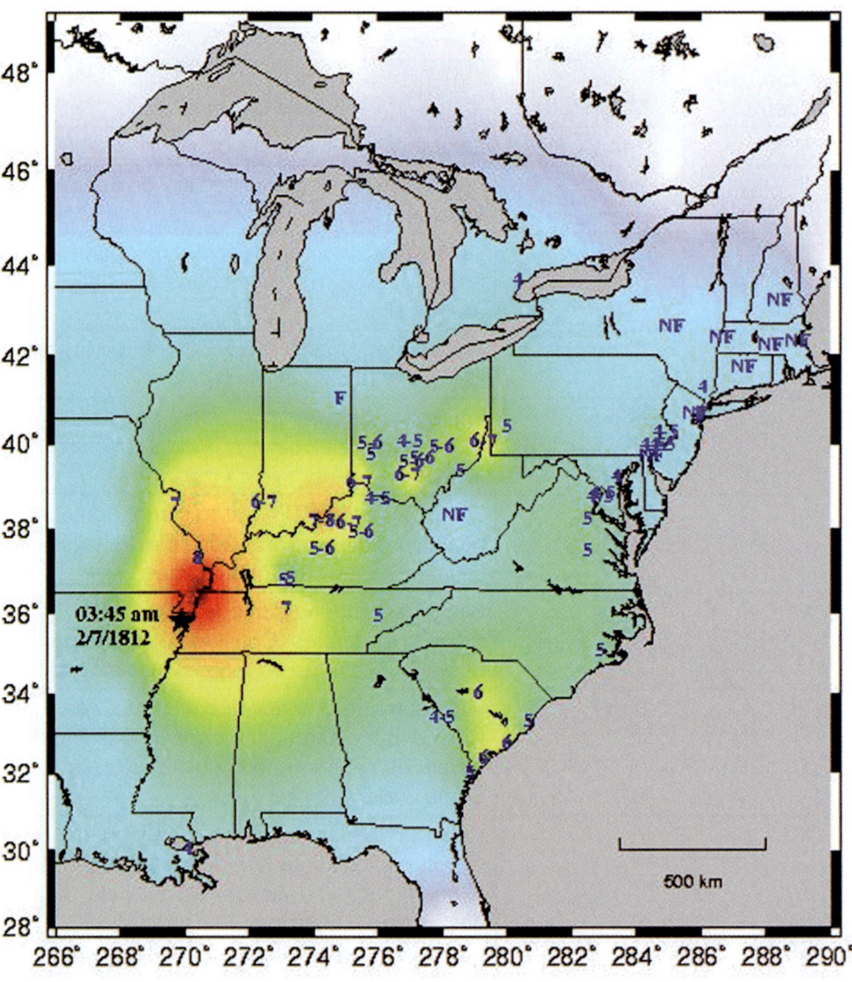

PERCEIVED SHAKING	Not felt	Weak	Light	Moderate	Strong	Very strong	Severe	Violent	Extreme
POTENTIAL DAMAGE	none	none	none	Very light	Light	Moderate	Moderate/Heavy	Heavy	Very Heavy
PEAK ACC.(%g)	<.17	.17-1.4	1.4-3.9	3.9-9.2	9.2-18	18-34	34-65	65-124	>124
PEAK VEL.(cm/s)	<0.1	0.1-1.1	1.1-3.4	3.4-8.1	8.1-16	16-31	31-60	60-116	>116
INSTRUMENTAL INTENSITY	I	II-III	IV	V	VI	VII	VIII	IX	X+

Courtesy Susan E. Hough, USGS

Figure 11-22 Intensity map of the 1811–1812 New Madrid earthquakes.

11.4 predicting earthquakes

Scientists are trying very hard to predict where, when, and how strong future earthquakes will be. They have been much more successful in predicting the size and location of earthquakes than when they will occur. By studying past earthquakes, using seismometers to monitor earthquake activity, and measuring how strain is accumulating in rocks, scientists can predict the potential size of future earthquakes and estimate the time between earthquakes. This information is very important for land-use planners and engineers taking measures to minimize earthquake damage. By understanding where earthquakes will occur and how strong they will be, people can be reasonably prepared for their impacts.

By combining historical records of earthquakes with monitoring of earthquake activity, scientists try to find recurrence patterns to help them predict earthquakes. Using this information, they define models for earthquake **recurrence intervals**. Earthquake recurrence intervals are the average time span between large earthquakes (≥ magnitude 7) at a particular site. The elastic rebound theory predicts that faults under continuous stress should slip at regular intervals. As an analogy, imagine slowly dragging a cement block along the sidewalk with a bungee cord. The cord will stretch against the weight of the block and the force of pulling but only to a certain point. The block will then slide, releasing the elastic strain that has accumulated in the bungee cord. If you keep pulling at the same rate, the block will slide at regular intervals.

Earthquakes in Alaska

Recall the Kodiak High School students gathering GPS data to predict earthquakes. Several of the Kodiak High School students' parents and grandparents remember all too well the dangers of earthquakes to their coastal community. Historical records show that the Kodiak region of the Aleutian Megathrust generates large earthquakes about once every 60 years (on average). By taking GPS measurements, the Kodiak High School students are measuring how the land is moving in response to the compressive forces of plate convergence. This information helps scientists measure the accumulating elastic strain (deformation) in the Earth's crust. Some faults can slip gradually without generating earthquakes; this is called aseismic slip, or creep. By measuring accumulating strain, the Kodiak students can help determine whether the fault is locked and likely to rupture again (producing an earthquake), or whether it is creeping without strain accumulating.

Earthquakes in the Pacific Northwest

Historical records are not always available in places where we have reason to believe there is potential for great earthquakes. The Pacific Northwest is an example of such a place. The Pacific Northwest sits above a subduction zone (Figure 11-23), similar to the Aleutian islands of Alaska. However, there are no historical records of great earthquakes in the Pacific Northwest, such as the great earthquake that hit Alaska in 1964. The pattern of great earthquakes in places such as Alaska has motivated scientists to try to assess the earthquake risk in the Pacific Northwest.

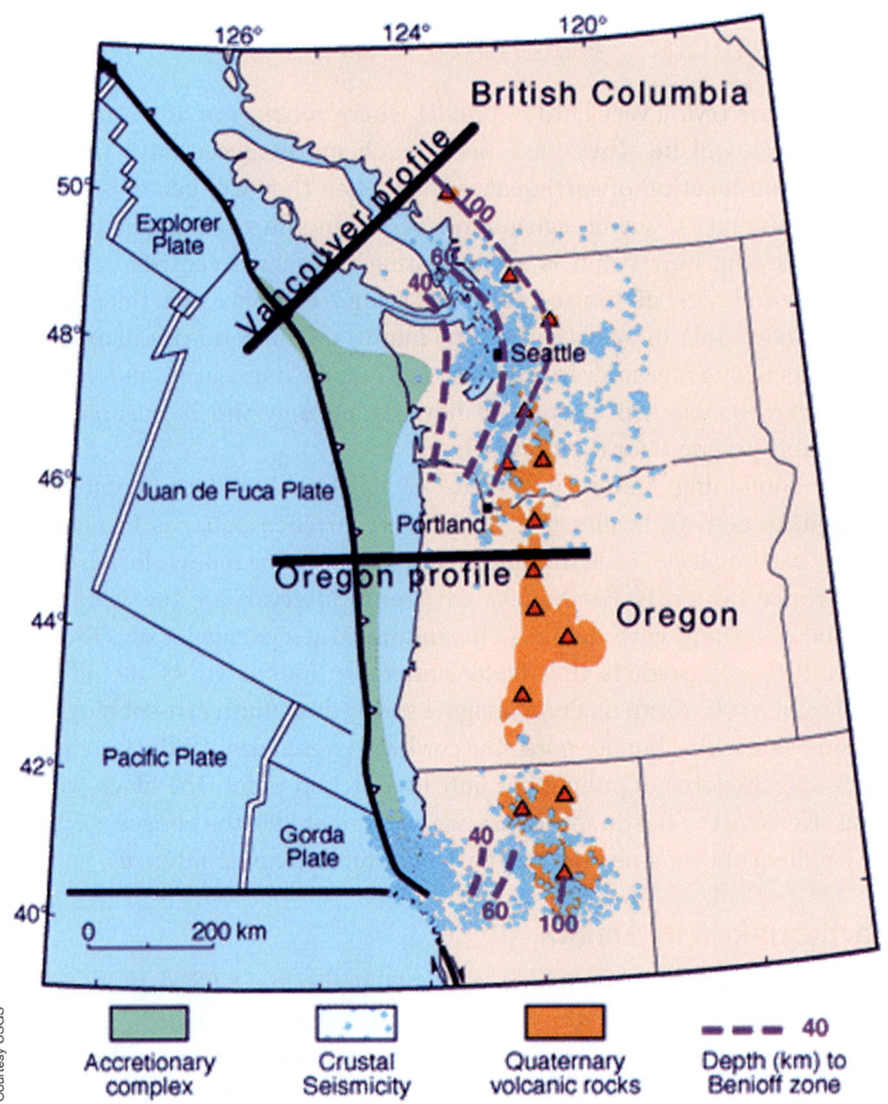

Figure 11-23 Pacific plate, Juan De Fuca plate, Cascadia subduction zone, and North American plate.

The lack of historical large earthquakes associated with the Juan De Fuca plate is puzzling for seismologists under pressure to assess the earthquake risk in the heavily populated regions of the Pacific Northwest. Geodetic measurements (surveys and GPS measurements) demonstrate that parts of the Washington coast are rising, and the coastal mountain ranges are being compressed. Geodetic measurements also show that the Juan de Fuca plate is subducting beneath North America at a rate of about 4 cm (1.5 inches) per year. This strongly suggests that strain is accumulating in the North American plate above the subduction zone that will likely be released by slip on faults, resulting in potentially great earthquakes.

Over the past 20 years, a new picture has emerged of the earthquake history of the Pacific Northwest, with startling new evidence for past great earthquakes. Geologists have found evidence of rapid sinking of the coastline in distinct episodes at least six times over the past 7,000 years, the most recent episode being only 300 years ago. Although there were no seismometers in the year 1700, evidence that a great earthquake (greater than magnitude 8) rocked the coastline from Vancouver to northern California, is preserved in the soils of the coastal lowlands. Scientists cut trenches several meters deep and exposed a peat layer, the remains of a former marsh, covered by layers of sand and mud (Figure 11-24). The marsh abruptly subsided 0.5 to 1 meter (1.6 to 3.2 feet) in a great earthquake about 300 years ago. In some locations, stumps of Sitka spruce and red cedar were found in the peat layer, and fossils of salt-tolerant plants were found in the mud overlying the sand.

Figure 11-24 Soil sample trench with layers of peat, sand, and mud.

What do these alternating layers tell you about the changing abiotic environmental conditions? Alternating layers of fossilized remains of saltwater plants and terrestrial plants are important clues to earthquake history. How do you think earthquakes radically change the ecological environment in coastal regions? Notice that 300 years ago, beach sands were deposited on top of peaty soil. This layer of beach sand is found on top of 300-year-old peaty sediments on the inland side of 9.1 meter (30-foot) high hills. How could beach sand have been deposited over hills? Imagine a giant wave, perhaps greater than 10 meters (32 feet high), hitting the coast. Scientists believe that tsunamis generated by the great earthquake of 1700 hit the west coast from southern British Columbia all the way to northern California. Historical documents from Japan record a tsunami the same year, possibly generated from the same earthquake! Given the evidence of great earthquakes in the Pacific Northwest, earthquake risk needs to be taken seriously in this region.

The Present Status of Earthquake Prediction

Now that you know from historical records and prehistorical evidence that earthquakes follow somewhat predictable patterns, why can people not better predict when earthquakes will occur? In terms of "long-term prediction," we can predict earthquakes with some success. In California, earthquakes are predicted in terms of probability based on the frequency and magnitude of past earthquakes. Forecasts for earthquakes in California, based on probability calculations, are projected over decades. For example, in the San Francisco Bay area there is a "62 percent probability of at least one magnitude 6.7 or greater quake, capable of causing widespread damage, striking the San Francisco Bay region before 2032" (Figure 11-25).

Attempts to predict earthquakes at precise locations, within even a few years, have not been successful. In the 1970s, scientists were confident that smaller earthquakes preceded large earthquakes. They thought that the smaller "precursor" earthquakes could be identified and used to predict when large earthquakes were likely to occur. In 1975, this method predicted a 7.5 magnitude earthquake to within a few days in China. The government successfully evacuated the city of Haicheng and saved thousands of lives. Unfortunately, this success was short lived. In 1976, 250,000 people died in another unpredicted China earthquake. As it turns out, not all large earthquakes are preceded by several smaller ones.

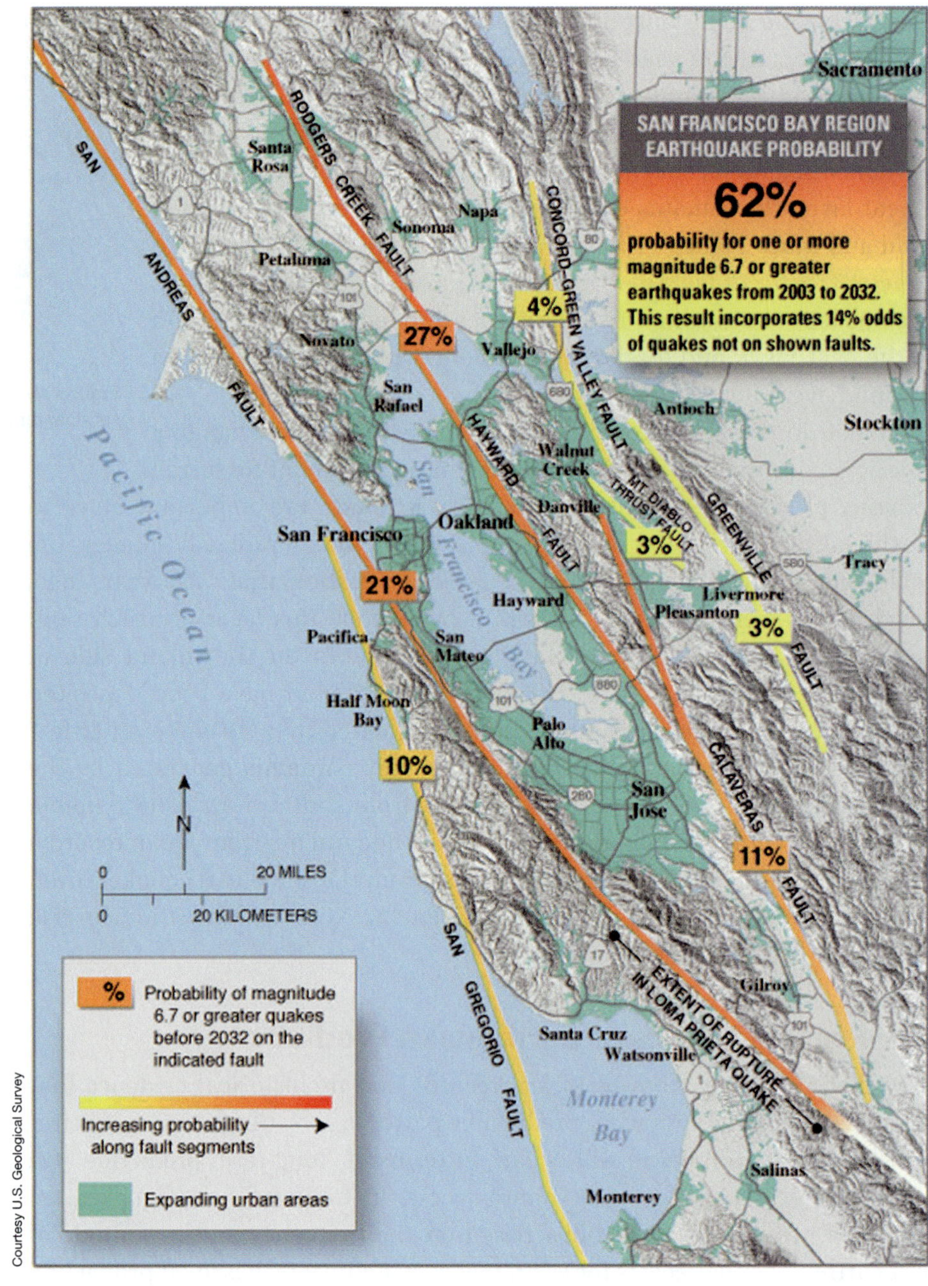

Figure 11-25 Seismic hazard map of the Bay area in California.

Meanwhile, during the 1970s to 1980s, several U.S. government agencies, also working toward accurate earthquake prediction, set up the Parkfield experiment in California. Seismic records showed that the Parkfield segment of the San Andreas Fault had generated magnitude 6 earthquakes approximately every 22 years from 1857 to 1966 (Figure 11-26). Therefore, this segment of the fault was chosen as a site to closely monitor in an attempt to "capture" information about the next strong earthquake when it occurred. An extensive array of instruments was set up in the 1980s and is still operating. Instruments including seismometers, strain meters, lasers to detect and measure fault movement, and ground water level monitors were placed.

The hope is to document precursors to a large earthquake that would then be helpful in predicting strong earthquakes elsewhere. However, the earthquake that was predicted to hit sometime between 1988 and 1993 has still

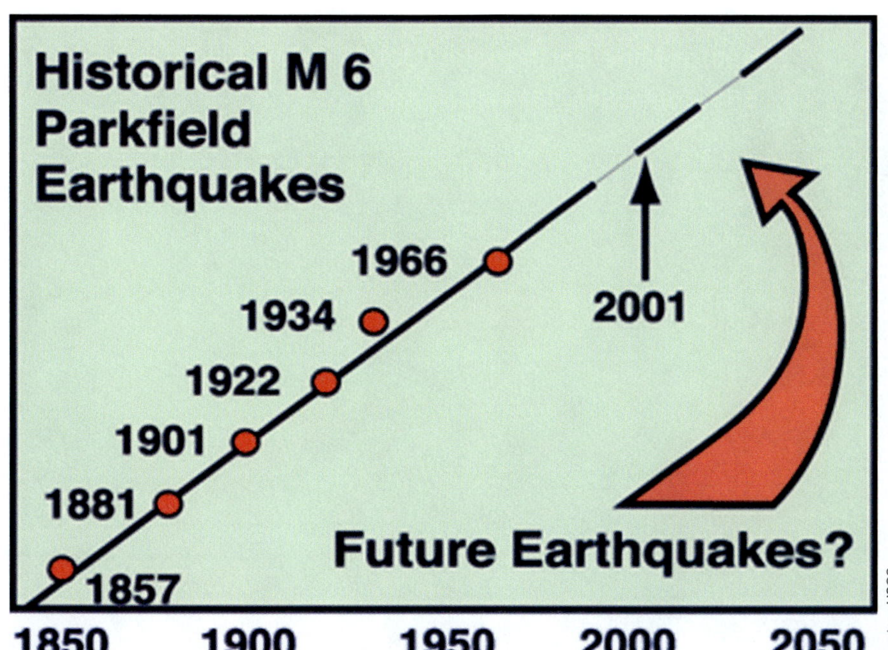

Figure 11-26 Frequency of M 6 earthquakes in Parkfield, California.

not happened. Some speculate that two large earthquakes in 1983—on different segments of the active fault in this area—changed the stress field in Parkfield and facilitated a seismic creep. If so, the creep may have reduced the probability of a near-term magnitude 6 earthquake along the Parkfield segment.

New Technologies

New techniques for mapping the Earth's surface using satellite technology hold tremendous potential for improving earthquake hazard analysis. In particular, GPS and interferometric synthetic aperture radar (InSAR) are two new techniques being used. InSAR measures the "quiet" or normal, nonearthquake motions associated with plate tectonics and rapid shifts that occur during a quake. Images are captured from space, using what is essentially an orbiting version of the speed gun state troopers use to clock speeders.

InSAR uses two radar images of the same area taken from the same point in space at two different times. The images are overlaid and show changes in the shape or position of the reflecting surface that occurred in the time elapsed between taking the two images. A spectacular example of InSAR images are those taken in the Mojave Desert in southern California that show surface displacement due to the 1999 Hector Mine earthquake (Figure 11-27). InSAR technology can image thousands of square miles in minutes, providing great potential for monitoring strain leading to earthquakes all over the world, including remote regions and in countries that cannot afford extensive ground-based monitoring.

At GPS stations, instruments are fixed to a point on the Earth's surface. These instruments use signals from GPS satellites orbiting the Earth to detect small displacements of the surface from which strain changes can be computed.

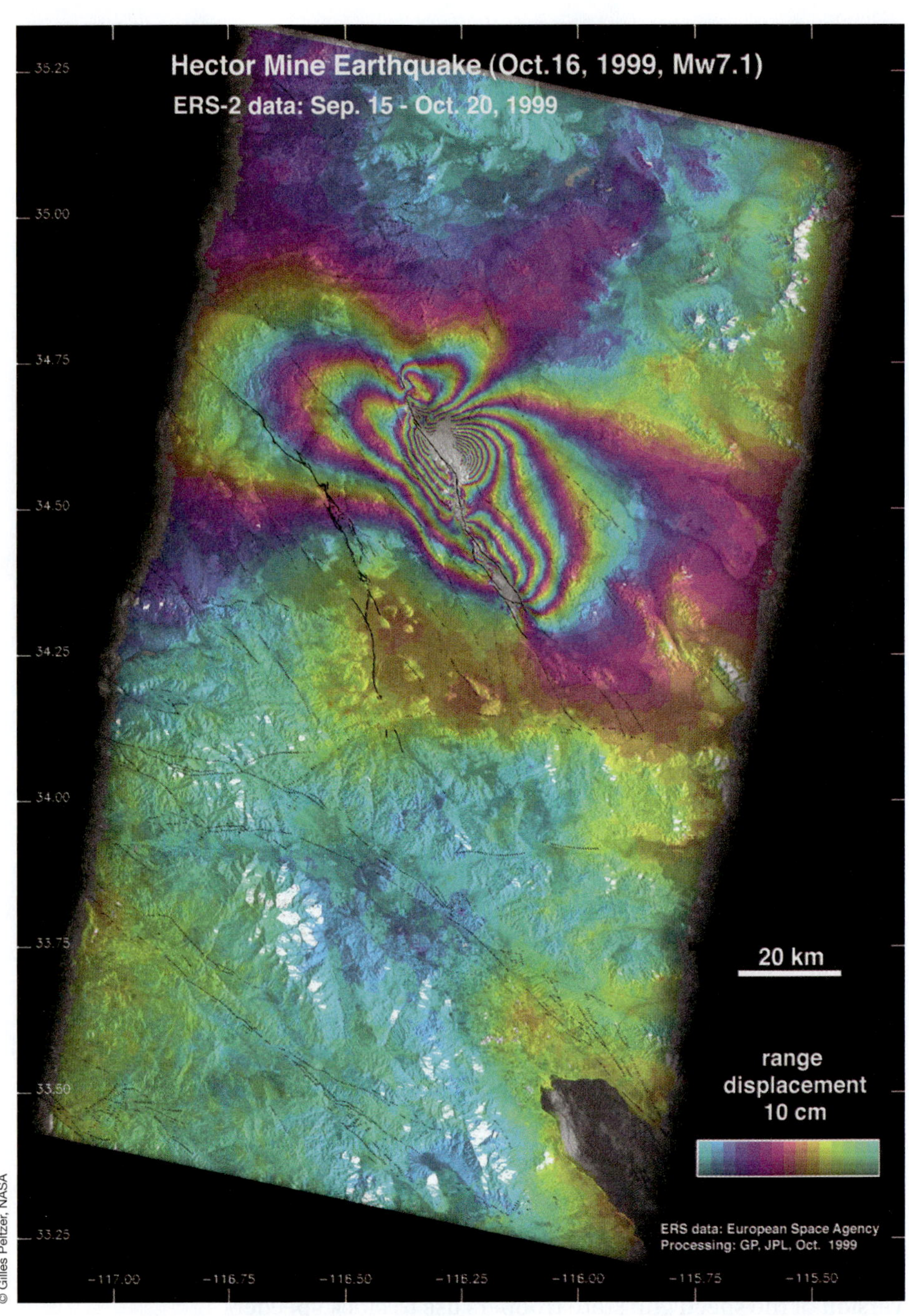

Figure 11-27 Surface displacement due to the Hector Mine Earthquake in the Mojave Desert, Southern California.

A network of permanent GPS stations located throughout Alaska, California, and much of the western United States monitors movement of the earth's surface. Unlike InSAR, GPS stations allow continuous monitoring of ground deformation but only at predetermined points. Scientists are using the combination of InSAR coverage and GPS stations dispersed over earthquake-prone regions in Alaska, California, and the western states. By monitoring earthquake prone areas, they hope to learn more about earthquake activity and the patterns of strain accumulation that lead up to earthquakes. InSAR and GPS data allow scientists to image and calculate the strain patterns associated with active faults before and after earthquakes over large areas.

11.5 people and earthquakes

Earthquakes have caused hundreds of thousands of deaths and billions of dollars in damages in the past century. Sadly, improvements in technology and understanding of earthquake hazards have only slightly reduced the death toll worldwide (Table 11-3). As recently as December 2003, more than 30,000 lives were lost when an earthquake destroyed the ancient city of Bam in southern Iran (Figure 11-28). The magnitude 6.3 earthquake that hit Bam, the home of a 2,000-year-old citadel, should not have claimed so many lives. Despite a history of devastating earthquakes in Iran, the people of Bam and many other places in the developing world live in nonreinforced brick and mud houses that collapse easily during strong ground shaking.

In countries with stricter building codes, similar size earthquakes claim fewer lives and cause less damage. The Northridge, California, earthquake struck at 4:31 a.m. on January 17, 1994. This earthquake was approximately the same size as the one that struck Bam, Iran, on December 26, 2003. Almost no houses collapsed in Northridge, and most people were in their beds rather than on the streets where falling debris could hurt them. Although the 1994 Northridge earthquake caused $40 billion of property damage, only 67 people died.

Table 11-3 *Most destructive earthquakes on record in the world, ≥ 50,000 deaths.*

Date	Location	Deaths	Magnitude	Comments
October 8, 2005	Pakistan, Kashmir	87,350	7.6	Estimated deaths: 100,000+
January 23, 1556	China, Shansi	830,000	~8	
December 26, 2004	Sumatra	283,106	9.0	Deaths from earthquake and tsunami
July 27, 1976	China, Tangshan	255,000 (official)	7.5	Estimated death toll as high as 655,000
August 9, 1138	Syria, Aleppo	230,000		
May 22, 1927	China, near Xining	200,000	7.9	Large fractures
December 22, 856+	Iran, Damghan	200,000		
December 16, 1920	China, Gansu	200,000	7.8	Major fractures, landslides
March 23, 893+	Iran, Ardabil	150,000		
September 1, 1923	Japan, Kanto (Kwanto)	143,000	7.9	Great Tokyo fire
October 5, 1948	USSR	110,000	7.3	(Turkmenistan, Ashgabat)
December 28, 1908	Italy, Messina	70,000 to 100,000 (estimated)	7.2	Estimated deaths are from earthquake and tsunami
September, 1290	China, Chihli	100,000		
November, 1667	Caucasia, Shmakha	80,000		
November 18, 1727	Iran, Tabriz	77,000		
November 1, 1755	Portugal, Lisbon	70,000	8.7	Great tsunami
December 25, 1932	China, Gansu	70,000	7.6	
May 31, 1970	Peru	66,000	7.9	$530,000,000 damage, great rock slide, floods
1268	Asia Minor, Silicia	60,000		
January 11, 1693	Italy, Sicily	60,000		
May 30, 1935	Pakistan, Quetta	30,000 to 60,000	7.5	Quetta almost completely destroyed
February 4, 1783	Italy, Calabria	50,000		
June 20, 1990	Iran	50,000	7.7	Landslides

(a)

(b)

Figure 11-28 Bam Citadel in Iran before (a) and after (b) the December 2003 earthquake.

Nonetheless, serious earthquake hazards exist in heavily populated regions of this country too. As Bam, Iran, and Northridge, California, illustrate, the size of the earthquake is not the only factor that determines the intensity of damage. Earthquakes of similar magnitude can cause devastation in one place and relatively little damage in another.

Land-Use Choices

Similar to the problem of large populations living near active volcanoes, many people live in earthquake country. Did population growth in California slow down after the great earthquake of 1906? On the contrary, the San Francisco Bay and Los Angeles areas are two of the most populated regions of the country; they are also the areas with the greatest risk of earthquakes. Unlike historic Bam, Iran, where people can trace their roots for hundreds, if not thousands of years, most people living in California have not lived there for more than one or two generations. They have moved to California by choice. Californians are well aware of the potential earthquake hazards and are reminded quite frequently by the rumbling of minor earthquakes.

For those who choose to live in earthquake country, there are also local land characteristics that contribute to earthquake hazards. Scientists observed that certain neighborhoods in the San Francisco area experienced stronger ground shaking and sustained more damage to buildings than other neighborhoods the same distance from the epicenter of the 1906 earthquake. They found that the areas that sustained the most damage were underlain by thick layers of soft soils rather than by bedrock. Soft soils tend to amplify ground shaking more than hard soils, or bedrock.

Unconsolidated material such as sand, silt, and clay that is saturated with water may liquefy during strong seismic events, causing buildings to collapse and sink; this phenomenon is referred to as **liquefaction**. In 1985, a magnitude 9.1 earthquake shook Mexico City. The most severe damage to buildings in the city was concentrated within just a few city blocks. Scientists looking for answers for the localized damage learned that this part of the city was built on sediments deposited in an historic lake. The lake was drained by the Spanish to accommodate expansion of the city after their conquest of the Aztecs. These sediments amplified seismic waves and became liquefied, causing the buildings above them to collapse. Adjacent buildings built on bedrock, or hard-packed soils, were undamaged. From this history, how would you use geologic maps of soils and rock to plan urban development? Areas of soft soils underlie many heavily populated areas in earthquake prone cities, including Los Angeles, parts of San Francisco, and Seattle.

Construction Technology and Design

Some people say that it is not earthquakes that kill people, but rather poor building construction. You just learned that a building's location is important too. But if buildings are located AND constructed properly, earthquake impacts can be minimized. Buildings that are not constructed to withstand shaking are the number one killer during large earthquakes. Nonreinforced buildings made of brick, mud, concrete block, stone, or similar materials, collapse during strong ground shaking. In places such as California, laws have been passed that require buildings to be designed and built to withstand ground shaking. Older buildings that do not comply are required to be publicly listed and modified to meet earthquake standards. Brick buildings throughout the Midwest and East coast are also vulnerable to earthquake damage. Most of the damage and deaths due to the 1886 Charleston, South Carolina, earthquake were caused by collapsing brick buildings. Countries without strict building codes have suffered the largest fatalities due to poor construction (see Table 11-3).

To illustrate the importance of construction design, consider the different outcomes of two earthquakes similar in size that both happened in close proximity to cities, the 1989 Loma Prieta earthquake near San Francisco (M_w = 7.1), and the 1999 Turkey earthquake on the North Anatolian Fault near Koceali (M_w = 7.4). Koceali is a modern industrial center with an educated workforce that had undergone a recent economic boom and constructed many apartment buildings to accommodate its growing population. The earthquake caused massive building collapse, including the new apartments, during 45 seconds of violent shaking. About 17,500 people died and 133,000 homes were destroyed. The Loma Prieta earthquake, on the other hand, caused 62 deaths and left 16,000 uninhabitable homes.

Tsunami Preparation and Warning

Earthquake-related hazards are not restricted to those that shake and destroy buildings or other infrastructure—they are the most common cause of tsunami. Tsunamis are large waves that can travel across entire oceans and affect communities thousands of kilometers (miles) from an earthquake's epicenter. Tsunamis are caused by fault movements that displace the ocean floor (Figure 11-29). These movements create large waves. Where these waves reach shallower water, their height can increase to over 25 meters (82 feet); they can engulf entire villages and crash down with disastrous effects.

The most recent example is the devastating tsunami caused by the December 26, 2004, earthquake off the shore of Sumatra, Indonesia (Figure 11-30). For many coastal earthquakes, the death toll from tsunamis far surpasses that from earthquake shaking. The 2004 Sumatra tsunami killed about 300,000 people.

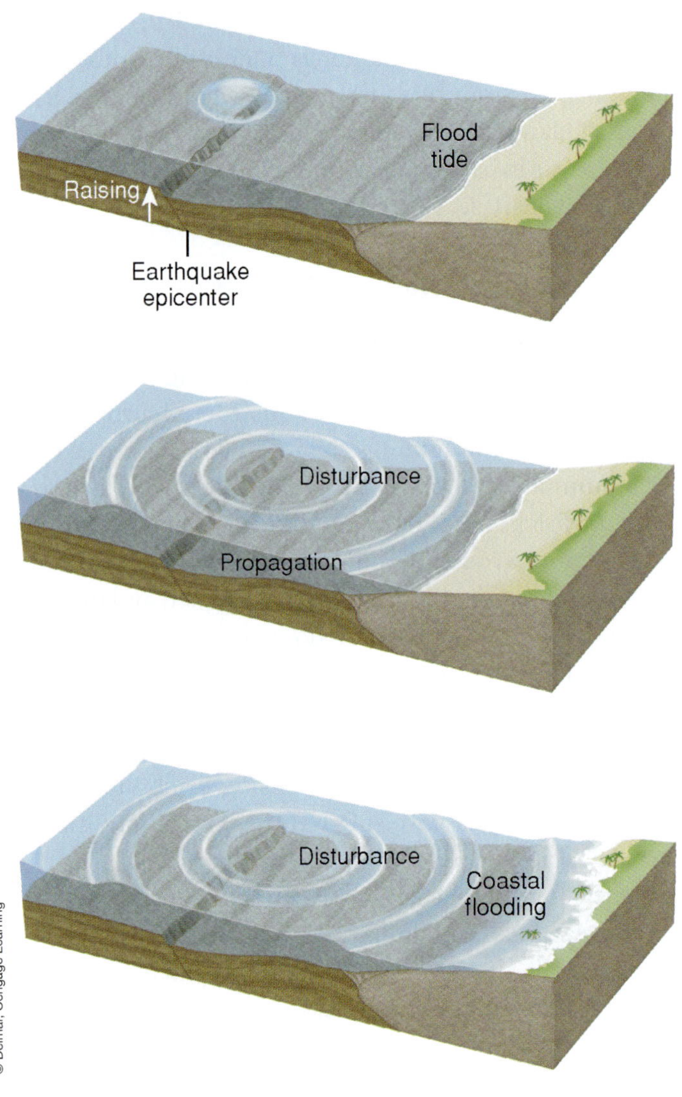

Figure 11-29 Three ways in which tsunamis can form: landslides, earthquakes, and volcanic eruption under water.

(a)

(b)

Figure 11-30 Before (a) and after (b) conditions at Banda Aceh Shore, Indonesia, from the December 2004 earthquake.

Tsunamis can travel in excess of 800 kph (500 mph) over the open seas where their wave height is less than 1 meter (3.3 feet). The speed decreases sharply as the tsunami reaches shallow water, but it is here that its height increases rapidly. Because of their great speed and small initial size, tsunamis are difficult to detect and can surprise entire coastlines if warning systems are not in place. A successful warning system for tsunamis was put in place for most communities around the Pacific after a devastating tsunami hit the Aleutian Islands in 1946.

The Tsunami Warning System is triggered when a seismometer in the region detects and locates a large earthquake. The system's headquarters in Hawaii receives the seismometer measurements and immediately transmits warnings of a possible tsunami to coastal communities. This type of warning system is now being developed in the Indian Ocean. Perhaps the tremendous loss of life caused by the 2004 tsunami will be avoided in the future.

Living with Earthquakes

Identifying earthquake hazards and understanding the risk to people and property is the first line of defense against earthquakes. Understanding earthquake hazard potential first and foremost requires identifying active faults. Faults are identified both by seismic monitoring and geologic mapping. The western states are the main areas with active faulting in the United States. These are the areas with the greatest earthquake risk. Although the western states have the highest risk, there is also risk in parts of the country that do not have active plate boundaries but do have ancient faults that act as planes of weakness that can be reactivated. The forces responsible for reactivating these old faults are not well understood, but the past history of earthquakes points to future earthquake potential even in places like South Carolina and Missouri. Both of these areas sustained major earthquakes in the 1800s.

To assess the earthquake hazard risk throughout the country, the United States Geological Survey (USGS), as well as other state and local government agencies, have constructed maps showing probable hazards from future earthquakes. Seismic hazard maps, such as in Figure 11-1, are contoured or shaded according to the peak ground acceleration that can occur during an earthquake (in other words how hard the ground will shake). The potential for ground shaking at any given place depends on several factors, including the distance to active faults, the geologic material (soft soil vs. firm bedrock), and the size and frequency of past earthquakes. Seismic hazard maps also include the probability of potential earthquakes. Probability is represented as an expectation that a parameter (such as ground acceleration) will be exceeded within a specified period such as 50 years. Seismic hazard maps are essential for appropriate land-use choices and engineering designs.

Building codes that take into account maximum potential ground shaking from earthquakes are now used in high-risk earthquake regions such as California. California has detailed seismic hazard risk maps for most counties, put together by both local and state agencies. Figure 11-31 is an example of a "shaking intensity map" for Oakland, California. It shows estimated shaking severity levels, in different parts of town, if a magnitude 7.9 earthquake were to hit the region. The San Francisco Bay area of California is one of the few places in the world where seismic risk maps have adequate detail for construction design and city planning purposes.

A new generation of highly sensitive seismometers and modern communication and computer networks may give people a few seconds warning of an impending earthquake. Even a few seconds warning could enable utilities

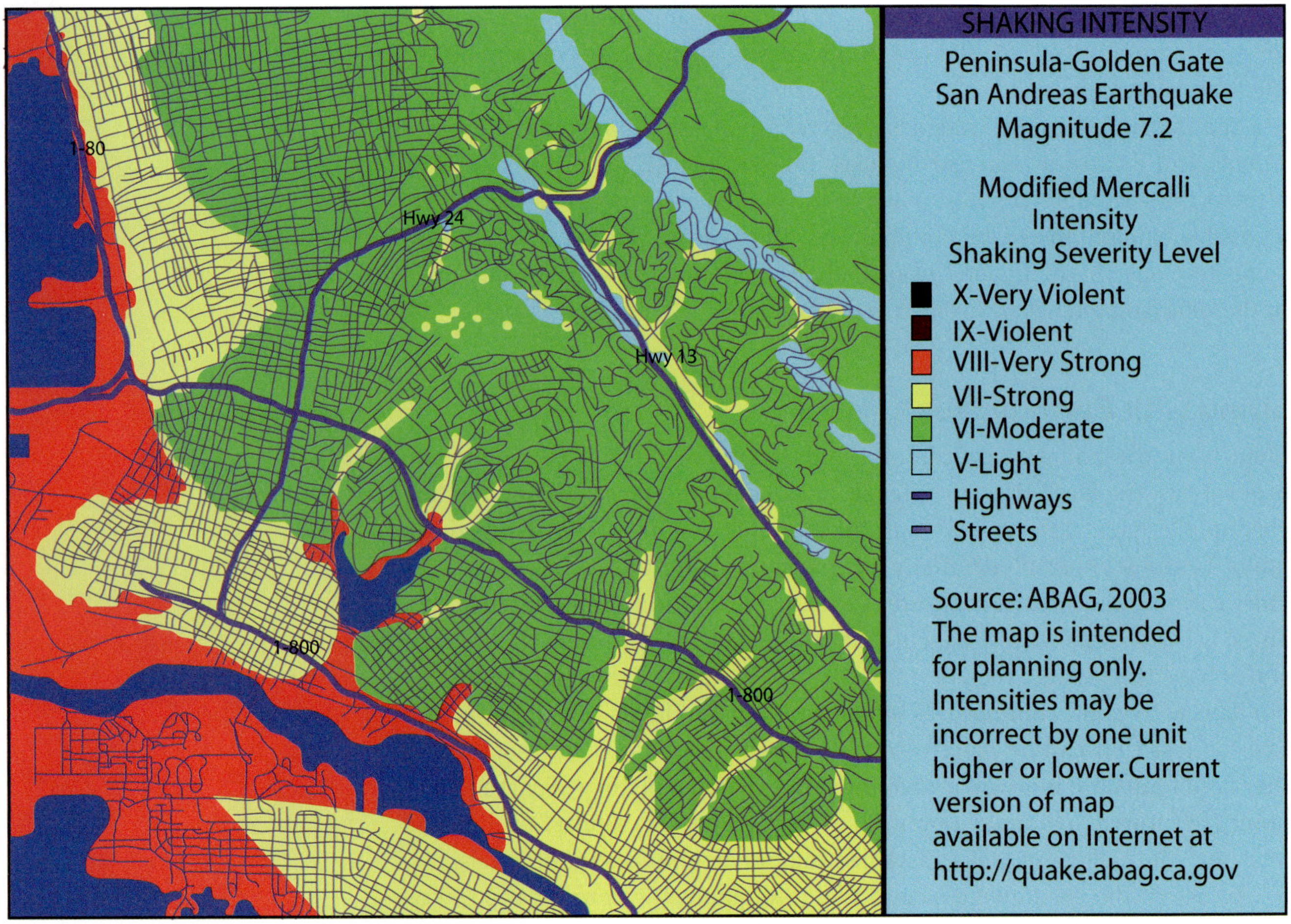

Figure 11-31 Earthquake hazard map for North Oakland, California.

to shut off electricity and natural gas. An example of a place that might benefit from this type of warning system is Mexico City. The 1985 earthquake that devastated Mexico City originated in a subduction zone more than 200 miles to the west. The time it took for the destructive surface waves to reach Mexico City from the focus of the earthquake might have allowed a couple of minutes warning if there had been a warning system in place. This type of warning is possible because radio waves can travel faster than seismic waves. A seismometer located close to the epicenter could trigger a radio signal that would then be transmitted to the surrounding cities and towns. How useful do you think this type of warning system would be for cities like San Francisco that are located very close to earthquake-generating faults?

11.6 conclusion

Earthquakes are direct evidence of the dynamic and changing Earth. They are a source of awe and wonder as well as devastating effects on people. As with most natural hazards, however, people's understanding of earthquakes can enable them to live more safely with them. Where people choose to live, and how they construct their homes and other structures, are important to living more safely with earthquakes.

IN THIS CHAPTER YOU HAVE LEARNED:

- Earthquakes are vibrations called seismic waves that travel both through the Earth and along its surface. P (or primary) seismic waves alternately compress and expand rock—they travel the fastest. S (or secondary) seismic waves move rocks from one side to another. P and S waves cause surface waves that move Earth's surface up, down, and sideways.

- Earthquakes are caused by the release of elastic energy that builds up along faults. When faults rupture—the block of rock on one side of the fault suddenly moves relative to the block on the other side—the stored elastic energy is released as seismic waves.

- Most earthquakes occur along plate boundaries because this is where the largest and most active faults are located. Faults at convergent and transform plate boundaries are particularly large—sometimes over a thousand kilometers (miles) long. They have caused the largest known earthquakes.

- Seismometers measure the ground motion caused by earthquakes. Scientists use seismometer measurements to determine earthquake size and location.

- An earthquake's size is determined by its magnitude and intensity. Magnitude is an estimate of the energy released by an earthquake. Magnitudes calculated today are "moment" magnitudes that take into account the area of fault rupture, the amount of movement on the fault, and the strength of the faulted rocks.

- Earthquake intensity measures the amount of damage caused by an earthquake and is based on people's observations. It is useful in determining earthquake size in areas with few seismometers.

- An earthquake's strong ground motion and shaking creates hazards for people and structures. They can also cause large and dangerous ocean waves, called tsunamis, if the fault rupture moves the seafloor.

- Both ground motions and tsunamis can devastate property and cause many deaths. Much earthquake damage comes from the collapse of poorly constructed buildings and other structures.

- Similar magnitude earthquakes will cause fewer deaths and less damage if buildings and other structures are appropriately located, designed, and constructed. This is the key way that people can live more safely with earthquakes.

KEY TERMS

Body waves
Elastic energy
Elastic rebound
Epicenter
Faults
Focus
Intensity
Isoseismal maps
Liquefaction
Magnitude
Mercalli scale
Mid-ocean ridges
Moment magnitude
Primary "P" waves
Recurrence intervals
Richter scale
Secondary "S" waves
Seismogram
Seismograph
Surface waves
Transform plate boundaries

REVIEW QUESTIONS

1. Describe how earthquakes happen.
2. Where on Earth are earthquakes likely to happen? Why is that?
3. How can scientists detect earthquakes?
4. How are the three types of earthquake waves different from one another?
5. What are three effects that earthquakes can have on the biosphere?
6. What is the relationship between earthquakes and tsunamis?
7. Why do you think that humans choose to live in earthquake-prone areas?
8. Why does the state of California have so many earthquakes?
9. How can humans best prepare for earthquakes?
10. If you were to design an earthquake-resistant building, what would you have to consider?

chapter 12

unstable land

High school students in Ventura County, California, have events right on their doorstep to investigate when they study unstable land (Figure 12-1). One of the most serious of these happened on January 10, 2005. After several days of heavy rain, a landslide hit the coastal community of La Conchita in Ventura County, destroying 36 houses and killing 10 people (Figure 12-2). Landslides are not new to La Conchita, which was built at the foot of a steep bluff and below an old landslide deposit. In fact, the 2005 landslide reactivated part of one that buried a few houses in La Conchita in 1995. You may wonder why the La Conchita community was built in such a precarious location in the first place. You are not the only one!

Figure 12-1 High school students investigating unstable land. (*Courtesy the Earth Science World Image Bank © Tim McCabe, NRCS*)

Courtesy AirPhoto USA

Figure 12-2 January 10, 2005 landslide, La Conchita, Ventura County, California.

Landslides and other effects of unstable land are among the most widespread hazards on Earth. They occur in all 50 states, even vertically challenged states such as Florida and Kansas. In the United States unstable land causes $1 billion to $2 billion in damages and more than 25 deaths (on average) each year. Globally, the situation is much worse—unstable land causes billions of dollars in damages and thousands of deaths and injuries every year. Losses due to unstable land are increasing in the United States. This is surprising given the examples of many poorly planned communities such as La Conchita that we have to learn from. You may wonder why losses are greater now than before. In large part, development of hazardous areas continues under pressures of urban growth.

Unstable land is most common where slopes are steep. To most people, the word landslide *brings to mind images of avalanches of rocks and debris roaring down a mountainside. However, unstable land can also be very slow moving and occur in flat areas or areas with gentle slopes. Sinkholes, caused by the collapse of underground caves, or creep, the slow movement of soils down gentle slopes, can be as destructive and costly as a rushing mountain landslide.*

Although problems from unstable land commonly catch communities by surprise, areas of land that are at risk can usually be identified. The timing of a particular landslide may not be easy to predict, but the areas prone to landslides, sinkholes, or soil creep can be identified. No matter which state you live in, you can be sure that certain areas have unstable land. Think of some reasons why you would want to know where the unstable land areas are in your state. Imagine that you are a real estate developer or a civil engineer about to construct a bridge or that you plan to buy land on which to build a house. It may seem unlikely that you would build in an area with unstable land hazards, but people do it all the time. In 1981, a sinkhole in Winter Park, Florida, swallowed a house, half of a six-lane highway, a parking lot, and a public swimming pool (Figure 12-3). Chances are the homeowners, as well as the pool and highway designers, thought they were building on solid ground.

Figure 12-3 Sinkhole in Winter Park, Florida *(Courtesy A. S. Navory, National Geophysical Data Center (NGDC))*

Landslides and other ground failures impose many costs on society. For example, landslides damage infrastructure systems that are expensive to repair. These include roads, tunnels, bridges, railroads, pipelines, electricity, and telecommunication lines, to name a few. Local, state and federal agencies are responsible for repairing these systems at great cost to all taxpayers. It is in everyone's best interests to minimize the impact of unstable land hazards.

AFTER COMPLETING THIS CHAPTER, YOU WILL UNDERSTAND:
What unstable land is and where it occurs.
How driving and resisting forces control unstable land.
How unstable land is a natural part of Earth system interactions involving the atmosphere, hydrosphere, biosphere, and geosphere.
How human activities can trigger or speed up unstable land processes.
How unstable land risks are evaluated and how people adjust to unstable land.
The role of land-use planning and engineering designs in managing unstable land.

 12.1 what is unstable land?

Ground that moves is **unstable land**. Steep shoreline bluffs that shed many landslides or a slowly slumping hillside are both examples of unstable land. Unstable land also occurs on the seafloor and can cause dangerous tsunami. In Chapter 9 you learned that over very long periods, all land is unstable and is continuously changing. In this chapter, however, unstable land is that part of the Earth's surface that changes and moves in ways that are disruptive to people's lives and property. These include several fast- and slow-moving types of land movement (Table 12-1). The different types of unstable land include landslides, mudflows and debris flows, soil creep, sinkholes, and subsidence.

Table 12-1 *Examples of fast and slow land movement.*

Type of movement	Example	Rate
Debris flow/avalanche	Yungay, Peru, 1970	270 mph
Earth flow (slow landslide)	Portuguese Bend, CA	1–10 feet per year
Soil creep	Humid environments	1–5 mm per year

Figure 12-4 1997 landslide that closed Highway 50 near Lake Tahoe, California.

Landslides

Landslide is a general term for the downslope movement of masses of rock, soil, or debris; they are also referred to as **mass movement**. "Landslide" is also used as a name for more rapid movements of fairly coherent rock, or unconsolidated masses, down slopes. Landslides can be very large or small. In many cases, landslides occur repeatedly in the same area. They are reactivated by storms or human modification of the land. New landslides can be triggered by earthquakes, heavy rain or snowmelt, volcanic eruptions, human activities, or simply by a steep slope collapsing under the force of gravity (Figure 12-4).

Mudflows and Debris Flows

Mudflows and debris flows (flows of relatively coarse material) are landslides that are saturated with water. They form rivers of mud, rock, soil, and other debris that take on the consistency of liquid cement and can level just about anything in their path. Mudflows and debris flows are among the most destructive natural disasters.

Figure 12-5 Armero, Colombia, destroyed by a debris flow on November 13, 1985. (*Courtesy: U.S. Department of the Interior, U.S. Geological Survey Cascades Volcano Observatory, Vancouver, WA*)

The conditions for debris and mudflows occur when water quickly saturates the ground either from heavy rainfall or rapid snowmelt. They commonly accompany heavy rains in mountainous regions or volcanic eruptions. Mudflows on the flanks of volcanoes can cause far more damage and fatalities than the eruption itself (Figure 12-5; see also Chapter 10). For example, take Casita Volcano in Nicaragua. On October 30, 1998, a rockfall near the steep summit of this volcano became a fast-moving flow of mud, rock, and debris that roared down a valley within minutes. The wall of debris, up to 3 meters (10 feet) high and 1,500 meters (4,900 feet) across, devastated everything in its path, including two villages that were completely buried. Almost 2,000 people died with little warning other than a brief loud rumbling that sounded like thunder. The initial rockfall became such a large and fast-moving flow because the surface materials in its path were exceptionally wet. Hurricane Mitch had been dumping huge amounts of rain on Nicaragua for days. On October 30 alone, about 50 cm (20 inches) of rain fell in the area of Casita Volcano. The weight of the water-saturated material, the steep slopes on the sides of the old volcano, and the fractured and altered nature of the bedrock near the volcano summit all combined to create this devastating disaster—it was not due to a volcanic eruption. Of course it did not help that the destroyed villages, established only decades earlier, had been located directly in the flow's path.

Volcano flanks are not the only place where debris and mudflows are common. Mountains in semiarid climates, such as southern California, are also particularly prone to mudflows during periods of heavy rain. When these dry regions receive heavy rain, very little of the water percolates down into the ground. Most of the rain runs off, collecting loose material with it. Mud and debris fill channels, replacing streams and dry riverbeds with mudflows.

Soil Creep

Less dramatic than landslides, debris flows, and mudflows, but still destructive, are the very slow, downhill movements of soil called **soil creep**. Evidence of soil creep is commonly so subtle that it requires detective work to identify. It happens too slowly for people to see, but it will gradually tilt trees or fence posts and can tilt and break building foundations (Figure 12-6).

Figure 12-6 Soil creep, as shown by bent tree trunks. *(Courtesy the Earth Science World Image Bank © Marli Miller, University of Oregon)*

Soil creep takes place in the top several feet of the soil and is caused by differences in soil expansion and contraction. In northern climates, water in the soil freezes and the ice pushes soil particles upward. When the ice thaws, the soil is pulled down slope by gravity. Over many cycles of freezing and thawing, soil moves slowly, or creeps, downslope. In other climates, especially where clays are present in the soil, wetting and drying cycles can cause soil creep. These clay-rich soils expand upward when wet and contract and gradually fall downslope when they dry.

Sinkholes

Crater-like holes that form when the land surface collapses are called **sinkholes** (see Figure 12-3). Sinkholes form on land that is underlain by a network of underground channels and caves. This type of landscape is called **karst** (Figure 12-7). Karst regions are characterized by caves, sinkholes, underground streams, and other features formed by the slow dissolving of rocks. Rocks such as limestone, dolomite, salt, and gypsum dissolve more easily than other rocks and are typical of karst regions. Even the weak acids that form in soils are effective at dissolving these rocks over time. At first the rocks dissolve along cracks, but these cracks gradually enlarge and produce caves and networks of underground channels (Figure 12-8).

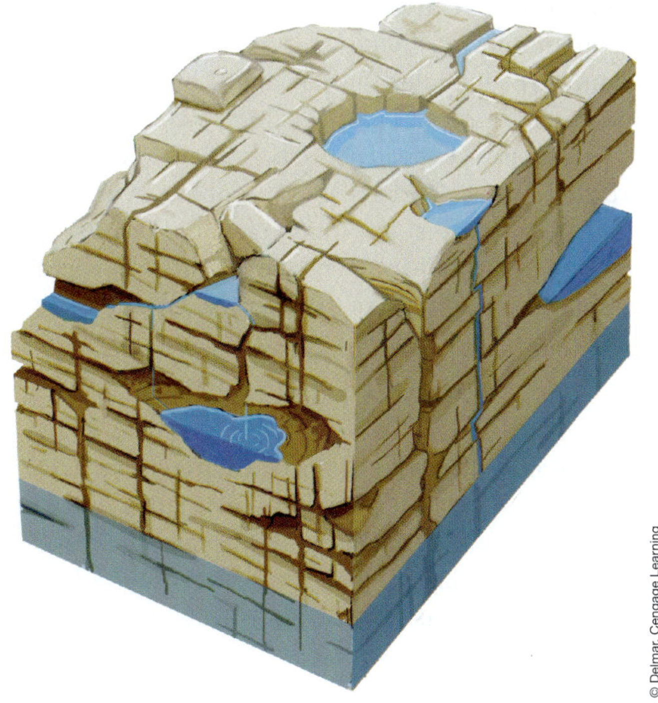

Figure 12-7 Bedrock surface in karst terrain.

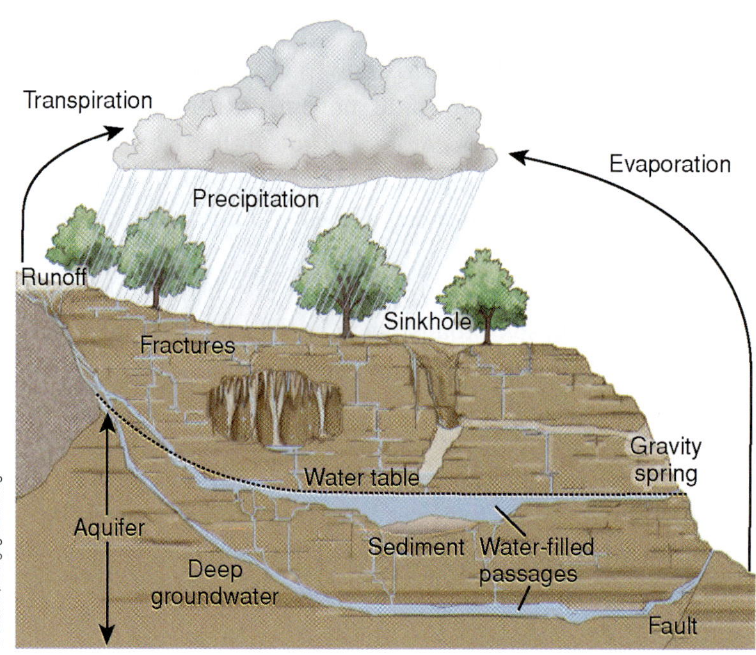

Figure 12-8 Hydrologic cycle in karst areas.

Figure 12-9 Sinkholes abound in this typical West Virginia karst landscape. (*Courtesy the Earth Science World Image Bank © R. Ewers*)

Sinkholes commonly form in what seems to be a gentle and stable landscape (Figure 12-9). Karst areas include regions of relatively low topographic relief such as Florida and much of Georgia and South Carolina (Figure 12-10). Sinkholes form suddenly when the roofs of rock caves within karst networks become weak and collapse, creating a large circular pit. Sinkholes also form by collapses within the soil and other unconsolidated material that overlie the karst bedrock. The washing of the overlying material into underlying caves can eventually cause catastrophic collapse (Figure 12-11).

The problems caused by sinkholes are pretty obvious when they open up and devour homes, cars, or streets, but there is another problem with sinkholes that is caused by people. In some places, people use sinkholes as garbage dumps and dispose of everything from refrigerators to dead cows in them. Volunteers recovered piles of tangled wire, metal sheets, broken glass, appliances, and auto parts from a Kentucky sinkhole during just one day of cleanup. Disposing of garbage in sinkholes can seriously degrade the quality of ground water because sinkholes provide a direct path for pollutants to enter the ground water system (see Chapter 13).

Subsidence

Subsidence is the gradual lowering of the land surface over large areas. In the United States, more than 17,000 square miles in 45 states, an area roughly the size of New Hampshire and Vermont combined, are affected by subsidence. Some of the more spectacular examples of subsidence are in the southwestern states, including Arizona, Nevada, California, and Texas. Some areas of California sank about 8 meters (25 feet) in the past 50 years. More than 80 percent of subsidence cases identified in the United States are caused by the

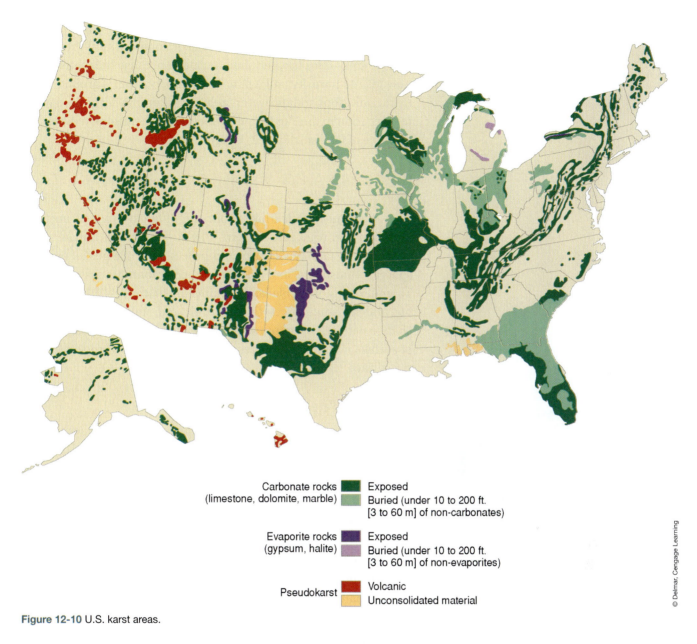

Carbonate rocks (limestone, dolomite, marble)
- Exposed
- Buried (under 10 to 200 ft. [3 to 60 m] of non-carbonates)

Evaporite rocks (gypsum, halite)
- Exposed
- Buried (under 10 to 200 ft. [3 to 60 m] of non-evaporites)

Pseudokarst
- Volcanic
- Unconsolidated material

Figure 12-10 U.S. karst areas.

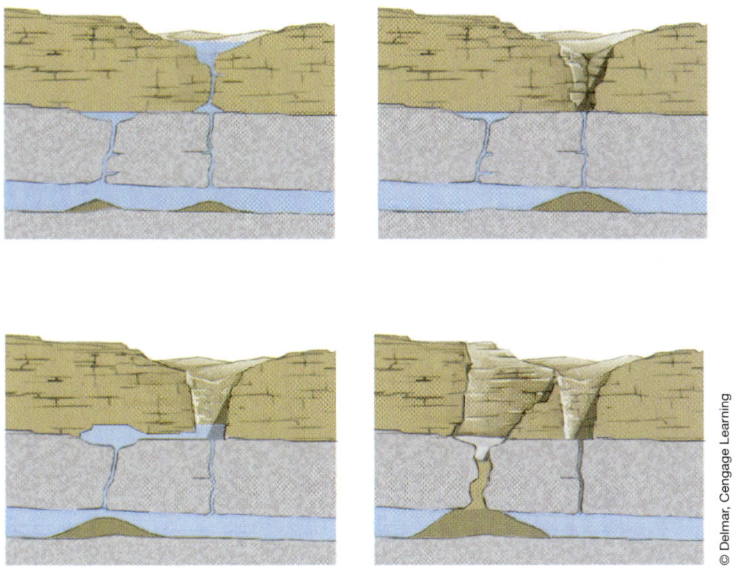

Figure 12-11 Sinkhole collapse.

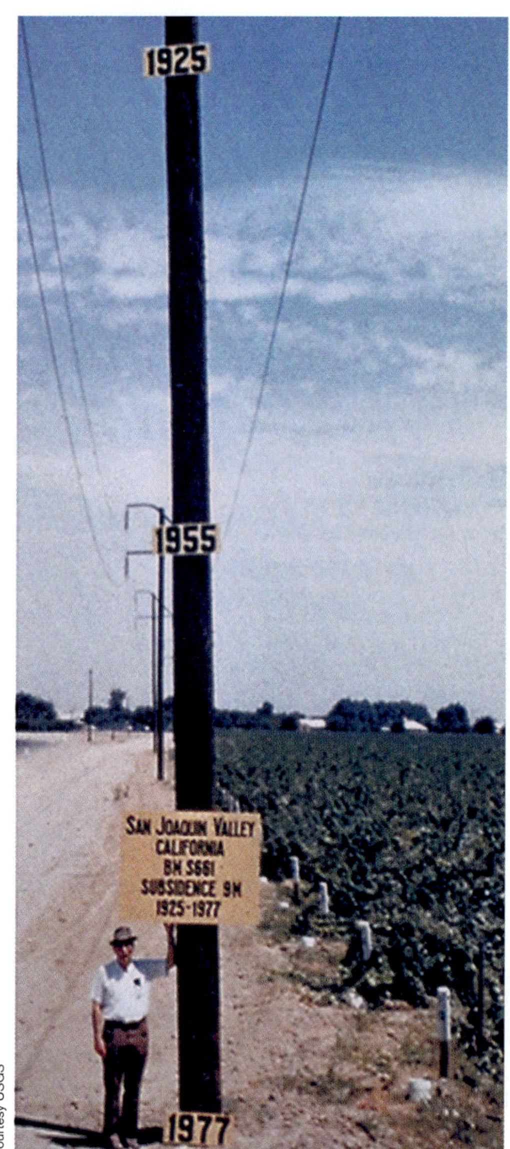

Courtesy USGS

Figure 12-12 San Joaquin Valley in California where the land has dropped due to overpumping of ground water.

withdrawal of ground water. Once water has been removed from compressible unconsolidated material, it compacts and the ground surface sinks. This is a permanent change because the material has now lost its porosity and will not decompact, even if ground water extraction stops and the water table rises. Pumping of ground water for irrigation in the San Joaquin Valley of California created a classic example of land subsidence (Figure 12-12).

12.2 causes of unstable land

Now that you have learned about the different types of unstable land, consider what causes them. Why do some slopes remain stable while others fail? What parts and processes of the geosphere, biosphere, atmosphere, and hydrosphere contribute to unstable land? To understand unstable land requires understanding how the driving and resisting forces that control land stability combine and change with other factors such as the steepness of slopes, the type of surface materials on slopes, and the properties of rocks beneath the surface.

Gravity—The Driving Force

Gravity is always operating—pulling downward on surface materials. Just as you feel the force of gravity pulling you downhill when you stand on a steep slope, a boulder or grain of sand is subject to the same force. If you stand on a flat surface, gravity's pull (your weight) is directly downward. But on slopes, gravity's pull is split into two parts—one pulling down and one pulling in a direction parallel to the slope (Figure 12-13). This is why the steepness (or angle) of a slope is such an important factor when evaluating land stability: the steeper the slope, the greater the gravitational component pulling surface materials downhill and the more likely it is that they will collapse, fall, or slide.

Friction and Cohesion—The Resisting Forces

Are there steep slopes where you live that appear to be stable? A steep slope is not enough to cause unstable land. This is because there are forces that resist gravity's pull on slopes. These are the resisting forces—friction and cohesion. Friction is the force resisting movement along a surface or boundary between

Figure 12-13 The two gravity components on a slope and how they proportionally change with slope angle.

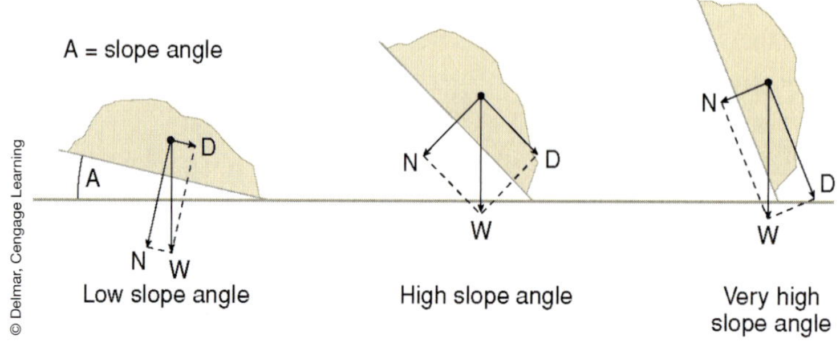

© Delmar, Cengage Learning

two objects. The more friction there is, perhaps because the contact surface is very rough and irregular, the harder it is for movement to occur along the surface.

Cohesion is the force that causes individual particles in a material to be attracted to one another—to tend to stick together. Loose sand has very little cohesion, but clay-rich soil can have a great deal.

Friction and cohesion combine to resist gravity's pull on slope materials. If these forces are small compared to gravity's pull, downslope movement is likely. Now, let us consider how driving and resisting forces interact to cause unstable land. Factors such as material properties, the water table, precipitation, earthquakes, volcanic eruptions, vegetation, and land use can lead to changes in driving and resisting forces that make land unstable.

Material Properties

The stability of a sloping surface depends on both the angle of the slope and the properties of the material covering the slope. Solid rock without fractures can have high internal friction and cohesion and be stable at high slope angles (Figure 12-14). On the other hand, highly weathered and fractured rock is likely to have low cohesion and friction and be quite unstable. The soil and loose material overlying bedrock usually does not have the strength of the underlying rock. Engineers and geologists dealing with slope stability problems refer to soil and this loose material as "**unconsolidated**." The unconsolidated material is usually the material that moves when land is unstable.

Figure 12-14 Indian dwelling in cliffs of the Verde Formation. (*Courtesy the Earth Science World Image Bank © Larry Fellows, Arizona Geological Survey*)

Unconsolidated material consists of loose, unattached particles such as gravel or sand. The nature of unconsolidated material varies widely. Unconsolidated material may include fragments as large as boulders or as fine as dust. Unconsolidated material may consist entirely of particles of similar size, such as sand in a dune, or it may be a mixture of a wide variety of particles from silt to boulder size, such as the material within a debris flow.

Examples of unconsolidated material include rock and boulder fragments that have accumulated at the base of a cliff or steep slope, called **talus**. Unconsolidated material deposited by rivers or streams is called **alluvium** and includes both well-sorted deposits, such as sand and gravel, and unsorted mixtures of sediment ranging from silt to boulder size. Your backyard, the beach, mudflats, and gravel pits are all examples of places where you can find unconsolidated material. Unconsolidated material is deposited by a variety of agents, including rivers, oceans, volcanoes, snow melt, wind, glaciers, landslides, and even heavy machinery.

The stability of unconsolidated material depends on many factors, including the steepness of the slope on which it lies. The maximum angle at which unconsolidated material is stable is called the angle of repose. The **angle of repose** is the angle between the slope surface and horizontal and varies for different materials (Figure 12-15). The angle of repose for dry sediments generally ranges from ~30° to 37°. The size of the unconsolidated material (sand, gravel, cobble, or boulder), as well as the shape of the material (rounded versus angular), affects the angle of repose. There is more friction between angular particles than rounded grains, so they will remain on a steeper slope. Imagine how the angle of repose would differ for a pile of marbles as opposed to a pile of pennies.

Figure 12-15 Angle of repose with protractor and granular material.

The nature of unconsolidated material plays an important role in the stability of the land surface. For example, volcanic regions such as the Pacific Northwest typically have clay-rich soils because volcanic ash (see Chapter 10) readily breaks down to form clay. Clay has physical properties that contribute to unstable land problems on steep slopes. Clay layers in soil absorb water easily because they are very porous. However, clay is not very permeable and water does not flow through it easily. Layers of clay in soil act as a barrier to water percolating downward. Imagine a clay layer at the base of a thick layer of unconsolidated material on a steep slope. If there is a large amount of rain, the clay layer acts as a barrier and the material above becomes saturated with water. This adds to the weight of the material (increasing the driving force of gravity) and may trigger a landslide.

In addition to clay acting as a water barrier, it is also slippery when wet. If you have ever made anything out of clay, you probably noticed how slick and slippery it felt when it got wet. This is because wetting clay reduces its cohesion. Refer to the diagram in Figure 12-13 and think about how reducing the cohesion of material on a sloped surface would influence its stability. How do you think the combination of reduced cohesion (from wet clay) and increased weight (from water in the material above the clay) affects slope stability?

Role of the Water Table

Land subsidence is caused by changes underground. These changes can include lowering of the water table, underground mining, and thawing of permafrost in northern climates. The most common cause of land subsidence in this country is the lowering of the water table as a result of excessive withdrawal of ground water.

Below the water table, water fills the open spaces in fractured bedrock and unconsolidated materials. When water is removed from sand and gravel, most of the open spaces remain. This is because the grains are rigid and retain their internal structure (Figure 12-16a). In unconsolidated materials that contain clay or silt, the open spaces are not as well preserved when water is removed because some of the individual grains are not rigid. When water is removed, the open spaces between the grains are compressed and the material compacts, becoming thinner (Figure 12-16b). Once unconsolidated material is compacted, it is unlikely to regain its original thickness, even if water is applied. Compaction occurs when ground water is pumped faster than it can be replenished or recharged (see Chapter 13). Many areas in the arid southwest pump large quantities of ground water for irrigation purposes. Aquifers in these regions may be slow to recharge, taking several years to be replenished.

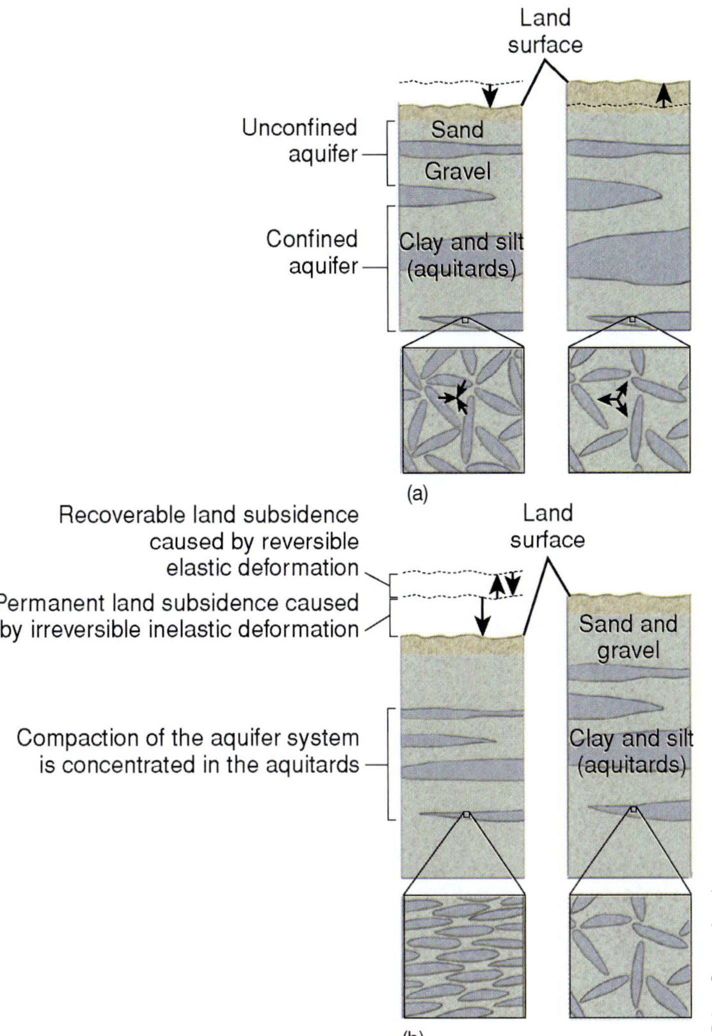

Figure 12-16 (a) Magnified grains touching with pore spaces, both with and without water. (b) Magnified representation of clay grains touching with pore spaces with and without water.

If the water table has been drawn down too quickly and the aquifer material has compacted, the ground surface may subside and never regain its original elevation.

Most observed sinkhole collapses occur in unconsolidated materials overlying bedrock, typically limestone. Although many natural sinkholes cannot be prevented, those caused by human activity, especially those caused by the overpumping of ground water, can frequently be avoided. Overpumping typically lowers the water table and decreases the buoyant support of the unconsolidated materials. The combination of gravity, loss of buoyancy, and water pressure can activate a collapse and result in the formation of a sinkhole. Overpumping, however, is not the only way to lower the water table. During a drought, for example, the water table is lowered naturally. The combination of a drought and rapid ground water extraction frequently results in conditions favorable for the formation of sinkholes.

Precipitation

Saturation of unconsolidated materials, either by rain or snowmelt, is a common cause of landslides. The added weight of the water, combined with the lessening of friction and cohesion, contributes to destabilizing slope material. In southern

Figure 12-17 1995 and 2005 landslide scarps; La Conchita, California.

California, infrequent heavy rains trigger numerous hazardous landslides and mudflows. In 2005, during the months of January and February, heavy rain seemed like a welcome relief to drought-stricken areas of southern California. However, the steep, dry slopes and drainages of the coastal ranges quickly became avenues for mudslides and slope failure (Figure 12-17).

Earthquakes

Strong shaking of the ground from earthquakes can also cause unstable land to fail (see Chapter 11). For example, during the Loma Prieta earthquake that shook the San Francisco Bay area in 1989, thousands of landslides were triggered throughout an area of 13,986 square km (5,400 square miles). Most of the damage caused by earthquakes commonly results from landslides triggered by ground shaking. After the Loma Prieta earthquake, tens of millions of dollars in damages to houses, other buildings and structures, and utilities were caused by landslides. In addition, landslides blocked many transportation routes, hampering rescue and relief efforts. How do you think strong shaking affects the driving and resisting forces on surface materials? Shaking primarily decreases friction and cohesion of the surface materials for a brief but long enough period to cause slope instability.

Figure 12-18 Stream undercutting a bank. (*Courtesy the Earth Science World Image Bank © Tim McCabe, National Resources Conservations Service, USDA*)

Volcanic Eruptions

Volcanic eruptions can cause unstable land in a number of ways. The heat from an eruption can quickly melt large amounts of snow. The snowmelt fills river and stream channels, potentially causing debris flows, or saturating slopes of clay-rich soils to the point of failure. Volcanic eruptions may also be accompanied by earthquakes, which, together, may trigger landslides. Recall from Chapter 10 how the 1980 eruption of Mount St. Helens caused a massive debris-avalanche. Often the most damage and loss of life associated with an eruption results from debris flows and mudflows.

Steepened Slopes

Erosion at the base of a slope can cause unstable land; it undercuts the slope and removes critical support for the unconsolidated overlying material (Figure 12-18). Erosion, in effect, makes slopes steeper and increases the driving force of gravity parallel to the slope. People can also have the same impact as erosion when they undercut and steepen slopes with roads or other excavations (Figure 12-19).

Figure 12-19 Highway cut causing a landslide. (*Courtesy the Earth Science World Image Bank © Bruce Molnia, Terra Photographics*)

Vegetation

Removing vegetation from slopes can also lead to unstable land. Vegetation on slopes protects soils from erosion during heavy rains and absorbs some of the water. Root systems help anchor unconsolidated materials to slopes, indirectly increasing cohesion. Steep terrain is more prone to landslides and/or mudflows when it loses its vegetative cover. Loss of vegetation can happen either from wildfires or from human activities, such as deforestation.

There are several cases of denudation from wildfires leading to landslides. In the summer of 1994, for example, a forest fire burned Storm King Mountain in Colorado. A few weeks later, heavy rains resulted in several debris flows, one of which blocked a major highway and almost dammed the Colorado River.

Figure 12-20 Abandoned homes in the Portuguese Bend region of the Palos Verdes Peninsula in California. (© *Bruce Perry, Department of Geological Sciences, California State University Long Beach*)

People's Changes to the Land

Land development can affect ground water levels as well as cause changes to slopes, loss of vegetation, and a variety of other factors that contribute to unstable land. The Portuguese Bend region of the Palos Verdes Peninsula in California is a classic example. Hundreds of homes were built here in the 1950s on an ancient landslide deposit; each of these homes had its own septic system (Figure 12-20). Water usage in the homes introduced a new source of water and eventually raised the water table. The raised water table increased the weight of the unconsolidated materials overlying bedrock, and decreased the cohesion in a near-surface layer of clay-bearing material. The material beneath the houses started to move very slowly during construction of a road at the top of the ancient slide. During road construction, excavated material was dumped onto the upper slope of the old slide, adding even more weight and increasing the driving force. In this case, the combination of road construction and water usage triggered movement of the unstable land.

 12.3 living with unstable land

Unstable land causes unsafe environments, loss of life, and economic hardships in all 50 states. What can we do to minimize the risk of living with unstable land? First, we can identify hazardous areas. Land-use planning of any kind should always take into consideration unstable land hazards, and land-use choices should minimize the risks. The following section shows how some communities have identified and adjusted to unstable land hazards. However, there are communities like La Conchita, California, that did not assess landslide risk and make satisfactory adjustments before the community was developed. In some cases, there are engineering and design responses that people can use to mitigate or minimize problems from unstable land.

Identifying Hazards

To find unstable land areas, scientists and engineers use a combination of tools. For example, to identify landslide hazards in a given area they evaluate slope steepness, surface materials, underlying bedrock characteristics, and history of past landslides. One way to integrate this information is with maps—especially topographic maps, geologic maps, and maps showing past landslide deposits. In Leavenworth, Kansas, all three types of maps were used to predict landslide hazards and guide land management choices. You may be surprised that even a "vertically challenged" state like Kansas has areas with landslides (Figure 12-21).

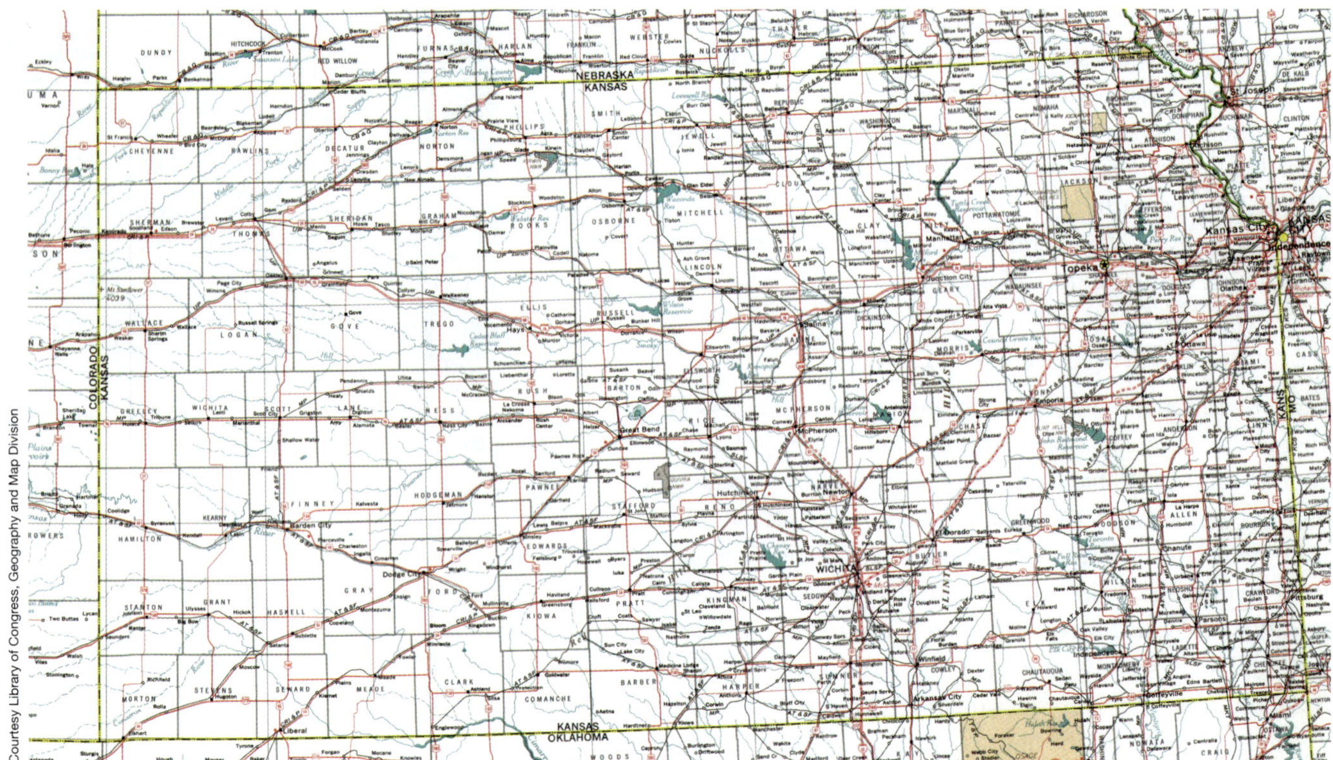

Figure 12-21 William Smith 1815 geologic map.

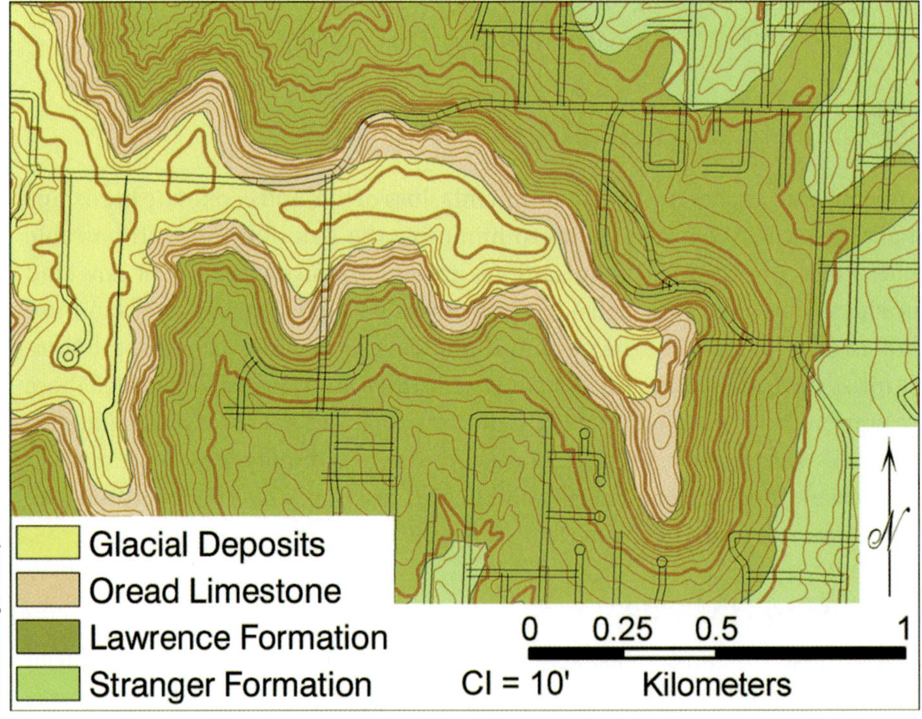

Figure 12-22 Geologic map of Leavenworth County, Kansas.

Landslide-Hazard Mapping in Kansas

The map in Figure 12-22 shows both topography and geology in part of Leavenworth County, Kansas. This map shows the surface distribution of glacial deposits (yellow) and rock units (pink, brown, and green). The glacial deposits cover upland areas with low slopes and are not very susceptible to

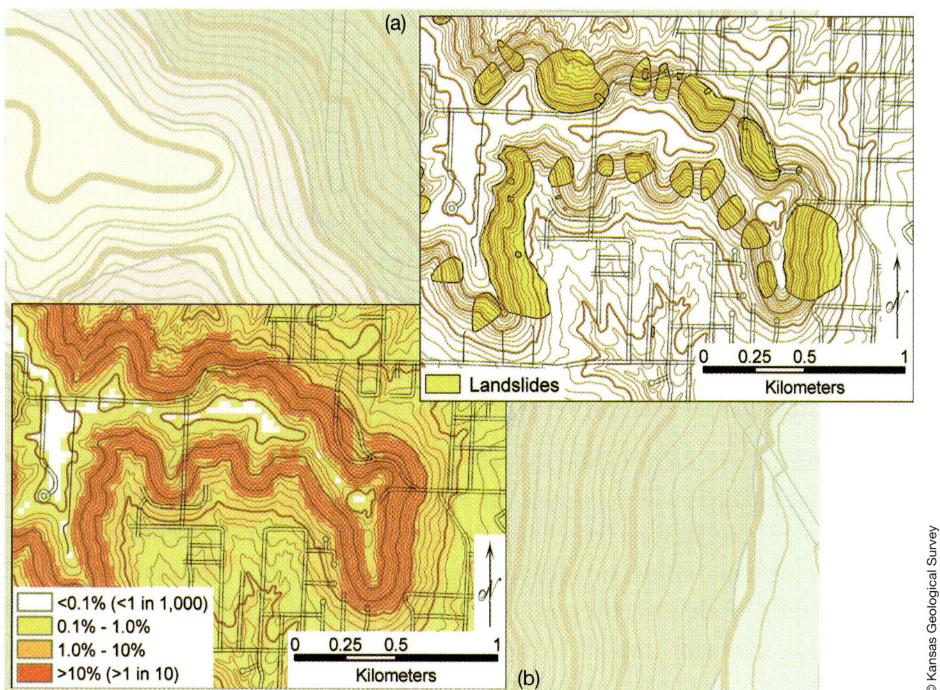

Figure 12-23 (a) Landslide map of Leavenworth County, Kansas. (b) Landslide probability map of Leavenworth County, Kansas.

landslides. Beneath the glacial deposits, the Oread limestone, a strong rock unit with much cohesion, supports the hilltops. Beneath the limestone, however, are the Lawrence and Stranger Formations, which are composed of weaker rocks including shale, siltstone, and sandstone. These rocks easily break down and weather to become sediments with low internal friction and cohesion; in other words, they are quite susceptible to landslides.

In addition to the geologic and topographic map of Leavenworth, geologists also constructed a map that shows where landslides occurred in the past (Figure 12-23a). They then developed a "landslide-hazard map" from a statistical analysis that takes into account slope steepness, bedrock characteristics, and the distribution of previous landslides (Figure 12-23b). The map shows the probability of a future landslide at any point on the map, given the slope steepness and underlying bedrock. Their analysis shows that slope steepness is the primary factor determining slope stability, but bedrock type also provides important information on landslide susceptibility. Geologists found that where the slopes are steeper and underlain by the weaker rocks of the Lawrence and Stranger Formations, landslides are likely. Landslide-hazard maps, like that shown in Figure 12-23, are important tools for local government officials, planners, developers, engineers, and landowners to assess risk and take appropriate actions. In Leavenworth County, Kansas, land-use planners can now identify and zone landslide-prone areas for uses such as open space rather than housing developments.

Assessing Risk

Development in areas with unstable land cannot always be avoided. If this is the case, then assessing risk and delineating areas that are most and least at risk can help minimize property damage and injury. Risk assessment is not simple for any geologic hazard because of the many variables involved; however, it is important. Risk is the chance, high or low, that damage will be caused by a geologic hazard, in this case unstable land. How do you assess risk in unstable land areas? Assessing risk involves systematic evaluations that include:

- Identifying where hazards are within a given area.
- Evaluating the probability or chance that the hazard will occur.
- Producing "hazard maps" such as the landslide-hazard map for Leavenworth, Kansas.
- Determining whether existing safety precautions and zoning restrictions are adequate.
- If safety precautions are not adequate, decisions are needed to adjust to the hazard such as relocating roads or buildings and making design changes.

Scientists and engineers use mathematical models to compute the probability that hazards will occur within a certain time frame. These models take into account the factors that contribute to unstable land problems such as slope, geologic materials, soil properties, precipitation conditions, and so forth. In developing models to compute probability, scientists try to accurately weigh each factor that contributes to causing unstable land. There is always some uncertainty in calculating risk, so probability is usually given as a range or as a minimum. For example, landslide risk assessment would be reported as "there is a greater than 60 percent chance that there will be a landslide in the Oakland area within the next five years," or "The Live Oak Canyon has a probability of greater than 60 percent that a debris flow will occur in response to a one hour rainstorm." Areas with different hazard probabilities are shown on maps like the one in Figure 12-24, which shows the probability of debris flows in a region of southern California that was burned by wildfires.

One way that scientists can assess the risk that a landslide may occur is to perform slope stability analyses. Engineering geologists often use the relationship between resisting forces (shear strength) and driving forces (shear stress) to perform slope stability analyses. The ratio of resisting forces to driving forces is called the factor of safety (FS), where FS = resisting forces/driving forces. If the factor of safety is less than or equal to 1 (i.e., FS $\leq$ 1), the slope will fail (mass movement will occur) because the driving forces equal or exceed the resisting forces. If the factor of safety is significantly greater than 1, the slope is considered stable. However, if the factor of safety is only slightly greater than 1, small disturbances may cause the slope to fail. For example, if FS = 2, the slope has resisting forces that are twice as large as the driving forces and is, therefore, considered stable. On the other hand, if FS = 1.05, the slope's strength (resisting force) is only 5 percent greater than the driving forces and slight undercutting or steepening, seismic shaking, or even heavy rain could easily cause it to fail. In California, construction of new buildings, or additions to buildings, typically requires that the factor of safety be at least 1.5.

Assessing Sinkhole Risk in Maryland

Another example of an area where hazard probabilities are calculated is Frederick, Maryland. In Frederick, a city plagued by sinkholes, geologists developed a risk assessment tool specifically for the urban area. Unlike landslides, it is generally impossible to predict where sinkholes might form based on topography. Instead, what geologists found in the Frederick region is that sinkholes

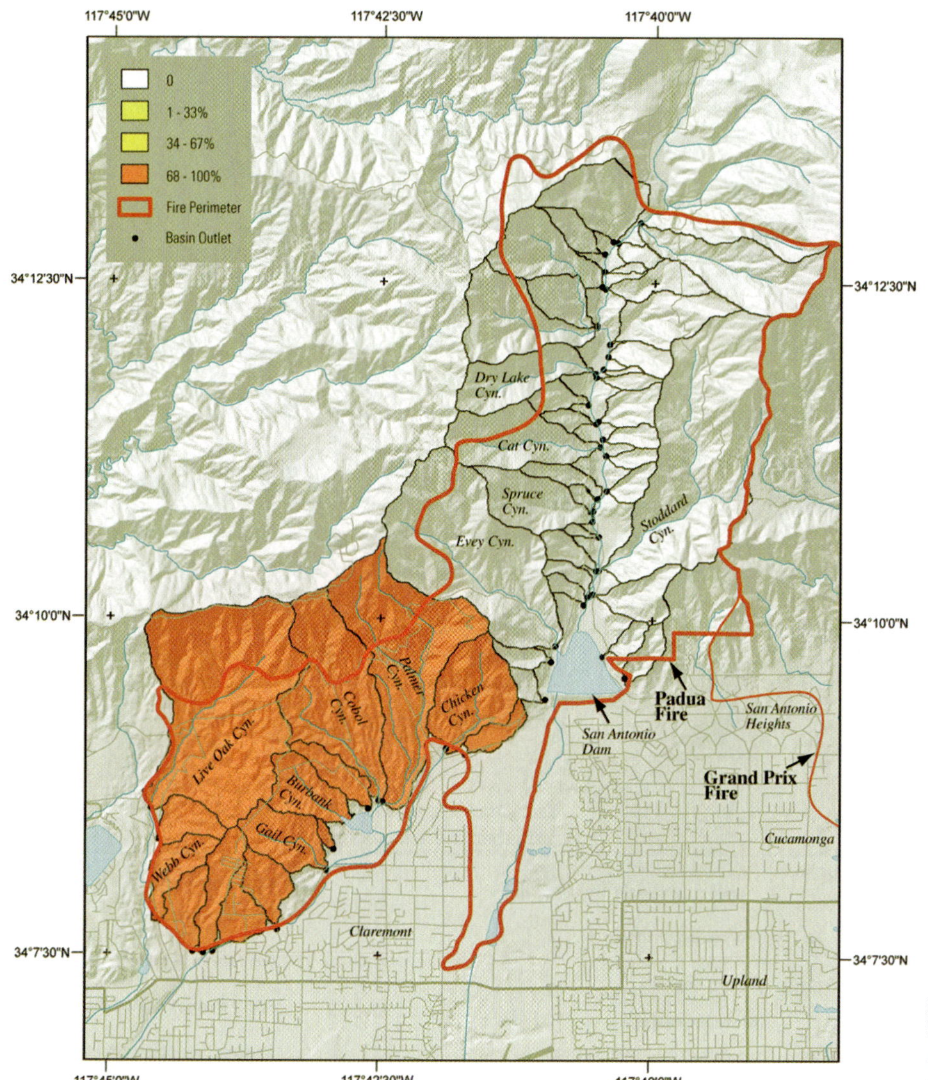

Figure 12-24 Probability of debris-flow occurrence from basins burned by the Padua Fire in response to the 25-year, 1-hour storm of 1.46 inches.

formed in specific bedrock units. Carbonate rocks (limestone in particular) underlie west central Maryland and are the types of rocks that develop karst features, including sinkholes. The geologic map of the Frederick Valley shows two limestone formations, the Frederick and the Grove (Figure 12-25). At first, it was not obvious which formation was most likely to be associated with sinkholes. However, after the formations were subdivided into rock types, the increased detail revealed that some of the subdivided bedrock units had more sinkholes than others. The more detailed geologic map with subdivided rock types allows the map to be used as a predictive tool for potential sinkhole development.

Examine the map of rock units and sinkholes included in Figure 12-25 and see if you can find a correlation between sinkhole distribution and rock type. The data in Figure 12-26 show that most sinkholes are present in the upper part of the Frederick Formation. Previously, it was thought that the Grove Formation was the unit most susceptible to sinkhole development. However, an increased level of detail in both the rock descriptions and mapping of different rock types allowed geologists to evaluate the susceptibility of rock units to sinkhole formation. By subdividing and mapping units as precisely as possible, and by accurately locating sinkholes with a global positioning system (GPS), geologists were

Figure 12-25 Geologic map of Frederick County, Maryland.

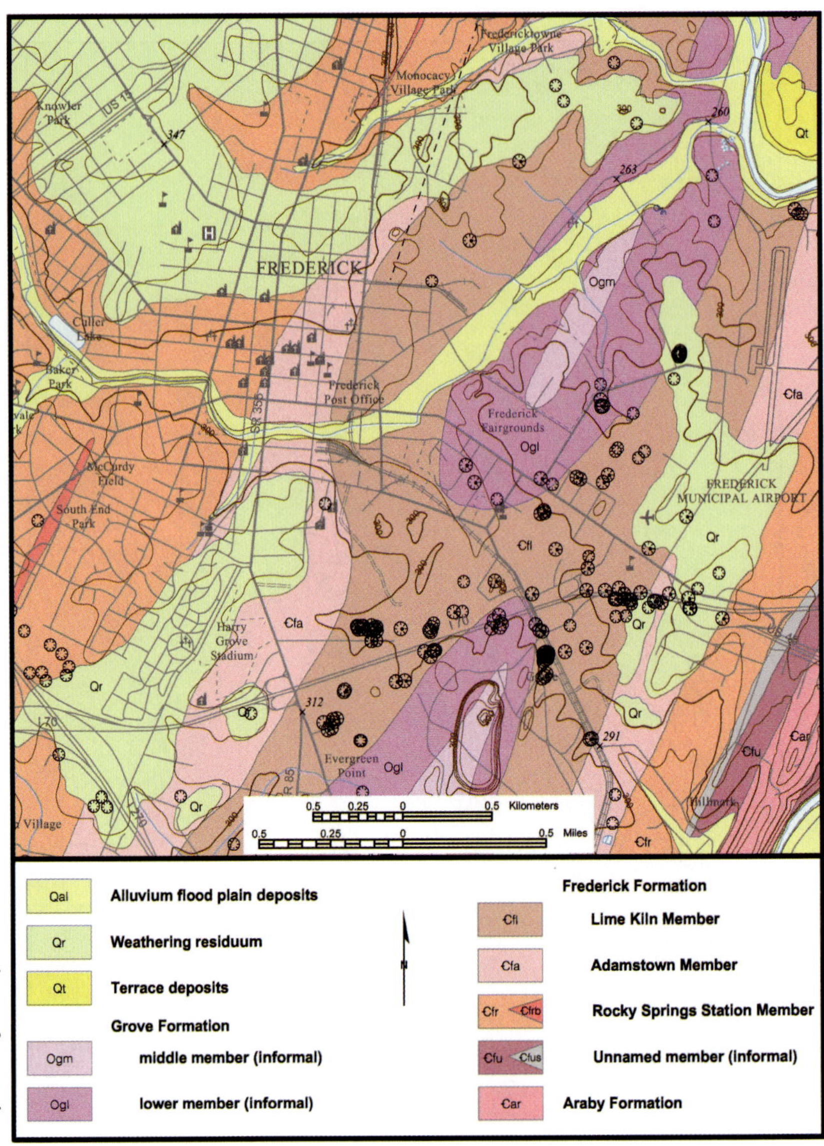

Sinkholes

Stratigraphic unit	Number of sinkholes
Grove formation middle member	0
Grove formation lower member	20
Frederick formation lime kiln member	83
Frederick formation Adamstown member	17

Figure 12-26 Number of sinkholes occurring in various stratigraphic units.

able to develop a new tool, the susceptibility index (SI), that shows the relative likelihood of sinkhole occurrence/formation for each unit (Figure 12-27). Planners and developers can now use the SI as a tool to evaluate the risk of sinkhole occurrence in areas considered for development.

Geologic maps are the main tools for conveying data important to understanding sinkhole distribution. Although sinkhole development in susceptible areas cannot be completely prevented, policy makers and the public can develop assessment strategies that can avoid most property damage and personal injuries. The same is true for landslides, subsidence, and other types of unstable land. Careful assessment of topography, geology, and history of surface movement can go a long way toward avoiding problems caused by unstable land.

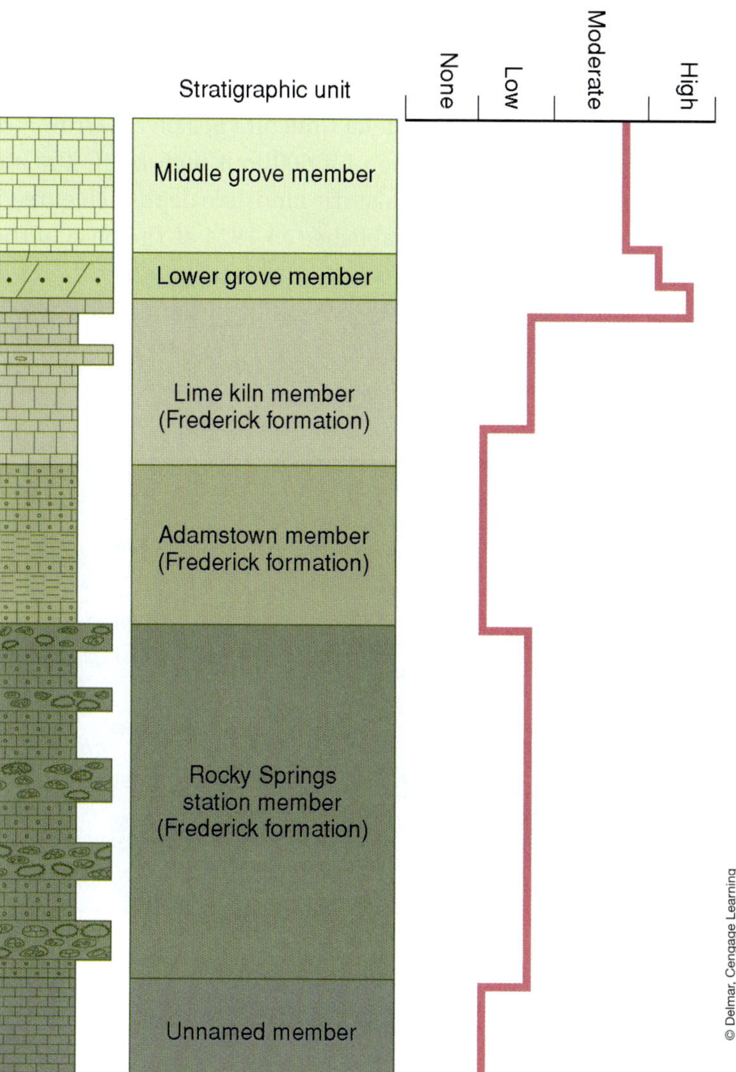

Figure 12-27 Sinkhole susceptibility in Frederick County; stratigraphic units.

Land-Use Choices

There are many examples of communities, and even cities, built in geologically hazardous zones. Whether it is a city built near an active fault such as San Francisco, or a small town at the base of a landslide deposit, like La Conchita, where and how we choose to build should always take into consideration the history of unstable land. Land-use planning is one of the most effective and economical ways to reduce landslide losses by avoiding the hazard and minimizing the risk. In California, extensive restriction of development in landslide-prone areas has been effective in reducing landslide losses. For example, the Los Angeles area achieved a 92 to 97 percent reduction in landslide losses for new construction by implementing appropriate zoning regulations. In many other states, however, there are no widely accepted procedures or regulations for landslides.

La Conchita, California—A Worst Case Choice?

Recall from the beginning of the chapter that a landslide buried part of the seaside town of La Conchita in January 2005. The heavy rain of January 2005 reactivated part of a landslide that had previously buried part of La Conchita in 1995 (Figure 12-28). No one was injured during the 1995 landslide, but nine houses were severely damaged or destroyed. La Conchita lies on a narrow strip of land between the shoreline and the base of a 600 foot-high bluff. The steep slope above La Conchita is part of a large historic landslide deposit (Figure 12-28). The community of La Conchita was established in 1924 at the base of the historic landslide. It is not clear if the developers who first built homes there were aware of the risk, but geologists had previously identified the landslide deposit. Had the town planners consulted local geologic experts, they would have realized that

Figure 12-28 Reactivation of La Conchita landslide due to heavy rain in January 2005.

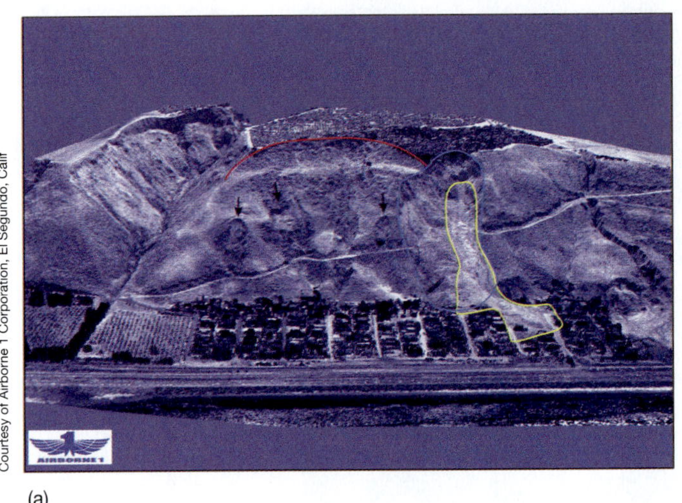

Courtesy of Airborne 1 Corporation, El Segundo, Calif

(a)

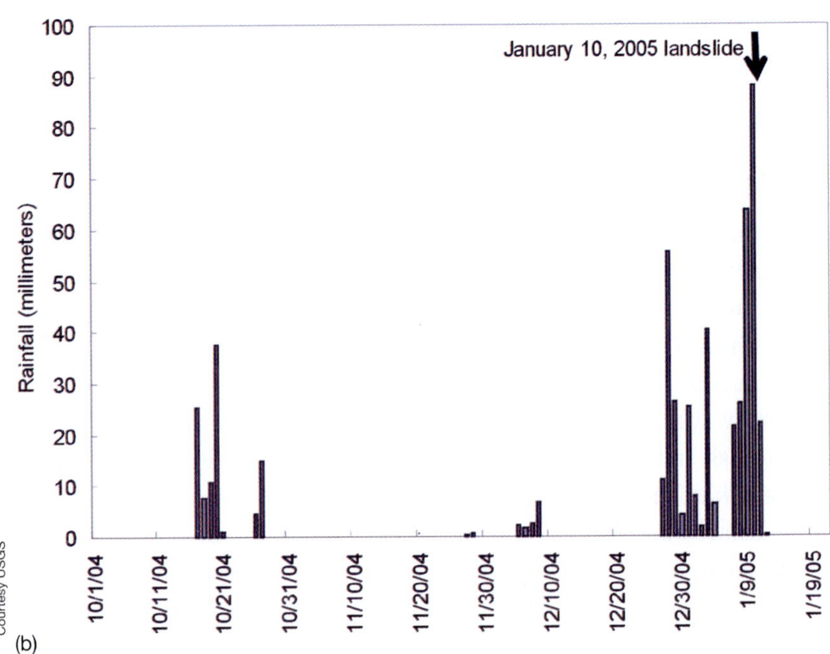

Courtesy USGS

(b)

they were building in the path of potential landslides. In some cases, it seems that people choose to build in scenic locations, or expand urban areas, either because they downplay the risks or simply ignore them.

Oakland, California—A Best Case?

Oakland, California, has steep hillsides and unstable rock and soil that generate damaging landslides during severe storms and wet winters (Figure 12-29). The landslide hazard is made even worse by the earthquake-generating Hayward Fault, which runs through Oakland. A large earthquake generated by the Hayward Fault could trigger many landslides in the surrounding hills. The challenge for city planners is to predict what areas are more likely to suffer future landslides. With this information, engineering designs can be applied that mitigate the landslide hazard or, if necessary, zoning restrictions applied to the most susceptible undeveloped areas. Landslide susceptibility maps, similar to the sinkhole susceptibility maps produced in Frederick Maryland, have been produced for the Oakland area. Geologists have combined geologic maps and landslide deposit maps with calculations for expected ground shaking from an earthquake to produce landslide susceptibility maps (Figure 12-30). Future developers in the Oakland area can use this information to avoid building new communities in dangerous locations; prevent potential injuries and deaths; and save the city, state, and taxpayers millions of dollars.

(a)

Courtesy USGS

(b)

Courtesy J. Coe, USGS

Figure 12-29 (a) Oakland, California's steep hillsides, can result in landslides during severe storms and wet winters. (b) Other contributors to California landslides are unstable rock and soil.

Design and Engineering Responses

When unstable land threatens developed areas, engineered solutions can sometimes help avoid problems. However, avoiding development in areas threatened by unstable land is much safer and more cost effective in the long run. In some cases, construction of pipelines and/or roads cannot avoid crossing unstable land areas and needs to be carefully designed to minimize risk.

The following section provides two examples of engineered solutions to unstable land problems. In the first example, the community of Portuguese Bend goes to great lengths to continue to live on an active landslide. In the second example, engineering designs helped to build a safe oil pipeline through hundreds of miles of land, including landslide-prone areas and areas with unstable soils.

Living on a Landslide at Portuguese Bend, California

In the case of the Portuguese Bend landslide, several different engineered responses have allowed people to continue to live in their homes on slowly moving earth. Houses that remain are equipped with wall-supporting jacks that can be adjusted to compensate for the rising or sinking of the earth beneath them (Figure 12-31). Beginning in the 1980s, movement of earth-flow has been slowed down in some areas by pumping ground water from wells and by improving surface drainage. The community is trying to stop the movement all together, but this is unlikely because the toe of the slide is continually

Figure 12-30 Oakland landslide hazards.

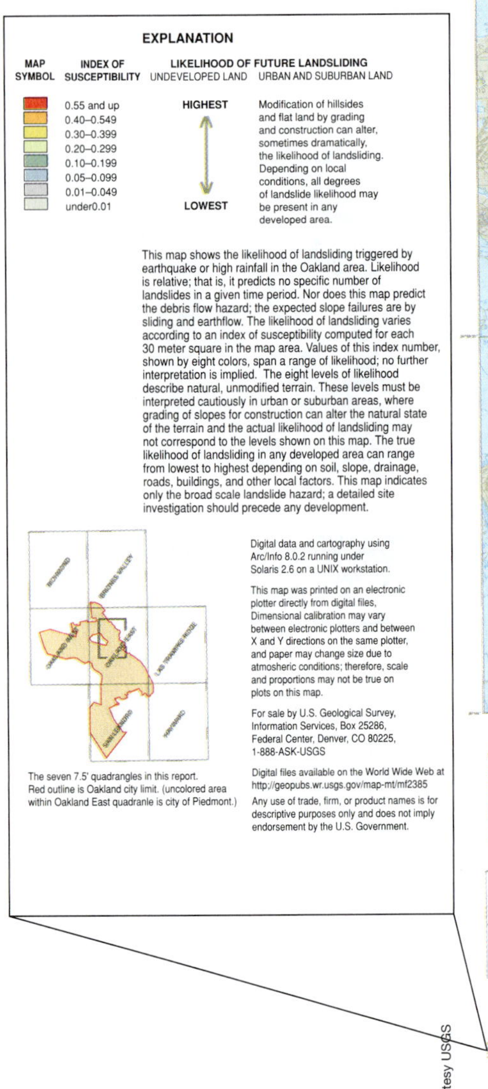

EXPLANATION

MAP SYMBOL	INDEX OF SUSCEPTIBILITY	LIKELIHOOD OF FUTURE LANDSLIDING UNDEVELOPED LAND	URBAN AND SUBURBAN LAND
(red)	0.55 and up	HIGHEST	Modification of hillsides and flat land by grading and construction can alter, sometimes dramatically, the likelihood of landsliding. Depending on local conditions, all degrees of landslide likelihood may be present in any developed area.
(orange)	0.40–0.549		
(yellow)	0.30–0.399		
(light yellow)	0.20–0.299		
(green)	0.10–0.199		
(blue)	0.05–0.099		
(gray)	0.01–0.049		
(light gray)	under0.01	LOWEST	

This map shows the likelihood of landsliding triggered by earthquake or high rainfall in the Oakland area. Likelihood is relative; that is, it predicts no specific number of landslides in a given time period. Nor does this map predict the debris flow hazard; the expected slope failures are by sliding and earthflow. The likelihood of landsliding varies according to an index of susceptibility computed for each 30 meter square in the map area. Values of this index number, shown by eight colors, span a range of likelihood; no further interpretation is implied. The eight levels of likelihood describe natural, unmodified terrain. These levels must be interpreted cautiously in urban or suburban areas, where grading of slopes for construction can alter the natural state of the terrain and the actual likelihood of landsliding may not correspond to the levels shown on this map. The true likelihood of landsliding in any developed area can range from lowest to highest depending on soil, slope, drainage, roads, buildings, and other local factors. This map indicates only the broad scale landslide hazard; a detailed site investigation should precede any development.

The seven 7.5' quadrangles in this report. Red outline is Oakland city limit. (uncolored area within Oakland East quadranle is city of Piedmont.)

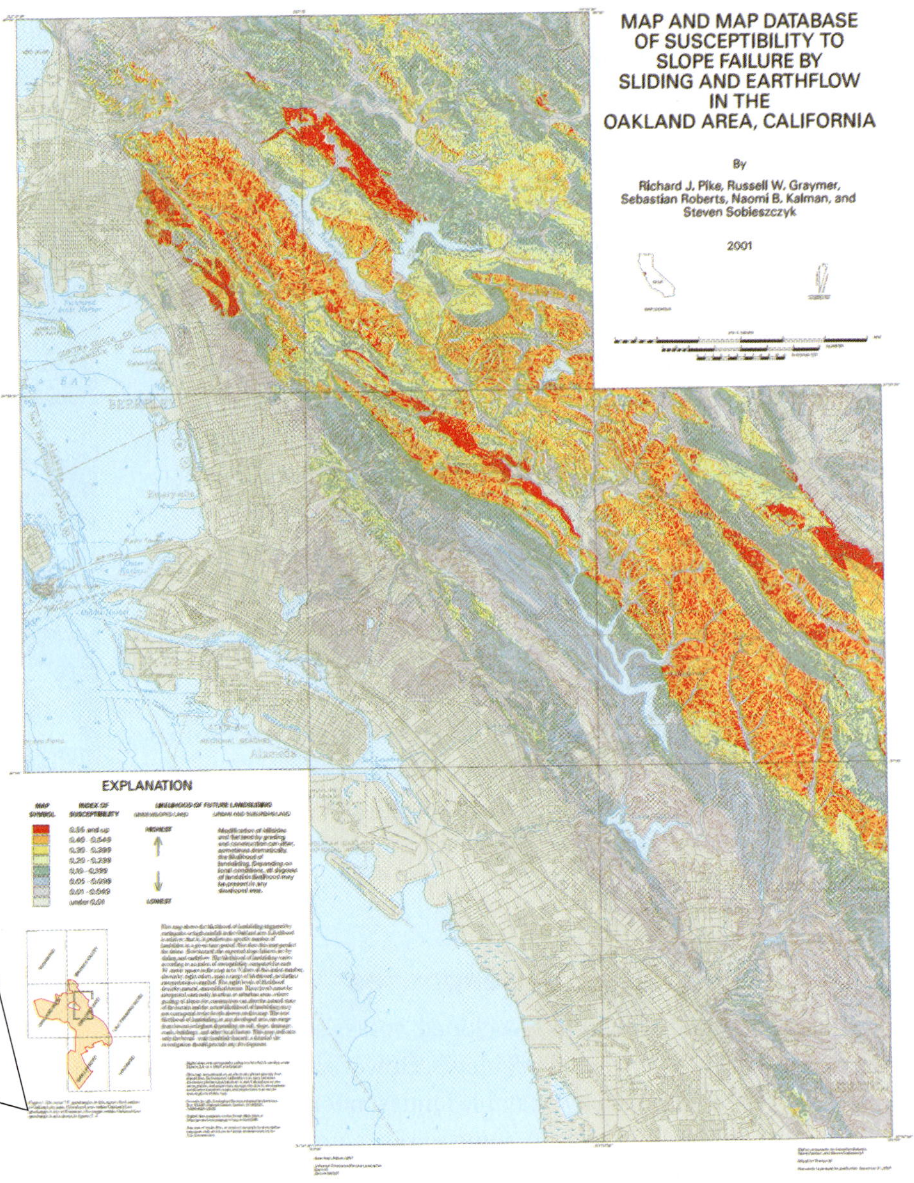

Digital data and cartography using Arc/Info 8.0.2 running under Solaris 2.6 on a UNIX workstation.

This map was printed on an electronic plotter directly from digital files. Dimensional calibration may vary between electronic plotters and between X and Y directions on the same plotter, and paper may change size due to atmosheric conditions; therefore, scale and proportions may not be true on plots on this map.

For sale by U.S. Geological Survey, Information Services, Box 25286, Federal Center, Denver, CO 80225, 1-888-ASK-USGS

Digital files available on the World Wide Web at http://geopubs.wr.usgs.gov/map-mt/mf2385

Any use of trade, firm, or product names is for descriptive purposes only and does not imply endorsement by the U.S. Government.

MAP AND MAP DATABASE OF SUSCEPTIBILITY TO SLOPE FAILURE BY SLIDING AND EARTHFLOW IN THE OAKLAND AREA, CALIFORNIA

By

Richard J. Pike, Russell W. Graymer, Sebastian Roberts, Naomi B. Kalman, and Steven Sobieszczyk

2001

Courtesy USGS

(a)

(b)

Figure 12-31 Remaining houses on the Portuguese Bend landslide equipped with heavy-duty, wall-supporting jacks. (© *Bruce Perry of the Department of Geological Sciences, California State University Long Beach*)

oversteepened by coastal erosion (Figure 12-32). Erosion of the toe of the landslide removes the lower supporting material and triggers further movement. What would you do if you lived in a house on the Portuguese Bend landslide? Keep in mind that if you wanted to move to a different location, it might be difficult to sell your house.

Figure 12-32 Part of the Portuguese Bend landslide showing coastal erosion of the toe of the slide. (*© Bruce Perry of the Department of Geological Sciences, California State University Long Beach*)

Protecting the Trans-Alaska Pipeline System

The Trans-Alaska Pipeline System crosses 1,300 km (800 miles) of land, including three mountain ranges and over 800 rivers and streams. It was designed and constructed to move oil from the North Slope of Alaska to the port of Valdez, where tanker ships can load year-round. When traversing this much mountainous and frozen terrain, encountering unstable land becomes inevitable. The mountainous regions of the pipeline are prone to landslides and avalanches. Over half of the pipeline runs through country with permafrost soils that become unstable when thawed. Because permafrost soils can either contract or expand, depending on temperature conditions, they are considered unstable. Contracting and/or expanding soils put large stresses on pipes and can rupture them, causing oil spills.

The pipeline crosses 680 km (420 miles) of frozen ground that is unstable if thawed, but not prone to landslides. In these regions, the pipes are located above ground. The above-ground pipeline has specially designed vertical supports placed in drilled holes. The vertical supports contain pipes with anhydrous ammonia, which vaporizes below ground, rises and condenses above ground. This process removes ground heat whenever the ground temperature exceeds the temperature of the air.

The other 620 km (380 miles) of pipeline crosses areas with a high risk of landslide. The pipeline is buried in these sections to protect it from landslide or avalanche damage. Six kilometers (4 miles) of the landslide-prone region also have unstable permafrost soils. This section of the pipeline is refrigerated to prevent heat from the pipeline from thawing the soils. Refrigeration plants, at each of these sections, circulate chilled brine through loops of pipe to maintain the soil in a stable frozen condition.

The total cost of the privately funded pipeline construction was about $8 billion. This successful project shows how engineering design can guide construction and enable people to use unstable land areas if they must.

Landslide Mitigation Strategies

Landslide mitigation strategies and techniques include a wide range of technologies applied to a wide range of situations. Different strategies are applied depending on the type of landslide and the geologic setting. Strategies may include avoidance (relocation of construction), stabilization, control, prevention and, even taking no action (living with recurring maintenance).

If the strategy is to stabilize, control, or prevent landslides, then a variety of landslide mitigation techniques may be evaluated. These techniques include modifications to existing surface and subsurface drainage, increasing resisting forces, and/or reducing driving forces. Evaluation of landslide mitigation techniques for effectiveness and cost sometimes leads to reevaluation of the landslide mitigation strategy.

Table 12-2 *Examples of fatalities due to unstable land.*

Year	Location	Type	Fatalities
1916	Italy, Austria	Landslide	10,000
1920	China	Earthquake triggered landslide	200,000
1945	Japan	Flood triggered landslide	1,200
1949	USSR	Earthquake triggered landslide	12,000–20,000
1954	Austria	Landslide	200
1962	Peru	Landslide	4,000–5,000
1963	Italy	Landslide	2,000
1970	Peru	Earthquake related debris avalanche	70,000
1985	Columbia	Mudflow related to volcanic eruption	23,000
1987	Ecuador	Earthquake related landslide	1,000

12.4 conclusion

As populations expand into unstable land areas, a growing number of people must contend with the hazards and costs of unstable land. The cost of property damage from unstable land is huge and the loss of life and injury devastating. Table 12-2 shows the impact of some of the major landslides on human life over the last century.

As an example of the high cost of living with unstable land revisit Oakland, California. During the 1997 to 1998 rainy season, the area surrounding Oakland experienced more than 200 landslides, leading to financial losses in excess of $47 million. The Oakland, California, region is densely populated and will most likely remain that way. The challenge for the future is to predict what areas are more likely to suffer future landslides so that proper engineering or zoning restrictions can be applied. Using unstable land anywhere would benefit from this approach. At a minimum, be prepared to investigate unstable land hazards wherever you choose to live.

IN THIS CHAPTER YOU HAVE LEARNED:

- Unstable land changes and moves in ways that disrupt people's lives and property. Landslides, soil creep, subsidence, debris and mudflows, and sinkholes are different types of unstable land.
- Unstable land occurs in all 50 states, causes large amounts of property damage, and injures or kills people every year.
- Gravity is the driving force that causes unstable land. It is resisted by the forces of friction and cohesion. The steeper the slope, the greater is the component of gravity that pulls parallel to the slope. This is why slope steepness is a major factor in unstable land.
- Changes in driving and resisting forces can lead to unstable land movements—mass movements. The driving force can be increased in several ways, as when rain water increases the weight of surface materials. Likewise, the resisting forces can be decreased in several ways, as when rocks become strongly fractured. Loose, unconsolidated materials have lower friction and cohesion (less resisting force) than solid rock.

- Loose, unconsolidated material has a maximum slope at which it is stable. This is called the angle of repose and is controlled by the material's resisting forces.
- Precipitation, vegetation, the water table, earthquakes, volcanic eruptions, and human impacts can all change the driving and resisting forces in ways that lead to mass movements. These changes to driving and resisting forces reflect interactions between Earth's systems.
- Unstable land can be identified and mapped, and the likelihood of future land movements—the risk of a future landslide, for example—can be assessed. Risk assessment and mapping is a key tool for characterizing unstable land hazards. Topographic maps that show slope steepness, geologic maps that show where different Earth materials are distributed, and maps that show where previous land movements occurred are important foundations for risk assessment and mapping.
- Understanding where unstable land movements are most likely can guide land-use choices. Land-use planning can use zoning restrictions as well as modified building and construction codes to adjust to areas with increased unstable land risk.
- If people need to use unstable land, as where highway or pipelines are located, engineering designs can help adjust to land movements and/or stabilize the land.

KEY TERMS

Alluvium
Angle of repose
Karst
Landslide
Mass movement
Mudflows and debris flows
Sinkholes
Soil creep
Subsidence
Talus
Unconsolidated material
Unstable land

REVIEW QUESTIONS

1. Define unstable land.
2. Give three examples of unstable land.
3. What are three factors that can cause landslides?
4. What causes debris flows and mudflows to occur?
5. What are the causes and effects of soil creep?
6. What is the relationship between karst and sinkholes?
7. Why is it not a good idea, environmentally, to dispose of garbage in sinkholes?
8. Explain the relationship between ground water and subsidence.
9. Why are all steep slopes not unstable? Explain your reasoning.
10. What are the steps in assessing risk from unstable land?

unit four

depending on the earth

chapter 13

water

Have you heard of the river that burned? Students at Strongsville High School in Strongsville, Ohio, live near this river. Native Americans called it the Cuyahoga ("Crooked River"). The Cuyahoga and its tributaries drain a 2,100 square km (810.8 square miles) watershed originally home to abundant fish and wildlife. But industrialization came to the Cuyahoga River valley in the 1800s and, like many rivers around the country, it became the region's waste disposal system—the place where industrial and community waste was dumped. Oil, chemicals, sewage, and debris clogged the river, destroyed its aquatic habitat, and made its water unfit to drink or swim in. Conditions were such that the river surface caught on fire from time to time and the Cuyahoga became the river that burned. The most

recent fire on the river, in 1969, was widely publicized and the Cuyahoga became one of the key examples of surface water pollution that led to new milestone environmental legislation, including the Clean Water Act passed by Congress in 1972.

Have Clean Water Act regulations helped the Cuyahoga River? Strongsville High School students tested the water quality of the river to find out. They collected water samples and conducted tests to determine the concentrations of dissolved oxygen, fecal coliform, nitrate, total solids, and turbidity and temperature, which are all important measures of water quality. Their tests showed that the overall water quality at their sample site was medium in 1997 and good in 1998. This is much better than it was before the Clean Water Act, but more progress is needed. Swimming is still discouraged in the 37 km (23 miles) of the river within Cuyahoga Valley National Park, and the lower river near Cleveland is targeted for continued cleanup because it still does not meet water quality standards. It appears that Strongsville High School students will have reason to monitor water quality in the Cuyahoga for years to come.

Is the water around you safe to drink or swim in? Most of the time the quality and the availability of our water is something we take for granted. We just turn on the faucet and take a drink, not thinking much about where the water comes from or whether it is polluted. We assume that it is unendingly available and safe to use. This chapter is about the significance of water for Earth systems and your life.

AFTER COMPLETING THIS CHAPTER, YOU WILL UNDERSTAND:
How the chemical nature of water makes it Earth's most important compound.
Why Earth, the water planet, is unique in our solar system.
The role of water in the biosphere, geosphere, and atmosphere.
The nature, distribution, and amounts of Earth's water resources.
People's needs and uses of water.
People's impacts on water quality and availability.
How water resources can be conserved and protected.

 13.1 earth—the water planet

Earth is a special place when it comes to water. Water's physical properties combined with Earth's position with respect to the sun produce a unique planet. Earth is the only planet in our solar system where water can exist as a solid, liquid, and gas at or near its surface. This physical setting produces the beautiful blue, cloud-enshrouded sphere, teeming with life that we call home (Figure13-1). The key to this unique outcome lies in the chemical nature of water.

Figure 13-1 Earth—the water planet. (*Courtesy NASA Goddard Space Flight Center Image by Reto Stöckli*)

The Nature of Water—A Better Earth Through Chemistry

Water is a molecule that combines two atoms of hydrogen (H) with one atom of oxygen (O) to form H_2O. The hydrogen atoms, small and positively charged, share electrons with the slightly larger and negatively charged oxygen atoms, but the hydrogen atoms are not symmetrically located around the oxygen atom. Instead the hydrogen atoms are both closer to one side of the oxygen atom (Figure 13-2).

This causes the water molecule to have a residual positively charged side (close to the hydrogen atoms) and a residual negatively charged side. Therefore, even though the molecule is overall electrically neutral, it still has a two-sided electrical character. One side acts a bit negative, and the other a bit positive. This type of molecule is called **polar** and water is the most important example.

Water's polar character is what gives it some very important properties.

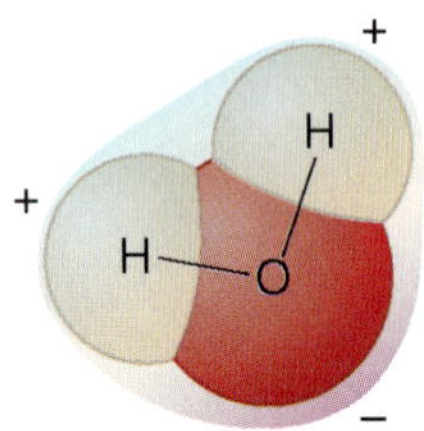

Figure 13-2 Arrangement of oxygen and hydrogen atoms in a water molecule. (*© Delmar, Cengage Learning*)

- Water molecules are attracted to one another and can form thin films and coatings; they have **cohesion**.
- Water molecules are attracted to even slightly and variably charged solid surfaces; they have **adhesion**.
- Water molecules are attracted to other variably charged molecules; they can break up and dissolve many substances.

These properties of water are very important in many of Earth's physical processes, such as the dissolution of minerals to form soils. These properties are also vital to biological processes where water is essential to cell functions, such as the breakdown and adsorption of nutrients. However, there are other properties of water that are also very important.

- Solid water, ice, is less dense than liquid water—it floats. This is a very unusual physical property for an Earth material. This is what causes rivers, lakes, and oceans to freeze downward from their surface.
- Water can adsorb heat without its temperature rising, as many other materials do. This makes water a wonderful modulator of climate as it adsorbs and stores heat from the sun.
- Water can naturally occur as a solid, liquid, or gas on Earth's surface. There are specific combinations of temperature and pressure at which all three phases of water can coexist. These define water's triple point and it commonly occurs at or near Earth's surface.

The **triple point of water** is the key property that helps make Earth unique—the water planet.

Third Rock from the Sun

Earth would be just a big rocky planet if it were closer to the sun. But its orbit keeps Earth at just the right distance from the sun's heat source for water to exist in all its phases—solid, liquid, and gas. Earth is also just the right size. Its gravity keeps water from escaping to space. This is why Earth is the unique water planet—the only planet we know of that can support life.

In its early stages of development, Earth was not the water planet it is today. Water was a component of the primordial material that aggregated to form the initial Earth. Primitive meteorites that fall to Earth are samples of some of these materials. Some of these meteorites have water-containing clay minerals. This, and other water from ice or gas particles in comets and other interplanetary space debris, was the original water that became part of Earth.

As you learned in Chapter 10, magmas transfer dissolved gases, including water, from the interior of the Earth to the surface and atmosphere during volcanic eruptions. Volcanic activity that transferred water to Earth's surface was widespread and common during Earth's early development when the internal structure of the planet was formed from about 4.5 to 3.8 billion years ago. The development of abundant sedimentary rocks by at least 3.5 billion years ago is evidence that large oceans comparable to those of today covered Earth by this time. As large parts of Earth became covered by water, one of its most important cycles began—the water cycle.

 13.2 water and earth's systems

It seems amazing that water can make up an entire Earth system—the hydrosphere. In Chapter 1, you learned how the hydrosphere varies and how it interacts with other Earth systems in the water cycle (Figure 13-3). You know a lot about the role of water in the atmosphere, but there is more to learn about the role of water in the biosphere and geosphere.

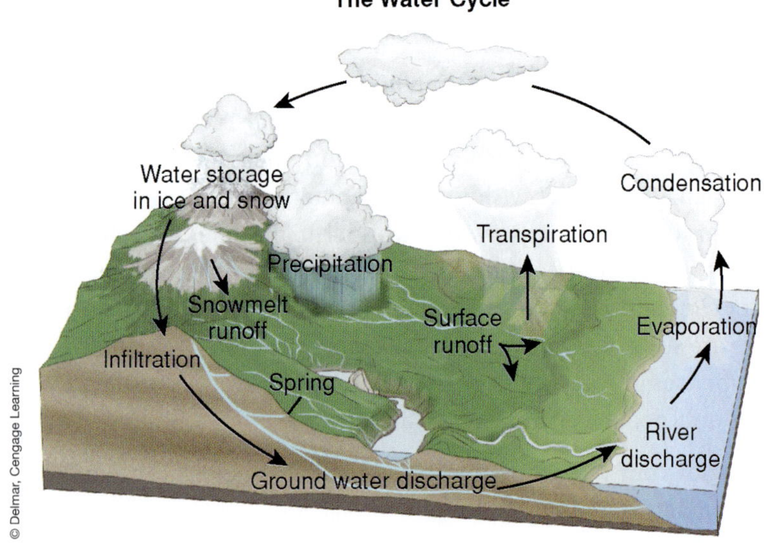

Figure 13-3 The water cycle.

Water and the Biosphere

Water and life are interconnected at all levels of the biosphere. It is a key ingredient within individual organisms where it is essential for cell functions. It is home, or habitat, to entire populations and communities. In addition, at the ecosystem and biosphere levels, it provides pathways for species migrations and the distribution of life-sustaining food and nutrients even over very long distances.

Water and Life

Water is the most abundant component of living matter. It makes up about 60 percent by weight of mammals' bodies, including people. Of this water, about two-thirds is within cells and one-third is in body fluids such as blood. Water is essential to every function in our body, including:

- Transport of nutrients in and waste products out
- Digestion and adsorption of nutrients
- Maintenance of body temperature

If you have monitored how much water you use each day, you will notice that it is much more than you drink. Do you think that the amount of water you drink is what your body needs? The amount of water your body needs depends on many factors, such as your weight and the physical activity you experience each day. In general, your body works better if you drink lots of water. Rules of thumb say that 8 glasses or 2 liters a day should work just fine.

How long can you live without water? Your body is constantly losing water through external respiration (breathing), perspiration (sweating), and excretion of wastes. If it is not replenished, you will start to feel lousy, not just thirsty. Headaches, poor concentration, and tiredness are some symptoms of low water levels in your body. You can get by for a few days on your body's stored water, but eventually you, and all organisms that need water, will die without it. People will die without water in about a week.

Not just any water, however, will save your life. Can you imagine being shipwrecked on a desert island, surrounded by water as far as you can see, but not a drop you can drink? Water's quality is very important to you. Your body needs fresh, clean water—water that does not have lots of dissolved salts, water without pollutants from wastes, and water that does not carry parasites and diseases.

It may surprise you, but only about 2.5 percent of Earth's water is even potentially fit to drink; 97.5 percent is salty ocean water. As you will see later, only a small part is fresh water available for use by people and other organisms. Therefore, water we can use is a relatively limited resource, even more limited if we degrade its quality or are not careful about how we use it. How we use water is important because almost everywhere it is on Earth, it is home, or habitat, for other organisms.

Water as Habitat

Because of water's essential life-supporting nature and its abundance (over 75 percent of Earth's surface is covered by water), it is not surprising that organisms have adapted to living in, on, and around water wherever it is present—even underground! Water-rich or aquatic habitats are a rich source of biodiversity that equals that of the land. You have learned about aquatic habitats along coasts in Chapter 5 and along streams in Chapter 6, but there are other important ones. There are many aquatic habitats in standing water bodies such as ponds, lakes, and the oceans. Wetlands are an especially important habitat that is transitional between terrestrial and other aquatic habitats. Some unique aquatic habitats even exist underground in caves.

Ponds and lakes are enclosed bodies of fresh water, ranging from those you can throw a stone across to those thousands of square kilometers in size. If streams enter or leave ponds and lakes, they can be connected to large surface water systems. A few lakes are not filled with fresh water. These lakes, like the Great Salt Lake in Utah, are within confined watersheds where long-term evaporation rates exceed the addition of water from precipitation. As the water volume in these lakes slowly decreases, they become saltier and saltier. Some are more salty than sea water.

Habitat in ponds or lakes is divided into three zones (Figure 13-4). The littoral zone is the shallow water along the shore. Here the sun's heat warms the water and lots of organisms like algae, aquatic plants, snails, clams, insects,

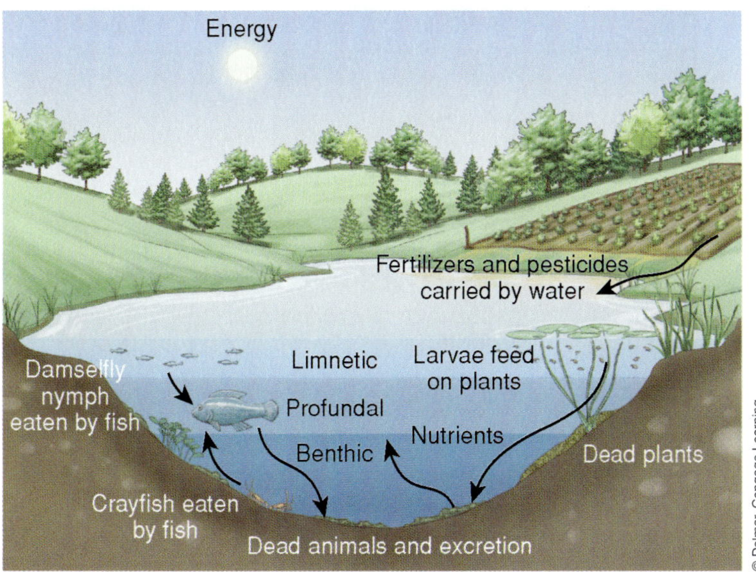

Figure 13-4 Three zones found in a lake or pond habitat.

crustaceans, fish, and amphibians find homes. These are very productive parts of ponds and lakes and good sources of food for land animals like fox and birds.

The shallow water in ponds and lakes away from shorelines is the limnetic zone. Here the sun's heat can also warm the water, at least down to certain depths. Floating organisms such as plankton are common here. Fish also live in this part of ponds and lakes as they come near the surface in search of food.

The deeper part of ponds and lakes, the profundal zone, is dark and colder. When organisms that live near the surface die, they sink to the bottom and become food for bottom-dwelling organisms. Mixing of shallow and deeper waters in ponds and lakes commonly evens out water temperatures and provides oxygen to the deeper waters. In some cases though, very deep waters can become stagnant, continue to collect decaying organic matter sinking down from above, and develop low oxygen contents. The stagnant water becomes anoxic and not many organisms can live there.

Oceans are the largest ecosystems on Earth and life can be found everywhere in these vast and deep, saltwater-filled basins. There are four basic ecological zones in the oceans: intertidal, pelagic, benthic, and abyssal (Figure 13-5). The intertidal zone is along the shoreline where waves and tides intermittently cover the adjacent land. You learned about this habitat in Chapter 5.

The pelagic zone is the open shallower part of the ocean offshore from the intertidal zone. Floating organisms, including some types of seaweed and especially plankton, live here. The great fishes and mammals of the oceans—tuna, marlin, swordfish, sailfish, shark, dolphin, seal, and whale—cruise wherever they want in the pelagic zone. A special habitat in the pelagic zone is coral reefs.

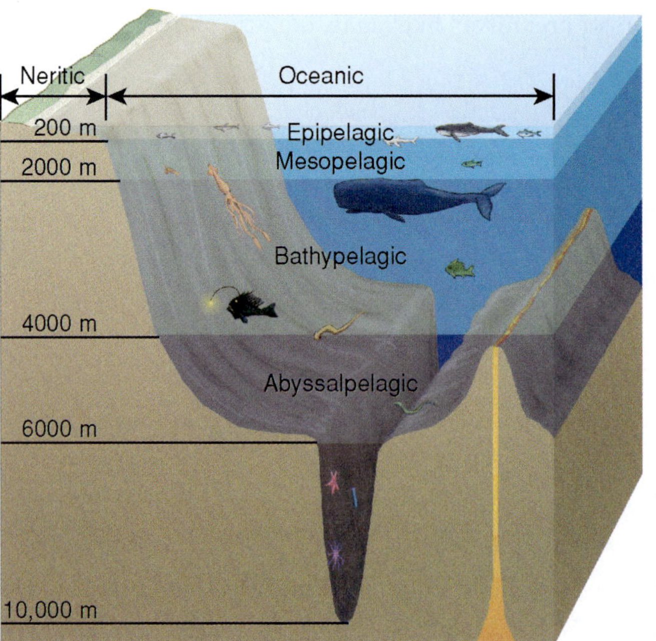

Figure 13-5 Four principal habitats in oceans.

Coral reefs are locally but widely distributed in warm ocean waters. They are places where corals can build up rigid and solid structures, the reefs, by precipitating calcium carbonate from the ocean water. Many fish find coral reefs a great place to live.

The benthic zone lies below the pelagic zone but above the deepest (abyssal) zone. Light does not penetrate through the benthic zone, and the seafloor, commonly sandy or muddy, collects fine sediment and organic debris that sinks to the bottom from the pelagic zone above. Even though the bottom of the benthic zone is dark and cold, the waters are rich in nutrients and support much life, including bacteria, fungi, sponges, worms, starfish, and fish.

The abyssal zone is the deepest, darkest, and coldest in the ocean. The waters are not nutrient rich but do commonly have high oxygen contents. But even here, invertebrates and fish can live (Figure 13-6). As in some lakes, some deep parts of the ocean can become stagnant, continue to collect decaying organic matter sinking down from above, and develop low oxygen contents. The stagnant water becomes anoxic and not many organisms can live here. The organic material that collects on the bottom of anoxic regions in ancient oceans created some of Earth's best petroleum source rocks.

Figure 13-6 Invertebrates that live in the abyssal zone. (*Courtesy of NOAA. Microphotograph taken by G. Carter*)

Wetlands are special aquatic habitats because they help improve water quality and because they support very diverse populations of organisms. Technically there are several types of wetlands, including those developed in brackish water along coastlines, but in general, they are all areas where water is present for all or most of the year and where aquatic plants are abundant. In fact, the distribution of specific types of aquatic plants is used to help define the distribution of wetlands in many places. Some different types of wetlands are:

- Swamps, where the vegetation includes trees and shrubs
- Marshes, where standing water is common and the plants, such as sedges, cattails, cypress, and gum, are adapted to living in water-saturated soils
- Fens, which are marsh-like areas with accumulations of plant debris that form peat and influxes of water from surrounding areas
- Bogs, which are places where peat has accumulated, but there is no inflow or outflow of water

Wetlands are developed from the coldest Arctic regions to the hottest of the tropics. They are teeming with life and have very high species diversity. In addition to the many aquatic plants that live in them, fish, reptiles, and amphibians live in the water, insects seem to be everywhere, birds of all kinds find good places to nest and much to eat, and mammals such as raccoons and muskrats are right at home.

The plant-rich character of wetlands helps them to be water purifiers (Figure 13-7). Water quality is improved as it passes through wetlands by vegetation that settles and filters impurities. The large amount of decaying vegetation can develop reducing conditions that help precipitate other impurities that

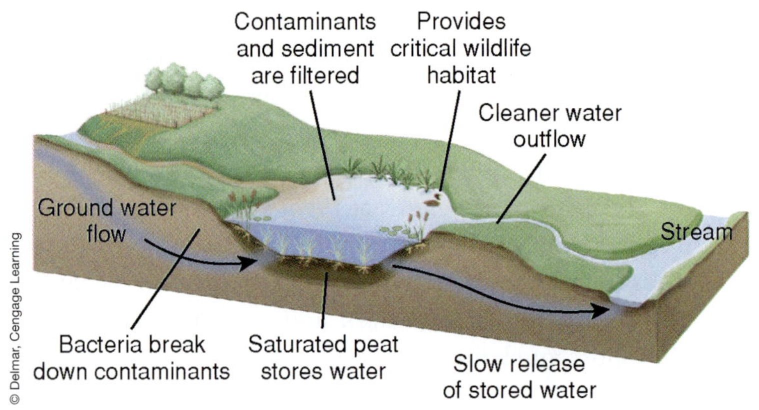

Figure 13-7 Water-purifying processes found naturally within a wetland.

are dissolved in the water, such as heavy metals. In fact, artificial or constructed wetlands have been created to remediate or help clean up metal-bearing water (see Chapter 15).

Wetlands such as salt marshes form in intertidal zones along the coast. These are important parts of the transition from fresh to saltwater habitats and have their own special cast of plants, insects, fish, and other animals such as shrimp and shellfish. They are common in estuaries. In tropical areas, mangrove swamps line the shore of estuaries.

Aquatic habitats also include places underground, especially in caves. Caves are simple but unique freshwater ecosystems. You learned how caves and underground passages between them develop in Chapter 12. Recall that water easily passes from the surface into underground systems where conditions have formed many caves. These areas are called karst terrain. In some places the underground water system is completely filled from the surface down. Although it is always dark underground, many organisms have adapted to living in aquatic habitats in caves. These organisms have developed special ways to live in the absence of light. In place of eyes, they have elongated legs and antennae to help feel their environment (Figure 13-8).

Fish, salamanders, spiders, beetles, crabs, and other animals have evolved such species. If light is not available, then who are the primary producers in these cave habitats? Life can exist here because food and nutrients are brought in the water that migrates downward from the surface. The sun is still the primary source of energy even in dark cave ecosystems.

Figure 13-8 Aquatic cave organism. *(Courtesy of the NPS. Photo by Rick Olson)*

The movement of food- and nutrient-bearing surface water into underground aquatic habitat is just one example of how pathways in the hydrosphere support life. These pathways are especially important at the ecosystem level. Streams and rivers physically connect entire watersheds and transport material from terrestrial to aquatic habitats, including the oceans (see Chapter 6).

Circulation of ocean waters is very important to the biosphere. Currents in the oceans are caused by wind and density differences. Density differences result from variation in both temperature (cold water is denser than warmer water) and salinity (saltier water is denser than fresh water). Global wind patterns mostly cause the easterly and westerly movements of shallow ocean water (down to

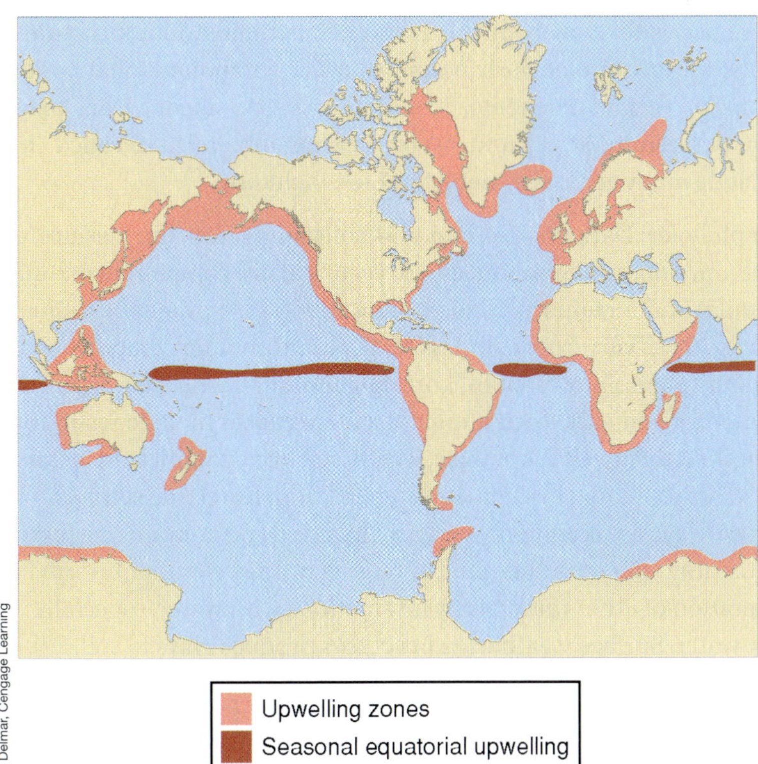

Figure 13-9 Four major upwelling zones around the world.

100 meters or 328.1 feet in depth). Cold water from the polar regions sinks and initiates deep ocean currents that carry it north and south into equatorial latitudes. There are some special places where winds blow warm surface waters away from adjacent continents and allow cold deep ocean water to rise or upwell (Figure 13-9).

Upwelling ocean water is rich in nutrients. The upwelling zones are incredibly productive regions where fish and birds flourish. They are some of the best marine fisheries in the world (Figure 13-10).

Figure 13-10 Peruvian fishing boat heading out to sea. *(Courtesy NOAA, Photographer: Shannon Rankin, NMFS, SWFSC)*

The physical pathways provided by the hydrosphere are essential to the survival and reproduction of many species in the biosphere. As you learned in Chapter 4, gray whales bear their young in warmer waters of lower latitudes then migrate northward to the food-rich Arctic Ocean in the summer. Other species that need to migrate in the oceans are sea turtles and tuna. Salmon born in cold, clear streams migrate to the ocean within a year, where they grow impressively before returning to spawn years later. Salmon die after returning to spawn, but other species, certain smelt, eel, and steelhead, for example, migrate back and forth between streams and oceans to reproduce. The stream habitat is essential for reproduction of these species, but the ocean habitat is where they find food plentiful.

Water and the Geosphere

Water plays many roles in the geosphere. It is a chemical component in minerals, it causes rocks and minerals to break down or become hydrated, and it transports rock and mineral debris—sediments. As you learned in Chapter 6, erosion by water helps shape the landscape everywhere on Earth. Many mineral families have water as a chemical component in their atomic structure. In these minerals, water is molecularly bound to other compounds that contain combinations of various elements, including oxygen, silicon, iron, magnesium, sodium, potassium, or calcium. Water-bearing minerals are called "hydrated" and include many general mineral families, including;

- Amphiboles: Dark-colored minerals containing iron, magnesium, calcium, sodium, silicon, aluminum, and oxygen that are common in certain granitic and metamorphic rocks. Amphiboles are prismatic in shape.
- Micas: Micas vary from colorless to black and contain combinations of sodium, potassium, calcium, iron, magnesium, silicon, aluminum, oxygen, and other elements. Their atomic structure causes them to readily break into thin sheets. Platy flakes of mica, weathered from granitic and metamorphic rocks, are common in sediments derived from terrestrial settings.
- Clays: Clays are commonly very small in size and contain mostly sodium, potassium, silicon, aluminum, and oxygen. They commonly form from the alteration of other minerals by interactions with water, especially warm or hot water. Surface weathering processes produce clays in soil.

Soil-forming processes are discussed in Chapter 14. Precipitation percolating through soil can dissolve minerals, interact with them to form new minerals such as clays, and carry mineral grains downward through the soil. Because water expands when it freezes—the property that makes ice less dense than water—fractures filled with freezing water expand and break apart rocks and mineral grains. Freezing water is a very powerful force that can break massive rocks (Figure 13-11). This is the process, sometimes called frost action, that creates much of the surface rock debris that then becomes carried away by runoff of surface water. The breaking of rocks and minerals by frost action commonly starts the erosion process.

Transport of rock and mineral debris by streams and ocean currents is a key step in the rock cycle (see Chapter 1). This process delivers sediment to low areas on Earth's surface called basins. Layer after layer of sediments accumulate in basins and eventually get buried deep enough to become consolidated into rocks—sedimentary rocks like sandstone and shale.

Water gets buried with sediments. It is in the spaces between mineral grains or pores and in the structure of water-bearing minerals. Water in pores is salty if it was originally seawater or it has dissolved lots of minerals. This water, with its various dissolved minerals, is commonly produced along with oil and gas. Salty water produced this way needs to be properly disposed of.

Figure 13-11 Exfoliation in granite. *(Courtesy Earth Science World Image Bank © Marli Miller, University of Oregon)*

As you learned in Chapter 1, sediments can be buried deeply enough that they change. These changes, caused by increased temperature and pressure at depth and called metamorphism, causes minerals to recrystallize into new minerals. These mineral changes very commonly involve dehydration. Dehydration reactions cause water included in mineral structures to separate into a coexisting fluid. These water-rich fluids migrate along permeable pathways and are responsible for transferring heat from the deeper crust, causing extensive alteration of rocks near the surface and creating valuable mineral deposits such as gold-bearing veins (see Chapter 16). Metamorphic recrystallization of seafloor sediments and altered, water-bearing basalt releases water in subduction zones (see Chapters 8 and 10). Migration of this water into the overlying mantle helps cause partial melting and magma generation that eventually feeds volcanoes at the surface (see Chapter 10). In addition, volcanoes emit water vapor as one part of the water cycle (see Figure 13-3). Therefore, water's role in the geosphere is just as diverse as it is in the other Earth systems.

Water and the Atmosphere

You have learned about the many roles of water in the atmosphere in previous chapters. The following reviews some of the more basic relationships:

- Water is a component of the atmosphere both as a gas (water vapor) and as tiny droplets in clouds. This water enters the atmosphere primarily by evaporation from the surface of the oceans.
- Water droplets can become rain, snow, and hail that precipitate onto Earth's surface. This is the step in the water cycle that replenishes fresh water resources.
- Water, particularly in the oceans, is a tremendous storage place for energy from the sun. It becomes warm and stays warm for long times. This warm water, in turn, can heat large air masses and cause them to move, even become violent weather such as hurricanes. This is a major impact of the hydrosphere on the atmosphere.

13.3 earth's water resources

Earth is a closed system when it comes to water: Water is not gained or lost. The water that first accumulated on Earth's surface has been here ever since. We can change its quality and we can move it around, but we do not physically use water up. The water we use today might be the same water used by creatures, perhaps dinosaurs, hundreds of millions of years ago.

How much water is there? Earth has a lot of water. The total amount of water on Earth is 1,358,827,280 cubic km (326 million cubic miles). This huge amount would cover the entire United States to a depth of 144.8 km (90 miles). It is 215,768,471,688 liters (57 billion gallons) for every person on Earth. This is so much water that you might be wondering why we should be concerned at all about its availability. We need to be concerned because:

- Only a small part of this water is usable by people.
- The usable water is not always available where it is needed.
- People decrease the amount that is available and useful.

All of Earth's water is important, not just the part that people can use. Most of this water, over 97.5 percent, is in the oceans. You learned above that the oceans are a critical part of the hydrosphere and its interactions with the other Earth systems. It provides habitat for diverse species, modulates climate by storing energy from the sun, and provides water to the atmosphere that eventually precipitates and replenishes water on land. However, because it is Earth's nonsalty, fresh water that people use and need, we will focus on this part of Earth's important water resources.

Although fresh water is less than 3 percent of the total water on Earth, there is still a lot. But there is a catch—not all this fresh water is available for our use. Most of Earth's fresh water is trapped as ice in the permanent polar ice caps and in glaciers (Figure 13-12). People use ground water and water from lakes and rivers (surface water), which together are only 0.62 percent of Earth's water. Actually the water people mostly use is from rivers that have only 0.0001 percent of Earth's water! It is a much more limited resource than you might have first thought.

Water Use

Each person, on average, uses about 302.8 liters (80 gallons) of water per day in their homes, up to about 757.1 liters (200 gallons) per day, if outdoor uses such as lawn watering or car washing are included. This may or may not seem like a lot of water to you, but imagine going somewhere, picking up your 302.8 liters, and carrying it back home for your use. What if you had to carry water home for your whole family, say four people? You would need to bring home 320 gallons (1,211 liters) weighing 1,204.7 kg (2,656 pounds), which is over a ton of water! This could take awhile. You might soon wonder how you and your family could get by on less water. However, what we use in our homes each day is just the tip of the iceberg.

Each of us uses over 4,921 liters (1,300 gallons) of water per day. How can this be? Water is used in many ways outside our homes to support our society and standard of living. Our biggest uses of water are actually for irrigation, raising livestock, thermoelectric power generation, and industrial and commercial activities. Each day over 1,000 gallons of water per person are withdrawn from our country's surface and ground water resources to supply these uses. This does not count what we call in-stream uses such as for hydroelectric power generation, which takes another 43,154 liters (11,400 gallons) per day for every person in the country. Therefore, the total water used each day in the United States to support you and your lifestyle is about 49,210 liters (13,000 gallons). That is a lot of water. Luckily, water resources can be replenished.

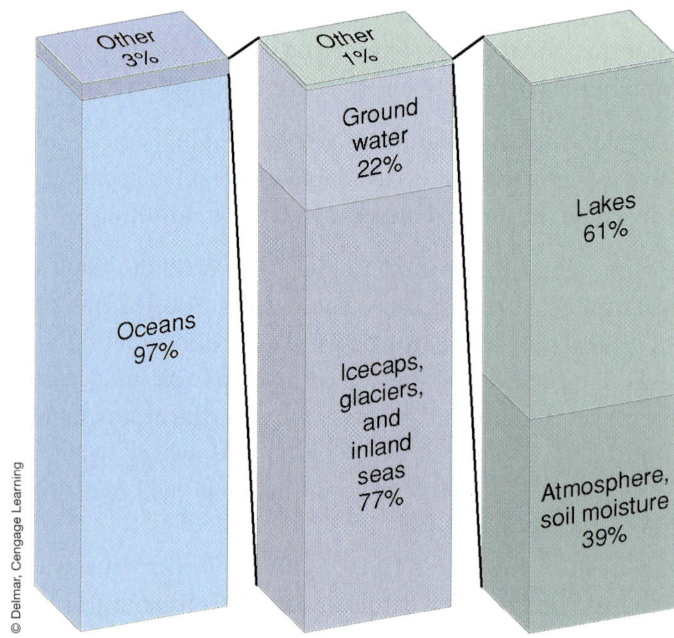

Figure 13-12 Earth's water distribution.

Water Sources

About 80 percent of the water that people use is from the surface, and 20 percent is from **ground water**. Significant amounts of rain and melting snow soak into the ground, fill up open spaces in soil, and percolate deeper to become part of a vast, interconnected ground water system. Areas where precipitation can soak into the ground and become part of the ground water system are called **recharge areas** (Figure 13-13). They are commonly in the higher elevations or headwaters of a watershed but ground water recharge can also take place at the bottom of lakes and streams. Because surface water resources have already been extensively used around the world, ground water is being used more and more to meet increasing needs. Because recharge of ground water resources can be much slower than the rate at which it is withdrawn, people can deplete ground water resources. This is a significant concern in some areas as we will see next.

Ground water is virtually everywhere under land. It completely saturates and fills all voids and fractures in rocks and other Earth materials below certain depths. The depth at which this saturation occurs, which varies from place to place, is called the **water table**. It is important to remember that water only forms what we think of as underground rivers and lakes in places where caves and other very large open spaces are present. Most water below the water table fills very small voids scattered through the Earth materials.

The amount of small open spaces in an Earth material is called its **porosity**. The small voids that make up a material's porosity are only connected by very small channels and fractures, most of which can only be seen with a microscope. Water cannot move very fast through these small openings. A measure of how well a fluid such as water can flow through a solid is **permeability**. The greater the permeability of a material, the easier it is for water to flow through it.

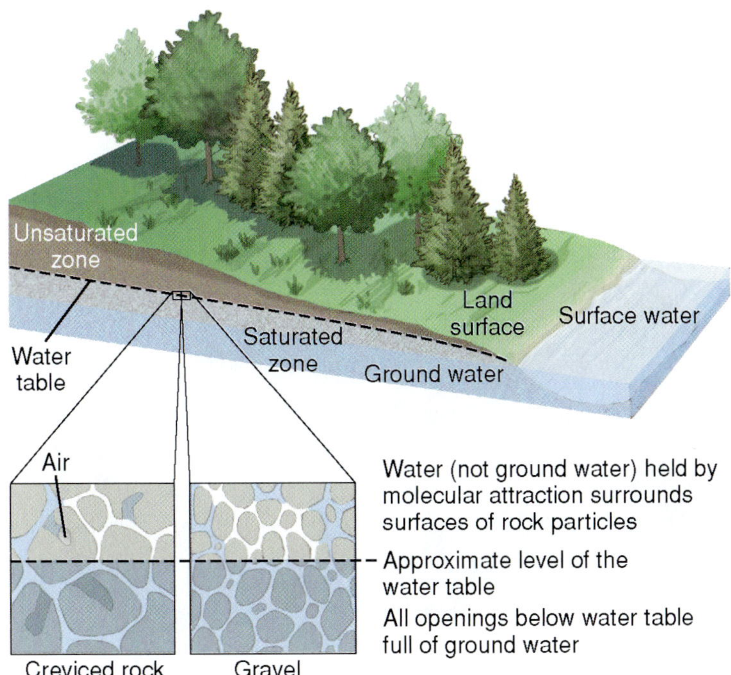

Figure 13-13 The unsaturated zone.

We commonly drill water wells into places below the ground water table that have good porosity and permeability. These good places to recover water from underground are called **aquifers** (Figure 13-14). Wells are not the only way to get water from underground. Some ground water naturally discharges to surface springs and rivers and some slowly migrates underground and discharges to the oceans along coastlines (see Figure 13-3).

Ground and surface water are not isolated from one another, and there are many ways they interact. Ground water can migrate and discharge into streams and lakes. In other places just the opposite can occur—surface water in streams and lakes can recharge ground water (Figure 13-15).

The interconnected nature of surface and ground water, in effect, makes them a single resource. This becomes readily apparent where people are involved with their interactions. For example, withdrawal from one can change the amount available in the other. Contamination in one can migrate and contaminate the other. People cause these and other impacts on Earth's water resources and many of these impacts are of increasing concern.

Replenishing Water Resources

Precipitation replenishes our surface and ground water resources. However, the amount of precipitation is not evenly distributed around Earth. For example, large parts of the mountainous western United States receive less than 38.1 cm (15 inches) of precipitation each year (an inch of precipitation falling on an acre of ground is about 102,206 liters or 27,000 gallons), but temperate rain forests in the Pacific Northwest can receive 10 times this amount, in places over 457.2 cm (180 inches) per year (Figure 13-16).

Tremendous variations in the amount of precipitation are also developed on a global scale (Figure 13-17). Large regions of Earth, like North Africa, receive very low amounts of precipitation, whereas equatorial regions are regularly drenched.

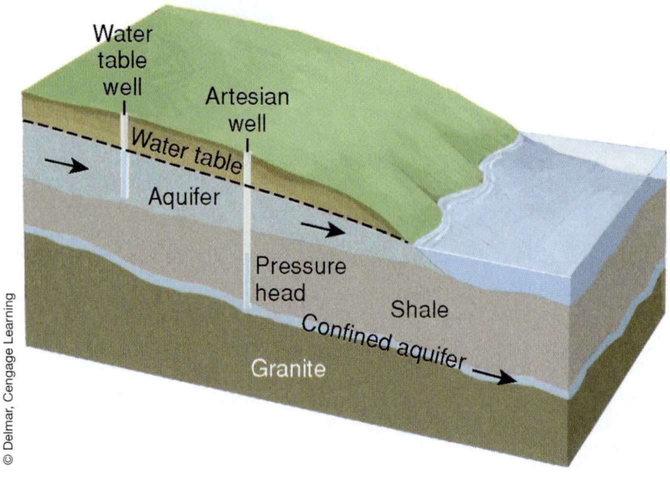

Figure 13-14 Aquifers.

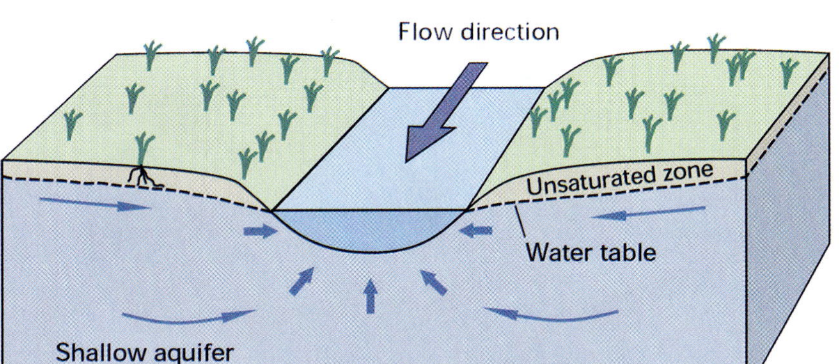

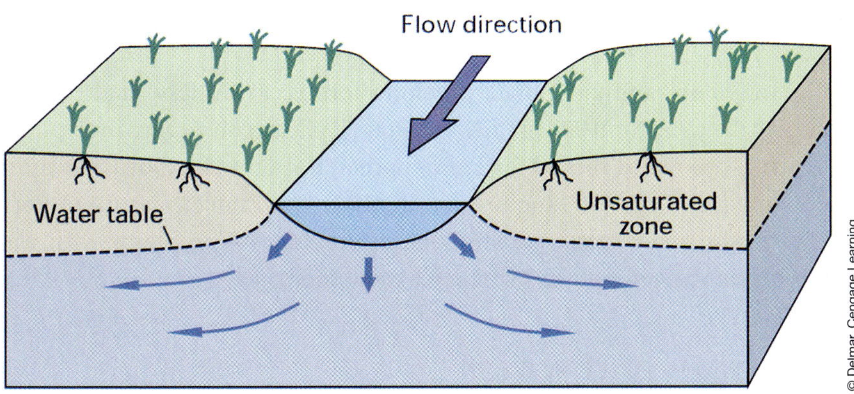

© Delmar, Cengage Learning

Figure 13-15 Gaining and losing streams.

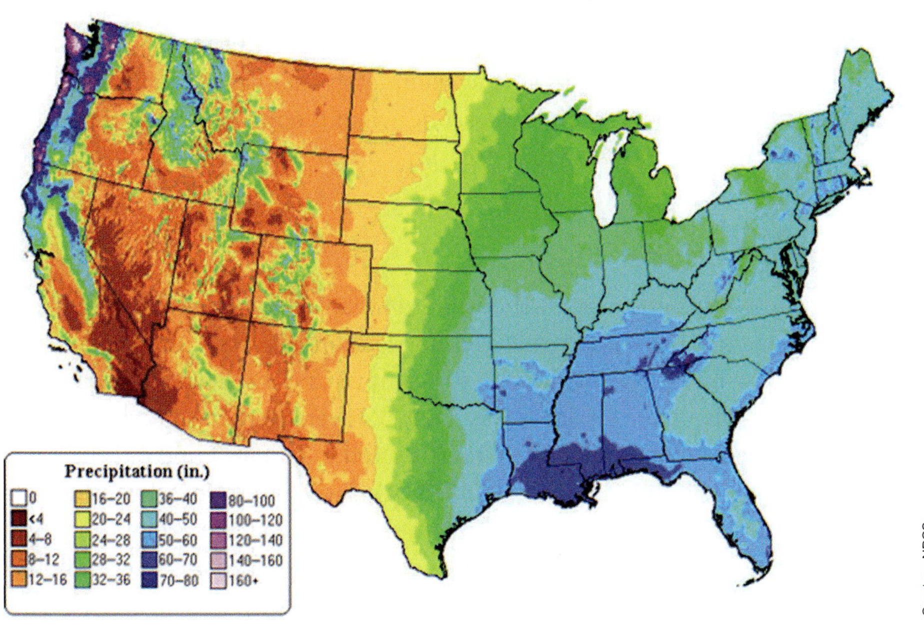

Courtesy NRCS

Figure 13-16 Precipitation map of the United States.

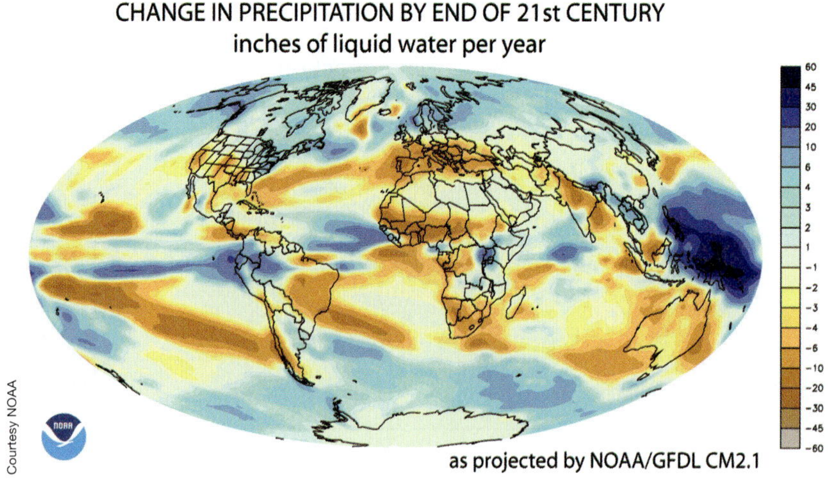

Figure 13-17 Projected global precipitation by the end of the 21st century.

The range in annual global precipitation is from less than 0.25 cm (0.1 inch) to over 2,286 cm (900 inches); Iquipui, Chile, received zero precipitation for 14 years. The world record for annual precipitation is held by Cherrapunji, India, where 2,299 cm (905 inches) fell in 1861. The uneven nature of Earth's precipitation is a key factor causing one of our biggest problems with water resources—they are not always where we need them.

13.4 people and water

Earth's water is a sustainable resource if our use is properly managed. For water resources, achieving sustainability requires surface and ground water management in ways that enable its indefinite use without unacceptable environmental, economic, or social consequences. The key question is: What is an unacceptable consequence and how do we recognize it? In most cases, this is not an easy question to answer. Our water use choices commonly have competing positive and negative impacts, and in other cases the significance of our impacts are not fully understood or completely manifested for many years. But we can learn from our experiences, and when it comes to water resources people have a lot of experience to reflect on.

Managing Surface Water

Since people first became farmers thousands of years ago, surface water has been managed for people's benefit. We have built irrigation systems, dams, canals, and viaducts—some on massive scales—to store, move, and distribute surface water from places it is available to places it is needed. Rivers were our first large highways. In the 1800s, many canals were constructed to enhance and expand river transportation, including canals that linked Lake Erie to the Ohio River through the Cuyahoga River valley. This development in the Cuyahoga drainage helped bring people and industry to the region—people who eventually contaminated the Cuyahoga River to the point that its surface could catch on fire.

Trains eventually replaced canals for transportation, but our efforts to control and manage surface water had just begun. The Reclamation Act of 1902 was passed by the federal government to foster settlement of the western states by constructing large-scale systems to store and manage surface water for the region's benefit. The next half century was an exciting time to be a civil engineer. The United States led the world in construction of large dams and irrigation systems—355 dams and 25,750 km (16,000 miles) of canals were built. In 1936, the four largest dams ever built to that time were under construction: Hoover, Bonneville, Grand Coulee, and Shasta.

In 1933, the middle of the Great Depression, another major federal program was passed—the Tennessee Valley Act. It established the Tennessee Valley Authority (TVA) whose grand purpose was the complete economic development of the Tennessee Valley region through management of its surface water. Fifty-eight dams eventually came to control the Tennessee for transportation, irrigation, flood control, hydroelectric power generation, and recreation. A comparison of the before and after situation in the Tennessee Valley region would declare the TVA a great success. Dams that control rivers now provide hydroelectric power that supplies 15 percent of our country's electricity needs. This efficiently produced, low-cost power does not create waste or pollute the environment. But, as we shall see, dams do change the environment and not all these changes are positive. Perhaps one of our best (and worst?) examples of managing surface water is the Colorado River.

The Colorado River system drains a 634,000 square km (244,789 square miles) region stretching from its headwaters in southwestern Wyoming to the Gulf of California (Figure 13-18). The principal use of Colorado River water within its sparsely populated drainage basin is irrigation. Water from the river is also moved outside the drainage basin for irrigation in places like the Imperial Valley of California. It supplies millions of urbanites in California with water too. If you have eaten fresh strawberries in winter, you may have eaten Colorado River water.

The 2,300 km (1,429 miles) long Colorado River has some characteristics that make sound management of its water particularly challenging. First, for the size of its drainage basin it does not carry much water—less than one tenth of the water carried by the Columbia River in the Pacific Northwest. Second, flow in the Colorado River varies significantly from year to year and it is hard to accurately estimate this critical parameter. Third, the river water is needed by people in seven states and Mexico. The competition for this limited surface water resource is almost unbelievable: All of the Colorado River's water has been legally allocated to specific users. Reaching these allocations has not been easy. Rights to Colorado River water may be the most regulated and litigated in the world.

Demands on the Colorado's water have not diminished. If all of the Colorado's water has been allocated, how will future needs be met? Is Colorado River water an example of surface water management in a sustainable way? We will have to wait and see but there are indications today that we have not managed the Colorado's water well.

To manage Colorado River water we have built 19 large dams and constructed hundreds of miles of diversions and aqueducts (Figure 13-19). These dams can produce enough electric power to meet the needs of up to 12 million

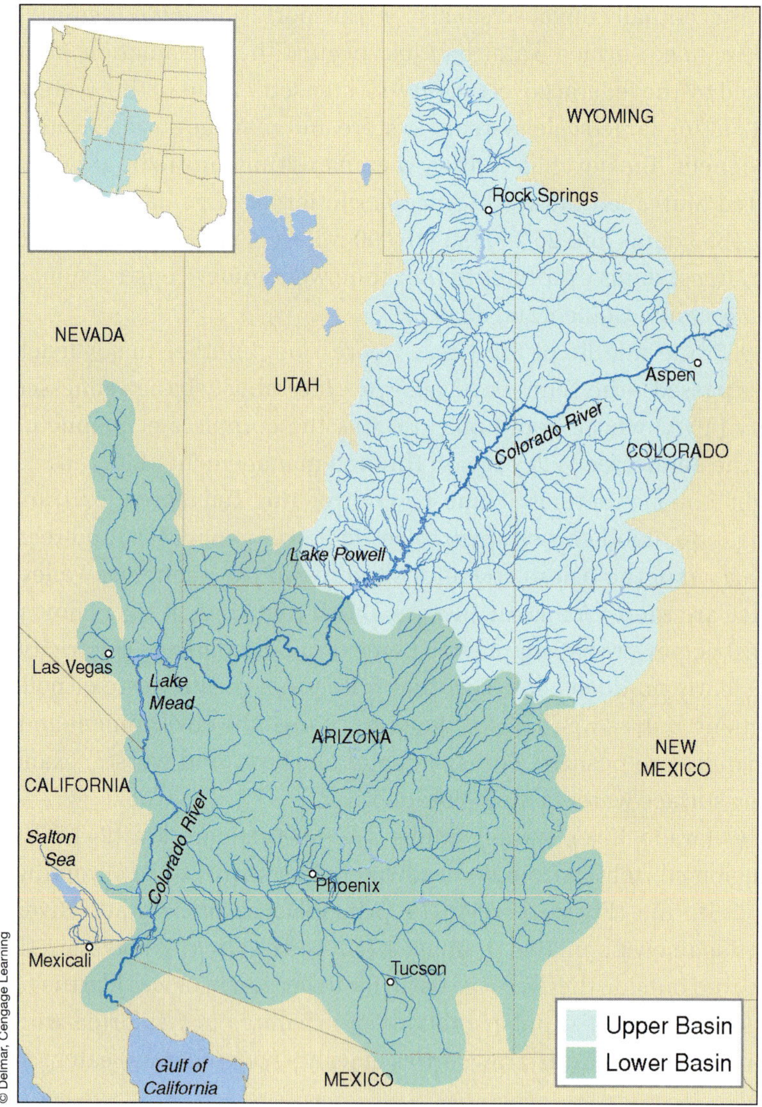

Figure 13-18 Colorado River basin.

people. The largest dams, Hoover and Glen Canyon, create reservoirs that store over three times the annual flow of the river. The reservoirs, Lake Meade and Lake Powell, are popular recreation areas. They are also places where sediment carried by the river gets trapped. Sediment deposition in reservoirs, called siltation, can completely fill them up. In the case of Lake Meade, it is estimated that its storage capacity was decreased 3 percent in 14 years due to siltation.

Figure 13-19 Laguna Diversion dam on the Colorado River. (*Courtesy Earth Science World Image Bank © Michael Collier*)

Dams, in general, disrupt a stream's natural sediment transport characteristics. Above dams, sediment gets trapped and deposited, as in Lake Meade. Below dams, the sediment-poor discharge may either scour the stream bed or have flows that are insufficient to carry away sediment brought to the river by downstream tributaries. These changes in flow and sediment carrying capacity can significantly change local habitat. For example, sand bars along the Colorado below the Glen Canyon dam were eroded and carried away after its construction—the sand bars originated during high flow periods that the dam effectively stopped. The sand bar ecosystems along the river were disappearing. To reverse this impact, the first flood intentionally caused for environmental purposes occurred on the Colorado in 1996. Water was purposely released from the Glen Canyon dam to create high flow levels. These high flows moved sand from the bottom of the river to new sand bars along its banks, which was a planned result in hopes of re-creating natural ecosystems along this part of the river. The results were encouraging and a repeat of this experiment took place in 2004.

But major changes in habitat along the Colorado still characterize its development. Today, four native fish—the Colorado River squawfish (a minnow that can grow to 6 feet long!), razorback sucker, boneytail chub, and humpback chub—are declared endangered species. The Southwestern willow flycatcher, a small bird that nests and once thrived in brush and trees along the lower river, is now endangered. The biological productivity of the delta at the mouth of the river is only 5 percent of what it was before its water became extensively used by people. In the case of the endangered fish, the effects of habitat change have been amplified by the introduction of very successful competitors such as pike and catfish. In some areas, native fish were actually poisoned to make room for introduced sport fish. In general, however, habitat disruption along the Colorado is a major factor in the depletion and endangerment of several native species.

Better understanding of the effects of siltation and habitat disruption can be gained from our experiences with the Colorado River, but this river also helps us understand how our use of surface water can have an impact on water quality. The extensive use and reuse of the river's water for irrigation as it moves downstream increases its salinity. Water in the upper reaches of the Colorado starts its downstream journey with a salinity of 50 parts per million (ppm). During irrigation, evaporation causes the salinity of the water to increase. Irrigation water that makes its way back to the river causes higher salinities downstream. In the lower parts of the river, salinities average 1,500 ppm and have been as high as 2,200 ppm! People are not supposed to drink water with more than 550 ppm salinity, and if the salinity is greater than 750 ppm, the water is not good for agricultural use either.

In effect, upstream users of Colorado River water are degrading the water that downstream users need. This was unacceptable to Mexico, an important downstream user. In 1973, Mexico reached an agreement with the United States that limited the salinity of the water delivered to them. To lower the salinity to acceptable levels, the United States had to construct a large and expensive desalination plant that processes Colorado River water before it is received by Mexico. Many of the people living in the southwest would probably conclude that the benefits of surface water management along the Colorado (power generation, irrigation, municipal use, and recreation) outweigh the negative impacts (siltation, habitat disruption, and nonsustainable use). The long-term consequences of these developments, however, are still being investigated and increasingly better understood. Perhaps more experiments like that tried in 1996 and 2004 when high flows were released from the Glen Canyon dam to re-create habitat will show us that we can get better at managing the Colorado and other river systems.

We seem to have nearly maximized our use of the nation's surface waters—we have not constructed many new dams in recent years (Figure 13-20). In some areas, we are actually proposing to remove dams. One of the most important

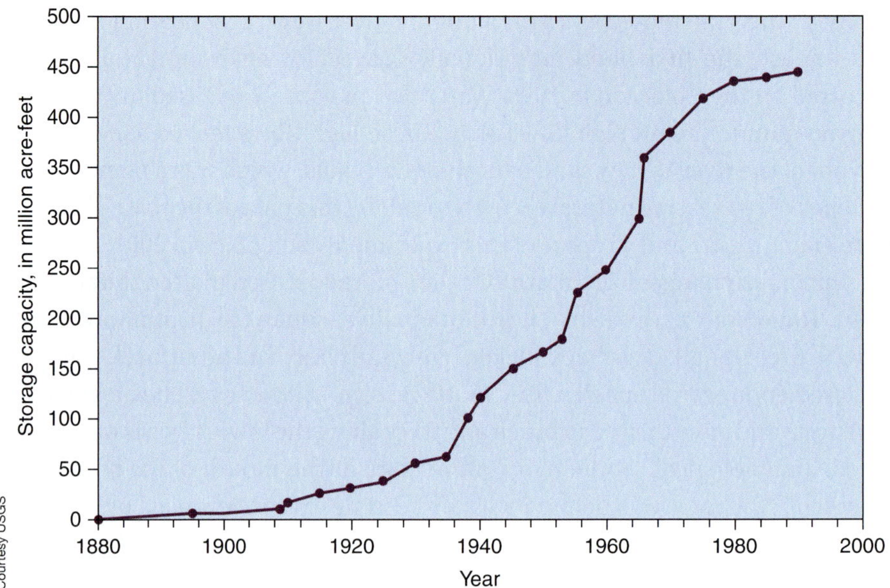

Figure 13-20 Surface water reservoir capacity in the United States.

implications of having maximized our use of surface water during the last century is that new water needs are being increasingly met by development of ground water resources. As you might have guessed, development of ground water has its own set of concerns.

Managing Ground Water

We obtain ground water from wells. As you recall, water moves through and is stored in porous underground zones called aquifers. Some aquifers can receive water percolating down from the surface throughout the area where they occur. These are unconfined aquifers. Others, confined aquifers, only receive water from specific recharge areas and are separated from the surface by impermeable zones over large parts of their area (see Figure 13-14). We especially like to use ground water for personal use because it is generally clean, locally available, and a dependable supply. Ground water provides about 40 percent of the water in our municipal systems and more than 40 million of us get our water from domestic ground water wells. As people maximized their use of surface water, they have increasingly turned to ground water for other purposes, including irrigation and industrial uses. People have used ground water in ways that have depleted this resource in specific aquifers, caused subsidence of the land surface, and decreased needed discharge to surface water bodies.

Even though ground water is a renewable resource, we can withdraw it at much faster rates than it can be recharged. In effect, we can mine ground water and seriously jeopardize its availability for future needs. This is not sustainable management of this valuable resource; an example is the High Plains Aquifer (Figure 13-21).

This aquifer covers 450,658 square km (174,000 square miles) in parts of eight states. It has been a major source of irrigation water since the 1940s; eventually over 56,633,693 cubic meters (2,000 million cubic feet) of ground water

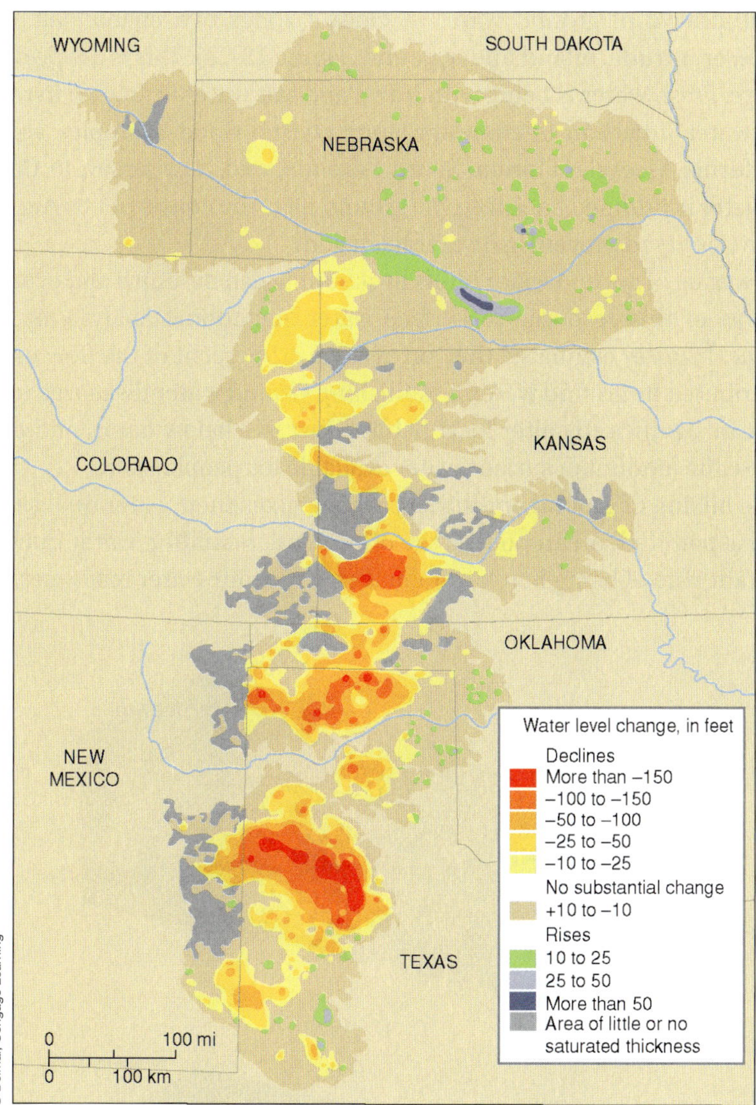

Figure 13-21 High Plains aquifer.

became withdrawn from the High Plains Aquifer every day for irrigation of farmlands. In areas where recharge could not keep up with withdrawal, the ground water levels became lower. By 1997, there were large areas where the ground water levels in the High Plains Aquifer had been lowered more than 45.7 meters (150 feet).

Withdrawing ground water can cause surface subsidence. The structure and water pressure in nearby underlying materials supports the land surface. If these materials contain significant amounts of silt and clay, they can compress when ground water is withdrawn from them. This causes a permanent reduction in the water storage capacity of the underlying material—the aquifer—and can cause land subsidence (Figure 13-22). Land subsidence caused by ground water withdrawal occurs in California, Arizona, Nevada, and Texas.

Figure 13-22 Land subsidence recording.

Withdrawal of ground water in coastal zones can enable salt water to migrate, or intrude, inland underground (Figure 13-23). This is a type of aquifer recharge. Fresh water is being withdrawn and salt water is flowing in to replace it. Saltwater intrusion is common along coasts. Good examples are known in California, Georgia, Florida, South Carolina, and New Jersey. In California, fresh water is pumped back into the ground near the coasts of Los Angeles and Orange County to prevent saltwater intrusion.

Lowering of ground water levels by pumping can diminish and even reverse discharge of this resource to surface water. A national analysis by the U.S. Geological Survey estimated that, on average, 52 percent of the flow in streams is contributed by ground water. Diminishing ground water discharge to surface water can significantly alter local habitat, just like below dams, and of course decrease the amount of surface water available for people's use.

The mixing of ground and surface water mixes their water qualities. If one source is polluted, it can pollute the other. Understanding water quality is an important part of being better at management of our water resources.

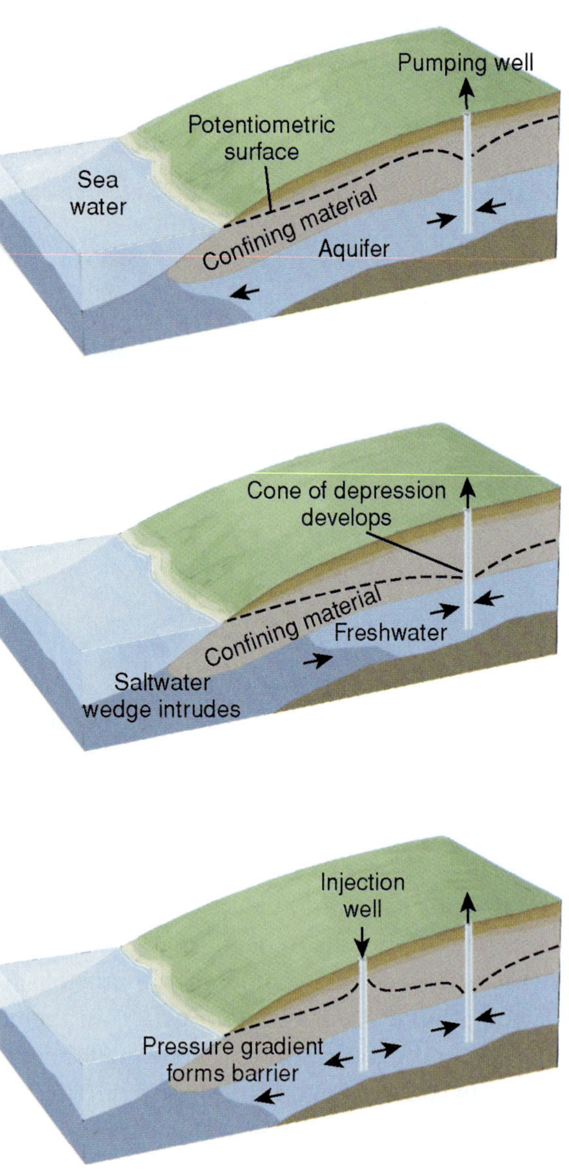

Figure 13-23 Saltwater intrusion caused by pumping of ground water.

If water's quality does not allow us to use it for our needs, it does not matter how much we have. We can understand water quality better by examining how the natural quality of water varies, what the principal contaminants in water are, and where they come from.

Understanding Water Quality

Natural water quality varies tremendously. In a sense, absolutely pure water does not exist on Earth; even rain has small amounts of dissolved constituents. As water percolates through soils and moves through the ground, it interacts with rocks and minerals. In this process, water dissolves mineral components. Natural waters can have high concentrations of iron, sulfur, and salt. They can be naturally carbonated and rich in calcium. Some waters are naturally acidic and metal bearing—they contain trace elements but significant amounts of metals such as copper and silver. Trace elements in natural water are minor constituents such as copper, silver, selenium, and arsenic that are present in very low but potentially significant concentrations—significant because even low concentrations of some trace elements can be harmful to aquatic organisms and people.

The Environmental Protection Agency (EPA) has identified about 90 contaminants (http://www.epa.gov/safewater/hfacts.html) in drinking water and established concentration limits or standards for their safe consumption by people. The contaminants of concern to EPA range from microbes such as *E. coli* to dozens of human-made organic and inorganic chemicals. They include many inorganic contaminants such as copper, mercury, and nitrate that can come from either natural or anthropogenic sources.

It is not common to evaluate water quality by analyzing samples for 90 different possible contaminants. Some contaminants are much more common than others and, as the students at Strongsville High School determined on the Cuyahoga River, can be sufficient parameters for evaluating water quality. These include acidity (pH), turbidity, and the content of dissolved oxygen, fecal coliform, phosphates, nitrates, and total solids.

Where do contaminants in water come from? These sources can be distinct sites, or point sources, such as industrial plants, wastewater discharge pipes, or personal septic systems. Sources that contribute contaminants from a large area, or nonpoint sources, include agricultural fields, urban developments, and even golf courses. Nutrients like nitrogen (in the form of nitrates) and phosphorus are commonly released to the environment by nonpoint sources, such as fertilized fields, and point sources, such as feedlots and septic systems. Bacteria, disease-causing organisms in human and animal waste, commonly come from point sources such as feedlots, sewer overflows, and leaking septic tanks. Feedlots have been the source of hormones in water (synthetic growth-enhancing hormones are common in cattle feed) that appear to have changed the sexual characteristics of fish in nearby streams!

Toxic substances, a broad category of contaminants that includes everything from trace metals to pesticides, are released to the environment from industrial sites, agricultural areas, storage tanks, and inappropriate disposal of industrial and household wastes. An example of a toxic substance is MTBE, a manufactured chemical that was added to gasoline to enhance its combustion and decrease air pollutants in the emissions from our cars and trucks. MTBE is soluble in water, carcinogenic to animals and possibly people, and now detectable in many urban wells and springs. It can be released to the environment from leaky underground storage tanks, accidental spills, and overflows during fueling at gas stations. This toxic substance is relatively stable and can travel long distances in ground water. What we are to do about MTBE and who is responsible for doing it is a major bone of contention as Congress struggles to develop and pass new national energy policy for our country.

Protecting Water Resources

What can we do about contaminated water? One thing we can do is be proactive about cleaning it up, and we have many systems that do just that. The best examples are our municipal sewage treatment plants (Figure 13-24a).

Flush the toilet—out-of-sight, out-of-mind? Not to the municipal engineers who are responsible for ensuring that this waste is properly treated before it is released to the environment or reused. Every community in the United States must meet certain water quality standards if they discharge wastewater to the environment—the same Clean Water Act that is guiding remediation of the Cuyahoga River has regulations that guide municipal wastewater discharge. To meet these standards, communities collect and treat our wastewater in large plants that use processes that accumulate solids and filter and disinfect water before it is reused or released (Figure 13-24b). When municipal wastewater is reused, it is mostly for irrigation but, technically, it could be treated sufficiently to be reused for drinking water. This happens every day on the Space Station, where complete recycling of wastewater is essential (Figure 13-25).

Preventing contamination of our water resources is a key to providing safe water for the environment and for our many uses. Prevention starts at home. We should be very careful about what we dispose of down our drains, such as cleaning solvents, prescription drugs, used motor oil, and so forth, which have no place in our wastewater systems. Better management of industrial wastewater will help too. Modern industrial facilities commonly recycle and reuse as much water as possible, but releasing even warm water can be a problem for local habitat. Much government regulation and monitoring is in place to ensure that industrial activities protect water quality, but accidents and other unfortunate events do happen. As we learned earlier, the Cuyahoga River is still not an acceptable clean river.

In many places, we need to conserve water better, and water conservation can also start at home. In California, installation of low flush volume toilets has saved million of gallons of clean water—and millions of dollars too. Mining of ground water—withdrawal at rates that far exceed those of aquifer recharge—is not managing this resource in a sustainable way. Therefore, understanding relationships between withdrawal and recharge of our ground water aquifers seems especially important as we come to depend more and more on them for our future water needs. Achieving sustainability in the management of our surface and water resources is an important, but so far unachieved, goal.

(a)

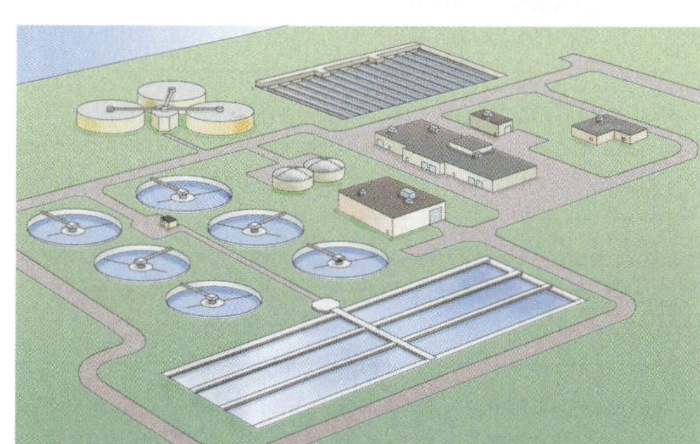

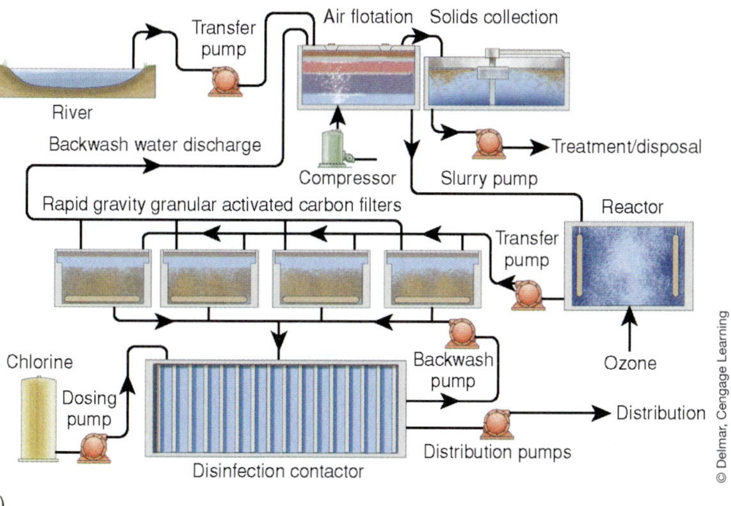

(b)

Figure 13-24 Sewage treatment plant.

13.5 conclusion

Earth, the water planet, is a unique and wonderful place for life. The chemical nature of water and Earth's place in the solar system combine to create conditions where life can thrive. Water is a resource that people depend on for many purposes. It is a sustainable resource if people carefully manage their use and impacts on water.

Figure 13-25 The International Space Station.

IN THIS CHAPTER YOU HAVE LEARNED THAT:

- Water is a polar molecule combining oxygen and hydrogen—H_2O. The polar character of water makes it an excellent solvent and gives it special properties of cohesion and adhesion.
- Earth's location in the solar system creates physical environments where water can exist as a solid, liquid, or gas. It is stable as a liquid over most of Earth's surface. This makes Earth the water planet.
- Water is essential to cellular processes, including the transfer of nutrients and wastes, digestion and absorption of nutrients, and maintenance of body temperature.
- Water creates habitat for organisms in places as diverse as the deepest oceans and underground caves. The biodiversity of aquatic habitats matches that of the terrestrial habitats.
- There are trillions of gallons of water for every person alive on the planet; however, 97.5 percent of this water is salty. About half of Earth's fresh water is tied up in polar icecaps and glaciers. Most of the remaining fresh water is underground.
- People mostly use surface water from streams and lakes, but ground water is also an important source. Because it takes many years—commonly decades, if not more—to replenish ground water, it can be depleted by human use.
- People use water for power generation, irrigation, transportation, and recreation, in addition to satisfying daily personal and household needs. Water is an incredibly important resource.
- People can contaminate water and degrade its quality in personal, agricultural, and industrial activities. Water pollution can be prevented, and polluted water can be treated to protect its quality.

KEY TERMS

Adhesion
Aquifer
Cohesion
Ground water
Permeability
Polar
Porosity
Recharge areas
Triple point of water
Water table

REVIEW QUESTIONS

1. Explain why water is such an important resource to humans and other life on Earth.
2. How does water's polar nature explain its properties of cohesion and adhesion?
3. Where did Earth's water first come from?
4. Why is water such an effective modulator of climate around the Earth?

5. Explain the role of heat in the water cycle.

6. Describe three ways in which the biosphere is dependent on water.

7. Describe three ways in which water affects the geosphere.

8. What negative effects have humans had on the water supply around the Earth?

9. What can humans do to conserve water?

10. What can humans do to improve water quality?

chapter 14

using earth's soil

Do you think soils are pretty much the same everywhere? Students at Moline High School in Moline, Illinois, tackled this question by investigating soils collected from 13 locations around the United States and one from the Caribbean region. They measured the acidity (pH), electrical conductivity, and buffering capacity of the soil samples (Figure 14-1). The results showed that the soils varied tremendously in their measured properties from place to place. For example, the soils ranged from very acidic (pH = 2.19) to very basic (pH = 9.08). No two soils were exactly the same. Variation of soil properties happens because of natural influences such as climate and the types of underlying rocks as well as human activities such as farming. What affects soil properties is important to people everywhere. People's lives depend on soil.

To engineers, soil is the surface material that can be excavated without blasting. Many soil scientists consider soil the loose (unconsolidated) surface material in which plants grow. Geologists think of soil as the natural surface mineral and organic layers that develop from weathering of underlying material such as rocks or sediments. **Soil** *is a blanket of loose material over the land surface. It is most commonly a porous mixture of minerals, rock fragments, and organic material with open spaces (voids) variably filled with air and water. Soil forms on all parts of Earth's land surface from the coldest Arctic regions to the hottest deserts.*

Figure 14-1 Students measure soil sample properties. (*Courtesy Earth Science World Image Bank. © Charlie Rahm, NRCS*)

Soil is a membrane between Earth systems. It connects air in the atmosphere with the solid rocks and minerals of the geosphere. The hydrosphere's water precipitates on soil, passes through it, and connects with underlying ground water. In addition, soil is home to literally millions of life forms of the biosphere. Earth's soil layer is where physical, chemical, and biological processes are constantly changing and transferring materials between Earth systems. It enables and completes many natural cycles of nutrients and organic materials at local to global scales. The Earth that has become such a wonderful and supportive place for life would not be so without soil.

AFTER COMPLETING THIS CHAPTER, YOU WILL UNDERSTAND:
Soil is a natural mixture of minerals, rock fragments, and organic material that make up a porous layer on Earth's land surface. All Earth's systems interact in soil.
Soils form from the physical, chemical, and biological weathering of rocks and minerals from the geosphere.
Weathering processes typically produce soil layers called soil horizons, and these horizons define the soil profile.
Soil is important because it sustains plants, cycles materials between Earth systems, provides a home for many living things, stores and cleans water, and underlies human-built structures.
People affect soil in many ways through farming practices, building structures, and disposal of wastes, to name a few.
Soil is a sustainable resource. If people carefully manage their use of soils, its quality and quantity can be maintained for the future.

14.1 how soils form

Everything on Earth's land surface is decomposing, disintegrating, and changing from one form to another—even rocks. The collection of interactions of air and water with the geosphere's surface that cause these changes is called **weathering**. Weathering includes physical, chemical, and biological processes.

Physical weathering causes rocks to disintegrate or break down into smaller and smaller pieces. When water freezes in rock fractures and small cracks, its expansion forces the rock apart. As this process continues for hundreds and thousands of years, rocks disintegrate into individual mineral grains. Plant roots can do the same thing as they grow along rock cracks and force them apart. Once rocks are broken, they can begin to move down slopes and be eroded by streams. Many rock fragments collide and break into smaller pieces. The end result is an accumulation of small individual mineral grains, especially quartz (SiO_2), the most common mineral in Earth's crust.

Chemical weathering changes minerals on Earth's surface. Water reacts with minerals such as feldspar to form clays. Oxygen in air and water reacts with metal-bearing minerals to form new minerals called oxides. Iron, a very common metal in minerals, reacts with oxygen to form iron oxides. Rust is a form of iron oxide and reddish orange, rusty stains on rocks are evidence that they are being "oxidized." Water soaking into the ground can become weakly acidic, most commonly by combining with carbon dioxide to form carbonic acid. Minerals such as calcite (calcium carbonate) are slowly dissolved by weak acidic water percolating through soil. Because chemical reactions speed up as temperatures increase, the changes brought about by chemical weathering occur faster in warmer climates than in colder climates. Chemical weathering, together with physical weathering, results in a mineral assemblage rich in clay, quartz, and oxides.

Biological processes aid physical and chemical weathering. Everything from microorganisms to burrowing animals plays a part. They mix the loose materials, add organic debris such as decaying plants, and evolve carbon dioxide that reacts to form weak acids. Biological processes are especially important in **humus**. Humus is a general term for the organic matter that imparts a dark color to upper parts of soil. It can consist of both plant and animal debris. Some of this material is acidic and insoluble in water (**humic acids**). The general function of soil organic matter is to supply nutrients such as nitrogen and phosphorus; help develop soil structure that assist aeration, permeability, and water-holding capacity; and serve as a source of food (energy) for soil organisms. Humus' impact on soil structure facilitates many chemical and physical soil processes. The combined result is a well-developed, mature soil that is a porous mixture of organisms, decaying organic debris, clays, silt- and sand-sized mineral grains (especially quartz), and oxides.

Well-developed soils are not a uniform, homogeneous mix of all these parts. Over time, the physical, chemical, and biological processes combine to stratify the soil materials from the surface downward into layers called **horizons**. Although climate and the nature of the underlying parent materials (rocks and minerals) are major influences on the rates and end results of soil development, the soil-forming processes work together to produce a generally similar set of layered horizons called a **soil profile**.

The Soil Profile

The horizons in a soil that form over many years define its profile. As Moline High School students learned, soils can vary a great deal from place to place. Soil profiles are very useful in characterizing soils and their differences.

Typical soil profiles (Figure 14-2) contain three or four master horizons. From the surface downward, the typical layers include the O, A, B, and C horizons.

- The **"O" horizon** is the layer of organic material, which is mostly decaying vegetation that accumulates on the surface.
- The **"A" horizon** is where organic and mineral materials are mixed. The organic material, humus, gives this horizon a dark color.
- The **"B" horizon** is the layer where material, especially clay minerals, that are moved downward by percolating water accumulates. It does not have structures inherited from underlying parent material.
- The **"C" horizon** includes weathered remnants of the underlying parent material. It grades downward into the parent material, which is called the "R" horizon.

The master horizons are a beginning framework for describing soil profiles. Soil scientists recognize many subhorizons and other internal characteristics that enable them to more carefully and completely describe a soil and how it has formed. Internal soil characteristics that help to describe and compare them are color, texture, and structure.

Soil color is important because it reflects a soil's internal changes and helps recognize the horizons that have developed.

- Organic material makes the upper horizon darker.
- Light-colored gray or white layers can indicate zones that have been leached.
- Mottled yellow, brown, and reddish zones can characterize "B" horizons where clays and oxide minerals are being accumulated.
- Strong reddish (rusty) colors indicate that the soil is well aerated and oxidized. Poorly aerated soils may be reduced and more yellow in color.

Figure 14-2 A well-developed soil profile.

Soil texture is determined by the size of mineral particles within it. The three key particle sizes are sand, silt, and clay.

- **Sand-size particles** range from 0.074 to 2.0 mm in diameter; they feel gritty when rubbed between the fingers.
- **Silt-size particles** range from 0.004 to 0.074 mm in diameter.
- **Clay-size particles** are tiny, less than 0.004 mm in diameter. Clay feels smooth and cohesive; when it is wet, it can be squeezed into forms.

Soil scientists use the proportions of these three particle sizes to classify soil texture. Figure 14-3 shows how the proportions of sand, silt, and clay can be displayed and used to classify a soil. Excellent soils, called loams, have subequant amounts of the three particle sizes. Loams are ideal for farming because they retain water well, are filled with organic matter that provides the nutrients needed by crops, and are loose enough to be easy to plow and till.

Soil structure reflects how individual soil particles are aggregated together. Four common soil structures are:

- **Granular**—the soil breaks up into subequant pieces 1 to 10 mm across.
- **Blocky**—the soil breaks up into subequant pieces 5 to 50 mm across.
- **Platy**—the soil breaks up into small tabular pieces.
- **Prismatic**—the soil is cracked into elongate blocks.

Figure 14-3 Soils have many textures.

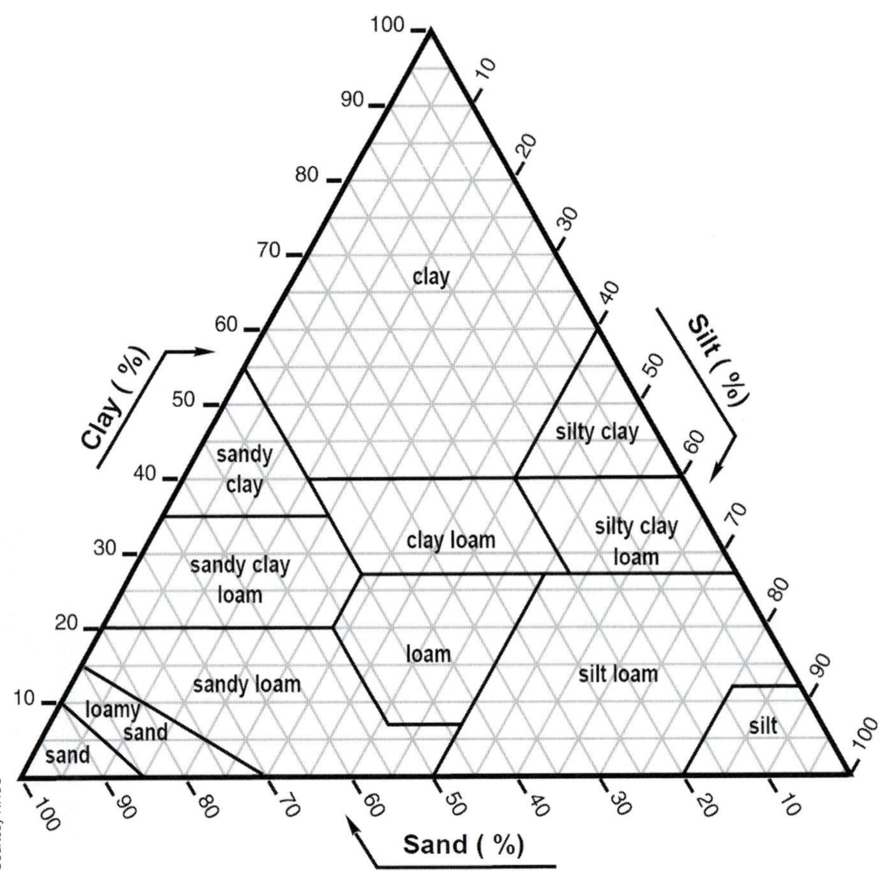

Not all soils have structure: Sand- or clay-rich layers can especially lack structure. Structure helps determine how easily water moves through a soil or how easily it is eroded. The more granular the soil, the easier it is for water to pass through it.

Stratifying Soils

How do soils become layered (stratified)? **Soil horizons** develop because of additions, changes, movements, and removal of soil materials. The most obvious material added to soils is organic matter, especially the decaying plant material that includes roots that get mixed into A horizons. However, dust, sediment from floods, and volcanic ash can add new material to developing soils over large regions (Figure 14-4).

Change in soil materials—what soil scientists call **transformations**—are caused by the physical, chemical, and biological weathering processes described earlier. They change original or added materials to clays, oxides, and remnant mineral grains (especially quartz). The development of acidic, neutral, or alkaline conditions by interactions of air, water, organisms, and mineral materials is an important part of transformation processes. They create conditions that enable water to dissolve soil components and nutrients to be absorbed by plants.

Movements of materials, called **translocations**, are key to developing soil horizons. Water moving downward through the soil physically carries minerals such as clays or precipitates minerals such as oxides and carbonates at deeper levels, especially in the B horizon. Soil material dissolved in water can

(a)

(b)

Figure 14-4 Volcanoes affect farmland.
((a) Courtesy Earth Science World Image Bank. © USGS Cascades Volcano Observatory (b) Courtesy NRCS)

also be completely carried away, or removed, from the soil. Removal is very effective in wet climates where soils get depleted as abundant water dissolves and carries away many components.

The processes that develop soils and their horizons take place very slowly. It commonly requires hundreds to thousands of years to make a well-developed soil. This is why people need to be careful how they use soils. Disturbing soils, depleting them of important components, or allowing them to be destroyed by erosion causes soils to be lost to future generations. These soils have not been used by people in a sustainable way. But soils are a renewable resource that can be properly maintained and sustained.

 14.2 why soil is important

For something people think of as just "dirt," soil is incredibly important (Figure 14-5). Except for fish and other types of aquatic life, most everything people eat depends on soil. This is a really important fact about soil: It is essential for terrestrial life.

In Chapter 2, you learned how plants use photosynthesis to convert energy from the sun, water, and carbon dioxide into simple sugars—food—for plants. This is the critical first step in making energy available for life. Plants are the basic foundation of the land's food webs and terrestrial plants cannot grow without soil. In the United States, agricultural lands produce at least 200 crops. Hay, wheat, corn, and soybeans are grown on about 80 percent of these lands, along with timber for homes and fuel, cotton for fiber, and feed for farm animals. Soils also recycle carbon and nutrients, store and clean water, effectively dispose of wastes, and provide a substrate for buildings and roads.

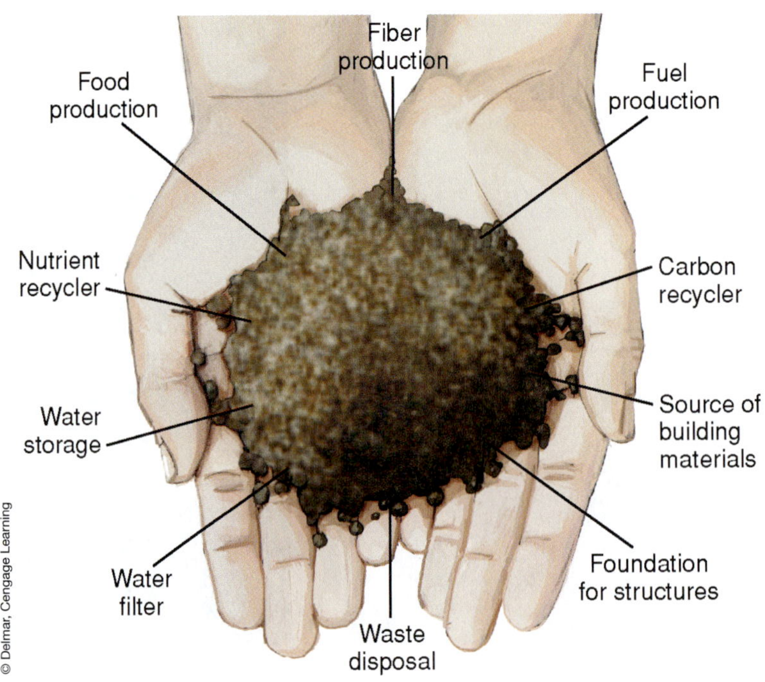

Figure 14-5 Soil uses.

Soil and Carbon

Carbon is a component of all life. Soil stores (reservoirs) a significant amount of this important element, changes it into different forms, and transfers it between parts of the Earth system. Soil stores about 1,500 billion metric tons of carbon or about twice the amount stored in the atmosphere. But the residence time of carbon in soil is only about 9 years, because it is constantly changed and transferred from soil to the atmosphere, biosphere, hydrosphere, or geosphere in the global carbon cycle (Figure 14-6).

Carbon is captured from the atmosphere by photosynthesis in plants rooted in soil. When plants and animals die, they add carbon to soils. This material is food for many life forms, particularly the decomposers that live in soil. Decomposers change organic carbon from decaying plants and animals into carbon dioxide. This carbon dioxide can then be transferred to the atmosphere through soil respiration or dissolved in ground water and transferred to other parts of the hydrosphere, even the oceans. Large amounts of carbon dissolved in water are eventually precipitated as carbonate minerals in soils, springs, or on the seafloor. Because soil processes link carbon in the different parts of the Earth system, it has a role in controlling the amount of carbon dioxide and other greenhouse gases in the atmosphere. Understanding soil's role in controlling carbon dioxide levels in the atmosphere is part of understanding the future of climate change (see Chapter 4).

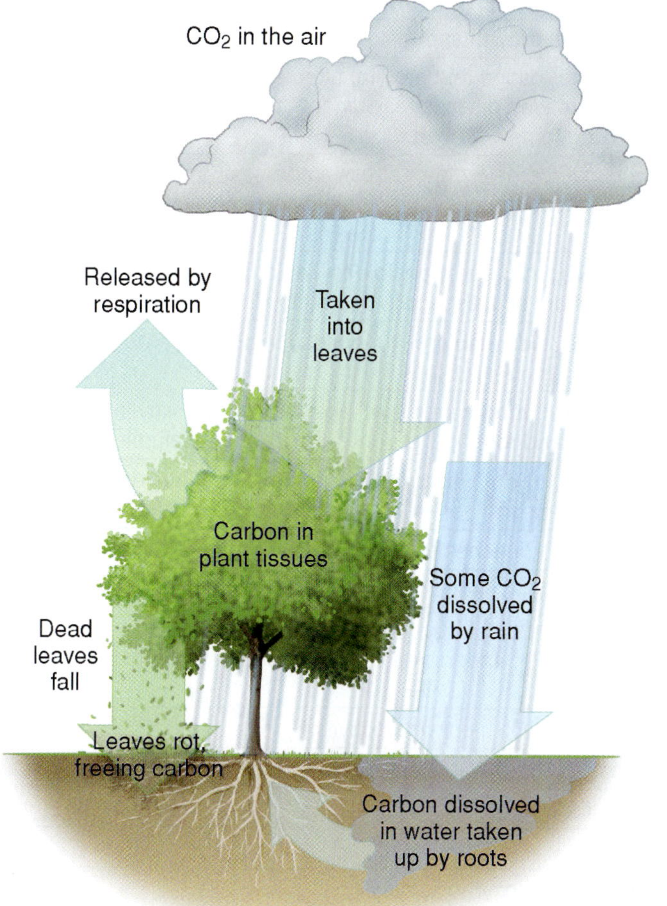

Figure 14-6 Soils are involved in the global carbon cycle.

Soil and Nutrients

Nutrients are the essential elements plants need for growth. All of the 17 essential nutrients (Table 14-1) come directly from soil except three—carbon, hydrogen, and oxygen. Examples of soil-based nutrients are nitrogen, phosphorus, potassium, calcium, magnesium, sulfur, and iron. Nutrients originally come from the geosphere, the parent material of soils. But once they are incorporated in plants, they can be recycled back into soil. When plants fall to the ground surface, they start disintegrating and decomposing and nutrients are released back to the soil from which they originally came. This is what makes soils a sustainable resource. Natural processes continually recycle nutrients in soil that plants need. People interrupt these natural processes when they farm and collect crops. Removing plants removes nutrients from the soil. If these nutrients are not replaced in some way, such as with fertilizers, the soil will gradually lose its ability to support plants. This is why it is important to manage the use of soil in a sustainable way.

Table 14-1 *The 17 essential plant nutrients.*

Element Symbol	Element Name	Element Action in Plants
C	carbon	Essential for plant growth and makes up 96 percent of the weight of a plant.
H	hydrogen	Essential for plant growth and makes up 96 percent of the weight of a plant.
O	oxygen	Essential for plant growth and makes up 96 percent of the weight of a plant.
P	phosphorus	Promotes root formation and growth; affects quality of seed, fruit and flower production; increased disease resistance; does not leach from soil readily; mobile in plant, moving to new growth.
K	potassium	Helps plants overcome drought stress; improves winter hardiness; increased disease resistance; improves the rigidity of stalks; leaches from soil; mobile in plant.
N	nitrogen	Absorbed as NO_3^-, NH_4^+; responsible for rapid foliage growth and green color; easily leaches from soil, especially NO_3^-; mobile in plant, moving to new growth.
S	sulfur	Absorbed as SO_4^-; leachable; not mobile; contributes to odor and taste of some vegetables.
Ca	calcium	Absorbed as Ca^{++}; moderately leachable; limited mobility in plant; essential for growth of shoot and root tips; reduces the toxicity of aluminum and manganese.
Fe	iron	Absorbed as Fe^{++}, Fe^{+++}; accumulates in the oldest leaves and is relatively immobile in the phloem; necessary for the maintenance of chlorophyll.
Mg	magnesium	Absorbed as Mg^{++}; leaches from sandy soil; mobile in plants.
Ni	nickel	Absorbed as Ni^+. This need by plants recently established; essential for seed development.
B	boron	Absorbed as $B(OH)_3^-$; important in enabling photosynthetic transfer; very immobile in plants.
Cu	copper	Absorbed as Cu^{++}, Cu^+; enzyme activity.
Mn	manganese	Absorbed as Mn^{++}.
Mo	molybdenum	Absorbed as MoO_4^-.
Cl	chlorine	Absorbed as Cl^-.
zn	zinc	Absorbed as Zn^{++}; enzyme activity.

How to remember the 17 essential elements: C HOPKINS CaFe is Mighty Nice, But Many More Prefer Clara's Zany Cup

Soil and Water

Soil both stores and cleans water. Because soil is a loose material, it is full of small voids called **pores**. Under most conditions, pores are filled with both air and water, but during periods of heavy precipitation, water can displace the air, completely fill the pores, and saturate the soil with water. This ability to soak up and store water helps diminish flooding and is very helpful to plants. Plants always need large amounts of water to grow—about 90.7 to 408.2 kg (200 to 900 pounds) of water for every 0.45 kg (1 pound) of new cellular growth. A soil's ability to store and provide this water, even in dry periods, is essential to keeping vegetation alive and healthy.

Soils that are almost always saturated with water develop wetlands, mushy and swampy areas (see Chapter 13), with thick O and A horizons. These types of areas are excellent water filters. Nutrients, sediments, and even chemicals and organic waste are removed from water as it passes through wetlands.

Soil and Waste

Ever wonder what happens to the stuff you flush down the toilet? Most goes to community wastewater treatment facilities through sewer systems, but more than 25 million households in the United States use private septic systems to dispose of people's waste products (Figure 14-7). Septic systems use soil to provide the final treatment of household waste. Wastewater from septic tanks percolates through drain fields in soil, leaving behind nutrients, bacteria, and remnants of organic matter. Sewage sludge from community treatment facilities can also be disposed of in soil. As much as one-third of the nation's sewage sludge—after appropriate processing—is applied to farmland as fertilizer. Other wastes can be added to farmland to amend soil. These include wallboard (gypsum), lime kiln dust, and materials collected in flues from electricity generating facilities (flue dust). Tilling these materials into soil recycles nutrients they contain.

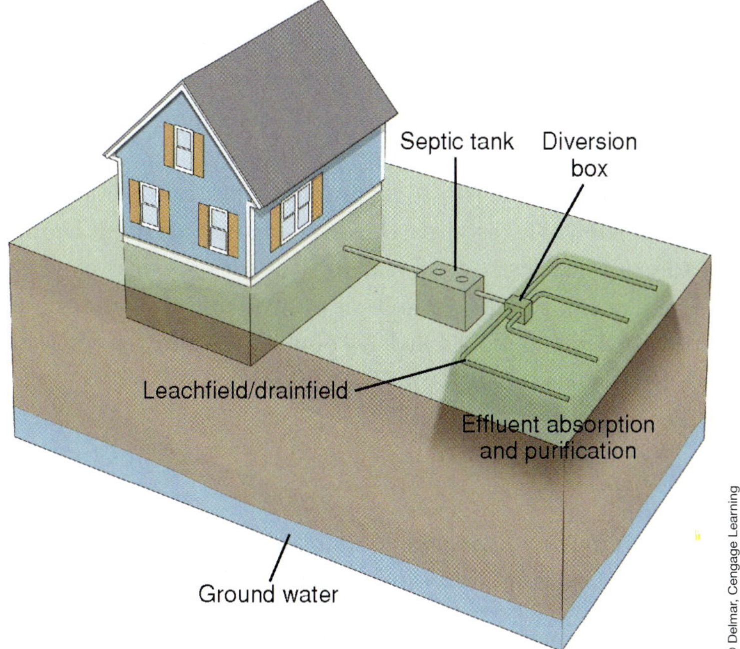

Figure 14-7 Septic systems fertilize soil.

Figure 14-8 Soil movement can damage homes. (*Courtesy Earth Science World Image Bank © Marli Miller, University of Oregon*)

Soil and Construction

Almost all buildings, roads, and other infrastructure on land is built on soil. Understanding soil's physical properties—how it compacts under weight, how it contracts or expands when wet, how it will react with other materials like cement, and how well water will pass through it—are important to safe and cost-effective construction. Engineers measure **soil's strength**—how well it withstands deformation—and classify them according to their physical properties at different water contents. Some soils are very good for foundations, but others, such as those that are very sandy or very rich in clay, can be problems. Excellent examples of poor soils for construction are clay-rich soils of the south-central United States (Figure 14-8). These soils will expand if wet and contract if dry; people in large parts of Texas need to literally water their home's foundations to keep them from cracking and tilting.

Courtesy NRCS, Photo by Tim McCabe

Figure 14-9 Mollisol cropland in Iowa.

Soil and Food

People's lives depend on soil resources and farming is how people use soil most directly. In the United States, 41 percent of all the land is used to produce food. Crops are grown on almost half this farmland. Almost 2 million farmers produce over $2 billion worth of farm products each year. Most of this value comes from cattle, dairy products, corn, and soybeans. United States farmland is so productive that about one-fourth of each year's agricultural products are exported. America's soil is being used to feed hundreds of millions of people.

Around the world, the most widespread and productive soils for farming are those that form in grassland or prairie biomes. Soil scientists classify the world's soils into 12 types or orders. The soil order that develops in grassland biomes is called **mollisol**. Mollisols are most extensive at mid-latitudes. They are characterized by a dark brown humus-bearing upper horizon and are rich in minerals (Figure 14-9). Highly productive mollisol regions grow grains so well that they become known as "breadbaskets."

If soils are properly managed, their productivity can be indefinitely sustained. They are a renewable resource. However, mismanagement can seriously deplete, degrade, or even destroy soil resources. If soils are destroyed, as when deserts take the place of productive soils, it may be thousands of years before good soils are again developed. People have done much harm to soils since first starting to extensively use them about 7,000 years ago.

 the soil ecosystem

Soil is full of life. In many places, soil contains more life and species diversity than exists on the surface. A shovel full of rich garden soil can contain more species than an entire rainforest! This should bring a whole new meaning to dirty fingernails! Life in soil is connected to the surface through plants. One tree can have 5 million root tips and the zone around plant roots is where much soil biological activity takes place. Soil-dwelling organisms include bacteria, fungi, nematodes, arthropods, earthworms, and even burrowing animals.

Bacteria

Bacteria are scattered throughout soil but are most abundant around plant roots (Figure 14-10). They are tiny one-celled organisms: A spoonful of soil can contain millions of bacteria. Some bacteria are photosynthetic (primary producers), some decompose organic matter, some "fix" nitrogen (take it from air and change it into forms plants can use), and some cause disease in plants. Actinomycetes are thread-like bacteria that are good nitrogen fixers (Figure 14-11). The earthy aroma people smell when digging or tilling soil is caused by actinomycetes.

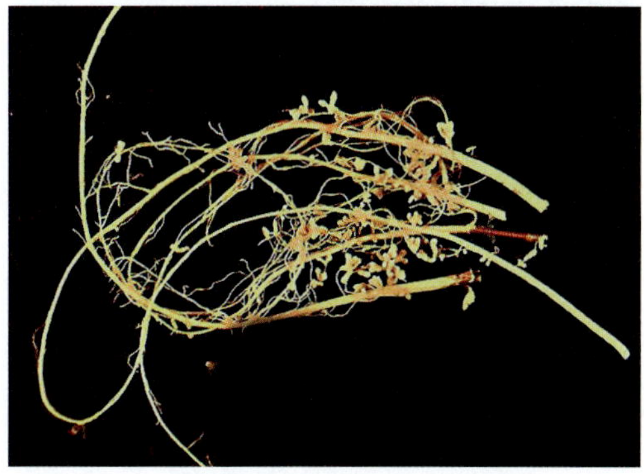

Courtesy ARS, USDA

Figure 14-10 Bacteria help keep soil fertile.

Fungi

Fungi, immobile multicellular organisms that are able to absorb food and grow, are one of the main subdivisions of life. Over 100,000 kinds of fungi have been described and many thousands more are expected to be living in soil. Most fungi are either too small to see, or hidden underground, but some visible kinds include mushrooms, the fungi that cause athlete's foot and ringworm, and the yeasts in bread and beer. In the ground, fungi grow webs of tiny filaments that can be concentrated around plant roots or extend in connected networks over many acres.

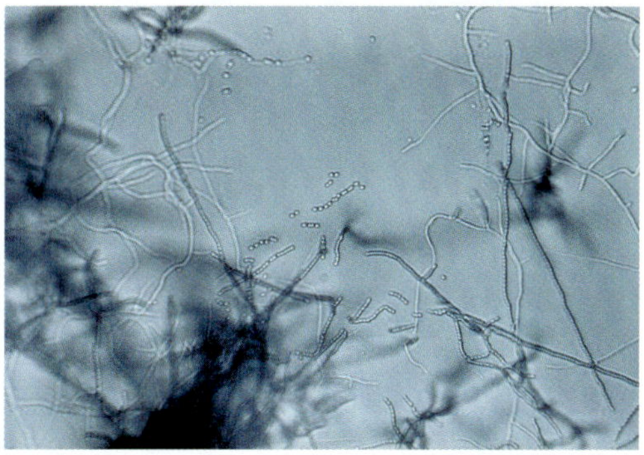

© CDC/Dr. David Berd

Figure 14-11 Actinomycetes help keep soil healthy.

The fungi around plant roots are especially important in soils. They are called mycorrhizal fungi. Some of these actually grow inside root cells. Mycorrhizal fungi have a symbiotic relationship with plants. This means that they extend the plants' contact with soil's water and nutrients while the plant provides sugars (through photosynthesis) that the fungi eat. But fungi are not picky about what they eat. They will absorb food from many sources in soil. The digestive enzymes they secrete break down decaying plant and other organic material so that it can be readily absorbed by the fungi.

Protozoa

Protozoa are single-celled animals that move around in soil and eat bacteria (Figure 14-12). Some protozoa can eat thousands of bacteria in a day. Although they are much bigger than bacteria, they are still microscopic in size and a spoonful of soil can contain many thousands of protozoa. Where there is food being eaten in soil, there is waste. Protozoa waste releases nitrogen and other nutrients into soils for plants and other organisms to use. Protozoa are in turn eaten by nematodes and other soil creatures.

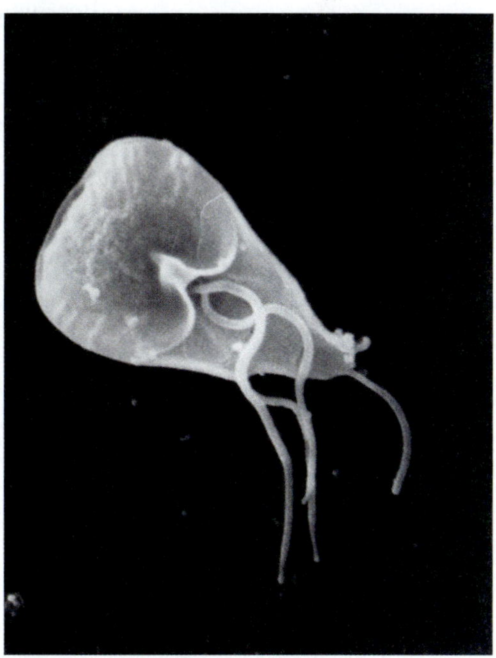

© CDC/Janice Carr

Figure 14-12 Flagellate protozoa live in soil but can cause diarrhea in animals.

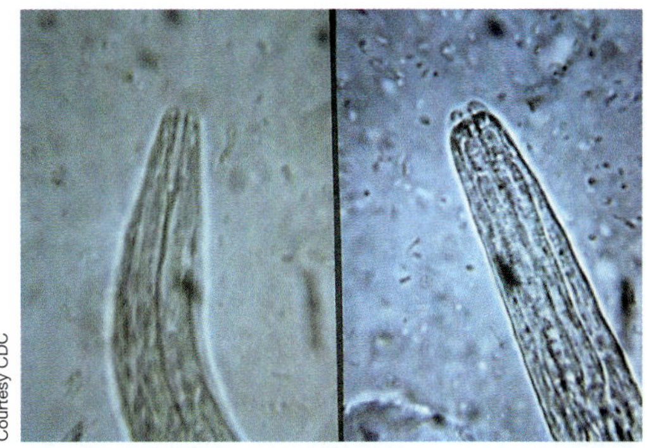
Courtesy CDC

Figure 14-13 Nematodes live in soil.

Nematodes

Nematodes are tiny roundworms that live in soil everywhere (Figure 14-13); millions can live in a shovelful of soil. Nematodes typically eat protozoa, although some eat fungi and bacteria, and others eat about anything they can find, including decaying plants. Some nematodes attack plant roots and cause damage to farm crops, but most recycle nutrients in soil.

Earthworms

Earthworms do much more than feed robins and help people catch fish. Earthworms are essentially eating tubes—plant debris and other organic matter goes in one end and worm feces, "castings," come out the other. Castings are rich in nutrients and help lower soil acidity. They are essentially fertilizer. Earthworms also help churn and mix soil. They burrow between soil horizons, helping create internal drainage, aeration, and mixing of minerals and nutrients. Earthworms are a very good thing in soil.

Figure 14-14 Antlions are beneficial to soil. (*© Clemson University—USDA Cooperative Extension Slide Series, Bugwood.org*)

Arthropods

Arthropods, organisms with jointed legs but no backbone, live in soils, too. Ants, sowbugs, spiders, centipedes, and scorpions are examples. Like earthworms, they help mix soil. They also chew up plant debris, control protozoa populations, and even eat each other (Figure 14-14). They help improve soil structure and make nutrients more available to plants.

Burrowing Animals

Burrowing animals make their homes in soil. Ground squirrels, gophers, and badgers are examples. Their burrows help mix surface and soil materials, create better soil drainage, and make homes for other animals such as snakes and owls. Small burrowers like shrews and moles eat bugs (arthropods) in the soil.

Soil, therefore, is so much more than "dirt." It is one important place on Earth that all parts of the Earth system—geosphere, biosphere, atmosphere, and hydrosphere—interact. It is critical to energy and material transfer within the Earth system.

14.4 people and soils

Civilization has roots in soil. Over 7,000 years ago, the great river floodplains of the Nile in Egypt and the Tigris and Euphrates in Mesopotamia were where people first learned to productively farm large areas. This enabled them to produce more food than they needed for themselves. This accomplishment freed people from the land—they could put energy and effort into new endeavors, trade food for other needs and resources, build cities, and eventually create civilizations. Ever since, people have been able to investigate the wonders around them, to develop the powerful tools of science and mathematics, and to challenge themselves to do more than simply provide for their own livelihood. Today there are about 6 billion people, and by 2030 there will probably be 8.3 billion. Although widespread agriculture led to great advances for the human race, many problems have arisen in the history of soil use.

People have caused extensive soil erosion, depleted soil of nutrients and productivity, displaced soil by land-use choices, and contaminated soil with pollutants. Fortunately, the causes of these problems are well understood and they can be prevented or mitigated.

Soil Erosion

Hundreds of thousands of acres of Earth's soil have been lost to erosion. Because soil is loosely aggregated surface material, direct exposure to wind and running water can cause it to be easily picked up and carried away, or eroded. Large areas of the midwest United States lose several tons of **topsoil** (the organic rich "A" horizon) per acre each year. It gets washed into rivers and carried away as mud and silt (Figure 14-15).

Heavy rain easily washes away bare soil. Raindrops start the process by moving soil particles on impact. As surface runoff accumulates, it flows down slopes, carrying soil particles with it. Where runoff becomes channeled, as along row crops or plow furrows, it cuts small **rills** (Figure 14-16). Eventually surface runoff can cut deeply enough to form **gullies** (Figure 14-17).

Soil erosion can occur wherever the protective cover of plants and plant roots has been removed. Throughout history, overgrazing, deforestation, and farming practices have been major causes of soil erosion.

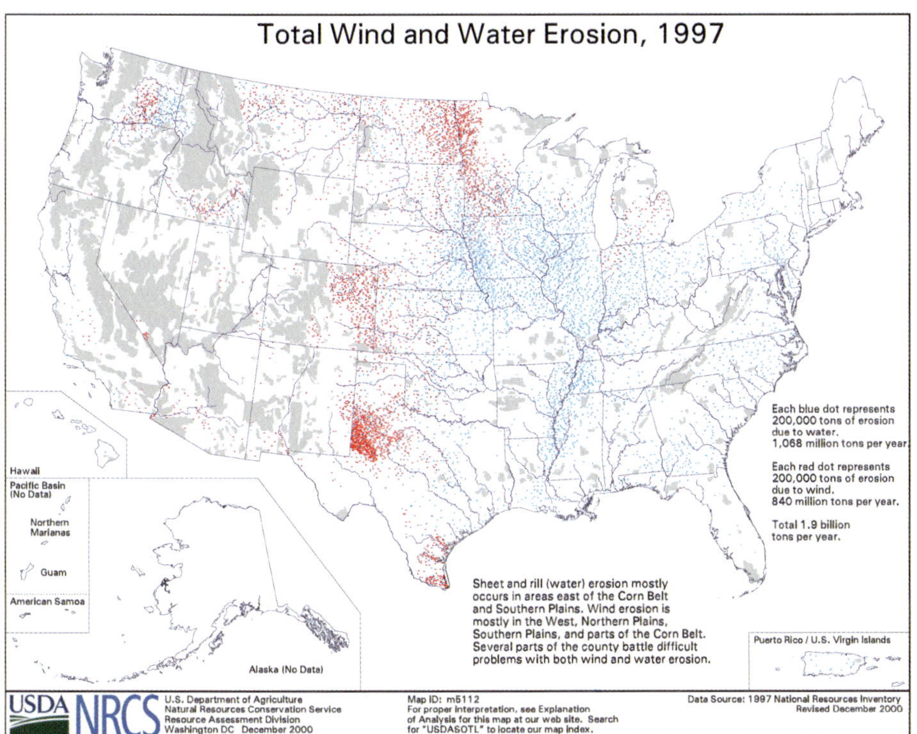

Courtesy NRCS

Figure 14-15 The United States loses topsoil to erosion.

Figure 14-16 Sheet and rill erosion in a field. (*Courtesy Earth Science World Image Bank © Lynn Betts, NRCS*)

Figure 14-17 Gully erosion results in soil loss. (*Courtesy Earth Science World Image Bank © Keith McCall NRCS*)

All grazing animals eat plant cover, and extended **overgrazing**—when there are more animals than the plants can sustain—essentially eats it all up. Overgrazing, especially by sheep and goats, is the cause of much soil erosion around the world. It has led to desertification in arid and semiarid regions. Soil that is washed off overgrazed areas contributes large sediment loads to rivers. The silt and mud that clouds the Tigris and Euphrates Rivers is soil washed from headwater regions overgrazed for thousands of years. This soil repeatedly filled and plugged the irrigation systems of ancient Mesopotamia.

Removal of trees for fuel and timber has caused serious erosion in many places. Global forest cover has been reduced at least 20 percent compared to that before agriculture became widespread. Over 100,000 square kilometers (38,610 square miles) of tropical rainforest are cut each year. Harvesting forests, or **logging**, decreases root density and strength, compacts soil where machines operate, bares soil where trees are dragged over the surface, and cuts roads and staging areas into soil. Since there are fewer trees after logging to use rain water, the soil becomes heavier and more unstable. This can cause channelized runoff. Landslides become more common and stream banks cave and erode. The increased soil erosion produces increased sedimentation in nearby streams, usually with negative effects on their ecology. In the Pacific Northwest, increased stream sedimentation after logging can make spawning grounds less productive and make water temperatures unhealthy for salmon and steelhead.

An ancient example of deforestation and its effects on soil are the once great cedar forests along the east coast of the Mediterranean Sea—the famous Cedars of Lebanon. These magnificent forests, fought over for thousands of years because of their importance for ship and building construction (timbers from Lebanon were used in the Temple of Solomon), once covered thousands of square kilometers in the Lebanon mountains. Now there are only about 300 trees in a few remnant patches (Figure 14-18). Their demise started with the Phoenicians about 4,500 years ago, who used the forest timber for ships and trade. The mountains of Lebanon are now stark and nearly barren, kept that way by overgrazing that followed on the heels of deforestation and soil erosion.

How people farm can cause soil erosion. Tilling and plowing churn soil, expose it to wind and rain, and at least temporarily remove its plant cover. If plowing occurs just before heavy rains, erosion can be extensive. Growing root and row crops can leave fields equally vulnerable. On all slopes the direction of plowing and planting is important. If it is not at right angles to the slope, the potential for erosion is great. Contour-parallel plowing and planting is important to erosion control.

One of the worst examples of erosion caused by farming practices occurred in the United States during the 1930s. Over 100 million acres in Texas, Oklahoma, Kansas, Colorado, and New Mexico became barren terrain where wind could gather small soil particles—dust—and carry them away (Figure 14-19). This region became known as America's "Dust Bowl." The expanded use of tractors and other farming machines allowed rapid conversion of the prairie grassland to plowed fields. When drought conditions affected the region, the new crops could not survive and bare soils became widespread. Winds and drought were a devastating combination for the region's soils. Resulting dust storms, black choking and churning clouds through which no one could see, carried soil thousands of kilometers away. This even darkened the skies and snow of New England and New York. One storm developed a dust cloud 2,000 kilometers (1,250 miles) long. The devastation drove people from the land and led to creation of the USDA's Soil Conservation Service in 1935. Soil conservation practices have since reduced soil erosion in the United States, but it still happens.

Conditions like those of the American Dust Bowl exist elsewhere in the world today. Rapid agricultural expansion in North China, combined with periods of drought, are creating dust storms that clog Chinese cities, close airports, and pollute air in neighboring countries. Dust from China made it difficult to see the Rocky Mountains from downtown Denver in 2001. In March 2002, NASA was able to take pictures of huge dust storms in northwest China from a satellite (Figure 14-20). Unfortunately, soil erosion is ongoing and extensive there. The lessons from past misuse of soil are not being heeded in many parts of the world.

Figure 14-18 Cedars of Lebanon.

Soil Quality

Soil erosion is not the only problem. A **soil's quality** is its capacity to sustain plant and animal productivity, maintain or enhance water and air quality, and support human health and habitation. High-quality soils:

- Support a diverse soil biology and productivity.
- Control water flow and character.
- Facilitate reactions between organic and inorganic materials.
- Store nutrients important to Earth system cycles.

Figure 14-19 Dust Bowl of the 1930s.

Figure 14-20 Dust storm in the Gobi Desert.

People change soil's ability to accomplish these functions by depleting nutrients, lowering its biodiversity, and contaminating it with mineral salts and other pollutants.

There are 17 key plant nutrients important to soil quality (see Table 14-1). Nutrients are depleted in soil by farming and irrigation. All plants take up nutrients from soil as they grow. By removing crops or trees from forests, or introducing grazing animals, people remove nutrients from soil. In natural settings, nutrients are recycled back to soil when plants and animals die. People's use of plants and grazing animals interrupts this cycle and causes nutrient depletion. Plowing also causes increased oxidation of organic matter in soil. This releases nutrients that can be dissolved and carried away by soil water.

Soil biology is important to soil quality. Bacteria, fungi, and the host of insects and animals that live in soil function to break down wastes and make nutrients available. Healthy and diverse biological activity may be the best indicator of general soil quality. Fertilizers and pesticides help specific crops grow, but at the same time they inevitably lower soil biodiversity and decrease its ability to process wastes.

Soil quality is decreased when soil becomes contaminated. Spilling or improper disposal of society's wastes is a common source of soil contamination. Another example is salt applied to roads during winter to guard against icy conditions. The salt dissolves in surface runoff, contaminates nearby soils, and kills vegetation.

Salt buildup in soil, **salination**, is more common and widespread in dryland areas. In these areas of low rainfall, take up and evapotranspiration of water by plants brings water with dissolved mineral salts to upper soil horizons. Evaporation of the salt-bearing water leaves behind deposits of the mineral salts in the soil and on the soil's surface. Over time, this process naturally degrades soils and makes them toxic to vegetation.

People cause salination too, especially where they irrigate in dryland areas. Water that is not used by crops, essentially left over from irrigation, evaporates, leaving behind salt deposits. This situation is made worse if the irrigation water is slightly salty to start with, but because the process is repeated over and over again during irrigation, it can have widespread degrading effects on farmland. Hundreds of thousands of acres are affected by irrigation-caused salination in the American Southwest. Ancient canal irrigation caused this problem in Mesopotamia (present day Iraq) about 2,000 years ago. One of the best examples of human-caused salination is in Egypt. After construction of the Aswan Dam in the 1960s, the annual Nile floods no longer flushed the fields of deposited salts. Irrigation became more intense, and salty ground water came closer to the surface. Now salination of farmland areas is a problem in the lower Nile River valley.

Soil Contamination

Soil contamination occurs when human-made chemicals are introduced into the natural soil environment. Major causes include the application of fertilizers and pesticides, the leaching of municipal and industrial wastes, and ruptured underground storage tanks. Common chemicals from these sources include solvents, heavy metals such as lead and cadmium, and other hazardous and potentially life-threatening substances.

One of the more widespread occurrences of soil contamination is through the application of fertilizers to agricultural lands. **Fertilizers** are chemicals that provide nutrients to help plants grow better. Farmers apply natural and chemical fertilizers to compensate for the loss of soil quality that occurs with erosion, irrigation, livestock grazing, and the harvesting of crops and trees.

Natural fertilizers include animal manure and human waste. More than 160 million metric tons of manure is generated by livestock in the United States each year. In many parts of the country this manure is added to the soil, helping to return nutrients and increase soil quality. Many countries around the world use human waste to increase soil quality. Sewage disposal plants convert their sludge into a product that can be piped or sprayed on low-quality soil. In addition to providing nutrients, natural fertilizers reduce erosion, help retain moisture in soil, and stimulate the growth of soil bacteria. Recycling of organic wastes from livestock and humans back to the farm fields also reduces the amount of waste that ends up in landfills or in streams, lakes, and rivers.

In many parts of the world, including the United States, soil depleted of nutrients is more often treated with industrially produced chemical fertilizers. These fertilizers are usually less expensive than natural fertilizers, at least over a few growing seasons. Chemical fertilizers include mixtures of nitrogen, phosphorus, and potassium. Nitrogen forms up to 30 percent by weight of organic rich topsoil. When such soil is subjected to intensive crop production or severe erosion, this nitrogen is rapidly depleted. For this reason, chemical fertilizers applied to soil are often largely nitrogen fertilizers.

The global production of chemical fertilizers has increased dramatically over the past four decades, because these fertilizers have helped farmers to increase crop yields. Since the 1940s, the United States and Canadian Midwest has become the breadbasket for much of the world with high yields of corn and other grains. Much of this increased production is the result of using fertilizers. Modern farmers all over the world are using chemical fertilizers in ever-increasing amounts.

Scientists are realizing, however, that the long-term effects from the intensive use of chemical fertilizers are cause for serious concern. Chemical fertilizers that are not taken up by plants may be washed into lakes, rivers, and coastal environments. Here, they stimulate the growth of water plants. When these plants die, they remove dissolved oxygen from the water, affecting fish and other aquatic life. Chemical fertilizers also increase soil erosion. Because they do not add complex organic matter to the soil, soil humus becomes continuously depleted. When humus is lost, soil retains less water, making the soil more susceptible to being blown and washed away. The loss of humus also changes the structure of the soil, reducing the space available for oxygen that plants' roots need. Finally, the production of chemical fertilizers requires enormous amounts of energy, mainly natural gas. During the petroleum crisis of 1973, gas prices soared and fertilizer became too expensive for farmers in many poor and developing countries. There was a significant decline in the manufacturing of fertilizers and widespread famine resulted.

Chemical fertilizers are not the only synthetic chemicals that are applied to the soil. Worldwide, over 2.5 million metric tons of chemical pesticides are used each year. **Pesticides** are chemicals that kill pests. They are sprayed on crops, orchards, swamps, pastures, forests, gardens, and lawns to control harmful insects, fungi, weeds, bacteria, birds, rodents, and others.

It is estimated that over 30 percent of agricultural crops are consumed and destroyed by pests worldwide. Pesticides reduce this damage and have helped increase crop yields. They have also helped save millions of lives worldwide by killing insects like mosquitoes, lice, fleas, and tsetse flies—all of which can carry fatal diseases.

Scientists are noting, however, that the effectiveness of chemical pesticides is decreasing. For example, many insect pests have become resistant to pesticides. Each time a field is sprayed, a small percentage of the insect population that is genetically resistant to the pesticide survives. Without having to compete with a larger population, these insects flourish. To kill these insects and their offspring, farmers must spray again at higher doses. This next application usually kills off much of the genetically resistant population, but again a small percentage of even more genetically resistant insects remain. These insects then flourish as a new and more troublesome pest. Farmers combat the newer strains with higher doses or more frequent applications. As a result, the use of pesticides rises and resistant pest strains increase. The cycle of pesticide application and generation of resistant organisms is not restricted to insects. Scientists are finding that weeds and a variety of other organisms that cause crop diseases are becoming resistant as well.

The intensive application of pesticides can severely affect soil quality. When pesticides are used for a period on the plants in an area, they eventually leach into the soil. Once in the soil, they kill microorganisms living there. These microorganisms break down organic material and aid in plant growth. Some pesticides can remain in the soil for years, keeping microorganisms from working in the soil for long periods.

Pesticides can also have other damaging effects on the environment. One serious problem is animals that are poisoned by eating foods that contain pesticide residues. Widespread application of pesticides can also eliminate food sources that certain types of animals need. Finally, poisoning from pesticides can travel up the food chain. For example, birds take in pesticides when they eat insects and worms that have consumed pesticides sprayed on crops or soil.

Another way that soil becomes contaminated is from poorly managed waste disposal practices. Septic systems are used by homes, offices, or other buildings that are not connected to a city sewer system. Septic systems are the largest source by volume of waste discharged to the land. A poorly maintained septic system can leak bacteria, viruses, household chemicals, and other substances into the ground, causing serious soil contamination problems. It is estimated that from one-third to one-half of existing systems could be operating improperly because of poor location, design, and construction.

Soil contamination also comes from municipal and industrial waste disposal. In the United States, hundreds of millions of tons of municipal and industrial solid waste are deposited in thousands of landfills each year. Today, this waste is carefully regulated and controlled. Landfills are supposed to have a protective bottom layer to prevent contaminants from entering into the surrounding soil. They are also covered with soil to minimize the penetration of rain and snowmelt into the waste. For many decades, however, dangerous municipal and industrial waste was deposited directly onto open land dumps without consideration of any long-term consequences. Rain and snowmelt trickled through the waste, sometimes carrying hazardous liquids and heavy metals into the underlying soil.

The last major source of soil contamination is from leaking underground storage tanks that contain gasoline, oil, chemicals, acids, solvents, and other types of liquids. Possibly as many as 10 million storage tanks are buried in the United States. The potential of these tanks to corrode, crack, and develop leaks increases with time. Estimates of the total number of tanks that are presently leaking substances into the ground and contaminating the soil are as high as 30 percent.

Figure 14-21 A city's development encroaches on farmland.

14.5 sustaining soils

Since agriculture's beginnings, over one-third of the almost 1,618,742,570 hectares (4 billion acres) of Earth's cropland has been degraded in some way (Figure 14-21). Since 1990, 2 to 2.4 million hectares (5 to 6 million acres) of cropland are lost each year to severe soil degradation. These are just cropland losses. They do not include soil degradation in forests and grazing lands. The sediment load of the Mississippi River is visual evidence of continued soil loss; muddy river water is soil down the drain.

New soil development is a slow process. In many places, only 2.5 centimeters (1 inch) of new topsoil can develop in a thousand years. But there is good news. People understand what causes soil degradation and how it can be prevented. Soil is a renewable Earth resource that sound management can protect and sustain.

Farming Practices

Because farming is the biggest and most direct way people use soils, sound farming practices are very important to sustaining soils. Farmers' and ranchers' "best practices" guide tilling, planting, grazing, nutrient management, and careful pesticide use. These best practices include tilling techniques that help decrease erosion and sustain soils, such as:

- Contour farming, which is tilling perpendicular to surface slopes. Plowed furrows catch soil and water rather than let it freely run off.
- Converting steeper slopes into a series of flat terraces. This has been used effectively for thousands of years to control surface runoff and erosion. They require maintenance and may not cover large areas, but they are effective at stabilizing soils and preventing erosion.
- Timing field tilling so that soils are not exposed for long intervals or during stormy periods.
- No-till cultivation techniques that use machinery to aerate soil, plant seeds, and weed fields without significant disturbance to soil.

Planting techniques that are important to stabilizing soil as well as managing its quality include:

- Planting different crops in parallel strips to ensure continuous vegetation cover.
- Planting crops that have different harvest times so that fields are never completely bare.

- Planting permanent plant barriers such as trees to provide protection from winds.
- Rotating different crops on a field through the years, which can help prevent selective nutrient depletion. Tobacco planting is a good example. Tobacco severely depletes soil nutrients, but if wheat is alternatively planted in the same field, the soil's productivity can be maintained.

Replenishing nutrients is very important to sustaining soil quality. Both natural (manure or treated sewage sludge) and manufactured fertilizers can be applied to soils to replace lost nutrients. Care must be taken to not contaminate soils with metals or, in some cases, with too many nutrients, especially nitrogen.

Pesticides can protect crops and help increase harvests, but their misuse can harm the natural soil biology or contaminate ground and surface water. Responsible pesticide use is very important to modern farming practices.

Just as farming practice choices can help sustain soil, so can grazing practices. Controlling the type of grazing animal, their quantity, and the time they are allowed to graze a specific area are all important. Grazing animals must be managed just like crops, and rotation of grazing through areas can be very helpful.

Other Actions

Other actions that help sustain soil quality include reforestation and judicious land-use choices. Reforestation can take up to a couple hundred years, but it is taking place over large parts of North America and other developed countries (see Chapter 4). Reforestation restores natural vegetation cover, soil biodiversity, water quality and retention, and nutrient cycling abilities.

Land-use choices are a major influence on soils. Controlling land use can be a very effective way of protecting and sustaining soils. For example, the Federal Agriculture Improvement and Reform Act of 1996 continued a federal government program that withdraws farmland that is at risk for soil loss. The government establishes long-term contracts with farmers to convert farmlands susceptible to extensive erosion to protective vegetation cover. Farmers are essentially paid to protect the soil. The program has reduced annual topsoil loss by 700 million tons or 19 tons per acre under contract.

How we choose to urbanize (see Chapter 8) affects soil. Urbanization either destroys or effectively removes soil from an active role in the Earth system. Over the years, about 1 million hectares (2.5 million acres) of highly productive farmland have been converted to urban uses. Have you witnessed farmland turned into subdivisions or malls in your community? It is a somewhat disturbing sight to see a mature orchard bulldozed down in order to make room for homes. It happens though; a lot of the big urban developments of California were once important farmland.

Understanding Desertification

Desertification could be a result of soil degradation in the 40 percent of Earth surface that is dryland. Today over 1 billion people live in dryland regions, but the starkest deserts, large expanses of wind-blown rock, sand, and marginal soils with little vegetation can support few animals and even fewer people. Desertification includes all the key processes that degrade soil: erosion, nutrient depletion, salination, decreased biodiversity, and decreased vegetation cover. Historically, it has been the interplay between natural stresses, especially drought, and stresses caused by people that have led to desertification. For example, in the eastern Mediterranean region, the coincidence of drought conditions with expanded agriculture, deforestation, and population growth may have contributed to the fall of the Mycenean civilization and caused people to move inland from the coasts between 4000 and 2500 B.C. But when scientists' attention first focused on desertification, people's impacts on land use were considered the key causes.

Scientists began to worry about desertification in Africa in the 1920s. The Sahara Desert was thought to be growing as a result of human mismanagement of the land. America's Dust Bowl of the 1930s increased scientific concern about desertification and its causes—clearly a result of farming practices in the case of 1930s America. The tie between people's land uses and advancing deserts has been a common theme since.

In the 1970s, severe drought and famine in Africa led to the United Nations Conference on Desertification in 1977. Here people's impacts were again considered the major cause of desertification. The responses in Africa that followed tried to limit those impacts. It was proposed that greenbelts be planted around the Sahara, grazing was diminished, and prohibitions on vegetation removal and burning were imposed. These changes created hardships for the indigenous people. They also did not consider more local land characteristics and the traditional knowledge gathered from many centuries of living on the drylands. As a result, the imposed measures have been largely ineffectual.

In 1996, another U.N. attempt was made, the Convention to Combat Desertification. Almost 160 countries have signed on and there are plans for combating desertification in most of the world's dryland regions (National Action Plans). But so far, there seems to be more talk than action.

The desertification example shows that even though causes and effects of soil degradation can become well understood, developing and implementing measures to sustain soils is complicated. So far people have gotten by with soil degradation because the hardest hit areas are poor and far away from the developed world. The developed world has bountiful soil resources, which, combined with technology to increase soil productivity, has made food available in unprecedented amounts. If people recognize soil's importance and manage their use of soil appropriately, soil resources can be sustained.

14.6 conclusion

The value of soils is difficult to overemphasize. As a membrane between Earth systems, soil provides the critical connections that make possible the material and energy transfers essential for life. The primary producers at the bottom of the terrestrial food webs are rooted in soil. In addition, humans advanced civilization due to their ability to produce food from soil. Soil is a sustainable resource if properly used, but soil resources are not being sustained around the world. Soil needs to be more appreciated and better protected.

IN THIS CHAPTER YOU HAVE LEARNED:

- Soil is a natural mixture of minerals, rock fragments, and organic material that makes up a porous layer on Earth's land surface. All Earth's systems interact in soil.

- Soils form from the physical, chemical, and biological weathering of rocks and minerals from the geosphere. Physical weathering causes rocks to break down into smaller and smaller pieces. Chemical weathering changes minerals in soil. Biological weathering aids physical and chemical weathering. Soil components are changed, moved, removed, and added to soils in various ways.

- Weathering processes typically produce soil layers called horizons. From the surface downward, the key master horizons include an organic-rich "O" horizon, a mixture of organic and mineral material in the "A" horizon, an accumulation of minerals that have moved downward in the "B" horizon, and remnants of underlying rocks are preserved in the "C" horizon.

- Soil horizons define the soil profile. Soil scientists subdivide and describe horizons in soil profiles in many ways. Soil color reflects a soil's internal changes and helps recognize the horizons that have developed. Soil texture reflects the size of mineral particles within it. Soil structure reflects how individual soil particles are aggregated together.

- Soil is important because it sustains plants, the foundation of many food webs in the biosphere. It also plays a key role in cycling of materials between Earth systems, it stores and cleans water, it helps people dispose of wastes, and it underlies most facilities that people construct. Food and fiber grown in soil feed, clothe and shelter the Earth's more than 6 billion people.

- Soil is full of life. There can be more species in a shovel-full of rich garden soil than in an entire rainforest. Organisms in soil, from bacteria to burrowing animals, play important roles in soil development, soil function, and plant sustenance.

- People affect soil in many ways. Exposure of soil can lead to soil loss by wind and water erosion. Inappropriate farming practices, overgrazing, and deforestation can cause soil erosion.

- Soil quality—its capacity to sustain plant and animal productivity, maintain and enhance water and air quality, and support human health and habitation—is degraded when people deplete its nutrients, lower its biodiversity, and contaminate it with pollutants.

- Soil is a sustainable resource. If people carefully manage their use of soils, its quality and quantity can be maintained for future use.

KEY TERMS

Chemical weathering
Clay-size particles
Fertilizers
Gullies
Horizons
Humic acids
Humus
Logging
Mollisol
Nutrients
Overgrazing
Pesticide
Physical weathering
Pores
Rills
Salination
Sand-size particles
Silt-size particles
Soil
Soil contamination
Soil horizon
Soil profile
Soil quality
Soil strength
Soil structure
Soil texture
Topsoil
Transformations
Translocations
Weathering

REVIEW QUESTIONS

1. What major components are there in soil?
2. How does soil form?
3. What are soil horizons and soil profiles? What do they tell us about soil?
4. What is the role of color in a soil horizon?
5. What is the role of soil in the biosphere?
6. What is the role of microorganisms in soil?
7. Describe three negative effects that human actions can have on soil.
8. What factors contribute to soil erosion?
9. How can soil erosion be prevented?
10. Explain how soil is a sustainable resource.

chapter 15

using earth's energy resources

As conventional energy sources such as oil and gas become more expensive, alternative sources such as solar energy can become more important. Students at Streamwood High School in Streamwood, Illinois, wondered how alternative energy sources work. They studied laboratory-size fuel cells that use hydrogen and oxygen to generate electricity and they made solar-powered stoves to cook food. Fuel cells are the subject of much research and they may someday power people's cars, but the solar cookers that the students built could be used right in the classroom. Solar energy is used for many purposes, but the sun needs to be shining for it to be available—cloudy days and night effectively turn off solar energy. Solar can provide for some future energy needs, but people need so much energy that other sources will continue to be needed.

Do you know how much energy you use and where it comes from? This chapter helps you understand the many sources of energy and the environmental concerns associated with both the production and use of energy.

Energy, like soil, water, and minerals, is an Earth resource everyone uses. As you learn about these resources, how people depend on them, and their related environmental issues, you will find that energy is like other Earth resources in some ways. But you will also come to understand that people's use of Earth's energy has some unique and special challenges. People have come to overwhelmingly depend on one source of Earth's energy—hydrocarbons, especially oil and natural gas. Hydrocarbon use has important environmental concerns; especially, its possible role in global climate change. Petroleum is a finite Earth resource that will become more and more scarce—and expensive—during your lifetime. Determining how people will adjust to smaller amounts of more costly petroleum is an important challenge you and your children will face.

AFTER COMPLETING THIS CHAPTER, YOU WILL UNDERSTAND:
Why energy is important.
What Earth's energy resources are.
What the environmental concerns associated with energy use are.
How the environment can be protected during energy production and use.
How energy conservation helps society and the environment.

 why energy is important

Humans have used energy to help them survive since they first controlled fire. As people became better at using energy, they have used it to help them prosper. Energy use is the best indicator of people's standard of living. A common definition of energy is the capacity to do work. Well, the more people use energy to do work for them, the more they prosper. The United States is the world's most striking example.

The United States consumes over 24 percent of all the energy used in the world. This is a huge amount of energy being used to do a tremendous amount of work for Americans. Is it all work? Technically yes, but the standard of living enjoyed by Americans provides for both work and play. Americans seem to expect this: to be able to travel anytime anywhere, to run every possible type of appliance or electronic gadget, and to be comfortable regardless of the weather. It is a lifestyle shared by a small part of the world's inhabitants. The availability of abundant, inexpensive energy is a key reason that the United States' standard of living leads the world.

It is difficult to comprehend how much energy people in the United States use or how much they have come to depend on it. Perhaps contrasting your use with that of a person in India or Africa who does not have access to abundant low-cost energy would help. Can you imagine living in a home without lights or appliances? Many people do. They cook their meals with wood or animal dung and simply stop doing many activities when it becomes dark at night. Perhaps a local village center has electricity and television or a telephone, but otherwise there will be no keeping up with the news, movies to watch, or calls to neighbors and friends for a chat. Can you imagine this situation?

Could you and your family commit to reducing your use of energy? If so, you have discovered ways to conserve energy. A mere 10 percent reduction in United States' annual energy consumption is equivalent to not using 1,880 million barrels of oil. At a price of over $100 a barrel, this reduction would save the United States' economy over $200 billion.

Do you think all Americans could conserve at least 10 percent of the energy they use? It would sure make a lot of money available for other purposes. A little bit of energy conservation can go a long way. However, to do anything about people's energy dependence, it first helps to understand what Earth's energy resources are.

15.2 earth's energy resources

You know what energy is, right? Sometimes you have a lot of it, sometimes you do not. Defining energy in a simple way is not so easy. The most common definition is "the capacity to do work (or produce heat or overcome resistance)." But how energy accomplishes work can be fairly complicated. Here are two important characteristics of energy:

- Energy has several forms.
- Energy can be converted from one form to another, and the total amount of energy in the universe never changes. (Scientists say that energy is *conserved.*)

The common forms of energy are **kinetic energy** (the energy of motion), and **potential energy** (the energy of position or stored energy). Kinetic energy includes heat energy produced by the motion of submicroscopic atoms and molecules; mechanical energy provided by the motion of larger physical objects such as a hammer; electrical energy provided by the movement of electrons; radiant energy that comes from the electromagnetic spectrum, including visible light and solar energy; and even sound energy that comes from the compression and expansion of spaces between the molecules that make up air.

Potential energy includes chemical energy provided by the position of electrons within atoms; gravitational energy provided by the position of objects in a gravitational field; electrostatic energy provided by the relative position of charged particles; or nuclear energy provided by the nuclei of all atoms.

How is energy converted from one form to another? In Chapter 2 you learned how energy from the sun, in the form of radiant energy, is converted into chemical (food) potential energy by the process of photosynthesis in plants. This energy conversion is the foundation of almost all Earth's numerous food chains. People's most direct use of energy, eating foods, ultimately depends on this conversion of the sun's energy. Explore some additional examples of energy conversions in the following activity.

Every time a person cooks food, heats a home, or burns gasoline in an automobile he or she is using—directly or indirectly—a conversion of chemical energy to heat energy. The chemical bonds in a fuel (wood, coal, natural gas, or petroleum) are converted to heat energy by burning. This conversion takes place in homes when a natural gas range is used to cook dinner, but it is also the principal way that electricity is generated at power plants for people's use.

There are four basic energy conversions involved in producing electricity from a fuel: (1) energy from chemical bonds in the fuel are converted to heat energy; (2) the heat energy produces steam, which has kinetic energy; (3) the steam's kinetic energy becomes mechanical energy as it turns turbines; and (4) the turbines' mechanical energy generates electricity.

Energy conversions are happening everywhere that work is being done for people. In this sense, energy, like soil and water, is essential for life. However, people use energy in many other ways. It is these other ways that make energy available to do work for people. Through these other uses, people improve their standard of living and create modern industrial society. Energy use has evolved from that necessary to sustain life to that sustaining the world's most advanced economies. People can

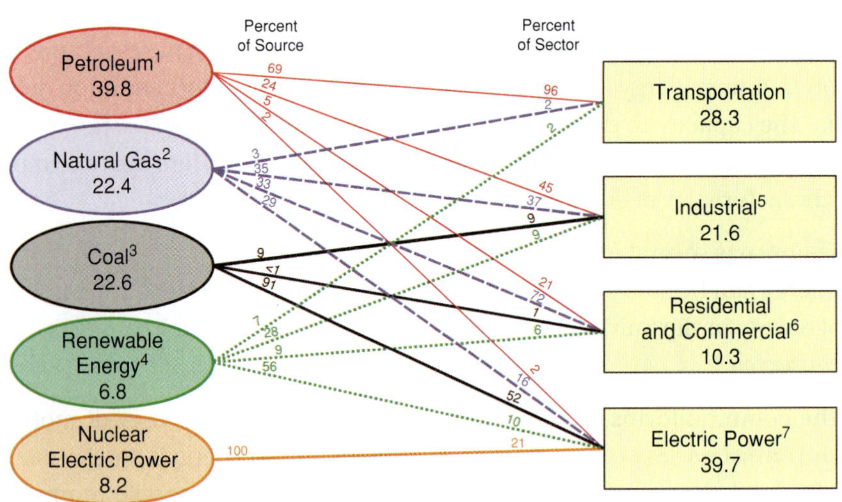

[1]Excludes 0.5 quadrillion Btu of ethanol, which is indicated in "Renewable Energy."
[2]Excludes supplemental gaseous fuels.
[3]Includes 0.1 quadrillion Btu of coal coke net imports.
[4]Conventional hydroelectric power, geothermal, solar/PV, wind, and biomass.
[5]Includes industrial combined-heat-and-power (CHP) and industrial electricity-only plants.
[6]Includes commercial combined-heat-and-power (CHP) and commercial electricity-only plants.
[7]Electricity-only and combined-heat-and-power (CHP) plants whose primary business is to sell electricity, or electricity and heat, to the public.

Note: Sum of components may not equal 100 percent due to independent rounding.

Sources: Energy Information Administration, *Annual Energy Review 2008*, Tables 1.3, 2.1b-2.1f, 10.3, and 10.4.

Courtesy DOE Annual Energy Review 2006

Figure 15-1 Energy use in the United States.

directly use the sun's energy to do work, but they mostly use Earth resources, including petroleum, coal, nuclear, and movements of the atmosphere and hydrosphere, to accomplish this. Figure 15-1 shows the role the different sources of energy play in the United States' economy. Petroleum (oil and natural gas) and coal are the principal sources of energy for the United States, but water and nuclear power generation are also important. Wind, solar, and geothermal power generation are providing small amounts of the United States' energy needs.

Petroleum

Oil and natural gas together contribute about 60 percent of the energy used each day in the United States. These are fuels containing only hydrogen and carbon and are called **hydrocarbons**. Oil is a mixture of complex hydrocarbon molecules that contain about one to two hydrogen atoms for every carbon atom. Natural gas is composed of simpler hydrocarbon molecules—the most common one, methane, contains four hydrogen atoms and one carbon atom. When hydrocarbon fuels burn, they are reacting with oxygen in the air to convert the chemical energy of the hydrocarbon bond to heat energy (Figure 15-2).

A series of geologic processes operating over millions to even hundreds of millions of years form the petroleum that people use. The formation of petroleum starts with development of plant and animal life, especially in oceans. The small plants and animals that float in oceans, called plankton, die and sink to the ocean floor. The remains of these organisms, along with other organic debris,

hydrocarbon + oxygen = > carbon dioxide + water

e.g., word equation: methane (natural gas) + oxygen = > carbon dioxide + water

and the symbol equation: $CH_{4(g)} + 2O_{2(g)} => CO_{2(g)} + 2H_2O_{(1)}$

Figure 15-2 Chemical reactions release the heat energy stored in the bonds of hydrocarbon fuels.

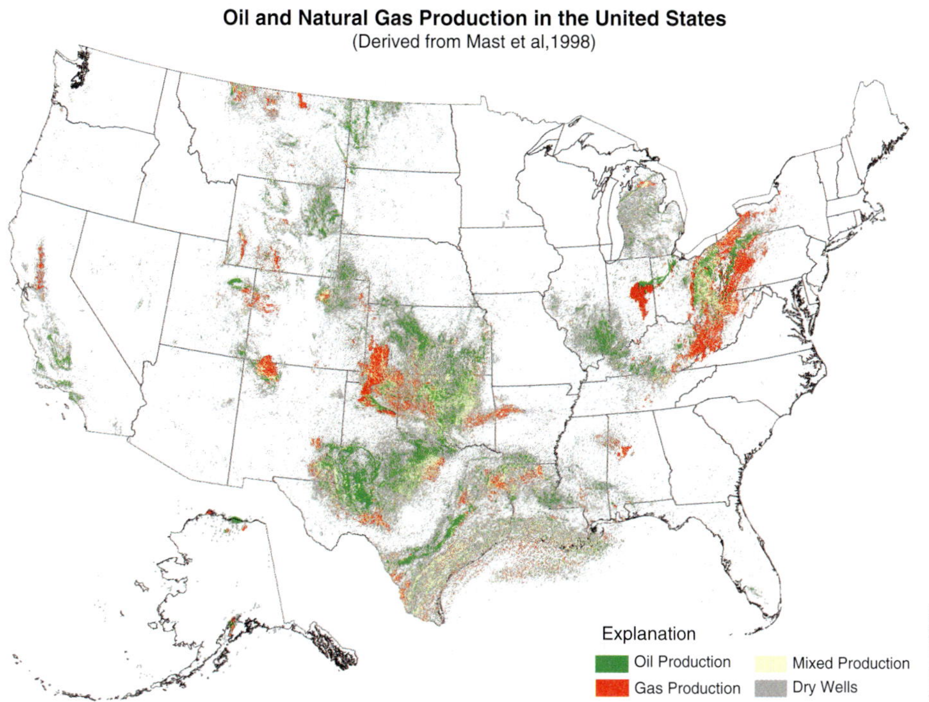

Figure 15-3 Sedimentary basins, where petroleum is generated and trapped, are widely distributed and are of various geologic ages.

accumulate and become incorporated in the ocean-bottom sediment. Eventually the organic-rich sediments are buried under other sediments and become part of the geosphere. The places where these sediments accumulate are parts of oceans or depressions in the Earth's crust called **sedimentary basins** (Figure 15-3).

Eventually the organic-rich sediment layers get buried within sedimentary basins to depths where they begin to change due to increased temperature and pressure. Higher temperatures at depth especially affect the organic matter in the sediments—it becomes converted to oil. If these sediments become even more deeply buried and heated to higher temperature, natural gas can form, either directly from organic matter or by breaking down the large carbon-rich oil molecules into small natural gas molecules. Because the material that is converted to oil or natural gas is organic, petroleum is a descendant of energy from the sun. Also, because of its organic origins, oil and natural gas (along with coal as you will see later) are known as **fossil fuels**.

Generating oil or natural gas in sedimentary rocks is just part of the process that makes them available for people's use. Next, it is necessary for the petroleum dispersed through the source sedimentary rock to migrate because of its natural buoyancy to places where it can form concentrated accumulations. These places are called **traps** because they prevent the petroleum from continuing to migrate (Figure 15-4). Without traps, the petroleum would eventually find its way to the surface, where it would form natural seeps (Figure 15-5). When petroleum migrates, it moves through cracks, fractures, and most commonly through small openings and voids between mineral grains. As you recall from Chapter 14, the volume of void space in a substance is its porosity. Rocks that have high porosity (30 percent is very high porosity) can hold large amounts of petroleum and are called reservoir rocks.

The petroleum people use comes from accumulations in traps; these are oil and gas fields. Geologists and geophysicists explore for oil and gas by using a combination of science and technology to predict the location of reservoir

Figure 15-4 Petroleum exploration has the goal of finding traps, which are geologic features that contain an accumulation of oil or natural gas.

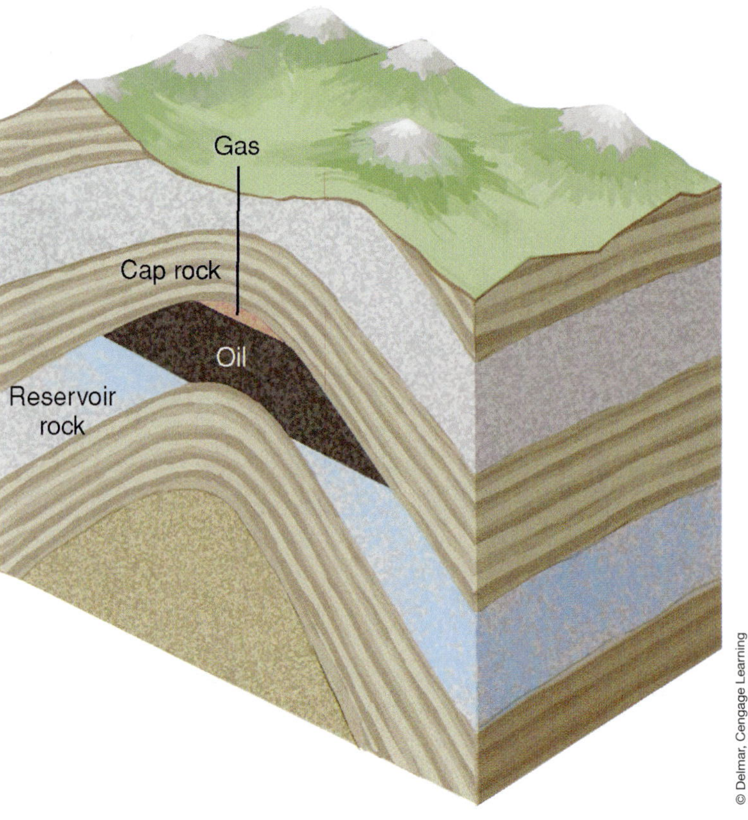

Figure 15-5 Natural oil seep.

rocks and traps in the subsurface. Although their ability to understand subsurface oil and gas fields has progressed tremendously, the real test of their predictions comes from drilling exploration wells. Even today it is common to drill 10 or more exploration wells before a new oil or gas field is discovered. Exploring for new oil and gas fields is a very challenging (and exciting) job.

Oil and gas fields that can be economically produced are not uniformly distributed around the world. The ingredients that eventually make good oil and gas fields—sedimentary basins with good organic-rich source rocks, good reservoir rocks, and traps that develop before the oil migrates—are unevenly distributed on Earth (Figure 15-6). Most of the world's economically recoverable oil and gas **(reserves)** is located in the Middle East, especially in Saudi Arabia, Iran, and Iraq. The Middle East contains 64 percent of the world's known oil reserves. This is over 10 times the oil reserves known in North America. Virtually all the developed world in North America, Europe, and Asia is dependent on Middle East oil.

Because it is not commonly found near to where it is used, extensive production, processing, and transportation systems for petroleum and its various products are needed. It is the handling and using of petroleum that develops environmental concerns. These concerns basically evolve around two issues: (1) direct release of petroleum or its products, most commonly from accidental spills, to the environment, and (2) emissions to the atmosphere of by-products produced during the burning of hydrocarbons as a fuels. You will learn more about these concerns and how they are addressed next.

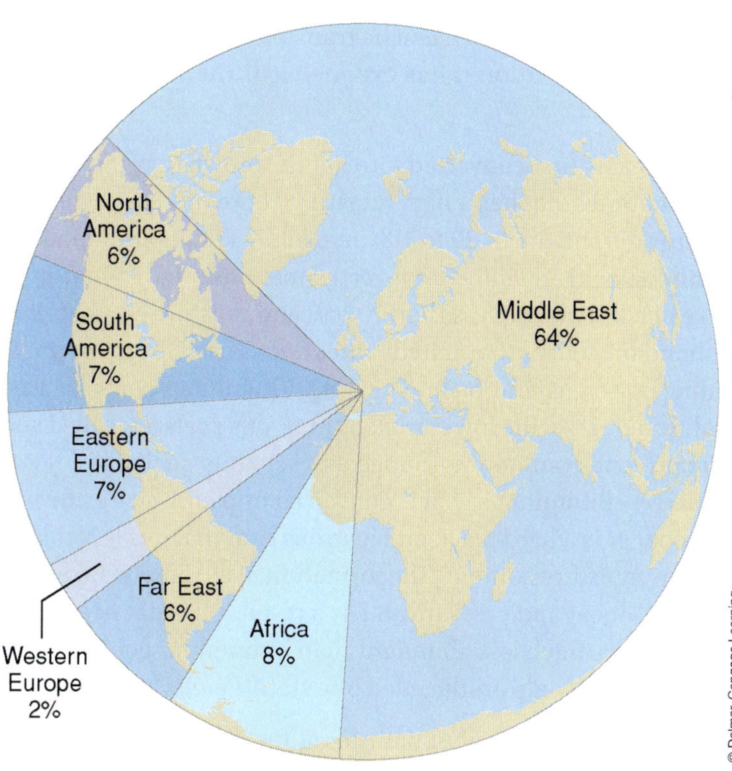

Figure 15-6 The United States, Canada, and Mexico together contain less than 6 percent of the world's estimated petroleum reserves.

Coal

Coal is a rock that burns. It burns because it is mostly composed of carbon and hydrogen. It contains other minor components such as water and minerals, but it is essentially a hydrocarbon fuel. Because this hydrogen and carbon originally came from ancient land plants, coal is a fossil fuel like petroleum.

Coal forms from **peat**, the accumulations of partly decomposed plants that grow in wet swampy or boggy areas. Peat is water-logged decaying plant debris, but if it is dried out, it can be burned. If it is turned into coal, then it can easily be burned. Coal has greater carbon content and provides more energy when it is burned than oil or natural gas. Because coal is basically recycled plants, the energy in coal was originally energy from the sun.

Peat deposits that become coal have formed at many times in Earth's history. For example, about 300 million years ago, forests full of ferns and large trees flourished in great expanses of warm and wet lowlands, especially along river floodplains and deltas. The tremendous peat deposits that formed under these forests eventually became extensive coal deposits. The geologic era during which these deposits formed is named the Carboniferous after the ancient forests and resulting coal deposits.

So how does peat turn into coal? It is a process called (you guessed it!) "**coalification**." As peat becomes buried by sediment, commonly sediment carried by nearby rivers, it gradually becomes compressed and heated. The weight of overlying sediment combined with heat from Earth's interior causes the peat to change or metamorphose into coal. The changes decrease the moisture content and increase the relative amount of carbon in coal compared to the original peat. As the process continues at higher and higher pressures and temperatures, the coal itself can be transformed. Coal is ranked according to the degree of coalification it has experienced. The three most general coal ranks are:

- **Lignite**—Peat is first converted into lignite, brown to brownish platy to massive material that has a high moisture content (up to 45 percent) and low energy output, about 20 to 100 percent less than bituminous coal.
- **Bituminous coal**—Lignite is converted into bituminous coal, a massive black coal with less than 20 percent moisture content. It is a very abundant coal in the United States and is used for many purposes, including power generation and heating. Bituminous coals are variable in quality and rank. The lower rank bituminous coals are called sub-bituminous and commonly grouped as a separate coal rank.
- **Anthracite**—Bituminous coal is converted into anthracite, the highest rank of coal. It is a hard black and lustrous coal that has less than 15 percent moisture content. The formation of anthracite takes considerable heat and pressure and as a result it has not been widely developed. It is much less abundant than bituminous coal, and in the United States it is only produced in one state, Pennsylvania.

Coal is the world's most abundant fossil fuel. The United States has more coal reserves than any other country, about 30 percent of the world's known coal reserves. Russia also has abundant coal reserves (about 16 percent of the world total) and the United States, Russia, China, Australia, and India combined have about 75 percent of all the world's coal reserves. The global distribution of coal reserves is very different from that of petroleum—the Middle East has very little coal. At the present rate of consumption, Earth's coal reserves will last at least 200 years.

Although people have used coal to heat homes for thousands of years, the current big use of coal is to generate electricity. Power plants burn coal to make the steam that runs huge turbines and electricity generators. Today, 51 percent of the electricity used in the United States comes from burning coal.

There are significant environmental concerns associated with people's use of coal. These concerns come from two main sources; (1) mining operations that produce coal, and (2) emissions to the atmosphere that accompany the burning of coal in power plants. The atmospheric emissions from power plants disperse materials and gases that have caused acid rain, global-scale pollution by such elements as mercury, and increases in carbon dioxide, a greenhouse gas, in the atmosphere. Because coal is such an abundant, low-cost source of needed energy, developing environmentally sound ways to use it is very important. As an example, mechanisms for preventing power plant emissions that foster acid rain have been widely applied in the United States. The environmental challenges of using coal are addressed more completely next.

Nuclear Energy

Why should you learn anything about nuclear energy? There are currently 103 nuclear reactors in 31 states that generate electricity. Combined, they produce 8 percent of the United States' energy needs. France produces 75 percent of its electricity with nuclear energy. Although there has not been one new nuclear power plant constructed for 25 years in the United States, it is very possible that nuclear energy will be more important to you and your children in the future.

Nuclear energy comes from the dense internal nucleus of atoms. Atoms are the fundamental building blocks of chemical elements. There are only slightly more than 100 different atoms. Each type of atom defines a unique chemical element. The atom's nucleus is made up of protons and neutrons surrounded by orbiting electrons. The number of positively charged protons equals the number of orbiting negatively charged electrons so that the atom as a whole is neutrally charged (neutrons are not electrically charged). The number of protons is unique for each type of atom, but for some the number of neutrons varies within a small range. Some atoms temporarily have too few or too many protons in their nucleus. These are unstable atoms that can spontaneously change into stable atoms. These changes occur when protons and neutrons interact or escape from the nucleus. The escape of protons and neutrons splits the nucleus in a process called **fission**. These changes in the nucleus release energy in the form of radiation and heat. For example, unstable (or radioactive) uranium atoms that have a dense proton- and neutron-rich nucleus change (decay) through several steps into stable lead atoms. Energy is released at every step of this change of uranium into lead. A pound of one type of uranium, an amount smaller than a baseball, releases the energy that 3,785 liters (1,000 gallons) of gasoline would provide as it changes into stable lead atoms. It is no wonder that uranium-fueled nuclear reactors power modern aircraft carriers and submarines.

The type of uranium that is used for fuel is called U-235 because it has 92 protons and 143 neutrons in its nucleus (92 + 143 = 235). U-235 is only about 0.7 percent of naturally occurring uranium so it needs to be concentrated, or enriched, to make good uranium fuel. Even though it needs to be concentrated, U-235 is good for fuel because it can be induced into undergoing fission. If a neutron hits a U-235 nucleus, it causes it to immediately split into two or more atoms, emit two or three new neutrons, and release lots of energy in the process. Induced fission can become a self-sustaining reaction as the newly emitted neutrons collide with nearby uranium atoms and keep the reaction going. This is the reaction that provides energy in nuclear reactors.

Nuclear reactors control induced fission reactions. Fuel rods enriched in U-235 are immersed in water. The rate of induced fission is controlled by moving rods of neutron-adsorbing material between the fuel rods. The energy released by the fission reactors heats water and generates steam. The steam, in turn, drives large turbines that spin generators to make electricity. Nuclear reactors produce electricity and radioactive wastes. There has been considerable concern about the safety of nuclear reactors, but the disposal of radioactive waste is the key environmental issue associated with the use of nuclear energy.

Energy from the Atmosphere and Hydrosphere

The kinetic energy provided by movements of air and water can be used to do work and generate electricity. Movements of air in the atmosphere—winds—are a form of kinetic energy developed when the sun variably heats the air, land, and oceans and causes differences in atmospheric temperatures. Atmospheric temperatures are the chief control on atmospheric pressure and the resulting movements of air from high to low atmospheric pressure areas.

Windmills have been used for centuries to do such work as pump water, grind crops, and saw timber. In 1888, a large windmill was constructed in Cleveland to produce electricity, but it was not until the last part of the

Figure 15-7 Wind farms produce energy. (*Courtesy Earth Science World Image Bank © Michael Collier*)

20th century that technological advances made wind-generation efficient and potentially widely useful. Although the United States only gets 0.1 percent of its needed electricity from wind, some electricity is generated from wind in 30 different states. California is by far the biggest producer of electricity from wind (Figure 15-7). Here large wind farms, areas containing hundreds of wind machines for generating electricity (there are 13,000 wind machines in California) produce more than 1 percent of California's electricity. This is about the amount of energy produced by one nuclear power plant. Many new wind plants are being developed in the United States and around the world, and California may be able to provide as much as 5 percent of its electricity by 2015. Denmark supplies 21 percent and Germany 12 percent of its electricity needs with wind-generated power.

The environmental concerns associated with **wind power** generation are: (1) wind farms need very large areas, about 0.8 hectares (2 acres) per machine; (2) some people consider wind machines unsightly additions to the landscape; and (3) in places they may be harmful to bird populations.

The use of moving water to do work goes back thousands of years to the first paddle-wheel mills that helped grind grains. The first use of moving water to produce electricity in the United States was in 1880, when some lamps in a Grand Rapids, Michigan, chair factory were run by a water-driven turbine. Today electricity generated by water-driven turbines provides 6 percent of the United States' needs. This power comes from conversion of the kinetic energy of flowing water into electrical energy. In many places, this process starts by storing water behind dams and developing gravitational potential energy. Release of the dammed water converts this gravitational potential energy into the kinetic energy of flowing and falling water. The flowing water moves through turbines that drive generators to produce electricity. The dams and their power plants can be huge, impressive facilities (Figure 15-8). This type of energy generation is commonly thought to be relatively environmentally benign, but there are some environmental concerns. As you learned in Chapter 13, dams invariably change habitat for animals and fish along rivers, they can become filled with sediment, and in some places they change ground water availability and character.

Geothermal Energy

Heat energy in the Earth can be used by people. This energy, called **geothermal energy**, comes from hot rocks in Earth's crust. The Earth is constantly producing heat at depth, primarily from the decay of natural radioactive elements. This heat gradually moves to Earth's surface (heat flow), where it is lost to the atmosphere. This is why rocks gradually get hotter with depth in the crust. In some places, rocks are extra (anomalously) hot. Volcanic activity is a major cause of anomalously hot rocks. In the United States, especially hot rocks are located in western states (Figure 15-9).

Hot rocks in the crust heat ground water and in some places make underground steam. The heated water or steam can be made to flow to the surface in wells where it is used to drive turbines and generators to produce electricity. There are now about 70 geothermal power plants operating in California, Nevada, Utah, and Hawaii. Together, they produce enough electricity to supply

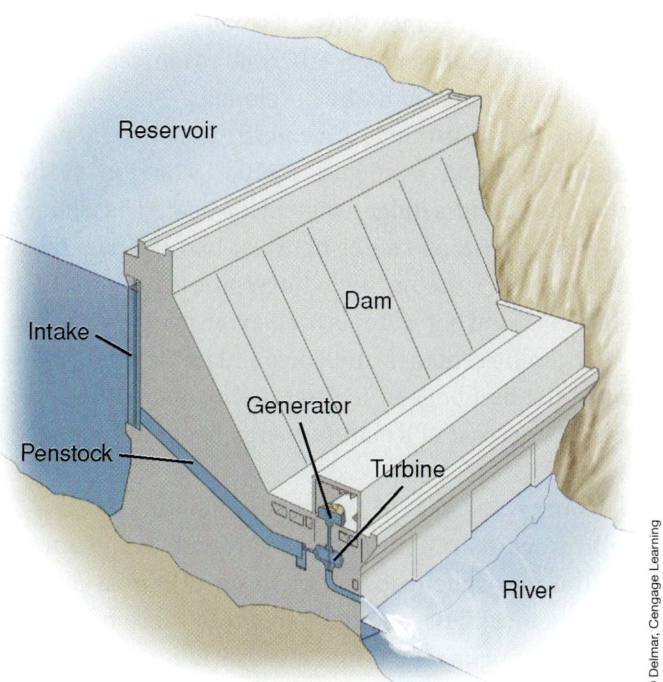

Figure 15-8 Water flows through a turbine.

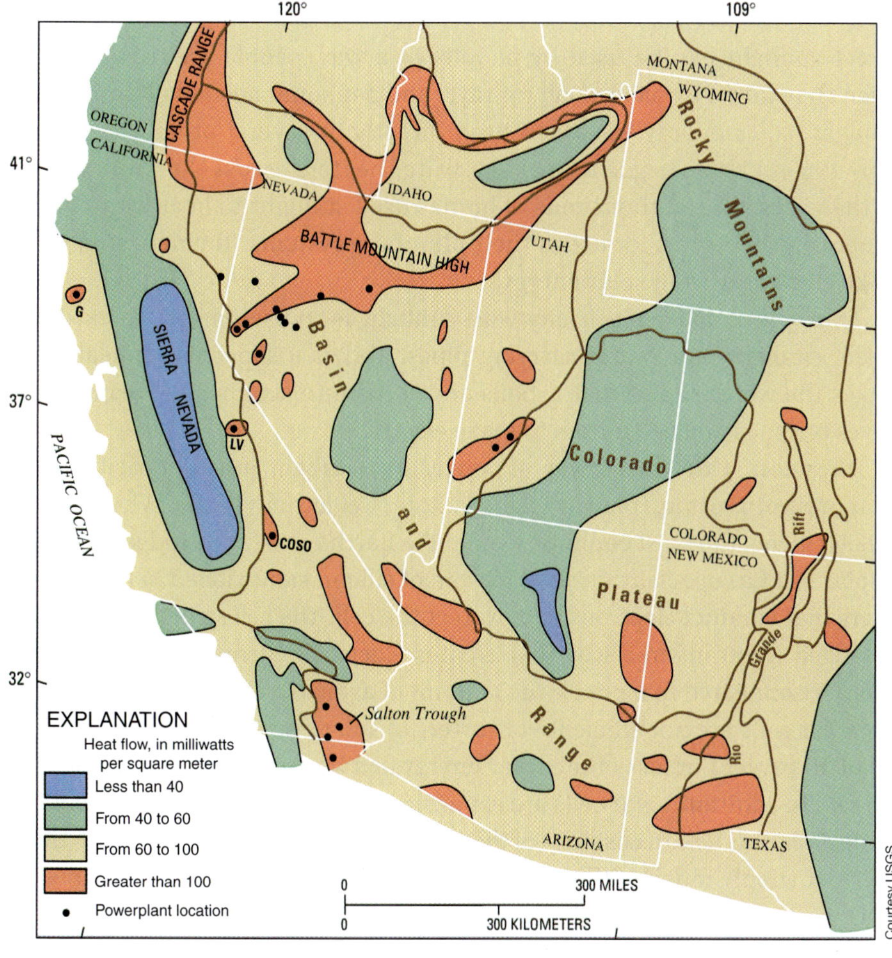

Figure 15-9 High heat flow areas in the western United States.

NPS, Photo by Ron Cooksy

Figure 15-10 Iceland power plant.

about 1.5 million homes. This is a small part of the electricity needed by the United States, but geothermal resources are widespread and additional resources may be developed.

Iceland is an example of a country that has extensively developed geothermal resources (Figure 15-10). Iceland is on the volcanically active Mid-Atlantic Ridge, where deep-seated basalt magma wells up along the spreading center between two tectonic plates. Many hot ground water systems, including some with geysers, are active in Iceland. Natural hot water is used extensively, as it has for thousands of years, to heat homes and cook and bathe with. Eighty-seven percent of the buildings in Iceland are heated with geothermal water. Overall more than 45 percent of Iceland's energy needs are supplied by geothermal resources.

Geothermal energy is less of an environmental concern than energy produced from hydrocarbons. Emissions of carbon dioxide, nitrous oxide, and sulfur dioxide to the atmosphere do accompany geothermal energy production, but the amounts are much less, about 1 percent of those released by a comparable coal-fueled power plant. The sound disposal of produced water, which can be salty or contain other dissolved minerals, has also been a concern, although now it is almost all pumped back into the ground to help maintain the heated ground water system. In some places, production of hot water or steam changes the local ground water system in ways that may not be wanted or helpful.

Solar Energy

Direct sunlight can be used by people. In a way, people are using sunlight when they lie on a beach, soak up rays, and get some needed vitamin D. But sunlight, solar energy, has been used directly in several other ways. In the early 1900s, people began using solar water heaters across the United States. By the 1920s, tens of thousands of homes used sunlight to heat water in containers on their roofs for use in their plumbing systems. The next step was to make electricity from solar energy.

One way to make electricity with sunlight is to use mirrors to catch and reflect sunlight onto water-carrying pipes. This is a super water heater that makes the water so hot that it boils and turns into steam. The steam in turn drives turbines and generators to make electricity.

The most widely used way of converting sunlight into electricity is with solar or **photovoltaic** (photo = light, voltaic = electricity) **cells**. When sunlight strikes silicon, the most common element in Earth's crust, it dislodges electrons. A solar cell has specially designed layers of silicon sandwiched between other layers that conduct electrons and protect the cell. The cell structure guides the dislodged electrons in a way that creates electrical current. Many individual cells are connected to increase the amount of available electricity.

Solar cells are not efficient converters of solar energy into electricity. Only about 15 percent or less of the solar energy can be absorbed and converted by the cell. In addition, solar cells are expensive to make. An installed system for a modest-size home can cost tens of thousands of dollars. As a result, solar energy does not supply much of the electricity needed in the United States. However, there are some very important uses of solar energy.

In remote areas where conventional electricity sources are not available, solar power can be very helpful to people. This is especially the case in less developed regions where all other available sources of energy—electricity or hydrocarbon fuels—are difficult to get and are expensive. For example, native people in the Amazon region of Ecuador have recently acquired electricity from solar cells to power communication radios and local facilities, including community movies. The solar-powered communication radio provides much needed communications for medical help and emergencies.

Another very helpful use of solar energy is in instruments. Many calculators are powered by solar cells—they seem to run forever and never need a battery change. Perhaps you have also seen solar cell panels running temporary highway signs or roadside instruments. In general, there is an important role for solar energy today and with advances in solar technology it can come to help many more people around the world.

Of all Earth's energy sources, solar has the least environmental concerns associated with its production and use. Progress is needed in making it less expensive and in overcoming the challenges of cloudy weather and the darkness of night, but it has an important role as an energy resource.

 15.3 the environmental concerns of energy use

Environmental concerns accompany both the production and consumption of energy. Energy production takes place in large industrial facilities; requires handling, processing, and transport of environmentally toxic fuels; and generates wastes that need to be safely disposed of. Some of these wastes are emitted to the atmosphere (emissions) where they can have significant environmental impacts. Energy consumption, especially the use of petroleum fuels for transportation, also releases emissions to the atmosphere.

Several steps in the production of energy are industrial operations that carry some risks of mechanical failure or accidents that can harm people or the environment. For this reason the safety of energy operations is an environmental concern, along with physical disturbances, waste disposal, and pollution.

Energy Production and Safety

The safety concerns associated with energy production are focused on petroleum, coal mining, and nuclear plant operations. Those associated with petroleum production deal with drilling, pipeline, and refinery operations. In many cases, the public and government leaders have become sufficiently concerned about these safety issues that new developments with perceived safety and related environmental risks have not been allowed. For example:

- Petroleum drilling is not allowed in many places, including offshore California, offshore Florida, and the Arctic Wildlife Refuge in Alaska.
- No new nuclear power plants have been built in 25 years.
- No major new oil refineries have been built in 25 years.

Examples of accidents and problems can be found in all industrial operations, not just energy operations. The recent safety record of the energy industry is a good one, as you will see next.

NOAA Restoration Center

Figure 15-11 Blowout at Spindletop oil field in Texas in 1902.

Petroleum Drilling

Petroleum exploration and production requires drilling wells. These wells vary from a few thousand to over 6,096 meters (20,000 feet) in depth and are drilled onshore and offshore. In offshore settings, drilling platforms or drill ships are used that are either anchored or directly connected to the seafloor. They can now drill in water over 3.2 km (2 miles) deep. In the spring of 2004, 12 Gulf of Mexico drilling rigs were operating in water depths of 1,524 meters (5,000 feet) or more—one in 2,749 meters (9,020 feet) of water. Offshore production from the Gulf of Mexico supplies a lot of the United States' oil and gas. There were 14 new oil and gas discoveries here in the 3 years between 2001 and 2003.

The principal concern associated with petroleum drilling is accidental spills caused by failure of drilling equipment. In the past, such failures were most common when drilling encountered unexpected high-pressure petroleum accumulations and oil or gas escaped uncontrolled up the well to the surface. These accidents are called **blowouts**. They have surprised drillers, created spectacular fires, and released large amounts of oil and gas to the environment (Figure 15-11). A not very spectacular but important blowout occurred offshore Santa Barbara, California, in January 1969. It released 80,000 to 100,000 barrels of oil to the marine environment. This oil eventually covered hundreds of square kilometers of ocean and 64 km (40 miles) of California coastline, including city beaches and marinas. The California Department of Fish and Game estimated that 3,600 birds died—all affected species recovered after a few years.

The cause of most of the problems at Santa Barbara was the lack of **well casing**, commonly installed under standard permit guidelines but allowed to be omitted in the blowout well by government regulators. Well casing isolates the well bore from adjacent geologic formations. When the subsurface pressures started causing oil and gas to flow to the surface (blowout) the operators worked hard to stop it. However, as they slowed the flow at the surface, the oil and gas forced its way into the adjacent geologic formations below where well casing should have been. From there it escaped upward through the formations and flowed to the seafloor.

Blowouts used to occur more frequently, especially in early petroleum drilling. They were called "gushers" when the oil squirted high into the air over the drill rig. The largest single oil release to the environment was from the 1979–80 Ixtoc I well blowout in Mexico. At least 3 million barrels of oil were released and estimates are as high as 10 million barrels. Although examples of blowouts can be found from around world, they have become much less frequent. This is because petroleum well drillers now:

- Have technology that allows them to better predict where high fluid pressures are present at depth.
- Monitor drilling conditions and quickly make adjustments needed to control flow in wells.
- Mechanically respond to increased flow by closing a series of specially designed valves that shut in the well and stop blowouts.

The valve systems that are installed on drilling wells are called "blowout preventers." When uncontrolled flow is detected by well monitors, the valves are automatically closed and fluid in the well cannot flow to the surface. They are an important safety device on all drilling wells. Today, very little petroleum is released to the environment from blowouts. In fact, all releases from offshore petroleum operations, which is mostly from discharge of impure water produced during oil and gas production (as you will learn in the waste disposal section later) and not blowouts, are less than 2 percent of the amount entering the marine environment from other sources.

Coal Mine Subsidence

The collapse of the ground into openings in underground coal mines is coal mine subsidence. This subsidence can happen suddenly and results in crater-like pits or surface sags and troughs. In places, the subsidence significantly affects property such as buildings and roads, although injuries to people are very uncommon. Mine subsidence caused a portion of the eastbound lane of a major thoroughfare—Interstate Route 70 (I-70)—in Guernsey County, Ohio, to collapse in March 1995. This subsidence event and the ensuing repair work closed all lanes of I-70 for several months, causing serious disruption to individuals and industries that used that route, and costing an estimated $3.8 million to repair.

Coal mine subsidence is most likely in older, abandoned mine settings. The subsidence is generally caused by mine designs that do not leave enough support (pillars of unmined coal) in place to prevent the overlying ground from collapsing. The nature and timing of the collapse is controlled by such factors as the size of the underground opening, the thickness and competency of the overlying material, and ground water conditions. The best remedy for coal mine subsidence is sound initial mine design. Modern mines are required to have subsidence control plans in place that fully characterize the underground conditions and define operating procedures and mine development that will prevent future subsidence. In old abandoned mine settings, where mining was not done in ways that safeguarded against subsidence, this hazard can be lessened by using maps that show the location and size of underground workings. These maps can guide surface land-use choices. For the many old mines that do not have maps of underground workings, the general areas of previous mining can be delineated to guide land-use choices. However, in many areas the land uses were decided many years ago and subsidence may unexpectedly affect surface property. In general, the public should consider old mine openings and facilities as hazardous areas that should not be casually visited or explored.

Petroleum Refineries

Petroleum refineries, where crude oil is converted into useful products such as gasoline, are large industrial complexes that operate 24 hours a day. At night, they are conspicuous brightly lit complex structures. By day, they are still large complex structures, but their most conspicuous features are billowing white clouds of steam rising above them. In 2002, there were 159 operating refineries in the United States with the capacity to process over 17 million barrels of crude oil each day. This processing requires the handling and storage of large amounts of volatile and potentially hazardous fluids and gases. The abundant pipes, connections, valves, and tanks in refineries are all possible points of mechanical failure.

Historically, refineries have been the site of tremendous explosions and fires. In 1955, refinery processing equipment, called a hydroformer, then the largest one of its kind in the world, unexpectedly exploded in Whiting, Indiana. Metal debris, some weighing many tons, flew high into the air and fell to the ground almost a kilometer (about half a mile) from the center of the blast. Fires quickly spread through the refinery and eventually engulfed 67 storage tanks in addition to the refining equipment. It took 8 days for the fires to be completely extinguished. Homes in a residential development adjacent to the refinery were damaged and one person was killed by falling debris.

The Whiting, Indiana, disaster is one of several that have occurred in places like Texas, Louisiana, and California, where many petroleum refineries are located. The good news is that every historical accident has enabled people to identify ways to change processes and facility designs that help prevent future disasters. For example:

- Residential communities are not developed immediately adjacent to refineries.
- Storage tanks are located separately from each other and other refining facilities.
- Parts of facilities are constructed with fireproof materials.
- Changes in facility designs have decreased the places where equipment is most likely to corrode, break, or inadvertently become disconnected.
- Processes and operations have become more efficient and streamlined, so that many fewer people are required to operate refineries.
- Extensive fire control systems that enable quick and effective response to accidents have been put in place.

Refinery safety is hugely important to their operators. Every accident or near accident (called incidents) are extensively investigated. Personnel are actively involved in developing safety procedures and ongoing safety training is required. In fact, refineries can have their own internal permitting requirements—work is not done that does not have an approved permit that enables all parts of the facility to know when, where, and how the work is being done.

The safety consciousness and fire control systems at refineries seem to be working, although these complex facilities continue to have accidents. There were at least six fires or explosions at refineries during the first half of 2004, but they were quickly controlled and damages and injuries were minimized by effective responses. The likelihood of a major disaster like that at Whiting, Indiana, in 1955 is significantly reduced today.

Pipeline Safety

There are about 2.2 million miles of pipelines that gather petroleum, transport it to refineries, and distribute its products such as natural gas and gasoline. These pipelines safely and economically transport millions of barrels of petroleum or petroleum products each day. Pipelines can transport gasoline from Texas to New Jersey for about three cents a gallon. Only the world's biggest marine tankers can transport petroleum more cheaply. For comparison, pipelines are much safer than truck transport of petroleum—truck transport causes 87 times more accident-related deaths, 35 times more fires and explosions, and twice as many accident-related injuries than transport in pipelines.

Even so, pipeline safety is an important issue. There were nine oil and gas pipeline accidents that caused people's deaths between 1989 and 2000. For example, 33 people died when a leaking natural gas pipeline in Puerto Rico caused a six-story building to explode. Twelve people camping by a corroded natural gas pipeline in New Mexico were killed when it exploded. Old, corroding pipes and valves; incomplete inspections and maintenance; and improper installation are causes of pipeline accidents. However, the largest single cause of pipeline accidents is digging by nonoperators where pipelines are buried. Accidental damage to buried pipelines causes accidents throughout the world.

Pipeline safety is so important that the federal government passed the Pipeline Safety Improvement Act in 2002. This act establishes rules and regulations that guide ongoing pipeline operations as well as repair and improvement of the aging national pipeline system. The Department of Transportation's Office of Pipeline Safety is responsible for ensuring that the requirements of the Pipeline Safety Improvement Act of 2002 are fully implemented.

Pipeline operators work to make pipelines safe by:

- Fitting pipelines with devices and coatings that minimize corrosion
- X-raying pipe welds to ensure their integrity
- Using monitoring and control devices that detect leaks and automatically shut down pipelines
- Examining the inside of pipelines with robotic devices (called pigs)
- Constructing pipelines so that weak, corrosion-prone parts are eliminated

Preventing accidents caused by digging requires cooperation by everyone working around pipelines. To help people know where buried pipelines are located, signs are posted to warn that they are nearby (Figure 15-12). More importantly, wherever a buried pipeline is suspected and people are digging, they should contact the pipeline operator or a special call center that is set up to help determine where buried pipelines are. A call should be made at least 72 hours in advance of digging. According to the Office of Pipeline Safety, the location information provided in this call will be forwarded to the pipeline operator and the owners of nearby underground utilities. The operator will then identify and mark the location of the pipeline in the area of the planned digging.

Nuclear Power Plant Safety

Radiation can be scary. Ultraviolet radiation from the sun is a leading cause of skin cancer and large doses of radiation from uncontrolled nuclear reactions, as in atomic bomb blasts, can kill or irreparably harm thousands upon thousands of people. On the other hand, controlled radiation is a very valuable tool. For example, in medicine, radiation provides internal images of people and helps treat cancer. As you learned in the previous section, controlled nuclear reactions are also an important source of electricity in the United States and many other countries.

Courtesy of USDOE

Figure 15-12 Signs used to identify the location of buried pipelines.

New nuclear power plants have not been constructed in the United States since 1979, when the Three Mile Island accident occurred, the most serious nuclear power plant accident in U.S. history. The Three Mile Island nuclear power plant is located near

Middletown, Pennsylvania. On March 28, 1979, a combination of equipment malfunction, design problems, and operator error led to a melting of fuel in a reactor—a "**meltdown**." Meltdowns are one of the most serious safety concerns at nuclear power plants. Meltdowns can occur if the cooling or reactor shutdown systems fail. If meltdowns are accompanied by other problems, such as steam explosions, gas generation, and failure of the containment system, then releases of radioactivity to the environment can occur. Although conditions were serious at Three Mile Island, there were no injuries to plant workers or people in surrounding communities. Radioactivity was released and about 2 million people were exposed to about one-sixth the amount of radiation they would have received from a standard chest X-ray. The Three Mile Island accident has caused major changes in how nuclear power plants are regulated, designed, and operated. It is unlikely that accidents like Three Mile Island will be repeated in the United States. Accidents like that at Chernobyl in the former USSR (now Ukraine), the worst nuclear power plant accident in history, are not thought to have been possible in the United States.

At Chernobyl, flawed reactor design, poorly trained personnel, and disregard for safety procedures led to a meltdown and explosions that blew the reactor apart. It took 9 days to put out the resulting fire. Tremendous amounts of radiation were released to the environment by this accident. Thirty-one people were killed, mostly from radiation exposure, during the accident. The radiation release contaminated large areas around the power plant and some even traveled through the atmosphere as far as Scandinavia.

Three Mile Island and Chernobyl are the two most serious accidents in the history of nuclear power plant operations. The Three Mile Island experience has guided the U.S. nuclear power industry and regulators toward much safer designs and operational procedures. Even early U.S. nuclear power plant designs were inherently safer than those in the USSR, including the failed reactor at Chernobyl. Because of its design, Chernobyl is not an analog for U.S. nuclear power plants or their safety.

Future energy needs will likely raise the possibility that new nuclear power capacity will be developed in the United States. Incentives for developing nuclear power capacity have been included in various energy legislation proposals in Congress, but new energy policies have not yet been passed. Eventually, however, nuclear power will need to be revisited as an expanding component of U.S. energy production.

Physical Disturbance

Many of the facilities that are needed to make energy available for people are large industrial complexes or other developments that change the landscape. These changes, or physical disturbances, may be considered unsightly and be unwanted by many people. In remote places, they may be developments that change natural settings that some people think should be undisturbed and preserved. Physical disturbances are one reason people have argued against new energy-related developments in many places. The most common energy-related developments that have physical disturbances of concern are oil and gas fields, petroleum refineries, surface coal mines, dams on rivers, and wind farms.

Oil and Gas Fields

Onshore development of oil and gas fields requires a few to many hundreds of wells, access roads, and pipelines. The fields can cover a few acres or very large areas up to hundreds of square miles. Drilling technology now enables oil and gas field developments that use much smaller facility areas, for drill sites and roads, for example, than ever before. The area used for a structure or device such as an oil and gas field facility is called its "**footprint**." Over the years, the footprint of these developments has steadily decreased.

Evolving drilling technology has been the key to decreasing the footprint of new oil and gas fields. Today the drill sites (pads) are at least half the size they were in 1990. A good example is the Alpine field developed in 2000 in the Alaskan Arctic. The 60 square km (22 square miles) of the Alpine oil field required a footprint of only 39.3 hectares (97 acres)—about half that needed for the same developed area at the nearby, but 19 years older, Kuparuk oil field. Technology that enabled the Alpine wells to be located only 3.1 meters (10 feet) apart on the drill pad (the wellhead spacing, originally 36.6 meters (120 feet) at Kuparuk) was very helpful. But the most important help was the ability to drill the wells horizontally at the 1,829-meter (6,000-foot) depth of the Alpine oil reservoir. Drillers can now drill down to the oil reservoir, turn the drill bit, and drill horizontally through the reservoir—even where it is only 6.1- to 18.3-meters (20- to 60-feet) thick—for distances of over 1.6 km (1 mile)! Horizontal drilling and close wellhead spacing means that fewer wells, fewer and smaller drill pads, and fewer roads are needed to develop new oil and gas fields.

Oil and gas fields have been developed in some of the largest cities and some of the remotest regions on Earth. As drilling and production technology has evolved, the ability to decrease the amount of physical disturbance, to decrease (and camouflaged if needed) the required facilities and infrastructure, and to anticipate and provide for the needs of neighbors, both people and wildlife, have been tremendously advanced (Figure 15-13). Today, developing an oil or gas field can have much less physical disturbance than a new sports coliseum, housing subdivision, or shopping mall.

Figure 15-13 A disguised urban oil field adjacent to a California marina. The tall white structure on the island conceals a drilling rig. *(Courtesy Earth Science World Image Bank © Bruce Molnia, Terra Photographics)*

Petroleum Refineries

There is no way to get around it. Petroleum refineries are a conspicuous feature on the landscape—a maze of pipes, tall columns, and other structures (Figure 15-14). They are brightly lit at night, like a mass of Christmas trees covered with white lights. The white billowing clouds rising above them are mostly condensed water vapor, but they include emissions (as you will learn later) that can pollute the atmosphere. Some emissions are also smelly. All in all, refineries are clearly a physical disturbance, but they are one that people need to deal with as long as they want to use petroleum products.

The best way of diminishing the impact of petroleum refinery disturbances is to locate them in areas where they do not conflict with other land-use choices, such as where people want to live. Refineries are commonly best located where petroleum transportation facilities,

Figure 15-14 A petroleum refinery. *(Courtesy Earth Science World Image Bank © Marcus Milling, American Geological Institute)*

ports or pipelines, are also well located. If a new refinery is proposed for an area, it is good to make sure that other land-use choices have been fully considered. Once a refinery has been built, it could be around for a long time.

Surface Coal Mines

Many large coal deposits exist at shallow depths and can be mined from the surface. Because they are most economical if the coal lies at shallow and uniform depths below the surface, they are most often developed where both the coal layer (seam) and the land surface are fairly flat lying. These surface mines first remove the overlying rock and soil (**overburden**), stack it nearby, and then start removing the coal. As the overburden is removed, mining can advance along the coal seam. The newly removed overburden is stacked in the back of the surface pit as the mining moves forward. Soil is kept separate from overburden rock so that it can be used in reclamation of the area after it is not needed for mining. This is the good news. Whereas there is considerable physical disturbance during mining, when mining stops, reclamation recontours the stacked overburden, covers it with soil, and plants new vegetation to help the disturbed area return to a natural setting. As in minerals mining, **reclamation** is the key to dealing with the physical disturbances of coal mining. Like mineral mining, coal mining eventually depletes its coal resource. Effective reclamation is a needed part of closing all surface coal mines.

Hydropower Dams

Rivers have been dammed to help produce electricity in many parts of the United States. In general, once large dams are constructed, they are not considered particularly unsightly. They even become tourist destinations and water sport recreation centers. But dams are a physical disturbance and they change both upstream and downstream landscapes and environmental conditions (see Chapter 13). Since 1999, 145 small dams have been removed from America's streams to help restore natural conditions. The most significant dam removal project so far attempted is now starting on the Elwha River in western Washington.

Two dams, one 32.9 meters (108 feet) tall and the other 64 meters (210 feet) tall, were constructed in 1911 on the Elwha River to produce electricity. The dams have prevented wild salmon and steelhead migrations on 112.7 km (70 miles) of the upper river for 93 years. In addition, the upper dam has trapped 13.8 million cubic meters (18 million cubic yards) of sediment. Removing the Elwha River dams is essentially a $182 million experiment. How the removal actions mobilize and distribute the trapped sediment has been carefully evaluated, but until the actions are actually carried out, the results will not be clear. Biologists and geologists have extensively studied the river and its present fish populations so they will be able to closely monitor the ecological changes that result from removal of the dams. The dam removal experience on the Elwha will be an excellent example for other places considering similar actions. Although removal of large dams and dam systems is not likely in light of America's need for the electricity they provide, some dams and their physical and ecological disturbances will be removed. The Elwha River removal experience will help guide other similar efforts in the future.

Wind Farms

Wind farms use rotor-driven generators perched atop tall towers. The towers are scattered across the landside about a half mile apart. A wind farm can cover very large areas but their footprint—the area needed for the towers, roads, and other required facilities—is only about 5 percent of the total wind farm area. Some of the best areas for wind farms are places like mountain passes and ridges where sustained winds are common. Such places are "well exposed" to the wind and to people's view. As a result, people are commonly concerned about the visual intrusion that wind farms make on the landscape. Because generator towers in wind farms can be over 122 meters (400 feet) high, it is hard to make them visually inconspicuous.

An example comes from Kansas where the Flint Hills rise from the western Kansas plains. They do not rise particularly high, because Kansas is a very vertically challenged state, but they are high enough to be a windy place. The hard underlying flinty bedrock has inhibited good soil development (and because of its resistance to erosion helped make the hills themselves) and farming. As a result, the Flint Hills are one of the largest areas of untilled tall grass prairie left in the United States.

A wind farm development proposed for the Flint Hills has been vigorously opposed by many residents of the area. The area is already developed by roads, homes (about 5 miles apart), and electrical transmission lines, and the wind farm "footprint" will be small, but the 35-story generator towers with blinking lights at the top will be anything but inconspicuous. Although many arguments have been raised against the Flint Hills wind farm development, the opposition is basically trying to prevent the visual impact of the wind farm on the landscape. Wherever wind farms are developed, this is the primary concern raised against them. Nonetheless, wind farms are a very environmentally sound energy source.

Waste Disposal

The disposal of waste materials produced by energy operations is a significant source of environmental concern. These wastes can be outright dangerous and toxic to plants, animals, and people. Diminishing the amount of generated wastes and recycling them is helping to overcome the related environmental concerns, but environmentally sound disposal of these wastes is still needed. Radioactive waste from nuclear reactor operations, petroleum drilling fluids, water produced during petroleum production, refinery wastes, and coal mine waste rock are the wastes that cause the most environmental concern.

Radioactive Waste

There are many sources of radioactive waste in the United States. Such wastes can be solids, liquids, or gases and vary from being barely radioactive to highly radioactive and dangerous. Any activity that uses or handles radioactive material generates radioactive waste—from workers' clothes and protective gear to spent fuel rods from nuclear reactors. Radioactive wastes are generated by mining, industrial, medical, defense, and scientific activities as well as electricity generation at nuclear power plants. Sources of radioactive waste are well known and many mechanisms are in place for safely disposing of waste from mining, industrial, medical, and scientific activities. The key approach is to physically isolate the waste in ways, commonly by burial, that ensures it will not expose people or release radioactivity to the environment. This waste will gradually lose its radioactivity as the radioactive elements in them decay, but for some elements it can take hundreds to thousands of years to decay to safe radioactive levels. The waste from nuclear reactors takes this long to lose its dangerous radioactivity. Safely disposing of radioactive waste from nuclear reactors is the biggest and most controversial environmental challenge facing nuclear power generation.

Radioactive waste from nuclear reactors includes used (spent) fuel and leftover material from reprocessing spent fuel (high-level radioactive waste or HLW). HLW is radioactive liquid created when spent fuel is dissolved to recover uranium and plutonium. In the United States, almost all (more than 99 percent) of the HLW results from the processing of spent fuel from the defense program reactors, not from commercial nuclear power reactors.

The amount of spent fuel in the United States continues to increase each year (Figure 15-15). Radioactive fuel contains uranium oxide-bearing pellets embedded in metallic rods. After most of the uranium in fuel rods has decayed, the rods are removed and stored in water-filled containment structures at the nuclear power plant site. The water serves as both a coolant and as a shield from the remnant radioactivity. These containment structures are temporary holding places for spent fuel, but it is where most spent fuel is now kept in the United States. This is because a satisfactory long-term repository for spent fuel and HLW has yet to be developed in America. As you will see, this is not for lack of trying.

Developing a spent fuel and HLW repository in the United States has become very controversial and both politically and environmentally challenging. Environmentally, the challenge comes from the need to develop a repository that can safely store radioactive waste for 10,000 years. Politically, the challenge is to select one site for all the nation's spent fuel and HLW. No one seems to want such a radioactive waste disposal site in their area, let alone

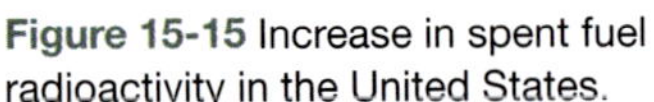

Figure 15-15 Increase in spent fuel radioactivity in the United States.

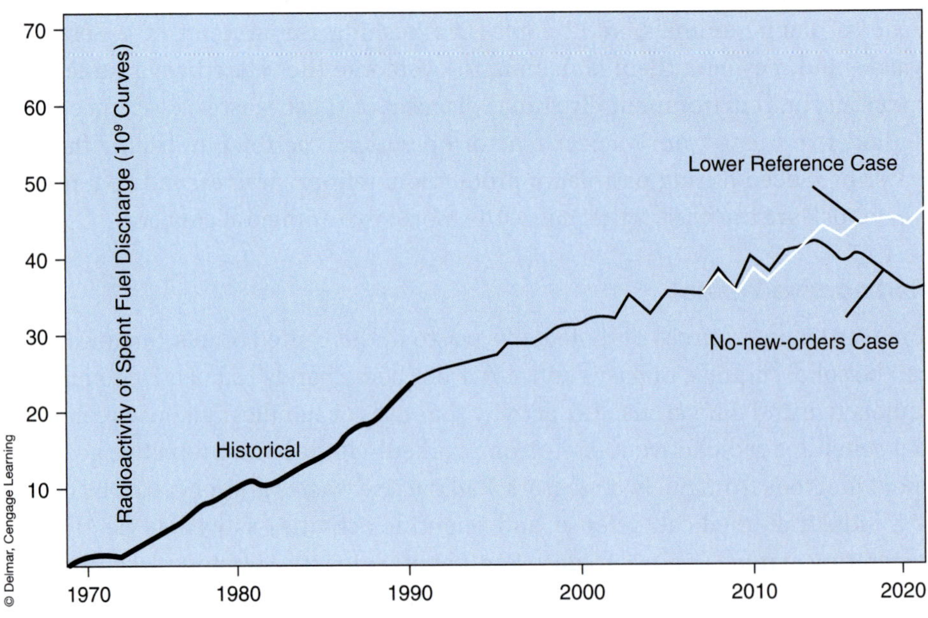

their state. However, one has been selected. The Department of Energy has selected the Yucca Mountain site in Nevada to be America's spent fuel and HWL repository.

Yucca Mountain is a **geologic repository**—a place where an underground facility will safely store radioactive waste. A geologic repository has to have large, homogeneous, and stable geologic formations; a deep ground water table; and other characteristics that make it suitable. At Yucca Mountain, a 1,829-meter (6,000-foot) thick layer of volcanic rock is the proposed host for the repository. The underground storage facilities in the volcanic rock would be about 304.8 meters (1,000 feet) below the surface and about 304.8 meters (1,000 feet) above the local water table. This site has been studied since 1979; billions of dollars have been spent in an effort to fully understand Yucca Mountain, and Department of Energy scientists have concluded that it will be a very satisfactory repository. But other scientists, people in Nevada, and people along the highways and rail lines who would be used to transport radioactive waste to Yucca Mountain have vigorously opposed its development. The controversy over developing the Yucca Mountain repository continues. While it does, America will continue to locally and "temporarily" stockpile spent fuel and HLW.

Petroleum Drilling Fluids

Finding and producing petroleum requires drilling wells. Drilling uses very hard metal drill bits, some with embedded diamonds that cut through rocks by breaking them into small pieces called **cuttings**. The drill bit is lubricated by a dense circulating fluid called **drill mud**. The drill mud flows to the bottom of the drill hole where it picks up the cuttings and carries them to the surface. At the surface, the cuttings are separated and the drill mud is circulated back down the drill hole.

Most drill mud is a mixture of water, clays, and chemicals. During drilling, it can pick up small amounts of petroleum from the rocks being drilled through. When drilling is done, both the mud and the cuttings must be disposed of. Early drilling just pumped the drill mud and cuttings into nearby surface pits or overboard into the water during offshore operations. The Environmental Protection Agency (EPA) classifies most drilling mud as nonhazardous, but if it contains oil and certain chemicals, it can cause environmental contamination. Some old disposal pits were simply abandoned when drilling was done. Today proper disposal of drilling mud and cuttings is required at all drilling sites. The more common ways that these materials are safely disposed of are:

- Mud and cuttings can be ground into a slurry, pumped down a well, and injected into selected subsurface rock formations where they will not contaminate ground water. This is an especially helpful approach in sensitive surface environments, such as wetlands, where surface pits are not feasible.
- If the drilling mud and cuttings are environmentally benign, they can be spread on nearby land and tilled into the soil or placed in surface pits, covered with soil, and revegetated.

Produced Water

Oil and gas wells commonly produce some water along with the petroleum. Initially, the amount of water is minimal, but as production continues, the amount of produced water increases. Because most petroleum wells in the

United States have been in production for many years, today's production averages six barrels of water for every barrel of oil. Produced water commonly contains salt and traces of petroleum, so it is not usable for drinking or other purposes like irrigation. Once it is separated from the accompanying oil or gas, it is necessary to properly dispose of the produced water. Because of its saltiness, it cannot be safely discharged to surface drainages.

Produced water comes from "conventional" oil and gas wells and from gas wells drilled into coal seams. Coal seams all have natural gas (methane) in them. This gas, coal bed methane (CBM), has caused unexpected underground explosions in coal mines that historically have killed many miners around the world. But in some settings, where the coal seams are not too deep, wells can be drilled that produce the gas contained in the coal. The methane-bearing coal also contains much water that is produced along with the gas. Concerns about not depleting ground water resources accompany CBM water production, and, if it is salty, it is like the water from conventional oil and gas wells and cannot be safely discharged to the surface.

There are three ways that produced water can be properly disposed of:

- If the produced water is not too salty, it can be treated and used for purposes such as irrigation.
- Produced water can be injected into deep porous rock formations where it cannot contaminate fresh ground water resources.
- Produced water can be injected back into the rock formations from which it is obtained. This is a common disposal procedure because it helps maintain pressures in the petroleum-bearing rocks and recover more oil or gas.

Produced water was not always properly disposed. Before the 1950s, it was commonly discharged to the surface. This salty water degraded surface water and soils. Today, these old degraded areas are being reclaimed by adding soil amendments, such as gypsum and organic fertilizers, and planting vegetation that tolerates the salty conditions and helps protect the soil from erosion.

Refinery Waste

The refining of petroleum removes impurities from oil and its products such as gasoline. These impurities include sulfur and metals. In addition, the refining process uses caustic chemicals, microorganisms for treating used water, and various catalysts. These materials eventually become part of the waste stream along with thick oil-water emulsions that are not easily broken down. Altogether about 3 million tons of waste are generated each year at U.S. refineries. This is about one-half ton of waste for every 1,000 barrels of oil that are refined. Refineries recycle 60 percent of these wastes back through the refining process and treat another 20 percent to make them inert. The less than 20 percent of refinery waste that remains to be disposed of (about 600,000 tons each year) is placed in landfills that are designed to prevent their release to the environment.

Another waste commonly developed at refineries, and wherever oil is stored, is a thick sludge composed of tar and sediment that sinks to the bottoms of oil storage tanks. These "tank bottoms" are collected when the storage tanks are cleaned and either added to the refining feedstock or treated to remove the petroleum components. Treated tank bottoms are spread on nearby ground or placed in repositories.

Coal Mine Waste Rock

Coal mining always includes removing at least some rock that is adjacent to or interbedded with the coal seams. This is called **waste rock** (or spoils) and it is placed back in the mine openings or stacked in piles (waste rock dumps) nearby the mine. In many cases, waste rock contains the mineral pyrite. As you will learn in Chapter 16, oxidation of pyrite is the principal cause of acid rock drainage. Acid rock drainage is an environmental concern at any mine, coal or otherwise, where pyrite is contained in waste rock. Proper disposal of waste rock in stable repositories that prevent water from infiltrating and migrating through them is very important to preventing acid rock drainage. As with metal mining, historically, coal mines have not all safely disposed of pyrite-bearing waste rock. If acid rock drainage has developed in old waste rock dumps, steps can be taken to mitigate or remediate the negative environmental consequences. The approaches for controlling acid rock drainage at coal mines are the same as those for metal mines (see Chapter 16).

Pollution

When chemicals or other materials are released to the environment and air, water, or soil is degraded, pollution has occurred. The production, transportation, and use of energy fuels can pollute the environment in several ways. Some of this pollution is accidental, and some accompanies people's consumption of energy fuels. Every person who uses energy in America contributes some pollution to the environment—even you. The energy-related pollution of most concern is caused by oil spills and emissions produced by burning fossil fuels. Oil spills are mostly accidental and degrade both soil and water. Emissions from the burning of fossil fuels accompany the use of every internal combustion engine (in lawnmowers, boats, and cars, for example) and the generation of electricity at coal-fired power plants. Emissions degrade air quality and contribute greenhouse gases to the atmosphere.

Oil Spills

Oil is released to the environment in several ways. The most publicized releases are oil spills caused by transportation accidents, especially the breakup or grounding of large marine tankers. Oil spills also accompany truck and rail accidents, releases of produced water from offshore production platforms, and from the improper disposal of used motor oil. It may surprise you, but in the United States, the improper disposal of used motor oil is the single largest source of oil pollution.

Oil in the marine environment has serious impacts on sea life and shorelines. Figure 15-16 shows the sources and amounts of petroleum released to the marine environment in North America. Natural seeps are by far the biggest source, over 62 percent, of oil pollution in the marine environment.

The consumption of petroleum leads to other large sources of marine oil pollution. River runoff that contains oil from streets and other onshore developments and atmospheric deposition of petroleum, derived from

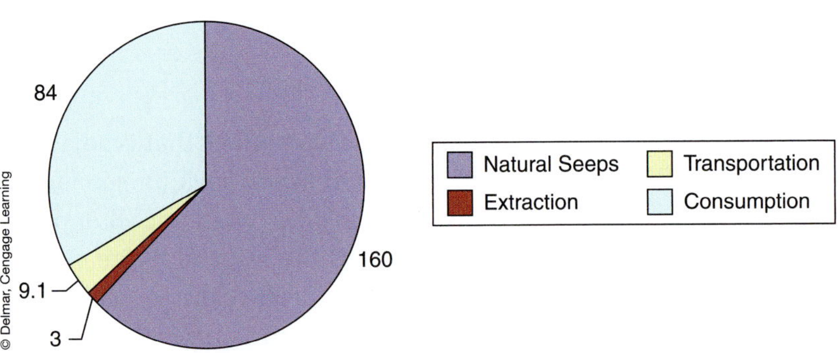

Figure 15-16 Petroleum released to the North American marine environment.

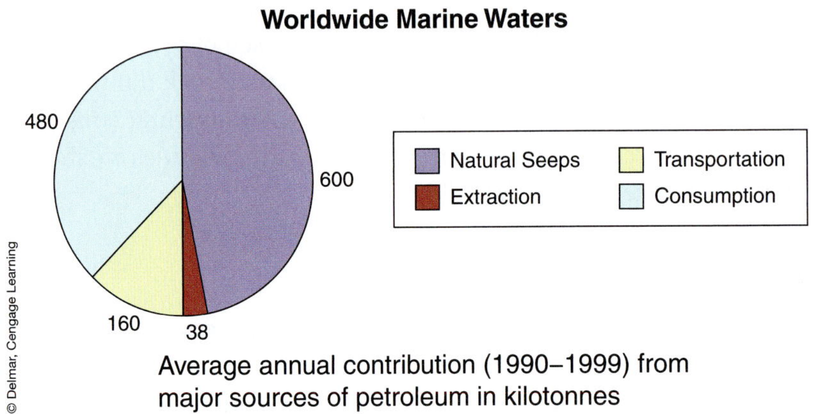

Figure 15-17 Petroleum sources in the worldwide marine environment.

volatile organic compounds (VOCs) that are emitted by internal combustion engines, power plants, and industrial operations, are the biggest sources of consumption-related oil pollution in the marine environment (Figure 15-17). Other consumption-related sources include outboard engines, aircraft fuel dumping, and discharges such as oily ballast water from ships. These many sources make the consumer the principal source of people-caused (anthropogenic) oil pollution of the marine environment. Offshore oil and gas operations, including drilling, production, and transportation (even in big tankers), cause only about 4 percent of the marine oil pollution in North American waters.

Even though the total amount of oil from natural and consumer-related sources overwhelms those from marine tanker spills, such spills are still devastating to the coastal environment. The largest in U.S. history occurred in 1989 when the oil tanker Exxon Valdez, carrying crude oil from Alaska, ran into a submerged rock soon after leaving port. The 41.6 million liters (11 million gallons) of oil released to the ocean covered large parts of Prince William Sound and its shorelines—a largely unspoiled marine ecosystem with sea otters, many types of birds, and important commercial fisheries including salmon and herring. In Prince William Sound, tremendous numbers of seabirds, perhaps thousands of sea mammals (seal and otters), and unknown numbers of shore life such as clams and snails were killed by spilled oil. In addition, remnants of the oil still remain in some places, directly or indirectly affecting wildlife.

Around the world, there have been over 50 marine oil spills larger than that of the Exxon Valdez, but none have received a more intensive cleanup effort or more thorough scientific studies to understand the environmental impact. The cleanup efforts included skimming to remove oil floating on the ocean surface, burning of floating oil, application of dispersants to help break up the floating oil, and extensive efforts to remove oil from shorelines—including some steam-cleaning techniques that had their own negative effects on shore life. Even so, the great majority of the spilled oil evaporated was dispersed in the water, or naturally degraded. These natural processes are what eventually overcome the negative impacts of all big oil spills: Nature heals itself.

What is natural degradation of spilled oil? It is partly just the evaporation of the oil that leaves behind tarry globs, but one of the most important natural processes is **biodegradation**. Oil is a mix of natural organic hydrocarbon compounds and there are many thousands of different microorganisms, like bacteria, that like to eat these compounds. Microorganisms are so effective at eating pollutants that an entire environmental cleanup industry has developed to use them.

Bioremediation is the managed application of microorganisms in the treatment of polluted soil or water. Figure 15-18 shows one bioremediation approach for cleaning ground water contaminated by diesel fuel that leaked into the ground from a storage tank. Bioremediation helps clean up thousands of polluted sites around the country. It can help clean up many different kinds of pollutants—oil, gasoline, and hazardous compounds like benzene and pesticides, for example.

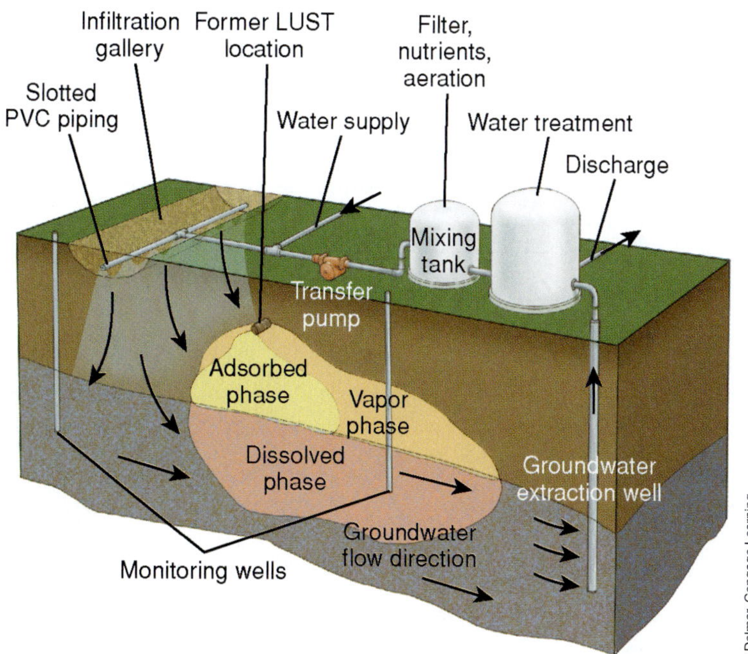

Figure 15-18 Process of in situ bioremediation of shallow contaminated ground water.

It is tough to remediate millions of gallons of oil spilled on miles and miles of shoreline. The best remedy for large accidental oil spills is to prevent them from happening in the first place. The Exxon Valdez oil spill has had a major impact on how oil is transported in North American seaways. Congress passed the Oil Pollution Act of 1990 in response to the Exxon Valdez accident. This act requires that all oil tankers in U.S. waters have double hulls. Double hulls significantly reduce the chance of major marine oil spills. Other factors that help prevent marine oil spills include:

- Tanker personnel are licensed, specially trained, and have random drug screening (like airline crews). The training includes simulations of real world emergency situations.
- Navigation technology now allows constant determination of a tanker's location within 4.6 meters (15 feet). Alarms sound if the tanker is even slightly off course.
- Tankers continuously transmit identification and location information to help avoid collisions with other ships.
- Powerful tug boats are permanently located in areas where tankers may ground if they lose power. Tug boats also routinely guide tankers through difficult to navigate waters.

Marine tankers are transporting oil more and more safely. The frequency of marine spills and the amount of oil spilled has significantly decreased (Figure 15-19).

Emissions

Petroleum refining and fossil fuel burning release pollutants to the atmosphere. People's cars and trucks are one of the biggest sources of these pollutants. The burning of gasoline and diesel fuel causes serious air pollution in many cities around the world. The burning of coal to produce electricity also emits significant amounts of pollutants. In addition, all fossil fuel burning, of gasoline, diesel fuel, jet fuel, or coal, releases the greenhouse gas carbon dioxide to the atmosphere. Controlling emissions and protecting air quality is an important environmental challenge throughout the world.

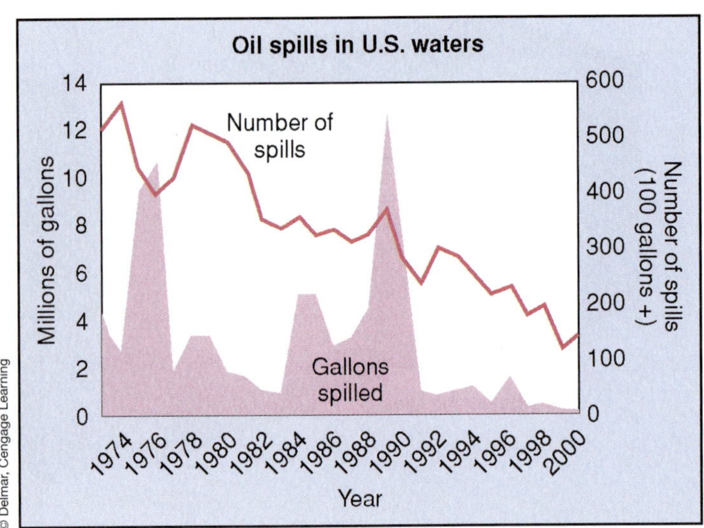

Figure 15-19 Oil Spill reductions in U.S. waters.

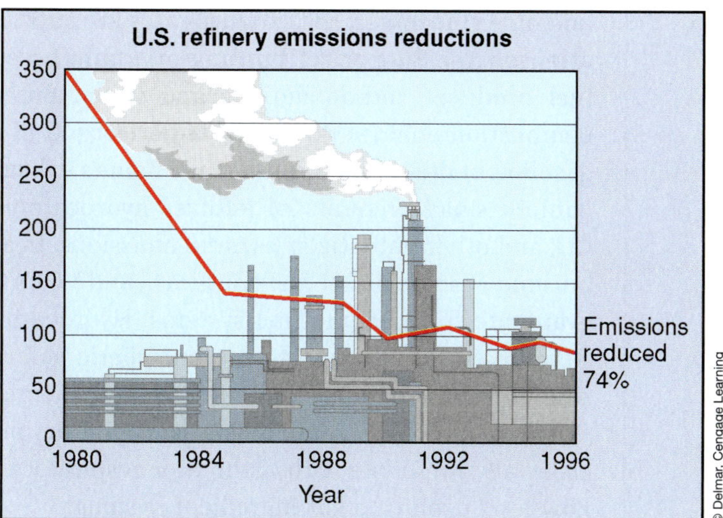

Figure 15-20 U.S. refinery emission reductions.

Refinery Emissions

Refineries emit particulate matter (a mix of tiny solid particles and liquid droplets), sulfur dioxide (SO_2), nitrous oxides (NOx), VOCs, carbon monoxide (CO), and other contaminants such as benzene. These pollutants come directly from the processing and from the burning of hydrocarbon fuels in processing equipment. Burning diesel fuel, for example, is a major source of the pollutants. Some VOCs come directly from petroleum processing, but fugitive VOC emissions are released by small leaks from valves, pipe connections, and storage facilities. Refineries produce a half pound of waste, which is 60 percent atmospheric emissions, for every barrel of processed crude oil.

The Clean Air Act authorizes the EPA to establish emission standards and regulate refinery operations that create emissions to the atmosphere (Figure 15-20). Increasingly stringent standards have required refineries to reduce their emissions of atmospheric pollutants. Since the 1980s, many billions of dollars have been spent to repair and upgrade refineries. Some of the steps that refineries have taken to reduce emissions include:

- Using processes that are more efficient and require less fuel
- Installing emission control devices such as catalytic converters and SO_2 recovery systems
- Using fewer and better valves and pipe connections to reduce fugitive VOC emissions
- Cycling excess gases back to fuel streams
- Installing vapor recovery systems on storage tanks

The effort to reduce refinery emissions continues. For example, in 2003, Chevron reached an agreement with the EPA to further reduce emissions at its refineries. Chevron is spending $275 million to repair and install new equipment that will reduce its refinery emissions by 10,000 tons each year.

Consumer Emissions

Using products produced by petroleum refineries, such as gasoline and diesel fuel, cause most of the air pollution in the United States (Figure 15-21). When these fuels are burned in the internal combustion engines of cars, trucks,

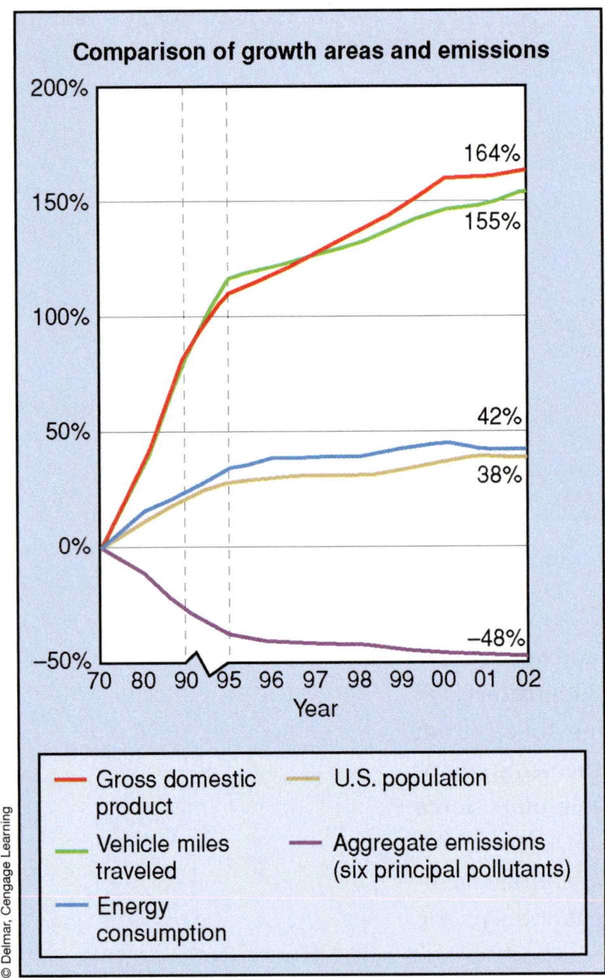

Figure 15-21 Comparison of U.S. growth areas and emissions.

and other machines, the exhaust gases are emitted to the atmosphere. The perfect burning of a pure hydrocarbon fuel produces carbon dioxide and water. But internal combustion engines do not do a perfect job of burning gasoline of diesel. As a result, the engine's exhaust gases contain small amounts of leftover hydrocarbons, NOx, CO, and other potentially harmful emissions. In addition, burning of some diesel fuel emits SO_2, and VOCs leak and evaporate from engine fuel systems. Hydrocarbons and NOx are especially important atmospheric pollutants. In sunlight, these pollutants are involved in chemical reactions that produce ozone. It causes respiratory problems, especially in people with asthma or respiratory disease. Low-level ozone is a key ingredient in smog.

Ozone and particulate matter makes smog a dense, hazy air pollution that obscures visibility (Figure 15-22). Wherever vehicles are abundant, smog can develop, but some conditions combine to make it worse. The weather can be a big factor, especially when it is sunny and conditions called a temperature inversion develop. During temperature inversions, vehicle exhaust and ozone-bearing smog can be very dense and dangerous. Increased death rates accompany serious smog events in major cities around the world. In addition the effects of intense smog are not all local. Air pollution from Los Angeles is blown east hundreds to thousands of miles—it causes the haze in the Grand Canyon! In the United States, the EPA's regulations authorized by the Clean Air Act have drastically reduced smog-forming vehicle emissions. Changes that have led to these reductions include:

Figure 15-22 Smog reduces visibility and harms health.

- Fuels have been reformulated to decrease their sulfur contents and increase their combustion efficiency.
- Catalytic converters have been added to vehicle engines that change exhaust CO, VOCs, and NOx into carbon dioxide (CO_2), nitrogen gas, and water.
- Vehicle fuel efficiency (miles per gallon) has increased.

A new car generates 1/28 of the emissions a 1960s car did. But continued reduction of emissions is needed. The EPA hopes to continue reducing NOx to levels 80 percent lower than they were in 1999. The sulfur content of diesel fuel is to be lowered by 97 percent between 2007 and 2010.

Coal Combustion

Air pollution existed before automobiles. A key source of air pollution, ever since it became a major energy fuel, is coal. A dense, foggy smog containing sulfuric acid and particulates (soot) emitted by household coal fires formed during a temperature inversion on December 5, 1952, in London. Before the air cleared days later, airports were closed, cars abandoned along roadways, and several thousand people died. London has always been known as a foggy place but the "pea-souper" of 1952 was a killer.

The replacement of coal as a household fuel by electricity and fuel oil significantly reduced air pollution in many cities. But today, the burning of coal is a major source of electricity throughout the world. As a result, emissions from burning coal continue to be an environmental concern.

In August 2004, ships collided in Hong Kong Bay because of poor visibility caused by smog attributed to special weather conditions and emissions from coal-fired power plants on the China mainland. The culprits of concern are some of the same ones produced by burning petroleum fuels—SO_2, NOx, and particulates—but because coal naturally contains more impurities and is not "refined" like petroleum, it contains other pollutants such as mercury. In addition, because the amount of coal burned by power plants is tremendous, so is the amount of emitted pollutants.

The SO_2 and NOx emissions from coal-fired power plants have been a cause of **acid rain**. These oxides react with water vapor in the atmosphere to form tiny droplets of sulfuric and nitric acid. This rain that includes these droplets is acidic and causes damage to vegetation, increased chemical weathering of some building stone (limestone and marble) and acidic streams and lakes that are harder for fish to live in.

SO_2, NOx, and particulates from coal burning can have the same smog-producing effect they do from vehicle emissions, but many coal power plants are not located where very smoggy conditions are easily developed. In addition, U.S. coal-fired power plants have been reducing their emissions. Regulations have required burning lower sulfur coal and more effective capture of SO_2 in exhaust gases. The exhaust (flue) gases are "scrubbed" by injecting calcium-bearing materials (limestone of lime) that reacts with SO_2 to precipitate solid gypsum (hydrated calcium sulfate). This gypsum dust is collected and used for making wallboard (sheetrock) and other purposes.

U.S. coal-fired power plants have also gotten better at decreasing particulates and NOx exhaust gases. The NOx reduction techniques involve preventing NOx formation in the first place by controlling the temperature and amount of oxygen available during burning, or converting it to elemental nitrogen before it is released to the atmosphere. Particulates are mostly collected from the exhaust gases by electrostatic precipitators or filters. This gets a lot of the larger particles, but some still escape.

Escape of particulate matter is the main way that other pollutants are released to the environment by coal burning. Coal naturally contains trace amounts of many elements. Fifteen of the 189 substances that the EPA has classified as hazardous air pollutants naturally occur as trace elements in coal. Of these, only mercury is considered by the EPA to be potentially hazardous to people. Mercury is an interesting element. This metal is a liquid at room temperature and it easily becomes a gas (volatilizes). This is why coal burning emits mercury. Even though present in trace amounts, the mercury that is there is easily volatilized and escapes with other gases up the smoke stack.

The mercury emitted to the atmosphere by coal power plants—over 40 tons per year in the United States—is an inorganic form that is not itself hazardous. Considering the billions of tons of coal that are burned each year this amount of mercury may not seem like much. But in the case of mercury, a little bit can go a long way. It is scattered by the winds across large parts of continents (and oceans) and settles onto landscapes. In wetlands, lakes, and streams microbial processes convert inorganic mercury into a form called **methylmercury**. Methylmercury is the most toxic form of mercury. When animals like fish eat material containing methylmercury it is not metabolized and gradually accumulates in their bodies. When people eat methylmercury-containing fish it will start accumulating in their bodies. Fetuses exposed to methylmercury because of their mothers' diets are at risk for neurological problems and, as young children, may experience learning disabilities. Because mercury pollution can be a serious problem, increased efforts to control the release of mercury to the environment are underway. The EPA's most recent plans are to lower mercury emissions from coal-fired power plants to 15 tons per year by 2018.

Greenhouse Gases

In Chapter 4 you learned about greenhouse gases and their role in influencing global climate. The greenhouse gas carbon dioxide is a major product of burning all hydrocarbon (fossil) fuels: gasoline, diesel, and coal. CO_2 is not a hazardous pollutant, but its role as a greenhouse gas makes its release to the atmosphere a significant concern.

The United States produces about 20 percent of the world's emissions of greenhouse gases—a direct result of its great dependence on fossil fuels. About one-third of these emissions come from coal-fired power plants; the rest come from internal combustion engines (especially in vehicles) and general "leakage" as when methane escapes to the atmosphere during coal mining. By far, the greenhouse gas of most concern is CO_2.

International discussions to define ways of controlling the amount of greenhouse gas emissions have been underway for many years. Because the United States is the world's single biggest CO_2 emitter, efforts here will be particularly important to adequately controlling them. Several possibilities exist for controlling CO_2 emissions, including:

- Increasing use of less carbon-rich fuels such as natural gas (methane).
- Becoming more fuel efficient, especially by using hybrid electric vehicles.
- Developing fuel cell technology that replaces hydrocarbon fuels with elemental hydrogen. The by-product produced by burning hydrogen is water.
- Capturing (sequestering) carbon dioxide in exhaust gases and storing it, perhaps in underground reservoirs or deep ocean settings where it can react with calcium to form stable compounds like calcium carbonate.
- Using captured CO_2. For example, it is very helpful in recovering petroleum from some underground reservoirs.

Controlling CO_2 emissions is an important challenge to the future of energy production. Although coal resources are sufficient to supply an increasing use of this energy fuel in the future, doing so will depend on how successfully the environment is protected from its emissions, including CO_2.

15.4 the energy future

The world's people need more energy every day. Population continues to increase and the economies of some very populous countries are expanding. For example, China became the world's second largest consumer of oil in early 2004 when its demand passed 6 million barrels per day. The history of U.S. energy use provides an example of what to expect as the world's demand for petroleum continues to grow (Figure 15-23).

In the history of U.S. energy resource use, the age of wood gave way to the age of coal, which in turn gave way to the age of oil. In general, the 1900s are considered the age of petroleum around the world. How long can the petroleum age continue? It depends on when global oil production "rolls over."

Rollovers happen when production of a finite natural resource such as oil cannot be produced in amounts needed to meet demand. Global production of oil had been predicted to peak as early as 2003 or perhaps sometime between 2010 and 2020 (Figure 15-24). Regardless of which prediction is closest, the most important point is that you will very likely live past the petroleum age.

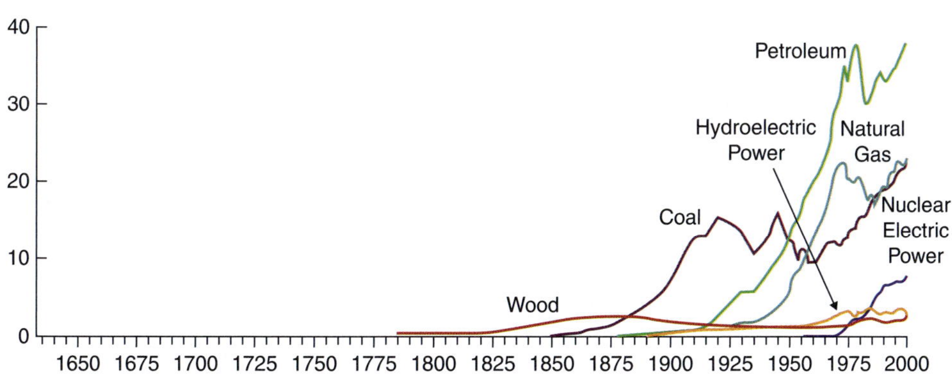

Figure 15-23 Energy sources and their contribution to U.S. energy demand from 1800.

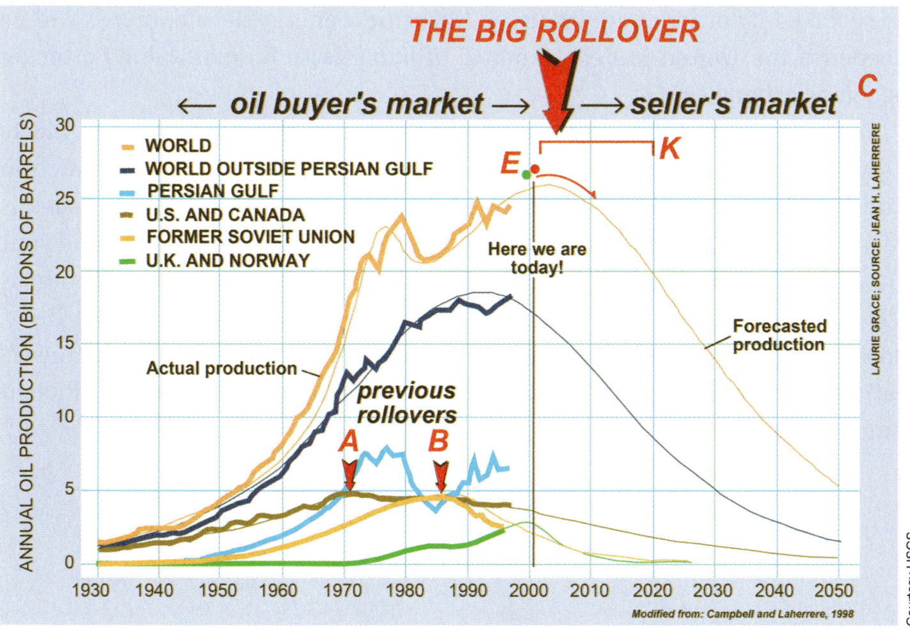

Figure 15-24 Rollover of global oil production.

Has global oil production already "rolled over?" Some people think so. The high oil prices in 2008 may at least in part reflect the inability of the world's oil producers to pump enough oil to meet demands, like China's, for this important energy resource. Increased prices for oil are to be expected after the "rollover" in global oil production. It will be a seller's market for oil producers after the "rollover."

The price of oil will be a major influence on the use of other energy resources. As oil becomes more expensive, energy providers will work to replace expensive oil with other less expensive energy sources. Everything from coal to solar may become more economically viable. Here are some of the things to expect after the "roll over" in global oil production.

- Natural gas will replace many oil uses. It will continue to be abundant and can be used in many ways. Some say the world will enter the natural gas age.
- Efforts will be made to expand the use of coal and nuclear energy to produce electricity.
- As gasoline becomes more expensive, alternative fuels and more efficient vehicles will be developed. Today's hybrid electric cars are an example. Hydrogen fuel cell technology may eventually replace internal combustion engines and the use of gasoline in vehicles. The hydrogen age may be coming along right behind the natural gas age.
- Increased efforts to conserve energy from all sources are likely.

Some of the energy alternatives in a life after the global oil production "rollover" will at least partly be determined by how well related environmental issues are addressed. The opportunity to expand natural gas and coal use will depend at least in part on satisfactorily controlling atmospheric emissions, including greenhouse gases. The increased use of nuclear energy will eventually depend on developing environmentally sound ways of disposing of spent fuel and HLW. People's responses to new energy developments—from nuclear reactors to wind farms—will be important to determining how oil will be replaced by other energy sources. These new energy developments will be needed if the United States' standard of living is to be maintained after oil becomes expensive.

How will you be involved in determining how the United States meets its energy needs in the future? A national energy policy has been debated in Congress for years. This policy, which covers the broad scope of energy issues, is very important and your congressional representatives should understand your views about them. In addition do not forget that your personal use of energy is something you make decisions about every day—what kind of car you drive, how much you drive, how long you leave lights on, and so forth. Energy conservation is important and your decisions will help determine how energy efficient America becomes.

conclusion

IN THIS CHAPTER YOU HAVE LEARNED:

- The United States consumes 24 percent of the all the energy used in the world, and per capita energy consumption in the United States is five times the world average. Energy use is the best indicator of a people's standard of living. The availability of abundant, inexpensive energy is the key reason the U.S. standard of living leads the world.

- Energy conservation is important. Reducing energy consumption by 10 percent could save the U.S. economy $75 to $100 billion per year. Even small amounts of energy conservation are significant to the U.S. economy.

- Energy has several forms and can be converted from one form to another. Physicists define energy as the capability to do work.

- Molecules that combine carbon and hydrogen—hydrocarbons—store chemical energy that is converted to heat energy when they are burned. Oil, natural gas, and coal are hydrocarbon fuels. Because they are derived from the remains of ancient life, they are also called fossil fuels. Chemical energy is converted into heat energy when a body consumes food or a stove heats a home.

- Nonrenewable sources of energy, all derived from the geosphere, are oil, natural gas, coal, and radioactive elements used for nuclear fuel. Oil and natural gas, petroleum, contribute about 60 percent of the energy used in the United States each day. Most of the rest comes from coal. The world is in the petroleum age of energy production. Over half of the petroleum used in the United States is imported from foreign countries.

- Renewable sources of energy include moving air and water, geothermal energy, and solar energy. These sources are used to generate electricity and altogether produce a little more than 7 percent of U.S. energy needs. Water flowing past dams provides 6 percent, and wind blowing through large fans produces about 1 percent of U.S. energy needs.

- There are many environmental concerns associated with energy production and use. These include safety, physical disturbances, waste disposal, and pollution concerns. Safety concerns accompany the operation of petroleum drilling rigs, petroleum refineries, nuclear power plants, underground coal mines, and the transport of petroleum products in pipelines.

- Physical disturbances accompany development of oil and gas fields, petroleum refineries, coal mines, hydropower dams, and wind farms.

- Wastes that cause environmental concerns include drilling fluids, produced water, refining waste, coal mine waste rock, and radioactive waste.

- Pollution concerns include oil spills and emissions. Emissions primarily come from petroleum refineries, coal combustion, and consumer combustion of hydrocarbon fuels. Emissions from burning of hydrocarbon fuels also release greenhouse gases that can significantly influence global climate. They remain a concern as the use of energy expands with increasing global population and growing economies.

- Environmental concerns associated with energy production and use are mitigated and prevented in many ways. Safeguards against accidents during production and transport of energy are well developed throughout the energy industry. Operational procedures, guided by regulatory standards, decrease the amount of wastes. Wastes that are generated can be disposed of in environmentally sound ways.

- When accidents and pollution occur, technologies have been developed to clean up, or remediate, the environment. A common and effective way of remediating hydrocarbon pollution is bioremediation.

- Petroleum is a finite resource that is becoming less abundant and more expensive. Other sources of energy will begin to replace petroleum in the future. People will decide how the transition from the petroleum age to the future is accomplished. Energy conservation and sound environmental stewardship will continue to be very important as the petroleum age fades away into Earth history.

KEY TERMS

Acid rain
Anthracite
Biodegradation
Bioremediation
Bituminous coal
Blowouts
Coal
Coalification
Cuttings
Drill mud
Fission
Footprint
Fossil fuels
Geologic repository
Geothermal energy
Hydrocarbons
Kinetic energy
Lignite
Meltdown
Methylmercury
Nuclear energy
Overburden
Peat
Photovoltaic cells
Potential energy
Reclamation
Reserves
Sedimentary basins
Traps
Waste rock
Well casing
Wind power

REVIEW QUESTIONS

1. Explain the difference between potential energy and kinetic energy.
2. What does "conservation of energy" mean?
3. Why are coal, oil, and natural gas called "fossil fuels"?

4. What are two advantages and two disadvantages of using fossil fuels?

5. What are necessary conditions for solar energy to be a viable energy resource in the United States?

6. Explain why Iceland depends so much on geothermal energy.

7. What are two advantages and disadvantages to using nuclear energy?

8. Choose one type of energy resource and describe its effects on the environment. Also, explain how humans are mitigating any negative effects on the environment.

9. Describe three ways in which humans could conserve their use of energy resources.

10. Discuss the role of alternative energy resources in the United States over the next 20 years.

chapter 16

mining

You probably know about recycling of paper, bottles, and cans. However, cans are mostly aluminum these days. What about materials that contain other metals from Earth's mineral resources? Students at All Saints College (actually a high school in Australia) investigated the metal content of dry cell batteries and how they could be recycled.

You know about dry cell batteries—they seem to be everywhere. They are in flashlights, of course, but also the television remote, the clock on the wall, and hundreds of other electronic devices. All Saints students collected every type of dry cell battery they could find, cut them open, and chemically analyzed their contents. They discovered that they

contained several metals, including manganese and zinc. Depending on the type of dry cell, they can also contain nickel, cadmium, lithium, mercury, and silver.

What do you do with used dry cell batteries? Just throw them in the trash? Most people do, but dry cells can be recycled. The All Saints students helped their local community set up ways to recycle these batteries, but, in general, recycling facilities for dry cells are not widely available. Most probably go into the local landfill. Dry cells are an example of how metals derived from Earth's mineral resources, although not fundamentally changed or destroyed by their use, wind up being discarded. Should we be concerned? The more metals are thrown away, the more people will need to mine mineral resources to meet their needs. If we want sustainable use of Earth's mineral resources and sound stewardship of the environment, then better recycling of metals could be important.

Mineral resources are the physical foundation of society. Everyone needs them and everyone uses them. In general, all mineral resources are mined and the environmental issues and problems that dominate concerns about mineral resources are those tied to mining and its related mineral processing operations.

AFTER COMPLETING THIS CHAPTER, YOU WILL UNDERSTAND:
What mineral resources are, how they form, and why they are important.
How mineral resources are found, mined, and processed for people's use.
The environmental impacts of mining, processing, and using mineral resources.
How the environmental impacts of using mineral resources are lessened or prevented.
Trends in mineral use and people's future needs of minerals.

16.1 mineral resources

The geosphere is the source of our mineral wealth. Many of Earth's materials, from the most common rocks to very unusual and rare minerals, can be useful. Most rocks are composed of minerals, which are naturally formed chemical elements or combinations of chemical elements that have specific chemical compositions, crystalline forms, and physical properties. The physical and chemical properties of rocks and minerals are the source of their utility and importance.

- Rocks can be very hard and durable substances that are useful by themselves, particularly for construction.
- Rocks can also have useful chemical properties such as limestone, an important source of the binding ingredient in cement, or phosphate rock, an important source of plant nutrients in many fertilizers.
- Minerals separated from rocks can also have useful chemical or physical properties. For example, gypsum is a mineral used to make wallboard sturdy and less flammable.
- Minerals are very commonly the source of useful chemical elements that are present within them. For example, most of the metals we use, such as copper, lead, zinc, nickel, and iron, are components of minerals.
- Minerals can also be a source of energy. The fossil fuels, such as petroleum and coal, are the subject of other chapters in this book, but minerals also provide uranium for nuclear reactor fuel.

Mineral resource distribution and availability varies tremendously within our country and around the world. The more common rocks and minerals such as those that make up crushed stone, sand, and gravel, are widely distributed. Most elements with useful properties, like the strength, malleability, and conductivity of heat and electricity of metals, are present in very small amounts in common rocks. However, it is necessary to have higher concentrations of these minerals for them to be recovered profitably. The geologic processes that concentrate metal-bearing minerals are uncommon and take place very locally. Therefore, the minerals that contain useful metals (Figure 16-1) are much less common, irregularly distributed, and challenging to find.

Why Are Mineral Resources Important?

Perhaps it seems unnecessary to explain the importance of mineral resources. After all, we all know that construction materials, metals, and other mineral resources are essential components in such everyday necessities as our

(a)

(b)

(c)

(d)

Figure 16-1 Metallic minerals.
((a) Courtesy Earth Science World Image Bank © StoneTrust, Inc.
(b) Courtesy Earth Science World Image Bank © United States Geological Survey
(c) Courtesy Earth Science World Image Bank © Dr. Richard Busch
(d) Courtesy Earth Science World Image Bank © Dr. Richard Busch)

homes, cars, appliances, and tools. Mineral resources are integral to the basic infrastructure of our society: transportation systems, power utilities, and food production and distribution. Indeed, we find ourselves becoming increasingly dependent on a vast array of new technologies—computer information systems and global communication networks—all of which need mineral resource components.

We have used more and more mineral resources through the years as our population has increased (Figure 16-2). Very importantly, our per capita consumption of many mineral resources has also increased. This means that we used more mineral resources than our parents and our children will use even more than ourselves. The Minerals Information Institute has estimated the amount of mineral resources that an American is expected to use in a lifetime, in effect our ultimate per capita consumption. Table 16-1 shows the physical amounts of various mineral resources that each of us is expected to use in a lifetime. It is difficult to comprehend the amount of mineral resources that people in the United States consume.

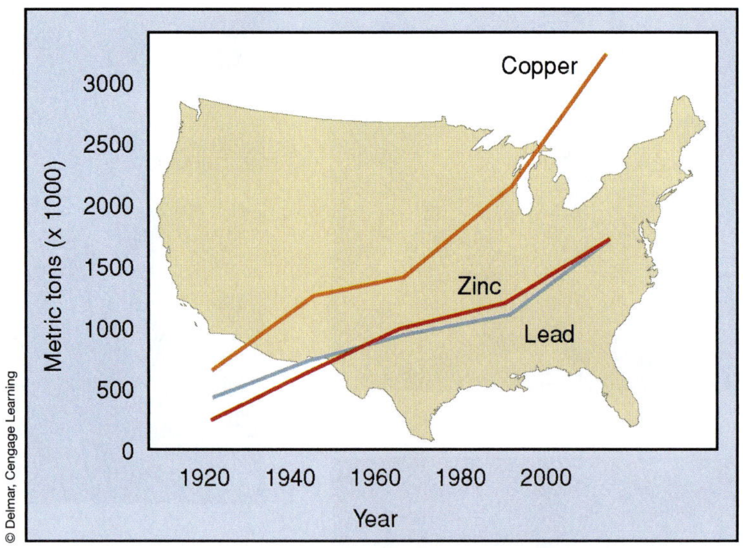

Figure 16.2 Historical consumption of copper, lead, and zinc in the United States.

Table 16-1 *Per capita lifetime consumption of minerals*

Mineral, Metal, and Fuel	Consumption
Stone, Sand, and Gravel	1.64 million lbs.
Coal	572,052 lbs.
Cement	69,789 lbs.
Iron Ore	34,045 lbs.
Salt	31,266 lbs.
Phosphate Rock	25,244 lbs.
Clays	22,388 lbs.
Bauxite (Aluminum)	6,176 lbs.
Copper	1,544 lbs.
Lead	849 lbs.
Zinc	849 lbs.
Gold	1.692 Troy oz.
Natural Gas	5.59 million cu. ft.
Petroleum	82,634 gallons
Other Minerals and Metals	> 28,564 lbs.
Total Minerals, Metals, and Fuels	3.6 million lbs.

After all, people do not go to a store and buy a few pounds of lead or zinc once in a while, yet over a person's lifetime he or she will in effect consume over 454 kg (approximately 1,000 pounds) of each of these metals (see Table 16-1). Well, maybe some people buy lead sinkers for fishing tackle, but individually, buying mineral resources is not a common experience.

The United States cannot supply all of its mineral resource needs. As shown in Figure 16-3, the United States is a net importer of 53 important mineral resource commodities; we import over 50 percent of our needs for 27 mineral resource commodities. From an environmental perspective, this high degree of reliance on imports means that the environmental consequences of providing these resources are taking place in other countries.

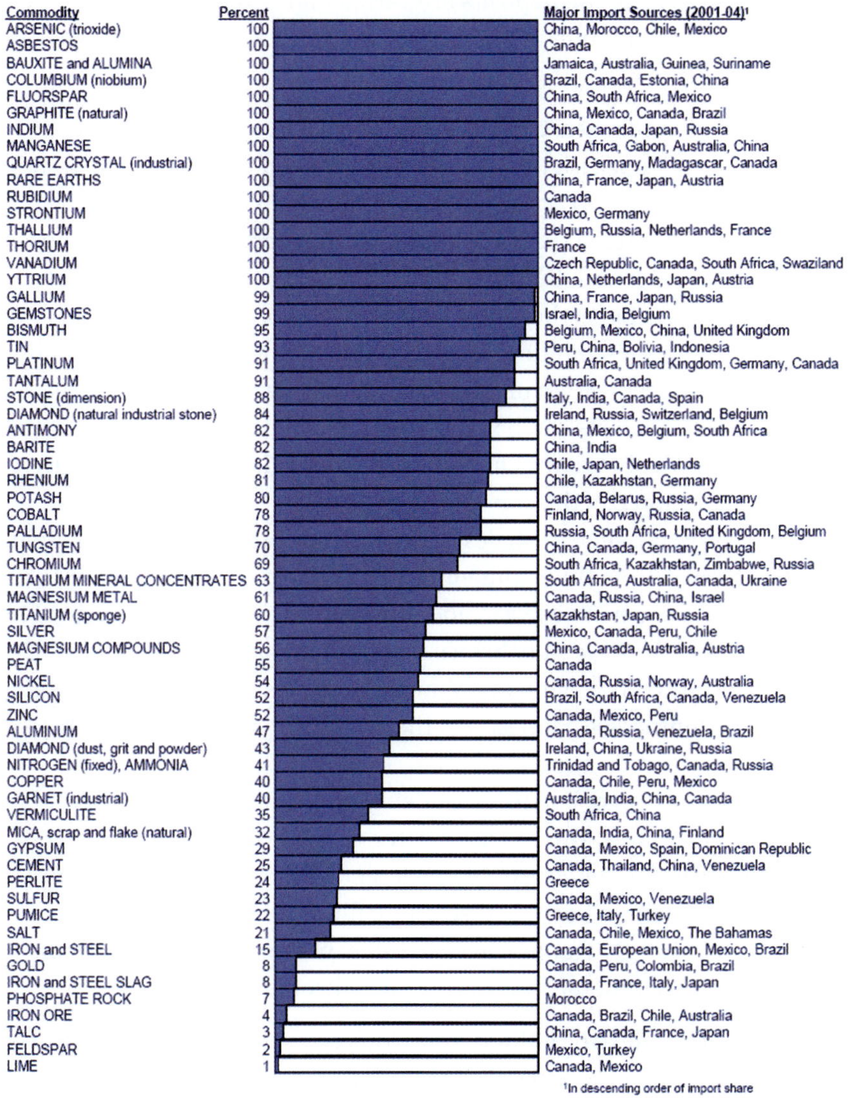

2005 U.S. NET IMPORT RELIANCE FOR SELECTED NONFUEL MINERAL MATERIALS
Commodity
Percent
Major Import Sources (2001-04)[1]
ARSENIC (trioxide) 100 China, Morocco, Chile, Mexico
ASBESTOS 100 Canada
BAUXITE and ALUMINA 100 Jamaica, Australia, Guinea, Suriname
COLUMBIUM (niobium) 100 Brazil, Canada, Estonia, China
FLUORSPAR 100 China, South Africa, Mexico
GRAPHITE (natural) 100 China, Mexico, Canada, Brazil
INDIUM 100 China, Canada, Japan, Russia
MANGANESE 100 South Africa, Gabon, Australia, China
QUARTZ CRYSTAL (industrial) 100 Brazil, Germany, Madagascar, Canada
RARE EARTHS 100 China, France, Japan, Austria
RUBIDIUM 100 Canada
STRONTIUM 100 Mexico, Germany
THALLIUM 100 Belgium, Russia, Netherlands, France
THORIUM 100 France
VANADIUM 100 Czech Republic, Canada, South Africa, Swaziland
YTTRIUM 100 China, Netherlands, Japan, Austria
GALLIUM 99 China, France, Japan, Russia
GEMSTONES 99 Israel, India, Belgium
BISMUTH 95 Belgium, Mexico, China, United Kingdom
TIN 93 Peru, China, Bolivia, Indonesia
PLATINUM 91 South Africa, United Kingdom, Germany, Canada
TANTALUM 91 Australia, Canada
STONE (dimension) 88 Italy, India, Canada, Spain
DIAMOND (natural industrial stone) 84 Ireland, Russia, Switzerland, Belgium
ANTIMONY 82 China, Mexico, Belgium, South Africa
BARITE 82 China, India
IODINE 82 Chile, Japan, Netherlands
RHENIUM 81 Chile, Kazakhstan, Germany
POTASH 80 Canada, Belarus, Russia, Germany
COBALT 78 Finland, Norway, Russia, Canada
PALLADIUM 78 Russia, South Africa, United Kingdom, Belgium
TUNGSTEN 70 China, Canada, Germany, Portugal
CHROMIUM 69 South Africa, Kazakhstan, Zimbabwe, Russia
TITANIUM MINERAL CONCENTRATES 63 South Africa, Australia, Canada, Ukraine
MAGNESIUM METAL 61 Canada, Russia, China, Israel
TITANIUM (sponge) 60 Kazakhstan, Japan, Russia
SILVER 57 Mexico, Canada, Peru, Chile
MAGNESIUM COMPOUNDS 56 China, Canada, Australia, Austria
PEAT 55 Canada
NICKEL 54 Canada, Russia, Norway, Australia
SILICON 52 Brazil, South Africa, Canada, Venezuela
ZINC 52 Canada, Mexico, Peru
ALUMINUM 47 Canada, Russia, Venezuela, Brazil
DIAMOND (dust, grit and powder) 43 Ireland, China, Ukraine, Russia
NITROGEN (fixed), AMMONIA 41 Trinidad and Tobago, Canada, Russia
COPPER 40 Canada, Chile, Peru, Mexico
GARNET (industrial) 40 Australia, India, China, Canada
VERMICULITE 35 South Africa, China
MICA, scrap and flake (natural) 32 Canada, India, China, Finland
GYPSUM 29 Canada, Mexico, Spain, Dominican Republic
CEMENT 25 Canada, Thailand, China, Venezuela
PERLITE 24 Greece
SULFUR 23 Canada, Mexico, Venezuela
PUMICE 22 Greece, Italy, Turkey
SALT 21 Canada, Chile, Mexico, The Bahamas
IRON and STEEL 15 Canada, European Union, Mexico, Brazil
GOLD 8 Canada, Peru, Colombia, Brazil
IRON and STEEL SLAG 8 Canada, France, Italy, Japan
PHOSPHATE ROCK 7 Morocco
IRON ORE 4 Canada, Brazil, Chile, Australia
TALC 3 China, Canada, France, Japan
FELDSPAR 2 Mexico, Turkey
LIME 1 Canada, Mexico
[1]In descending order of import share

Courtesy U.S. Geological Survey, Mineral Commodity Summaries

Figure 16-3 United States mineral resource import reliance.

People around the world aspire to a better standard of living and they will become bigger consumers of mineral resources as they come to achieve this. In general, the developed countries are using many times the amounts of mineral resources than less developed countries use. This differential consumption, combined with the uneven distribution of resources, is a significant influence on many global issues, including environmental ones.

Where Are Mineral Resources?

Mineral resources are where you find them—an apparently simple but exceedingly important characteristic. In this section, we will come to understand why this characteristic develops and why it is important to economic, social, and environmental issues that accompany the mining of mineral resources.

Mineral resources are produced by processes operating in the Earth. These processes mobilize useful elements such as metals from rocks in which they have low concentrations, and deposit them in places where they have higher concentrations. If these mineral deposits have high enough concentrations of the useful elements to allow profitable mining, they are called **ore deposits**. The **ore** in a deposit is that part that can be mined profitably.

Processes that form ore deposits are varied and complex. Many ore deposits form from the interaction of parts of the geosphere and hydrosphere. Some examples are:

- Heated seawater migrates through rocks below the ocean floor, dissolves metals such as copper, lead, or zinc, then emerges on the ocean floor in springs where the metals combine with sulfur and precipitate in sulfide minerals.
- Molten rock or magma that is emplaced within certain stratovolcanoes releases hot, water-rich fluids when it crystallizes that contain metals and other elements. As the metal-bearing fluid cools, it precipitates minerals, commonly sulfide minerals, which contain metals, especially copper. The concentration of copper is commonly less than 1 percent, but because these deposits can be huge they are a major source of this metal (Figure 16-4). The Bingham Canyon copper deposit near Salt Lake City, Utah, is an example of this type of ore deposit.
- Marine sediments such as shales and sandstones that fill basins contain trapped seawater. As the sediments become thicker, the trapped water is squeezed into permeable layers and it migrates toward the margins of the basin. These waters are salty brines that contain dissolved metals such as lead or zinc and other elements. Where the brines emerge on the seafloor or come in contact with limestone, the contained metals commonly precipitate in sulfide minerals.
- Rocks in deep parts of Earth's crust dehydrate when they become very hot and recrystallize. This releases water-rich fluids that contain metals such as gold and other elements, particularly silicon, which is the most abundant element in Earth's crust. These fluids migrate to shallower crustal levels and get trapped in structures where the contained elements precipitate with minerals such as quartz (silicon oxide).

These examples illustrate the tremendous diversity of Earth's ore-forming processes. The geologic settings that foster ore-forming processes are special and unusual circumstances that are not uniformly distributed around the world. Some areas will have experienced these processes and be relatively well endowed with mineral resources. An example is Arizona, where conditions were right about 70 million years ago to form many large copper deposits, but these types of deposits are lacking in nearby California (see Figure 16-4).

Figure 16-4 Bingham Canyon Mine near Salt Lake City, Utah. *(Courtesy Earth Science World Image Bank © Bruce Molnia, Terra Photographics)*

In general, regions that have undergone more complicated geologic histories with multiple episodes of deformation, mountain building, sedimentation, and magmatism are more likely to have experienced the processes that can form ore deposits.

The physical evolution of Earth has led to a varied, uneven history in which different ore-forming processes were active in different places at different times. The major result is that ore deposits are widespread but rare compared to other geologic features and really are "where you find them" (Figure 16-5). They are in such diverse environmental and geographic settings as high, mountainous rain forest in Indonesia, arid

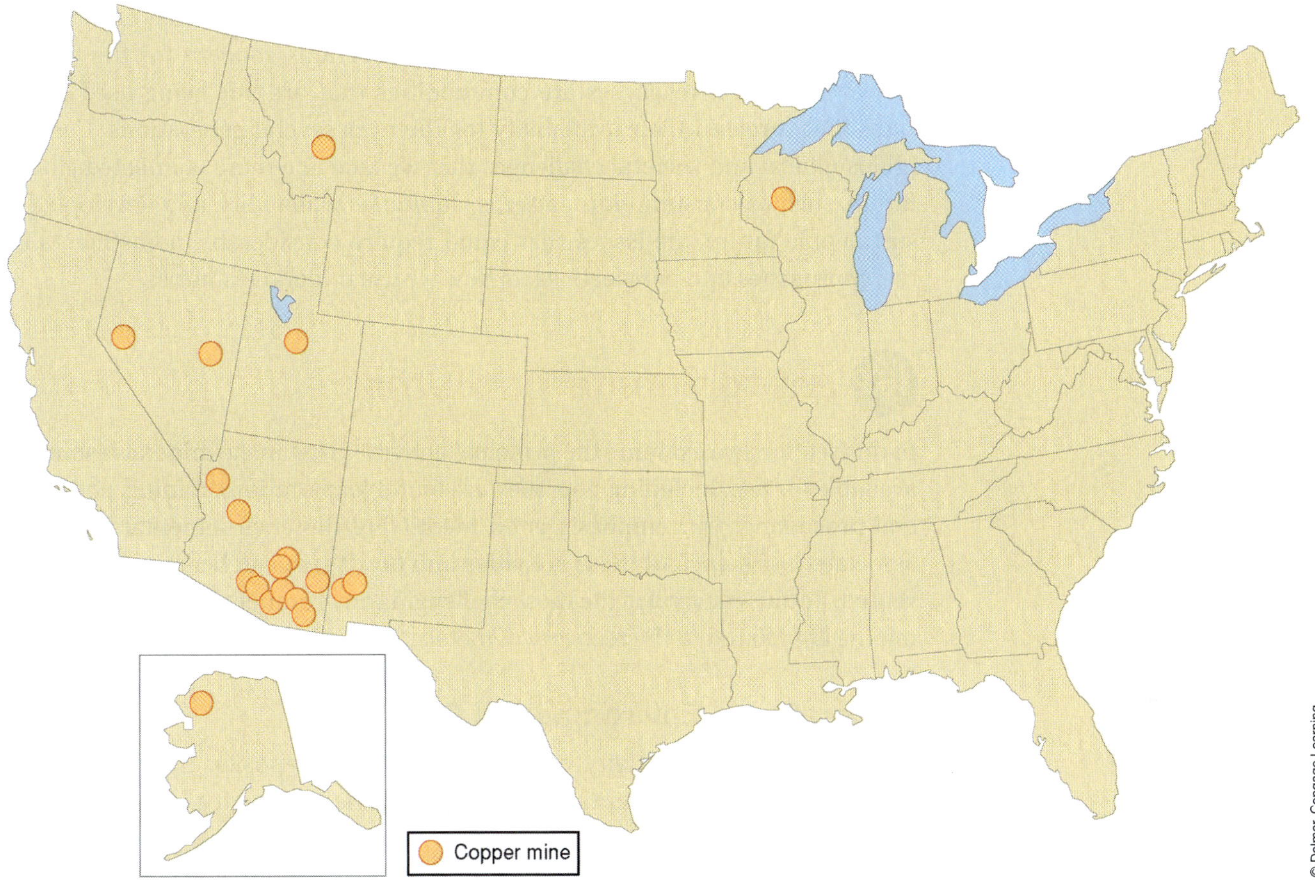

Figure 16-5 Arizona has many copper mines.

desert in Arizona, and the treeless Arctic tundra in Canada. The settings where ore deposits occur can play a significant role in determining the nature and the extent of environmental concerns at specific mine sites. For example, mining a type of ore deposit in the deserts of Arizona has different potential impacts on surface water and ground water quality than if the same mining occurs in areas of temperate climate, such as the Midwest.

A major implication of the uneven distribution of mineral resources is the uneven access to them and, therefore, to the economic, social, and environmental consequences of their mining. This is a key source of some challenging dilemmas for those people with mineral resource endowments and those who are the ultimate users of these resources. Mining is an industry that provides taxes, jobs, useful infrastructure, and other benefits to the community and country where it occurs. However, it can create cultural challenges in places that have not experienced previous industrial development, and it always presents certain environmental consequences, many of which have negative, longer-term implications if they are not satisfactorily addressed during mining.

Mineral resources are a finite resource. Geologic processes that form ore deposits do so over hundreds of thousands to millions of years of time. They are forming today in certain places but not at rates that can keep up with the need for mineral resources. Considerable discussion has centered on whether the finite character of mineral resources is a serious concern for the future. However, mineral resources are commodities that are not being used up at rates that threaten their availability for the next several generations. Certain philosophical and societal challenges that we face in a more connected global future, such as consumption patterns, equitable availability, and developing a sustainable future, are issues that could require considerable adjustments in our approaches and perspectives on how we use mineral resources.

16.2 environmental concerns

In this section, we examine the principal activities that make mineral resources available for use, including how they are found (exploration), mining, and mineral processing. Our emphasis is on identifying the environmental concerns associated with each of these activities and how these can be lessened or prevented. Because several of the most challenging environmental consequences of mining are related to the recovery of metals, this is an emphasis in this section.

Exploration's Environmental Concerns

Exploration is the activity of searching for ore deposits. Although many mineral deposits (**prospects**) have been identified through exploration, only a few are ore deposits with the size and mineral concentration sufficient to be mined at a profit. Mineral exploration is a challenging exercise that takes geologists to remote regions throughout the world and requires a variety of scientific and technical skills. Exploration geologists need exceptional perseverance, for they may examine dozens and dozens of prospects without finding one new ore deposit. On a worldwide scale, however, geologists find a few new ore deposits each year.

The exploration process begins with a geologist examining satellite images, geologic maps, and reports to identify areas favorable for mineral deposits. Once these areas are defined, the geologist conducts on-site field examinations to create more detailed maps and rock descriptions. Geologists commonly augment their field examinations with geochemical or geophysical exploration techniques that help them identify specific types of mineral deposits. **Geochemical techniques** are used to analyze samples of rocks, soils, water, vegetation, or stream sediments that may contain elements that are clues to the nearby presence of mineral deposits. **Geophysical techniques**, such as magnetic surveys, can help characterize rocks beneath the surface. Because these activities do not significantly alter or disturb the local area, little environmental concern is associated with this stage of exploration.

Once initial exploration has identified a mineral deposit of interest, more detailed efforts are needed to determine if an ore deposit is present. These continued exploration efforts most commonly involve evaluation of the subsurface character of the deposit by trenching and drilling. Trenching, drilling, and infrastructure such as field camps, trails, or leveled drill sites (pads) supporting these activities are the principal sources of environmental concerns during exploration.

Trenching

Trenching, with backhoes or other heavy equipment, exposes the near surface parts of a prospect for direct observation and sampling. Individual trenches can be up to hundreds of meters long and a few meters deep.

The principal environmental concern associated with trenching is physical disturbance of the landscape; these, in effect, are big ditches in the ground. If they are left as open ditches after exploration is completed, some may become hazards; in Alaska, open trenches could conceivably become moose traps! The good news is that trenches can be easily backfilled and the area returned to its original slopes and contours, or reclaimed. In time, the replaced materials can be expected to have a surface character similar to that before trenching took place.

Figure 16-6 Helicopters help explore for minerals by conducting aerial surveys and transporting people and equipment in remote areas. (*Courtesy Earth Science World Image Bank © United States Geological Survey*)

Drilling

Drilling uses machines that can grind their way downward and recover chips or cores of subsurface rock. A drill hole 1,000 meters (3,280 feet) deep would be a very deep example. Drilling machines (**rigs**) are portable pieces of equipment that can be transported by truck, airplane, or helicopter (Figure 16-6).

The rigs are run by gasoline or diesel engines and use air or water as their chief lubricant as they grind into the subsurface rock to retrieve chips or cores. The chief environmental concern with drilling is physical disturbance of the landscape. Drill sites need to be level places, but in most cases they do not have to be large. The number of drill sites may be a few to a few hundred depending on the nature of the mineral deposit that is being explored. The physical disturbance at the drill site is minimal and the primary need is to not spill fuel or lubricants and to remove miscellaneous equipment and debris after drilling operations cease. The drilling itself does not produce significant environmental consequences. In some places, roads or temporary trails may be constructed to enable drill rigs to be moved to drill sites with trucks or tractors. Roads used to access drill sites are generally the most common leftover, observable impact after exploration is completed. Even these, though, can be reclaimed.

Infrastructure in Support of Exploration

Roads or trails in support of drilling are examples of what is most commonly temporary infrastructure at exploration projects. Other examples are camps to house and feed the exploration team; fuel depots; work facilities for sample description, storage, and preparation; airstrips; local water systems; and waste management systems. In general, this infrastructure can be the most significant impact of exploration.

The exploration process—from initial office compilation to extensive drilling—is expensive and time consuming. It may take years of work and tens of millions of dollars to reach a development decision about a specific mineral deposit. The success rate is not high; hundreds of prospects may be evaluated before a new ore deposit is discovered. This means that infrastructure in support of exploration is most commonly temporary. Exploration projects need to guard against negative environmental impacts during operations, such as fuel spills and harmful waste management practices, but they also need to anticipate reclaiming physical disturbances upon project completion. Sound stewardship in these cases, essentially good housekeeping practices, is probably close to what most of us would expect if these activities were conducted on our own land.

To ensure that good housekeeping takes place, owners and managers of lands where mineral exploration occurs commonly require permits and, in some cases, bonding, to protect against long-term negative impacts of these operations. Permits commonly define the timing and scale of camp operations, how physical disturbances will be reclaimed, and how support facilities such as fuel depots and waste management systems will be built, operated, and closed.

Mining's Environmental Concerns

Once successful exploration finds a new ore deposit, the next step is mining. The mining process, from the surface in open pit mines or from underground, separates and removes ore from the surrounding rocks. Significant environmental concerns accompany mining. The environmental concerns vary from hazards associated with excavations to appropriate disposal of waste materials. In most cases, the most important concerns are those that directly or indirectly affect ground or surface water quality. This section explains the sources of these concerns and how they are lessened or prevented.

(a)

(b)

Figure 16-7 Open pit mining process. (*Courtesy Earth Science World Image Bank © Bruce Molnia, Terra Photographics*)

Infrastructure and Developments

Mines have facilities needed to support operations. Buildings for offices and equipment maintenance, power systems, fuel depots, water systems, roads, crushers, conveyors, and so forth—everything needed to support highly mechanized and technically complex modern mining operations. These are industrial facilities built for many years of use, but all ore deposits eventually become depleted and mining operations stop. Today's mining operations plan on eventual closure and common practice is to demolish, salvage, or otherwise remove all mine support facilities. In fact, permits to operate new mines commonly require that the operator have plans for closure that define what will be done with mining-related infrastructure.

Planning for closure is an important part of today's mining operations. Local communities and landowners around mines can be involved in determining how the mine site may be converted into other assets for the area after mining ceases. The challenges and opportunities associated with mine closure are a general theme that is further developed later. For now, remember that it is expected to happen and that it is a time of change that requires careful planning by all interested and affected parties.

Surface Mining

Surface or open pit mining is the most cost-efficient mining method, and the world's largest mines are **open pit mines**. The open pit mining process, especially for metals such as copper, includes blasting the ore loose, hauling it to a crusher, and breaking it into pieces small enough for milling (Figure 16-7).

Technology has evolved to handle tremendous volumes of material in the highly mechanized process of open pit mining. Mines like the one shown in Figure 16-4 produce up to 150,000 tons (136,000 tonnes) of ore daily. Typically, for every ton of ore produced, as much as 2 or 3 tons of waste rock is produced. Waste rock is the principal waste product developed by mining operations.

Waste rock, the name for rocks and minerals that enclose the ore and need to be removed to recover it, contains too few useful minerals to process. However, waste rock commonly contains some minerals that can lead to environmental problems, and environmental concerns accompany its disposal. Open pit mining creates large volumes of waste rock. For example, the waste rock disposal areas that develop at a surface mine like the Bingham Canyon mine sometimes cover hundreds or even thousands of acres (tens of km^2) and may be 100 to 200 meters (several hundred feet) high. Large piles of waste rock are commonly one of the most visible aspects of a surface mine.

Underground Mining

Figure 16-8 illustrates the underground mining process. Underground mines use vertical shafts as shown, or mine openings driven into mountainsides known as **adits** (horizontal openings) or **declines** (openings inclined downward from the surface). Figure 16-9 shows examples of relatively large underground openings and related mining equipment.

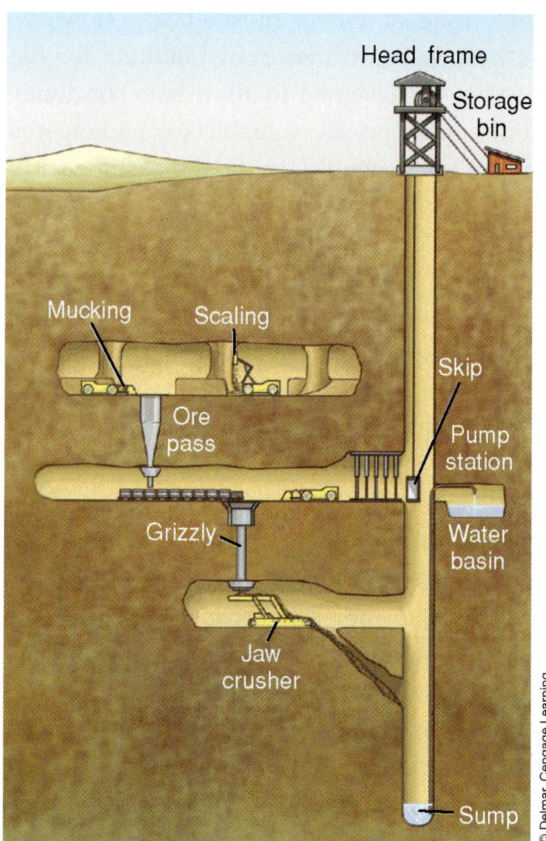

Figure 16-8 Underground mining process.

Figure 16-9 Underground mining operations. *(Courtesy Earth Science World Image Bank © Travis Hudson, American Geological Institute)*

Over the life of an underground mine, which may be several years to several decades, the volume of ore produced is most commonly a few hundred thousand tons to several million tons. This compares to production at larger open pit mines where 1 million tons of ore may be produced in just 1 week of operations.

Another big difference between open pit and underground mines is the volume of waste rock that is produced. It is common in underground mining for the volume of waste rock to be equal to or less than the volume of ore produced. In optimum situations, very little waste rock is generated and the waste rock can be used to fill underground areas where access is no longer needed. Where waste rock must be hauled to the surface, the resulting disposal areas, although much smaller than those at open pit mines, may still be highly visible. Because underground mining was the most common mining method before 1900, waste rock disposal areas (dumps) at the portals of mine workings are common in historical mining districts and at abandoned mines and prospects.

Excavations

Mine openings, either on the surface or underground, have both safety and environmental concerns. The safety concerns relate to slope stability and subsidence. The slopes in open pit mines are commonly steep and catastrophically fail if not properly designed. If underground mine workings come near the surface, then gradual or catastrophic subsidence can occur. Sound stewardship will leave these sites with reduced slopes not prone to collapse and at least surface markings, barriers, and warnings concerning hazards related to underground mine workings. The closing of openings to old underground mine workings may also be done as part of these efforts. However, some abandoned mine workings have become important habitats for bat colonies. Closure of mine openings can be designed to allow bats continued access and protection. This practice is especially valuable for endangered bat species. Because many old mine sites may not be safe, the casual visitor should avoid entering them.

Abandoned Mine Lands—The West's Legacy

Mining in the West became a very important contributor to its development from about the mid-1800s with the discovery of gold in California. The Mining Law of 1872 was especially important to westward development because it created incentives for mineral resource exploration and mining.

Early exploration in the west depended on direct observation by prospectors. Mineral deposits needed to be directly observable at the surface or in shallow surface or underground excavations. Tremendous physical effort was exerted in the search for exposed mineral deposits, and most uplands in the West have some evidence of the passing of the early prospectors and mine developers. Prospect pits, trenches, exploration shafts or adits, and small waste rock piles are widely distributed.

Mining efforts were also attempted in many of these same places, especially for gold and silver, but finding high-grade ore deposits was the key to their success. High-grade ore deposits do not tend to be large in their physical dimensions and most were mined by underground methods. Later, copper, lead, and zinc deposits became viable in many regions. These types of deposits were commonly mined out in a few years. The miners and prospectors moved on and left the disturbed areas, and in many cases whole towns—ghost towns—behind them. Prospects were also abandoned as soon as it was determined that they could not be profitably mined. As a result, there are many abandoned mine or prospect sites scattered throughout the West. Most are too small to be of significant environmental concern but some are. The environmental concerns at these sites are similar to those at other metal mines, but in the old days these concerns were not considered important. It was the exploration and development of the region that was a priority then.

Air Quality

Blasting, hauling, and crushing are dusty operations. Dust can cause respiratory illnesses and be harmful to workers and others who inadvertently breathe it. All mine sites that can produce dust as part of their operations must have active air quality protection programs to ensure safety of workers and the nearby community. Air quality monitoring and compliance with regulatory guidelines are required at all mining sites. Fortunately, effective dust controls are possible through design of processing equipment and mining techniques. Examples are water sprayers, sweepers, and vacuum systems.

Acidic Soils and Waters: Acid Rock Drainage

Although the character of waste rock varies with the type of ore deposit, those associated with metal mining present the most significant environmental concerns. Metals such as copper, lead, and zinc have a strong affinity for the element sulfur, and they combine with it to form minerals called **sulfides**. Probably the most familiar sulfide mineral is fool's gold or **pyrite**, which is composed of iron and sulfur. Pyrite is very common in waste rock from metal mines (and coal mines too). When pyrite is exposed to air and water, it undergoes a chemical reaction called **oxidation**. Oxidation of pyrite results in the formation of iron oxides that impart orange or red "rust" color to soils and waste rock. The oxidation process, which is enhanced by bacterial action, also produces acidic conditions that can inhibit plant growth at the surface of a waste rock pile. Bare, nonvegetated, orange-colored surface materials make some waste rock disposal areas highly visible, and they are the most obvious result of these acidic conditions. Figure 16-10 shows a natural example of pyrite oxidation.

Figure 16-10 Natural pyrite oxidation. (*Courtesy Earth Science World Image Bank © Larry Fellows*)

If water infiltrates into pyrite-laden waste rock, the resulting oxidation can acidify the water (Figure 16-11), enabling it to dissolve metals such as copper, zinc, and silver. This production of acidic water is called "**acid rock drainage**." If acid rock drainage is not prevented, and if it is left uncontrolled, the resulting acidic and metal-bearing water may drain into and contaminate streams or migrate into the local ground water. The acidity of contaminated ground water may become neutralized as it moves through soils and rocks, but significant levels of dissolved constituents such as sulfate can remain, inhibiting its use for drinking water or irrigation.

Figure 16-11 The oxidation process that produces acidic soils and waters. (*Courtesy Earth Science World Image Bank © Travis Hudson, American Geological Institute*)

Figure 16-12 Acid rock drainage from a mine. (*Courtesy Earth Science World Image Bank © Stuart Jennings, Montana State*)

Figure 16-13 Waste rock reclamation. (*Courtesy Earth Science World Image Bank © Stuart Jennings, Montana State University*)

Where acid rock drainage occurs, the dissolution and subsequent mobilization of metals and other elements of concern into surface and ground water is the most significant environmental impact associated with mining. Acidic and metal-bearing ground water occurs in abandoned underground mine workings and surface excavations such as open pits that are deep enough to encounter the ground water of a mineralized area. Such deep excavations are, in effect, wells that keep filling with water. This water must be removed during operations, commonly by pumping, and properly treated and disposed of to ensure that surface water quality is not negatively affected. Also, removal of mine water can at least temporarily change the local ground water levels in the mining area. However, after mining ceases, the mine workings will fill with water and some of the water may discharge to the surface through mine openings. Because these waters migrate through mineralized rocks that commonly contain pyrite before discharging, they can become a source of acid rock drainage (Figure 16-12).

If left unmanaged, significant volumes of acid rock drainage can form at large mine workings and degrade the quality of surface waters into which it flows. Preventing and treating acid rock drainage is a key environmental challenge of metal mining. Although the discharge of acidic drainage presents several challenges to protecting water quality, the significance and widespread occurrence of acid rock drainage warrants special efforts to prevent or minimize it. The common approach for preventing acid rock drainage is to physically isolate the sulfide-bearing materials from air and water. Standard reclamation practices (Figure 16-13) accomplish this to a degree by inhibiting water infiltration into waste rocks and other sulfide-bearing wastes produced by mineral processing (see next). Placing these wastes on impermeable bases such as constructed clay pads can prevent the migration of contained water away from them. Other possible measures to prevent or significantly reduce acid rock drainage include:

- Flooding of old mine excavations to cut off the oxygen supply necessary to sustain acid rock drainage
- Sealing exposed surfaces in mine excavations with a coating of material that is nonreactive or impermeable to inhibit the oxidation process
- Backfilling mine workings with reactive materials that can neutralize and treat waters that pass through them
- Adding chemicals to the water in flooded mine workings that can inhibit acid-generating chemical reactions and precipitate materials that can plug or seal off ground water migration routes

If acid rock drainage has not been prevented, then methods to treat this water are needed. The most common treatment for acidic and metal-bearing waters is the addition of a neutralizing material, such as lime, to reduce the acidity. This approach helped remedy acid rock drainage in the Savage River, investigated by the students from Beall High School. This "active" treatment process

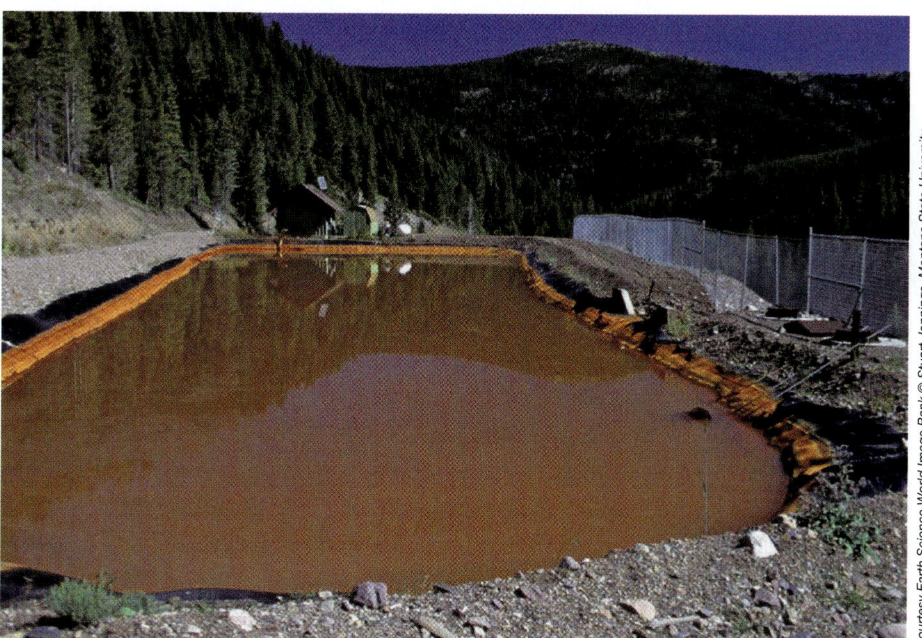

Figure 16-14 Active treatment plant for acid rock drainage.

precipitates calcium and metal hydroxides. Hydroxide precipitate can be a large amount of sludge that presents its own disposal problems. In fact, sludge management is one of the key challenges to the long-term effectiveness of this treatment method. The other challenge is cost. Because this approach is an active process, conventional acid rock drainage treatment facilities (Figure 16-14) require operation and maintenance that can go on indefinitely.

Alternatives to active treatment that may be appropriate in some cases are passive treatment systems such as wetlands, co-treatment, and reuse.

- Natural and constructed wetlands can satisfactorily treat some waters through combinations of oxidation, settling, anaerobic-based metal reduction and precipitation, and filtering processes.
- Co-treatment may be a viable option in some cases. Municipal wastewater treatment facilities may have pond systems and biosolid products that can be used to treat acid rock drainage.
- Reuse may be possible for some acid rock drainage. Highly acidic waters with high levels of dissolved constituents could be processed to recover valuable constituents.

In general, acid rock drainage is the most significant environmental concern associated with mining. It is a potential problem wherever sulfide minerals are mined for their metal content. However, it does not have to be an inevitable outcome of metal mining. Careful mine operations and closure plans can be expected to satisfactorily address this concern.

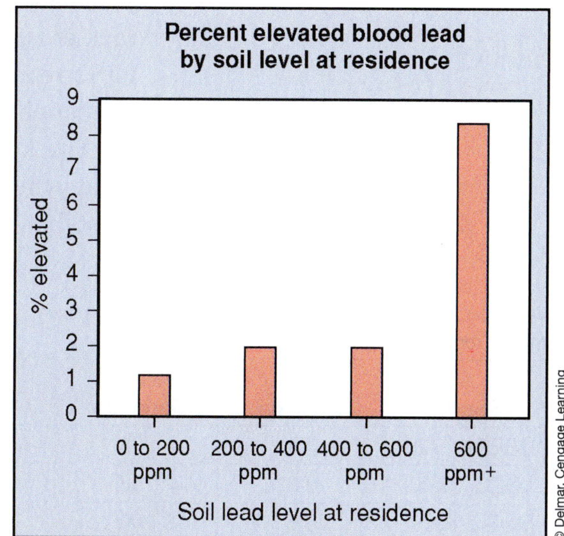

Figure 16-15 Percent of elevated blood lead level compared to soil lead level at a residence of children.

Soil Quality Concerns

High levels of metals and other elements, not just acidity, can be harmful to plants, animals, and, in some cases, people (Figure 16-15). In some places, particularly in old mining areas,

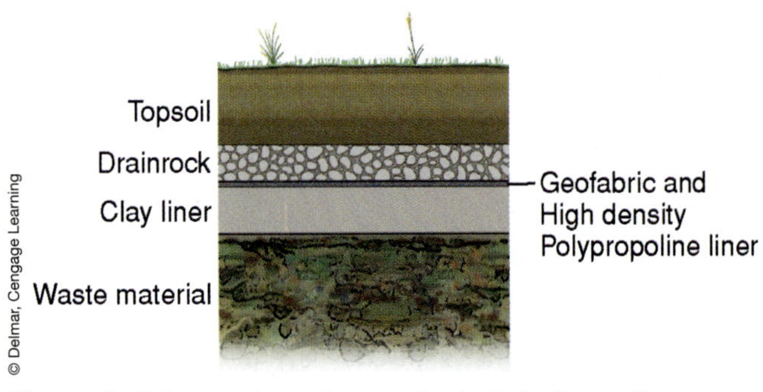

Figure 16-16 A cover design for a contaminated soil repository.

waste rock or mill tailings (see later) may have been dispersed in the environment rather than satisfactorily contained. A common approach used in dealing with contaminated soil is to move it to specially designed repositories that physically isolate it from all contact with the biosphere (Figure 16-16).

This approach can be very expensive. The volume and toxicity of the soil is not reduced; the soil is just physically isolated. Another approach that can be used in some cases is to add chemicals to the soil that react with the elements of concern, making them less mobile and biologically available.

Metal-bearing mining wastes have been considered potentially toxic and hazardous to people. Examples of metal concentrations that are considered potentially hazardous in surficial materials (soils) are listed in Table 16-2.

Though soil contamination is an important issue that needs attention, understanding the implication of it is much more complex than many people think. Examining the issue of soil lead levels will illustrate this point.

Water Quality Concerns

Waste rock disposal areas are located as close to the mine as possible to minimize haulage costs. Although the waste rock may contain useful minerals, the rock is still considered a waste, because the cost to process it would be greater than the value of the recovered minerals. If not properly managed, erosion of mineralized waste rock into surface drainages may lead to concentrations of metals or other undesirable elements in stream sediments. This situation can be potentially harmful, particularly if the elements of concern are in a chemical form that allows them to be easily released from the sediments into the stream waters. When this occurs, the elements are considered to be "mobilized" and "**bioavailable**." in the environment. In some cases, plants and animals may absorb bioavailable elements such as dissolved metals, causing harmful effects. Although current U.S. mining and reclamation practices guided by environmental regulations minimize or prevent waste rock erosion into streams, disposal of waste rock where it could be eroded into streams has occurred historically. These conditions still exist at some old or abandoned mines.

The key approach for preventing waste rock erosion is to place it in areas where it can be stabilized and reclaimed. Reclamation entails the reestablishing of viable soils and vegetation at a mine site, much like what

Table 16.2 *Examples of state and USEPA guidance levels (in ppm) for metals in residential soils. Concentrations at or above the guidance levels trigger regulatory concerns.*

Source	Arsenic	Cadmium	Copper	Lead	Manganese	Silver	Zinc
USEPA 1996	0.4	39	–	400	–	390	23,000
ADEQ 1995	0.91	58	4,300	400	580	580	35,000
MDEQ 1995	6.6	2,100	–	400	2,000	2,400	140,000

Source: Environmental Research Needs of Metal Mining, 1998, p. 5.

happens when your local landfill is closed. Figure 16-13 illustrates a simple but effective design for reclamation of waste rock. Many waste rocks can generate acidic conditions in their soils and in the waters that drain through them. Effective reclamation on these types of waste rocks adds lime or other materials that can neutralize acidity plus a cover of new topsoil or suitable growth medium to promote vegetation growth. Modifying slopes and other surfaces and planting vegetation as part of the process stabilizes the soil material and prevents erosion and surface water infiltration. Reclamation is now expected when waste rock disposal areas or mines are closed.

Protecting Surface Drainage—A Colorado Example

The Silver Swan Mine was a small, pre-WW II lead and silver producer in southwest Colorado. After operations ceased, it essentially became an abandoned mine site. The original mining practices deposited waste rocks from the mine nearby and along the banks of the Delores River.

The Delores River, a beautiful Rocky Mountain trout stream, was eroding into the waste rock pile and sending some metal-bearing material downstream in the stream sediments. Although the stream biota were not negatively affected, erosion of waste rock is an unnecessary risk to the environment in this area and can be prevented. In addition, the old mine workings discharged small amounts of weak acid rock drainage. The acid rock drainage supported wetland growth, even on the old waste rock.

The Silver Swan Mine waste rock dump was reclaimed in 1995 and 1996. The reclamation protects against (1) erosion of the waste rock by consolidating it to a position removed from the river's edge and placing a protective barrier against high flood waters; (2) infiltration of surface water into the waste rock by contouring and reclaiming the surface; and (3) direct discharge of acid rock drainage by routing it through a wetland pond system.

The placement, reclamation, and control of mine water discharge at the Silver Swan mine are small examples of approaches that can be used at all scales to stabilize and protect against erosion and acid rock drainage in waste rock disposal areas.

Mineral Processing's Environmental Concerns

Mining produces a product that is a mixture of useful and nonuseful minerals. The next step in providing mineral resources is the separation and processing of the useful minerals. **Beneficiation** is the process that crushes the ore and separates and concentrates the useful minerals. Beneficiation includes milling, leaching, flotation, and the creation of a waste product called tailings. The next processing step, called **metallurgy**, separates the desired elements such as metals from their parent minerals. Because the environmental concerns associated with mineral processing are dominated by those associated with the recovery of metals, these processes are emphasized next.

Figure 16-17 Milling and flotation. (*Courtesy Earth Science World Image Bank © U.S. Bureau of Mines*)

Milling

Milling breaks ore into individual mineral grains. Large rotating mills use metal balls or rods to grind the ore into tiny particles the consistency of silt, sand, and clay. The crushed and ground ore leaves the mill as water-rich slurry, which may be processed a variety of ways to concentrate the valuable minerals.

A concentration process commonly used for sulfide ores of copper, lead, and zinc is "**flotation**." In this process, the water-rich slurry from the mill is passed through large vats containing special bubble-making chemicals or "reagents." The vats are agitated and the metal-bearing minerals selectively attach themselves to the reagent bubbles and float off the surface of the vats—hence, the name "flotation" (Figure 16-17). Water is filtered from the bubble-rich liquid, and the resulting material is an ore concentrate that is rich in metal-bearing minerals.

Flotation leaves behind minerals, such as quartz and pyrite, that do not contain useful metals. The nonuseful minerals remain as part of the water-rich slurry in the agitated vats until almost all the useful metal-bearing minerals are floated off. After it has been stripped of useful minerals, the slurry is a waste product called **tailings**. The tailings are pumped into large ponds, called "impoundments," for disposal (Figure 16-18). Tailings are the primary waste material and the principal source of environmental concerns associated with milling. In some cases tailings have very high concentrations of pyrite—up to tens of percent—much more than is common in waste rocks. The most significant environmental concerns associated with beneficiation stem from the disposal of sulfide-rich tailings.

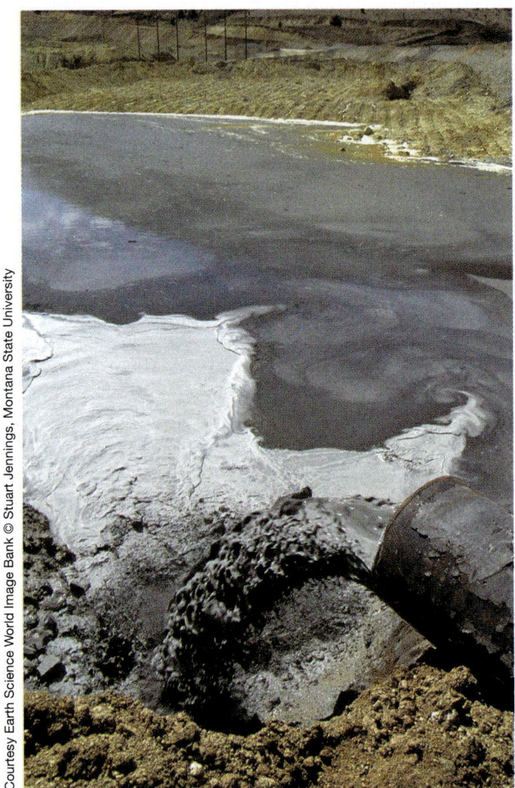

Courtesy Earth Science World Image Bank © Stuart Jennings, Montana State University

Figure 16-18 Slurry pipe discharging tailings.

Tailings Impoundments

Tailings storage areas, or "impoundments," can be very large. To save energy, tailings impoundments are commonly created somewhere downslope from the mill, so that gravity will help move the tailings slurry to the impoundment. Tailings impoundments can be located miles (kilometers) away from the mill where they are produced. The impoundments associated with the largest mills, such as open pit copper mines, can cover thousands of acres (tens of square kilometers) and be several hundred feet (about 100 meters) thick. Such impoundments must be soundly engineered to prevent their catastrophic collapse. Inadequately engineered impoundments have failed and have been sources of devastating releases of tailings to streams. Proper location of the impoundment so that it cannot be eroded or fail into environmentally sensitive areas is very important. Unfortunately, many old mines and some modern mines have discharged tailings to streams rather than satisfactorily impounding them.

Tailings produced from the milling of sulfide ores—primarily copper, lead, zinc, and nickel ores—may have concentrations of iron sulfides that are greater than those common in waste rock. Also, because tailings are composed of small mineral particles the size of fine sand and smaller, they can react with air and water more readily than waste rocks. Therefore, the potential to develop acidic conditions in pyrite-rich tailings is very high. The resulting acidic soil conditions give some tailings impoundments orange and buff-colored, nonvegetated surfaces, similar to some sulfide-bearing waste rocks. Tailings are saturated with water upon disposal and some parts of tailings contain high proportions of very fine-grained material. This clay-sized material, called "slimes," is relatively impermeable, and surface water can form ponds on it. The ponding of water on a tailings surface keeps the tailings saturated with water. If not prevented or controlled, such tailings are likely to produce acid rock drainage that can seep indefinitely from the base of the tailings impoundment. On the other hand, if tailings impoundment surfaces dry up, the fine slimes can be a source of wind-blown dust. Tailings dust may create health concerns if people or wildlife breathes it. Its migration into homes can increase human exposure to any harmful constituents that may be present. Therefore, reclamation of tailings impoundments is important to mitigating their potential negative environmental impacts.

Tailings can be reclaimed like waste rocks (see earlier). Addition of materials to neutralize acidity, such as lime, can be very important to successful reclamation, and slope modifications that inhibit surface water ponding and infiltration are necessary. Seepage of acid rock drainage from tailings can be prevented or minimized by placing an impermeable barrier, such as clay, at the bottom of the impoundment before tailings disposal. Many older tailings impoundments did not have such barriers. If not prevented or controlled, the acidic and metal-bearing waters that seep from tailings negatively affect stream habitats and ground water.

The Baia Mare Tailings Dam Collapse—A Romanian Example

Sulfide ores have been mined in and around the town of Baia Mare, Romania, for about 200 years. This mining has left various waste rock and tailings accumulations that are ongoing sources of metal-bearing dust, sediment, and acid rock drainage. Tailings slurry was pumped from old impoundments to a new one where modern cyanide-leaching techniques were used to recover gold and silver. This operation has been called an environmental cleanup by the operators, and if you were one of the local residents living close to the old tailings impoundment you would very likely agree. If appropriately designed, the new tailings impoundment could be a good long-term container that effectively removed the tailings from interactions with the environment.

On January 30, 2000, 25 meters (80 feet) of the new tailings impoundment dam failed and 100,000 cubic meters (3.5 million U.S. gallons) of water and tailings containing up to 100 tons of cyanide were catastrophically released to the environment; some leakage continued for about 60 hours. This was the first winter of an expected 7 years of operation. The cyanide-bearing material flowed down a road to a small tributary of the Tisza River. From there, the pollution made its way in a matter of days about 400 km (248.5 miles) downstream through Hungary to the Danube River in Yugoslavia. The cyanide caused a major fish kill in these rivers and was devastating to other biota as well. Pictures of dead fish floating on the Danube could be seen on the front pages of newspapers around the world.

What caused this environmental catastrophe? Weather conditions were bad and unusual—thawing snow cover and heavy rains filled up the tailings impoundment. The rising water saturated the dam and led to liquefaction and failure. However, even though the weather was unusual, it was not unheard of, and design problems clearly existed. A United Nations Environmental Program

(UNEP; http://www.unep.ch) assessment soon after the accident identified some of the key design problems. The water management system did not provide for needed adjustments such as transferring water to other locations and enabling treatment of waters before their release. Design inadequacies like this are obviously not examples of best practices in operations. In fact, the Baia Mare accident has become the impetus within a wide group of stakeholders to define a "Code of Practice" for the use of cyanide in the gold mining industry. By June 13, 2000, preparations were underway to recommence operations at Baia Mare. Many design adjustments in light of the UNEP and many other assessments and reviews were to be in place by reopening. The lesson, though, is that mining operations, regardless of their size and location, need to adhere to stringent "best practices" guidelines and drastically reduce the probability that accidents like that of January 30, 2000, at Baia Mare, Romania, will occur.

Leaching

Instead of milling, some metals—mostly from certain copper and gold ores—are concentrated through the process of leaching. After these types of ores are placed in large piles or "heaps" on specially designed pads (Figure 16-19), water solutions of sulfuric acid or dilute sodium cyanide are allowed to infiltrate and migrate throughout the ore heap. The solutions percolating down through the heap dissolve the desired metals before being collected from the base of the pile. Well-designed leach pads have synthetic or natural clay liners that prevent leakage of the chemical- and metal-laden fluids into the ground. The dissolved metals are precipitated in various ways from the collected solutions, which are then returned to the top of the heap to start the leaching process over again. Although leaching avoids milling and generation of tailings, it leaves behind large heaps of metal-depleted materials that contain residual chemicals from the leach water that have passed through them. Rinsing spent leach piles is done to ensure that the chemicals have been removed. Spent leach piles are nonetheless a source of environmental concern, and they must be properly reclaimed and closed.

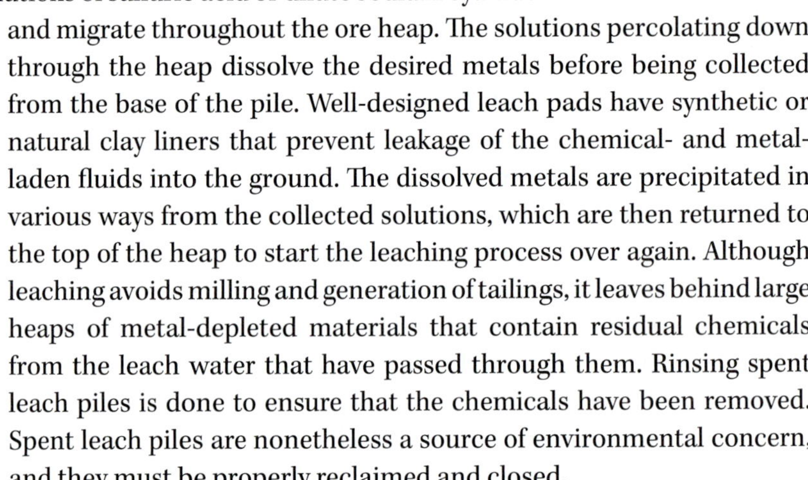

Figure 16-19 A heap leach pad in Nevada. *(Courtesy Earth Science World Image Bank © Stuart Jennings, Montana State University)*

Heap Leach Pads

Heap leach pads can also be very large. The key environmental challenges are to satisfactorily control the leach solutions during operations and to effectively clean up and close the leach pads after operations cease. If appropriate prevention and control measures are not taken, the leaching process, which uses sulfuric acid solutions to dissolve copper and dilute water/cyanide solutions to dissolve gold, can be a source of contamination that is harmful to plants, animals, and, in some situations, people. Cyanide degrades naturally in the presence of sunlight, atmospheric oxygen, and rainfall. Potential toxicity problems with this chemical are most likely to occur if concentrated cyanide solutions are accidentally spilled or released during leaching operations.

Controlling leach solutions during operations is accomplished by sound engineering designs that use strong, impermeable barriers at the base of the leach pads and effective collection systems for the leach solutions that percolate through the ore. Well-designed heap leach pads are closely monitored to ensure that leaks do not develop and that solutions are not released to the environment. However, inadequate designs are developed in cases and accidental releases of leach solutions can have devastating effects on streams and biota. Because some leaching chemicals may be left behind in the residual materials after the metals have been removed, a combination of rinsing, physical isolation, and detoxification of heap leach pads is common practice before they are reclaimed. Heap leach pads are most commonly engineered piles with regular shapes that facilitate leaching. They can be reclaimed like other areas of mine wastes, but they may still appear as constructed parts of the landscape.

Extracting Metals from Minerals: Pyrometallurgy

Mining concentrates ore and beneficiation concentrates useful minerals. The next processing step, called metallurgy, separates the desired elements such as metals from their parent minerals. The most common technique for separating metals has involved heating the minerals to their melting point. This type of "**pyrometallurgy**" is also the most significant with respect to environmental concerns. The heating process in pyrometallurgy is generally called "**smelting**." Historically, smelting facilities, called "smelters," have been large industrial developments located near mines or in other areas that can provide the necessary transportation facilities, water, and energy supplies (Figure 16-20).

Figure 16-20 Historical smelting operations. *(Courtesy Library of Congress Prints and Photographs Division)*

In the smelting process, the ore concentrate is mixed with other materials known as "fluxes" and then heated in furnaces until it melts. As the molten metals or the metal-bearing minerals separate from the other materials, they accumulate in the bottom of the furnaces and are removed. The other constituents, primarily iron and silica, float to the top of the furnaces. After they are removed, they cool to a solid glassy substance called **slag**. In some cases, the large piles of dark-colored slag near smelters make it the most visible solid waste product produced by smelting. In addition to slag, the other significant by-products from smelting are gases, which contain suspended particles. Historically, uncaptured sulfur dioxide was the greatest concern in smelter emissions, but other constituents, such as lead and arsenic, were locally important.

Smelter Slag Disposal

Slag, the principal solid waste from the smelting process, is an iron- and silica-rich glassy material that may contain elevated concentrations of metals such as lead and arsenic. The actual composition and form of slag will vary depending on the type of ore and smelting technology used. Most slag, because it is composed of oxidized, glassy material, is not very reactive and is significant as a potential source of metals released into the environment as waste rock and tailings. The main problems with slag are the physical disturbances and aesthetic impacts associated with large piles, generally dark-colored piles that have a fairly inert character and do not support vegetation. Slag piles can cover 0.1 to 1 km^2 (tens to hundreds of acres) and over 30 meters (100 feet) high. It is typical today for mining companies to reclaim these piles so that the areas can be used for other purposes. Slag has also been found to be useful in some cases; it has been used as an abrasive, to form railroad beds, and to fill sand traps at a golf course.

Smelter Stack Emissions

Smelter stack emissions have historically been a source of serious environmental concerns and negative impacts. Older smelting technology did not control either the particulate or the gas emissions very well. Sulfur dioxide has been the most common emission of concern because it reacts with atmospheric water vapor to form sulfuric acid. This form of acid rain was very harmful to vegetation and produced large areas of bare, dead soils downwind from active smelters. In addition, the particulates that were emitted could have potentially harmful constituents such as lead or arsenic, and their airborne nature increased people's exposure to them. For example, some people in communities nearby to active lead-zinc smelters were shown to have high blood lead levels. Living downwind from old-fashioned smelters was not healthy for people or the environment.

Figure 16-21 Kennecott smelter operation. (*Courtesy Earth Science World Image Bank © Michael Collier*)

Fortunately, the smelting process does not have to be as negative for the environment as it once was. Modern smelters use processes that drastically reduce particulate and sulfur dioxide emissions. An example is the modernized smelter built by Kennecott Utah Copper (Figure 16-21) that processes ore concentrates from the Bingham Canyon Mine near Salt Lake City. This smelter has reduced sulfur dioxide emissions by 95 percent of the previously permitted levels and captures 99.9 percent of the emitted sulfur. It came online in 1995, and is one of the cleanest operating smelters in the world.

 16.3 the mineral resource future

People's decisions concerning mineral resources—how they are used, how much are used, where they come from, how they are mined and processed—all have impacts on other people and on the environment. On top of all this, there is the challenge of how to foster a sustainable future as people use finite mineral resources. This final section brings many of these complexities into better focus.

Perspectives on Mineral Use and Availability

To better understand people's present and future roles in mineral resource use and its related environmental consequences, it is helpful to become reacquainted with the range of possible perspectives that one can have about mineral resource use and availability. Three perspectives are the most important and pretty much capture the range of possibilities: (1) the view from an economically advantaged society that consumes very large amounts of mineral resources; (2) the view from an economically less advantaged society that does not use a lot of mineral resources; and (3) the view from the industry that provides mineral resources.

As a student in the United States, you are part of a society that consumes more mineral resources per capita than any other in the world. You have become accustomed to having material things, from cars to computers, that depend on mineral resources. You use one of the best transportation and communication systems that can be built. You can obtain almost anything you want at the local store or in a few days by ordering things online. We grow up expecting this situation to get better and we work hard to provide even more for our children. It is actually amazing what we have come to take for granted when it comes to how our society provides for our physical needs or desires. But we are operating on a finite mineral resource budget and that others rightfully aspire to a better standard of living that depends on their receiving increasing proportions of the mineral resource pie.

Because so much of our mineral resources come from other countries, our consumption is playing an economic, social, and environmental role everywhere mineral resources are developed. We use the mineral resource, someone mines it somewhere for us, and environmental consequences occur somewhere in the world as a result. Our consumption fosters environmental consequences. Our daily decisions, as apparently inconsequential as using aluminum foil to help cook dinner, have environmental implications somewhere. The more we rely on imported mineral resources, the more we are using someone else's resources and the more environmental consequences we are passing on to them. Others may not be in near as good a position to deal with the environmental consequences as we are. The result is an inequitable distribution of the environmental consequences of mineral resource production.

It is hard for us to understand or relate to those in other societies who do not have our economic advantages, but our increasing global dependencies require that we do. Although several aspects of our high-consumption lifestyle are not considered desirable by many around the world, some degree of increased health and well-being for large parts of the world's population is needed. An increased access to mineral resources will be needed to achieve this in the future of these people. In addition, many first steps to better standards of living require the money that natural resource developments such as new mines provide. We have seen that there is a very uneven distribution of mineral endowment around the world. The natural inequities in original distribution combine with economic, social, and political barriers to limit resource availability. Is the concept of equitable mineral resource availability one that should be a guide in our future decisions? What is equitable mineral resource availability?

Mining supplies the world's mineral resource needs. For most of us, it is a remote activity run by people whom we seem to have little in common with. We may think of mining people as only concerned about their business and willing to cut corners to make a profit. Actually, it is hard to make a profit in this industry. The mining industry consistently has one of the lowest rates of return in the world; it is very competitive and high risk. It is a cyclic industry at the mercy of global commodity prices, and tremendous pressures are always in place to lower costs and improve profitability, in many cases just to stay in business. Competition has kept the prices we pay for the products of the mining industry low and allowed those who use them to develop robust businesses supplying manufactured products we have come to depend on. Mineral resources are such an important foundation of society that the industry providing them could be thought of as a service industry.

New Mine Development

The Raglan Mine is located on the Ungava Peninsula in the far north Nunavik Territory of Quebec, Canada. This is about 1,800 km (1,118.5 miles) north of Montreal in a treeless region where near-surface materials are perpetually frozen and the annual temperature is near −10°C (14°F). There are about 8,000 Inuit people living in 14 villages along the coast of Ungava Peninsula where they hunt and fish the nearby waters of Hudson Bay, Hudson Strait, and Ungava Bay. Transportation to and from this remote region can only be accomplished by air or water but there is open water only during a few months each summer.

Nickel- and copper-bearing sulfide deposits were first discovered in this region in the 1930s. Falconbridge Limited is a Canadian mining company that was founded in 1928, operates in 14 countries, employs more than 6,700 people, and is the western world's second largest producer of primary nickel products. Falconbridge started serious efforts to understand the Raglan nickel-copper deposits in the 1960s, but it was not until the 1990s that a combination of technical advances, exploration successes, and market conditions indicated that the Raglan deposits could be profitably mined. Today the Raglan deposits contain an estimated reserve of 22 million tons grading 3.06 percent nickel and 0.87 percent copper. However, the technical and economic challenges were not the only ones important to developing a new mine at Raglan. Falconbridge also needed to understand and adjust to the environmental and social implications of new mine development in this remote region, where their neighbors were Inuit people living a subsistence lifestyle. It may come as somewhat of a surprise, but the new mine that was started here in December 1997 was the first in Quebec to submit an environmental impact study and to sign a development and cooperation agreement with indigenous people.

The Inuit people of Nunavik Territory have lived in this region for at least 4,000 years. They have historically had a lifestyle that centered on fishing and hunting, particularly in the nearby marine environments where seals live and in certain rivers where fish such as Arctic Char can be caught. Their lifestyle fosters family and community relationships, shifts in priorities and activities that follow the seasons, and systems and values centered on consensus building. Important land claims were settled in 1975 when the Inuit people gained special rights of land ownership and use, including hunting, fishing, and trapping. Public institutions for Nunavik Territory were established, including Makivik Corporation (http://www.makivik.org) that represents the Inuit people and manages compensation funds provide to them as part of the 1975 land settlement agreement. The Inuit villages also have community organizations and two, Salluit, 100 km (62.1 miles) to the northwest, and Kangiqsujuaq, 60 km (37.3 miles) to the east, are fairly near the Raglan mine area. People from these villages could be expected to use areas near the new mine.

The Inuit people had several concerns about a new mine in their area. Environmental concerns focused on ensuring that the natural resources they depended on, such as the Arctic Char fishery, were not harmed. In addition, it was important to them to protect their traditional lifestyle, but at the same time participate in the economic and social benefits that the mine could bring. Falconbridge, on the other hand, has made corporate commitments to being environmentally and socially responsible miners. The two sides came together and reached the "Raglan Agreement," which provides for:

- Employment priority for qualified Inuit of the nearby villages and the territory
- Contracting priority for competitive Inuit enterprises
- Monetary compensation and profit-sharing for Salluit, Kangiqsujuaq, and Nunavik inhabitants

- Establishment of the six-member Raglan Committee, including a member from Salluit, Kangiqsujuaq, Makivik Corporation, and three from Falconbridge, that oversees implementation of the agreement and reviews outstanding environmental issues
- Environmental mitigation and monitoring beyond regulatory requirements and reporting of environmental monitoring results on a regular basis

Achieving the specifics and the intentions of the Raglan Agreement are challenging. Some of the biggest adjustments that have been made help to deal with the social concerns important to both sides. Some examples are development of mining-related education and employee training programs, housing employees at the mine site in order to not disrupt village life, establishing work schedules that allow Inuit employees to have more time with their families and communities, ceasing ore shipments at times it could disrupt hunting, and cross-cultural training for all employees—both Inuit and "Southerners." The cross-cultural training, which takes place at the mine site and in villages, provides practical exercises and tools for understanding the different cultures, communicating effectively, working well in cross-cultural teams, and recognizing and resolving cross-cultural conflicts in a positive manner. The keys to long-term success seem to be in place. These include direct involvement of the local people, respect for the inherent diversity of the setting, and a mechanism for working together that facilitates understanding and the evolutionary adjustments that will surely be needed as the project progresses. This modern example will be worth watching by everyone interested in the challenges of developing a new mine—wherever it is located.

Sustainability

Mineral resources are a finite, nonrenewable resource. Yet we are increasingly challenged to evolve to a sustainable future. At first glance, a sustainable future within the context of a nonrenewable resource seems impossible. We use the commodity, some may get recycled, but some is eventually lost to various waste streams and winds up in a landfill or worse. This is in fact what happens to a lot of mineral resources.

The payout from mining can be very direct. Wherever mines are, they are important sources of new wealth that can be used to directly benefit people and society. Mines can supply the financial capital needed for schools, hospitals, power systems, and other needed foundations of society in places where no other sources of funding are available. This conversion of natural resources into financial capital, and then into other forms of intellectual and social capital, is the legacy that needs to be provided by all natural resource exploitation. How well we manage this conversion is the key to developing a sustainable society.

The conversion of mineral resource capital to other forms of capital is one of the most important contributions that mineral resource development can make to a sustainable future. Historically, development of our mineral resources has sustained our society in this way although not necessarily in an organized, planned, and obvious fashion. The process starts with conversion of mineral resource capital to financial capital; it is this step that people are most familiar with. The financial capital that flows from mining provides every institution and individual in line to receive it with opportunity. But this is just a start; the big challenge is how we then come to convert the new financial capital into other forms of economic, social, and intellectual capital. Creating alternative and continuing sources of financial capital is one direct action that can be taken—replacing mining with another industry capable of carrying on and contributing to the economy. However, investing the mining-produced financial capital in social capital such as enhanced community institutions and intellectual capital through educational advancement of new generations is where future sustainability can really be fostered. If we are good at these types of conversions, is it possible that we really do not lose our mineral resources by using them?

We have many contemporary examples of these types of conversions. One that is still unfolding is centered on the State of Alaska's Permanent Dividend Fund. This investment fund, first established in 1968 to manage a capital windfall that accompanied oil and gas leasing in northern Alaska, has continued to grow through oil production royalties, lease sale contributions, and investment successes. Alaska's challenge is to convert this economic capital into other forms that can help Alaska have a sustainable future. The oil has already started to run out and the state's leaders are now facing decisions concerning how to put their financial capital to use. This is a contemporary example worth watching; it will provide lessons for all of us who become affected by future natural resource constraints and need to make adjustments. Another example, also with a short time fuse, is the situation faced by the Inuit people of Nunavik Territory, who have allowed the Raglan Mine to be developed on their land (see earlier). How will they invest their new financial, social, and intellectual capital being derived from the Raglan Mine development so that their future is what they want it to be after the mine is closed?

Becoming Involved

Because we import a large part of our minerals, we are passing on the related environmental consequences to others—people and countries that may not be able to afford the environmental stewardship we can. Can we do something about this? Should we? Can we really expect less advantaged societies to have the same environmental sensitivities we do? This is a complex issue, but there are some things we can do. The most important is to be involved in our own mining industry and help it evolve best practice standards and become more environmentally and socially responsible. We can support development of new mines in the United States and the related environmental technology that can foster sound stewardship of our mineral resources. Many folks seem to think that less mining in the United States is somehow better for the environment, but from a global perspective, just the opposite may be the case. How might you become involved in the mining industry and related decisions? You do not have to be a technical expert, just an interested and concerned stakeholder. An excellent example of concerned stakeholders coming together to support both mine development and sound environmental stewardship recently occurred at the Stillwater mine in Montana (see Chapter 4). The stakeholders sincerely worked together to understand each other's concerns and define common ground in an atmosphere of mutual respect. This sounds not only like good business but good society as well.

Individuals can also become more judicious and efficient consumers. Lower our mineral resource needs, substitute other more renewable or more recyclable materials for mineral resources, and of course get better and better at recycling mineral resources themselves. These are all just prudent measures that anyone might consider who was on a mineral resource budget. Becoming efficient consumers requires becoming involved in the related local to national decisions. Our country is not a recycling leader. Some countries, like Germany, are much more efficient.

Recycling—The Lead Versus Zinc Example

A possible solution to the dilemmas imposed by our consumption of a nonrenewable resource is to become very efficient at recycling it. Recycling is important to our use of many mineral commodities. For example, in 1990 the value of recycled metals in the United States totaled $37 billion, or only $2 billion less than the value of newly mined metal in 1990. In 1999, the results of recycling metals were about the same as in 1990. What can we expect in the future? A closer look at how we use and recycle lead and zinc provides some important insights.

Lead is the malleable, blue-gray metal that makes our car batteries heavy. It has a low melting point and has been used for many purposes for thousands of years: as a construction material, as part of solder alloys, in ceramic glazes, and as a constituent of crystal glass. For many years, lead was an important component of paint and an additive in gasoline. People can absorb the lead used in paints and gasoline and incur serious health problems as a result. The toxicity of the forms of lead in paint and gasoline is an important reason that these lead uses have been discontinued. Today, our principal use of lead is in the common lead-acid battery. In 1997, 87 percent or 1.4 million of our 1.6 million tons of lead consumption was in lead-acid batteries. Whereas some uses of lead are very dissipative, such as in ceramics and glasses, lead used in batteries is readily recovered. A large recycling industry (with its own environmental concerns) has developed around recovering lead from batteries. By 1997, 97 percent of spent lead-acid batteries were collected for recycling. This is a very important source of lead; through the 1990s, over 60 percent of our consumed lead has come from recycling of lead-acid batteries. Through all forms of lead recycling we are now able to supply about 70 percent of our annual lead needs. If we continue to decrease our dissipative uses of lead, which we have through discontinuing its use in paints and gasoline, for example, recycling may come to provide about 80 percent of our lead needs. This is an exceptionally high rate of recycling for such an important mineral commodity. We expect to only do half as well with the recycling of zinc.

Zinc is a metal with very diverse and beneficial uses. In 1999, we consumed 1.6 million tons, mostly in galvanizing steel to protect it from corrosion (56 percent), but also in alloys such as brass, in rubber tires, dry-cell batteries, dyes and pigments, and chemicals including fertilizers, animal feeds, fungicides, and pharmaceuticals. Many of these uses are inherently dissipative, even some of the zinc used to galvanize steel is sacrificed when it is preferentially oxidized instead of the steel. However, zinc recycling does occur. This is mostly accomplished through recycling manufacturing scrap, automobiles, alloys such as brass, and zinc construction materials such as roofing. This recycling produces about 25 percent of our annual zinc consumption. If we develop better zinc recycling technology, we may be able to increase the amount of recycled zinc to as much as 40 percent of our consumption, but we do not expect to be able to match the high level of recycling that we have accomplished with lead. Zinc's uses are just too diverse and inherently dissipative.

The examples of lead and zinc show that we can become very good, but not perfect, when it comes to recycling mineral resources. Technical factors controlling recycling efficiency and our consumption patterns indicate that finding and mining new ore deposits will still be needed in our future. However, continually becoming better at recycling of mineral commodities is a very worthy goal.

16.4 conclusion

This chapter has introduced you to mineral resources, why they are important, how we come to acquire them, and the environmental consequences that accompany their extraction from Earth. A very important message of this chapter is that there is a connection between people's individual actions and the corresponding mineral resource implication. We use products that require mineral resources, these resources are mined to provide them, and there are environmental consequences.

Another important message is that the environmental consequences of providing mineral resources for people need not be devastating. Mining operations are finite in time, they will all eventually be closed, and their most important environmental consequences can be prevented or lessened to various degrees. Ensuring that environmentally acceptable outcomes are in fact achieved is an ongoing challenge to the mining industry and to all other stakeholders. The concept of best practices is developing in this industry, but it has a way to go before it is a system of standards that are widely adopted.

Another message is that there is a role for everyone in addressing the environmental concerns associated with mineral resource extraction. Your first role evolves directly from your actions as a consumer of mineral resources. You have personal control over many decisions about what you use, how much you use, and what you do with the waste that comes from mineral use. In addition, you can come to think of the mining industry as a national asset in which you are a stakeholder. How well this industry evolves and becomes a sound environmental steward can depend on how each of us participates, as concerned stakeholders, in its future.

Finally, it is important to appreciate the complexity of many of the environmental issues associated with using and providing mineral resources. Many environmental concerns are only part of a bigger picture of economic and social needs and, in some cases, tradeoffs must be made. It is clearly an arena where simple answers are seldom right and first appearances are commonly wrong. However, you can now expect yourself to be more knowledgeable of the issues and much more capable of becoming involved. Your participation as a consumer and a concerned stakeholder is important to solving the environmental problems associated with providing the mineral resources people need.

IN THIS CHAPTER YOU HAVE LEARNED:

- Minerals or elements acquired from minerals are essential components of most physical products people use. Mining provides the mineral resources people need.

- Mineral deposits are local concentrations of useful minerals. They form from a variety of processes that mostly involve interactions between parts of the geosphere and the hydrosphere. Mineral deposits that can be profitably mined are called ore deposits.

- Ore deposits are fairly unique geologic features and they are not uniformly distributed around the world. Some countries have lots of mineral resources and some do not. The United States is a net importer of 53 mineral commodities and imports more than half of its needs for 27 mineral resource commodities.

- Exploration for ore deposits may involve trenching, drilling, and various facility developments that can physically disturb the landscape. Because most mineral exploration projects are temporary, sound environmental practices anticipate reclaiming disturbed areas, removal of facilities, and appropriate waste disposal. Exploration operations, commonly guided by permit requirements, need to guard against fuel spills and other sources of water and soil contamination.

- Mining produces minerals for people's use. It commonly disturbs the landscape and produces waste rock that must be properly disposed. Sulfide-bearing mineral deposits and waste rocks (especially those with pyrite–iron sulfide) can oxidize and produce acidic soils or acid rock drainage. Where it occurs, acid rock drainage is the most significant environmental consequence of mining. Modern mining operations protect against acid rock drainage and release of dust and waste rock to the environment. All mines deplete their ore deposits and must close. Environmentally sound mine closure, guided by permit requirements, removes mine facilities, stabilizes and reclaims waste rock, and takes measures to prevent surface and ground water contamination.

- Mineral processing separates and concentrates useful minerals or their components from ores. Processing of ore from sulfide-rich ores provides the most significant environmental concerns. Milling of sulfide ores produces a waste material called tailings that can contain significant amounts of pyrite (iron sulfide). Tailings disposal in impoundments must be done in ways that are safe from collapse and that prevent water infiltration and acid rock drainage.

- Leaching of ores to extract metals, especially copper and gold, must carefully control toxic fluids and flush and reclaim depleted ore materials.

- Separating metals from sulfide minerals requires melting in facilities called smelters. The primary environmental concern at smelters is emissions to the atmosphere. Such emissions can create acid rain and release other toxic materials. Modern smelters can capture potentially harmful emissions before they are released to the atmosphere.

- People's need for mineral resources is increasing. Developing new mineral resources has a number of social, economic, and environmental consequences. New mines can anticipate and control these consequences. Although minerals are a finite, nonrenewable resource, sound investing of the financial capitol created by mining can help society achieve a sustainable future.

KEY TERMS

Acid rock drainage
Adits
Beneficiation
Bioavailable elements
Declines
Drilling
Exploration
Flotation
Geochemical techniques
Geophysical techniques
Metallurgy
Milling
Open pit mines
Ore
Ore deposits
Oxidation
Prospects
Pyrite
Pyrometallurgy
Rig
Slag
Smelting
Sulfides
Tailings
Trenching
Waste rock

REVIEW QUESTIONS

1. What are three importance resources humans get from mining?
2. Describe one way in which ore deposits are formed.
3. Describe two ways in which rocks, just by themselves, are an important resource for humans.
4. Choose two properties of metals and explain why they are important to humans.
5. Explain how minerals can be a source of energy.
6. "Mineral resources are a finite resource." Explain what this means.
7. Describe two techniques that geologists use to locate mineral deposits.
8. Choose one type of mining technique and describe its advantages and disadvantages.
9. Describe two ways in which mining companies can reduce or eliminate acid rock drainage.
10. Describe two ways in which mining companies reclaim mining sites.

chapter 17

can you see the future?

What do you think high school will be like in the future? In places like Utah, there is an Electronic High School. Students can roll out of bed and essentially be at school. But what will energy cost and how clean will the air and water be for your children and grandchildren?

This book explains how Earth systems interact to create a dynamic and changing physical and biological environment. It has especially focused on helping you understand people's relationships to the environment, including their experiences along coasts and rivers and where natural hazards such as earthquakes, volcanoes, unstable land, and severe weather occur. In all these places, Earth's dynamic systems will have an impact on where people choose to live and how they build their homes.

You have also learned about how people affect Earth systems. People's impacts come from their use of Earth's resources—its water, soil, minerals, and energy resources. Both acquiring and using Earth's natural resources cause environmental impacts. These can range from physically changing the landscape to altering air quality on a global scale.

People make choices about how much of a resource they use and how they dispose of the waste they generate. You will use more energy by leaving lights on, driving a large car, or living in a poorly insulated home. You will contaminate soil and water if you discard waste oil or other trash inappropriately. These are examples of individual choices people make, but people also make choices about using Earth's resources at all scales.

Farmers make choices about how they will irrigate, use pesticides, and till their land. Energy companies make choices about how they will produce, process, and transport their products. Governments make choices about where and how resources can be developed. It is a complicated series of choices that combine to define people's affects on Earth systems. It is especially complicated because people need Earth's resources to survive, and conflicts can arise between resource use and environmental protection.

Meeting the basic health and security needs of the world's human population is the primary reason that people use Earth's resources. Providing for these needs while still being sound environmental stewards is an ongoing challenge. This challenge is even more significant when sustainable resource use is the goal. Do you think people will be able to meet their future resource needs and be sound environmental stewards too? Looking into the future is hard to do.

 constraining the future

A rosy future is possible, but it would take some major changes in how people interact with each other and the environment. How about a dire future? A dire future seems unlikely for several reasons.

- People are resilient and can learn from their experiences. Sometimes it seems that we are hard learners, but if we can come to understand a problem, we can usually figure out a way to deal with it.
- Economic factors can become strong promoters of change and innovation. There is probably nothing more powerful than a high gasoline price in making our nation's vehicle fleet more fuel efficient.
- The interconnected nature of Earth systems is being better understood and appreciated. People are becoming better at understanding how environmental effects spread and aggregate even at global scales.
- Communication and technology are making the world smaller. Lessons learned in one place can be quickly shared and applied elsewhere.
- The Earth is resilient too. It has undergone tremendous changes throughout its long and evolving history.
- When it comes to people and Earth systems, change, not catastrophe, is to be expected. So which future do you now expect?

 engaging for the future

You now have an understanding of how people affect Earth systems. Although these effects may at first seem far from your control, even small changes can become significant when millions of people participate. If millions of people stopped discarding used motor oil in the nearby storm sewer or ditch, the amount of oil entering the sea would be drastically reduced. If people drove cars that were more fuel efficient, the amount of emissions from vehicle exhausts would be significantly reduced. How do millions of people come to participate and share such goals? How do people become engaged and help make choices that provide for sound stewardship and sustainability? Your individual choices are an important start, but in your community, state, and nation there are mechanisms in place that will help you and others participate.

As you engage in discussions and decisions that affect the environment, be sure to appreciate and use the power of scientific inquiry. Remember that you do not need to be a scientist to solve problems like a scientist. You can:

- Accurately define problems.
- Collect relevant data and information about the problem.
- Analyze the data and design hypotheses about how to resolve the problem.
- Test your hypotheses and see which one works best.

You can also require others who are responsible for decisions in your community, state, and nation to be good problem solvers too. Support and accept sound analysis and decision making at all levels and in all relevant venues. This is a challenging responsibility, but as someone once said—only you can protect the environment.

glossary

a

abiotic Nonliving factors that affect the biosphere, such as air, water, and some parts of soil.

abundance The number of organisms in a particular area.

acid rain Rain with an unusually low pH.

acid rock drainage Production of acidic water.

adhesion Attraction of molecules to even slightly and variably charged solid surfaces.

adits Mine openings driven into mountainsides.

air mass A large body of air, where weather conditions are similar throughout its extent.

air pollutant Any gas or particle released into the atmosphere at a concentration that has a negative effect on human health or the health of the environment.

air pressure Force of the atmosphere pressing down on Earth's surface.

alluvium Unconsolidated material deposited by rivers or streams.

andesitic magma Magma with about 60 percent silica.

angle of repose The angle between the slope surface and horizontal and varies for different materials.

animalia Members of the animal kingdom.

anthracite A hard black and lustrous coal that has less than 15 percent moisture content.

aquifer Storage areas below the ground water table that have good porosity and permeability.

archaebacteria One-celled microscopic organisms are the primitive "old" bacteria.

atmosphere The envelope of gases that surrounds Earth.

b

basalt Dark-colored volcanic rock; the most abundant volcanic rock on Earth.

basins Places that accumulate sediments.

beaches Sandy areas that line coasts.

beneficiation The process that crushes the ore, separates, and concentrates the useful minerals.

bioavailable elements Elements available in the environment.

biodegradation Use of microorganisms to eat the natural organic hydrocarbon compounds in spilled oil.

biodiversity The variety of life forms in an ecosystem, biome, or biosphere.

biome Community that covers a large geographical area; the temperate forests of the northwestern United States are an example.

bioremediation The managed application of microorganisms in the treatment of polluted soil or water.

biosphere The life that inhabits Earth and evidence of past life.

biotic Living factors that affect the biosphere.

birth rate The number of live births per 1,000 people per year.

bituminous coal A massive black coal with less than 20 percent moisture content.

blowouts Accident that occurs when oil or gas escapes uncontrolled up the well to the surface.

body waves Earthquake waves that travel through the Earth's inner layers—primary waves and secondary waves.

braided stream A stream with multiple interconnected channels.

c

carbon cycle Transfer of the element carbon between Earth's systems.

carnivores Meat-eating organisms.

carrying capacity The number of people that Earth can sustainably support at some reasonable level of economic and social well-being.

cellular respiration Process by which organisms oxidize glucose to release energy, water, and carbon dioxide.

channel The cleft or indentation in the land through which streams and rivers flow.

chemical weathering Processes that change minerals on Earth's surface.

chlorofluorocarbons Chemicals containing chlorine and fluorine that are harmful to the ozone layer of the atmosphere.

clay-size particles Smallest particles in soil.

Clean Air Act Federal legislation that provided regulations for air quality in the United States.

Clean Water Act Federal legislation that provided regulations for water quality in the United States.

closed systems Systems in which only energy is transferred in or out.

coal A rock that burns because it is mostly composed of carbon and hydrogen.

coalification Process by which peat turns into coal.

coasts Long strips of land at the interface between land and water.

cohesion Attraction of molecules to each other that can result in the formation of thin films and coatings.

commensalism A relationship between two organisms that benefits one but does not help or harm the other.

community All the populations of various species that live in an area.

competition Organisms vying for the same resource.

conduits Underground channels in volcanoes.

consumers Organisms that rely on other organisms for food.

convergent plate boundaries Locations where plates move toward each other.

Coriolis Effect The westward deflection of air movement due to Earth's rotation.

crude rates Birth and death rates that are not specific to a particular age group.

crust Uppermost rigid layer of the Earth.

cuttings Rock pieces resulting from drilling with very hard drill bits.

cyanobacteria One-celled organisms that can make their own food.

d

death rate The number of deaths per 1,000 people per year.

declines Mine openings inclined downward from the surface.

demographic transition The evolution of populations.

discharge The total volume of water flowing past a point on the river in a given amount of time.

distribution Place where organisms are found.

divergent plate boundaries Places where tectonic plates move away from each other.

doubling time The amount of time it will take for a given population to double.

drilling Using machines that can grind their way downward and recover chips or cores of subsurface rock.

drill mud Dense circulating fluid in a drill.

drought An extended period of below normal rainfall.

dunes Hills of sand found on beaches.

dynamic systems Systems that are constantly changing.

e

earthquake Vibration due to the rupturing of rocks along fault zones that moves outward from the site of fault movement through the Earth and along its surface.

ecology The scientific study of the interactions between organisms and their environment.

ecosystem Community of organisms and its abiotic factors.

El Niño The interaction of the oceans and the atmosphere every few years that moves warm surface water from the west Pacific to the east Pacific Ocean.

elastic energy The energy released when a fault slips and the Earth's surface changes shape.

elastic rebound The change in shape of the Earth's crust after it breaks.

energy Ability to do work.

epicenter The location on the Earth's surface directly above the focus.

episodic An event that occurs periodically.

erosion Removal of sediment by the action of wind or water.

estuary Partly protected place along coasts where fresh water and saltwater mix.

eubacteria Common or "true" bacteria.

exploration Process of finding minerals.

exponential growth Growth in a population that occurs when there is a positive growth rate for many years.

f

faults Surfaces along which movement occurs between two large blocks of rock.

fertilizers Chemicals that provide nutrients to help plants grow better.

fission Process of splitting the nucleus of an atom.

flood Overspill of water onto the land. It is a naturally recurring event that results from unusually heavy rains or, in cold climates, from rapid snowmelt.

floodplains Areas along streams or rivers that have a tendency to be covered with water (flood).

floodwater Water resulting from flooding action.

flotation Process of passing water-rich slurry from the mill through large vats containing special bubble-making chemicals or "reagents."

flux The rate at which energy and matter transfers take place between systems.

focus The location within the Earth where an earthquake originates.

footprint The area used for a structure or device such as an oil and gas field facility.

fossil fuels Fuels with organic origins: coal, oil and natural gas.

fungi Organisms that grow embedded in a food-bearing medium.

g

geochemical techniques Methods used by geologists to analyze samples of rocks, soils, water, vegetation, or stream sediments that may contain elements that are clues to the nearby presence of mineral deposits.

geologic repository Geologically stable place where an underground facility will safely store radioactive waste.

geophysical techniques Methods, such as magnetic surveys, that are used to help characterize rocks beneath the surface.

geosphere The solid Earth and all its related parts.

geothermal energy Heat energy from hot rocks in the Earth's crust.

glaciation Covering an area with large sheets of ice.

gradient Slope.

gravity Force of attraction between all matter in the universe.

ground water Water below Earth's surface.

growth rate Percentage population change.

gullies Deep channels in soil.

h

habitat Physical environment in which organisms live.

herbivores Plant-eating organisms.

high pressure Areas in the atmosphere that are developed where air sinks.

horizons Soil layers.

humic acids Acidic material in soil that comes from humus.

humus The general term for the organic matter that imparts a dark color to upper parts of soil; it can consist of both plant and animal debris.

hurricane An organized ring of convective storms with circulating air due to the conservation of angular momentum.

hydrocarbons Fuels containing only hydrogen and carbon.

hydrosphere All water on Earth, including the water in the oceans, lakes, rivers, underground, the atmosphere, in living things, and in permanent ice accumulations.

i

igneous Rock formed from molten material.

indicators Representative factors that can be measured.

individual One organism.

inner core Solid, innermost layer of the Earth; mostly made up of metallic iron and nickel.

intensity The amount of damage an earthquake causes.

ions Charged particles.

isoseismal maps Maps that show the distribution of modified Mercalli intensities using contours of equal intensity.

k

karst Regions that are characterized by caves, sinkholes, underground streams, and other features formed by the slow dissolving of rocks.

kinetic energy The energy of motion.

l

lahar Giant, fast-moving mudflows.

landforms The shapes of the land surface.

landslide A general term for the downslope movement of masses of rock, soil, or debris.

lava Magma above the Earth's surface.

lignite Brown to brownish platy massive coal material that has a high moisture content (up to 45 percent) and low energy output.

liquefaction The process by which unconsolidated material such as sand, silt, and clay becomes saturated with water during strong seismic events.

lithosphere Rigid rocks of the crust and upper mantle.

logging Process of harvesting forests.

low pressure Areas in the atmosphere that are developed where air rises.

m

magma Molten material from deep in the geosphere.

magnitude Strength of an earthquake.

mangrove swamp Wet areas in tropical and subtropical areas of the world, including southern Florida that are commonly found on the fringes of estuaries and are dominated by mangrove vegetation.

mantle Layer of the Earth under the lithosphere.

mass movement Another term for a landslide.

matter All physical substances.

meandering Winding stream.

meltdown Melting of fuel in a nuclear reactor.

Mercalli scale Measure of earthquake intensity based on human observation of damage.

mesopause Top of the mesosphere.

mesosphere Atmospheric layer above the stratosphere.

metallurgy Separation of desired elements such as metals from their parent minerals.

metamorphic rocks Rocks that are new mineral aggregates formed from high pressures and temperatures.

methylmercury The most toxic form of mercury to the environment.

microhabitats Subdivisions of habitats.

mid-ocean ridges Underwater mountain ranges.

migration The movement of people from one area to another.

milling Process of breaking ore into individual mineral grains.

mineral Naturally occurring inorganic solid made up of an element or a combination of elements that has an ordered arrangement of atoms and a characteristic chemical composition.

mollisol Type of soil in grassland biomes.

moment magnitude Determination of the amount of energy an earthquake releases that is calculated from the area of the fault rupture generating the earthquake, the amount of movement on the fault, and the strength of the faulted rocks.

monsoons Very heavy rains that occur where very hot land masses in the summer are surrounded by much cooler ocean waters.

mudflows and debris flows Flows of relatively coarse material.

mutualism Relationship between two organisms that benefits both organisms.

n

natural growth rate For the Earth as a whole, the difference between birth and death rate for the human population.

niche Role of organisms in their habitats.

niche dimensions The physical conditions an organism can tolerate, the resources it uses, and other aspects of its lifestyle.

nonpoint source pollution A source of pollution that cannot be tied to a specific point of origin; pollution that occurs as pollutants are carried along in runoff and flow into a river or seep into ground water and are carried far away.

nuclear energy Energy produced by nuclear fission.

nutrients The essential elements plants need for growth.

o

oil spill Oil released to the environment by transportation accidents, releases of produced water from offshore production platforms, and from the improper disposal of used motor oil.

omnivores Organisms that eat both plants and animals.

open pit mines Mines at the surface.

open systems Systems in which both energy and matter are exchanged.

optimal range Range within which the condition of a factor is ideally suited for an organism.

ore Rock with recoverable minerals.

ore deposits Mineral deposits with high enough concentrations of the useful elements to allow profitable mining.

oscillation Shifts between conditions.

outer core Liquid, outer part of the Earth's core.

overburden Removal of the overlying rock and soil in surface mines.

overgrazing Process of allowing livestock to feed on plants in an area without giving the plants enough time to regrow.

oxidation Type of chemical reaction in which there is a net loss of electrons.

ozone Molecule consisting of three oxygen atoms.

ozone layer Ozone layer about 30 km above the Earth that serves as a protective shield as it absorbs and filters out some of the sun's harmful ultraviolet radiation.

ozonosphere Ozone layer of the atmosphere.

p

parasitism Relationship between two organisms that benefits one and hurts the other.

peat Accumulations of partly decomposed plants that grow in wet swampy or boggy areas.

permeability A measure of how well a fluid such as water can flow through a solid.

pesticide Substance applied to crops to kill unwanted organisms, such as insects.

photochemical smog Thick, brownish haze that forms when nitrogen oxides and hydrocarbons in the air react with sunlight.

photosynthesis Process by which plants make glucose from carbon dioxide and water in the presence of chlorophyll and light.

photovoltaic cells Devices that convert the sun's light to electricity.

physical weathering Processes that cause rocks to disintegrate or break down into smaller and smaller pieces.

physiography Description of Earth's landforms and other natural features on Earth's surface.

plantae Members of the plant kingdom.

plates Pieces of the Earth's lithosphere.

point source A source of pollution that can be tied to a specific point of origin.

polar Condition in a molecule where one side acts somewhat negative, and the other somewhat positive.

pollutants Harmful substances such as disease-causing organisms, toxic chemicals, and metals that have a negative effect on an ecosystem.

pollution Contamination of a part of the Earth system, such as air, water or soil.

population Group of organisms of the same species that live in a specific area.

pores Small spaces between particles in an Earth material.

porosity The amount of small open spaces in an Earth material.

potential energy The energy of position or stored energy.

predation Act of preying upon another organism.

predator Member of a species that eats a member of another species.

prey Member of a species that is eaten by a member of another species.

primary waves Earthquake waves that move through the Earth by alternately compressing and expanding rocks.

producers Organisms that make their own food.

prospects Mineral deposits.

protista One-celled organisms with internal cell structures, such as a nucleus, that separate them from bacteria.

pyrite Mineral common in waste rock.

pyroclastic flow A very hot, dense cloud of volcanic gases, rock debris, ash, and water.

pyrometallurgy Technique for separating metals that involves heating minerals to their melting point.

r

range of tolerance Adaptation of a species to live within a certain range of physical environmental conditions.

rate Pace at which something happens, such as environmental change.

recharge areas Areas where precipitation can soak into the ground and become part of the ground water system.

reclamation The process of reestablishing of viable soils and vegetation at a mine site.

recurrence intervals The average time span between large earthquakes (≥ magnitude 7) at a particular site.

reserves The world's supply of economically recoverable oil and gas.

reservoir Places where energy or matter are stored in systems.

residence time The average amount of time a substance is contained in a specific reservoir.

resource partitioning A situation in which two species competing for a single resource adapt so that parts of the same resource can be used by both species.

resources Requirements for organisms, such as food, water, and habitat. Something a species needs to survive and reproduce.

rhyolitic magma Magma with 70 or more percent silica.

richter scale Scale that measures earthquake size with a range from 1 to 10.

rifting Plate divergence that pulls apart the lithosphere.

rig Drilling machine.

rills Small, narrow furrows in soil.

riparian zone The bankside and banktop environments of a river taken together.

river Large stream of water.

rock cycle The process by which rocks are formed, changed and broken down in the Earth system.

rocks Aggregates of one or more minerals.

rock system The process by which rocks move through the Earth system.

roughness The friction between river water and the bed, banks, wood and large rocks in the channel.

s

salination Salt buildup in soil.

salt marshes Areas typically surrounding estuaries just above the intertidal zone that are characterized by an assemblage of salt-tolerant grasses.

sand-size particles Largest particles in soil.

scale An ordered reference standard, such as those used for time and space.

secondary waves Earthquake waves that can only travel through solid rock and move the rock in a shearing, or side-to-side, motion.

sediment Small rock particles.

sediment carrying capacity The amount of sediment a river can move.

sedimentary basins Places where sediments accumulate in parts of oceans or depressions in the Earth's crust.

sedimentary rock Rock formed from sediment such as sand, silt, and clay.

sedimentation Accumulation of small particles, or sediment, in an area.

seismogram Graphical output from a seismometer.

seismograph Instrument that detects vibrations from earthquakes.

seismometer Device used to detect earthquake vibrations.

self-regulates System that operates with minimal external intervention to maintain a steady state.

shield volcano Volcano with gentle slopes where fluid basaltic lava is erupted.

silica A combination of the elements silicon and oxygen and the most abundant component in Earth's rocks and minerals.

silt-size particles Midsize particles in soil.

sinkholes Crater-like holes that form when the land surface collapses.

sinks Places where matter gets trapped and essentially removed from system interactions.

slag The principal solid waste from the smelting process.

smelting Heating minerals to their melting point.

soil A natural mixture of minerals, rock fragments, and organic material that makes up a porous layer on Earth's land surface.

soil contamination Decrease in soil quality due to pollutants.

soil creep Very slow, downhill movements of soil.

soil horizon The stratification of soil materials from the surface downward into layers.

soil profile A generally similar set of layered soil horizons.

soil quality Soil's capacity to sustain plant and animal productivity, maintain or enhance water and air quality, and support human health and habitation.

soil strength The extent to which soil withstands deformation.

soil structure A reflection of how individual soil particles are aggregated together.

soil texture A reflection of the size of mineral particles within soil.

southern oscillation The phenomenon that occurs under El Niño conditions, when the westward-blowing winds decrease, or actually reverse direction, and blow eastward due to changes in air pressure.

species Groups of interbreeding populations that cannot make fertile offspring with other similar groups.

steady state Situation or system that appears to be unchanging.

storm surges Large amounts of water pushed by strong storm winds.

stratopause Top of the stratosphere.

stratosphere Atmospheric layer above the troposphere.

stratovolcano Cone-shaped volcano with steeper slopes that explosively erupts a wide range of materials.

subduction Process by which thin and dense oceanic plates move toward light and thick continental plates, and the dense plate tends to sink below and under the less dense plate.

subsidence The gradual lowering of the land surface over large areas.

sulfides Minerals that are a combination of sulfur with copper, lead, or zinc.

surface waves Earthquake waves that travel more slowly than primary or secondary waves and cause most of the earthquake damage to buildings and other property.

sustainability The act of providing for the present needs of humans in ways that do not jeopardize the needs of humans in the future.

symbiotic An interaction that involves two organisms from different species living in direct contact with one another.

system Group of things that regularly interact or are interdependent and that together form a unified whole.

t

tailings Slurry that is stripped of useful minerals.

talus Rock and boulder fragments that have accumulated at the base of a cliff or steep slope.

tectonic movements Motion of the geosphere.

temperature inversion A phenomenon that occurs where a warm air mass moves into a region of cold air, preventing the colder air from rising higher into the atmosphere and dispersing its pollutants.

tephra Material blasted into the air by a volcano.

thermosphere Uppermost layer of the atmosphere.

tides Regular rises and falls of sea level along coasts.

tiltmeter Instrument to detect ground deformation.

tonne A metric ton, equal to 1000 kg, or 2,204.6 pounds.

topsoil The organic, rich top horizon of soil.

tornadoes Strong vertical rotation of air that is focused into a small area and forms visible funnels at the base of the clouds that sometimes reach the ground surface.

trade winds Air masses that move regularly westward at low latitudes.

traits Distinguishing features of a species.

transform boundary Location where plates slide horizontally past each other.

transform plate boundaries See *Transform boundary*.

transformations Changes in soil materials.

translocations Movements of soil materials.

transpired Process in plants through which water gathered in plant roots is released to the atmosphere through the leaves.

traps Places that prevent petroleum from continuing to migrate.

trenches Deepest places in the ocean.

trenching Using backhoes or other heavy equipment to expose the near surface parts of a prospect for direct observation and sampling.

tributaries Streams that feed into larger streams.

triple point of water The specific combinations of temperature and pressure at which all three phases of water can coexist.

trophic level Particular level of a food chain or web.

tropopause Top of the troposphere.

troposphere Lowest layer of the atmosphere.

trunk stream Primary stream in a watershed.

u

unconsolidated material Soil and loose material overlying bedrock.

unstable land Ground that moves.

urban area According to the U.S. Census Bureau, any town or city with more than 2,500 residents.

urban heat island effect The condition in which the air in the center of the city, particularly at night, can be as much as 10°F warmer than surrounding rural areas.

urbanization Formation of cities.

v

viscous Quality of a fluid that is its resistance to flow or its thickness.

volcanic eruptions Outflow of matter (lava, ash, gases, etc.) from volcanoes.

Volcanic Explosivity Index (VEI) Scale to express the size of a volcanic eruption.

volcano Places on Earth where molten magma comes to the surface from deep in the geosphere.

w

waste rock Rock that is adjacent to or interbedded with the coal seams; the principal waste product developed by mining operations.

water cycle The process of water evaporating, condensing, precipitating, and reevaporating; transfer of the water molecule between Earth's systems. Also called the hydrologic cycle.

water table The depth at which ground water saturation occurs, which varies from place to place.

watershed The land areas that contribute surface runoff to a specific, connected set of streams and rivers.

waves Movements of near-surface water caused by wind.

weather Conditions of the atmosphere that include such aspects as air temperature, air pressure, wind, precipitation, relative humidity, and cloud cover.

weathering processes The collection of interactions of air and water that break down the geosphere's surface.

well casing Covering that isolates the well bore from adjacent geologic formations.

westerlies Characteristic winds of temperate regions.

wind power Windmills that produce electricity.

z

zero population growth (ZPG) Population that is stabilized at a specific size and composition.

index

f

g

y

z